Geotechnical, Geological and Earthquake Engineering

Volume 48

More information about this series at http://www.springer.com/series/6011

Dietlinde Köber • Mario De Stefano
Zbigniew Zembaty
Editors

Seismic Behaviour and Design of Irregular and Complex Civil Structures III

Springer

Editors
Dietlinde Köber
Reinforced Concrete Department
Technical University of Civil Engineering
Bucharest, Romania

Mario De Stefano
Department of Architecture (DiDA)
University of Florence
Florence, Italy

Zbigniew Zembaty
Faculty of Civil Engineering
Opole University
Opole, Poland

ISSN 1573-6059 ISSN 1872-4671 (electronic)
Geotechnical, Geological and Earthquake Engineering
ISBN 978-3-030-33531-1 ISBN 978-3-030-33532-8 (eBook)
https://doi.org/10.1007/978-3-030-33532-8

This Springer imprint is published by the registered company Springer Nature Switzerland AG.
The registered company address is: Gewerbestrasse 11, 6330 Cham, Switzerland

Preface

Buildings optimally designed against seismic loads are regular, symmetrically shaped structures. The design of such buildings is clearly described in student textbooks and in the first pages of any civil engineering code. However, such buildings are rarely met in reality for economical, functional and formal reasons. In densely populated urban areas there is a need for developing any piece of land; investors formulate various specific demands with respect to the shape of the building and its functions; architects want their works to distinguish from other buildings; and topographical and subsoil irregularities may require atypical design solutions. As an effect, most of the building structures are irregular both in plan and in elevation. Furthermore, ground rotations and wave passage effects also contribute to irregularity issues in the seismic response of structures.

It is a specific feature of the seismic design of irregular structures that they are always conservative with respect to regular structures. This means that the presence of any type of irregularity generates additional seismic loads, internal forces and strains compared to the regular structures. For these reasons, substantial engineering efforts are devoted to solve irregularity issues at the stage of research and during the design.

From 1996, every 3 years, a cycle of workshops devoted to study the seismic response of irregular structures takes place under the auspices of Working Group 8 'Seismic Behaviour of Irregular and Complex Structures' of the European Association of Earthquake Engineering. Starting from the 2011 Workshop, which took place in Haifa, Israel, the proceedings are published as monographs in the Springer Geotechnical, Geological and Earthquake Engineering series. This was volume 24 which appeared in 2013, after the Haifa Conference of September 2011[1] followed

[1]Lavan O., De Stefano M. (editors), *Seismic Behaviour and Design of Irregular and Complex Civil Structures*, Springer, Geotechnical, Geological and Earthquake Engineering, Dordrecht, 2013 (http://www.springer.com/gp/book/9789400753761).

by volume 40 which appeared in 2016, after the conference which took place in October 2014 in Opole, Poland.[2]

This volume presents the third monograph devoted to the irregularity issues in structural seismic response. It contains state-of-the-art papers presented during the 8th European Workshop on the Seismic Behaviour of Irregular and Complex Structures which took place in Bucharest, Romania, on October 19–20, 2017. Forty-five participants from 13 countries took part in this successful event.

The book contains 33 reviewed and edited chapters selected from the lectures presented during the workshop. The chapters are grouped in three parts:

- Seismic Load, Ground Motion, Rocking Excitations
- Seismic Analysis and Design of Irregular Structures
- Seismic Control and Monitoring of Irregular Structures

Bucharest, Romania Dietlinde Köber
Florence, Italy Mario De Stefano
Opole, Poland Zbigniew Zembaty

[2]Zembaty Z., De Stefano M. (editors), *Seismic Behaviour and Design of Irregular and Complex Structures II*, Springer, Geotechnical, Geological and Earthquake Engineering, Dordrecht, 2016 (https://www.springer.com/la/book/9783319142456).

Contents

Contributors

V. Alecci Department of Architecture (DiDA), University of Florence, Florence, Italy

Asimina Athanatopoulou Department of Civil Engineering, Aristotle University of Thessaloniki, Thessaloniki, Greece

William A. Avila Universidad de los Andes, Bogotá, Colombia

Athanasios P. Bakalis Institute of Structural Analysis and Dynamics of Structure, School of Civil Engineering, Aristotle University of Thessaloniki, Thessaloniki, Greece

F. Barbagallo Department of Civil Engineering and Architecture, University of Catania, Catania, Italy

M. Barnaure Faculty of Civil, Industrial and Agricultural Buildings, Technical University of Civil Engineering, Bucharest, Romania

D. Beben Faculty of Civil Engineering and Architecture, Opole University of Technology, Opole, Poland

Rita Bento CERIS, Instituto Superior Técnico, Universidade de Lisboa, Lisbon, Portugal

P. Bonkowski Faculty of Civil Engineering and Architecture, Opole University of Technology, Opole, Poland

M. Bosco Department of Civil Engineering and Architecture, University of Catania, Catania, Italy

Oana Carașca Popp & Asociații Inginerie Geotehnică S.R.L., Bucharest, Romania
Geotechnical and Foundations Department, Technical University of Civil Engineering Bucharest, Bucharest, Romania

Claudia Caruso CERIS, Instituto Superior Técnico, Universidade de Lisboa, Lisbon, Portugal

Ionuț Damian Technical University of Bucharest, Bucharest, Romania

Gabriel Dănilă "Ion Mincu" University of Architecture and Urbanism, Bucharest, Romania

Mario De Stefano Department of Architecture (DiDA), University of Florence, Florence, Italy

M. Dolšek Faculty of Civil and Geodetic Engineering, University of Ljubljana, Ljubljana, Slovenia

Alexandra Ene Popp & Asociații Inginerie Geotehnică S.R.L., Bucharest, Romania

P. Fajfar Faculty of Civil and Geodetic Engineering, University of Ljubljana, Ljubljana, Slovenia

Tomasz Falborski Faculty of Civil and Environmental Engineering, Gdansk University of Technology, Gdansk, Poland

Ioanna-Kleoniki Fontara Department of Civil Engineering, Technische Universität, Berlin, Germany

K. Fujii Department of Architecture, Chiba Institute of Technology, Narashino-shi, Chiba, Japan

T. Furuta Future Robotics Technology Center, Chiba Institute of Technology, Narashino-shi, Chiba, Japan

S. Galassi Department of Architecture (DiDA), University of Florence, Florence, Italy

George K. Georgoussis Department of Civil Engineering Educators, School of Pedagogical and Technological Education (ASPETE), Attica, Greece

A. Ghersi Department of Civil Engineering and Architecture, University of Catania, Catania, Italy

Adrian Iordăchescu "Ion Mincu" University of Architecture and Urbanism, Bucharest, Romania

Robert Jankowski Faculty of Civil and Environmental Engineering, Gdansk University of Technology, Gdansk, Poland

Jafar Kayvani Department of Civil Engineering, Faculty of Engineering, Kharazmi University, Tehran, Iran

Dietlinde Köber Technical University of Bucharest, Bucharest, Romania

Shinji Konishi Structure Maintenance Division, Tokyo Metro Co., Ltd., Tokyo, Japan

M. Kosič Faculty of Civil and Geodetic Engineering, University of Ljubljana, Ljubljana, Slovenia

Konstantinos Kostinakis Department of Civil Engineering, Aristotle University of Thessaloniki, Thessaloniki, Greece

M. Lapi DICEA – Department of Civil and Environmental Engineering, University of Florence, Florence, Italy

Oren Lavan Technion – Israel Institute of Technology, Haifa, Israel

Lj. Lazarov Faculty of Civil Engineering, University "Ss. Cyril and Methodius", Skopje, Republic of Macedonia

Kouichi Maekawa Department of Civil Engineering, School of Engineering, Tokyo University, Tokyo, Japan

Triantafyllos K. Makarios Institute of Structural Analysis and Dynamics of Structure, School of Civil Engineering, Aristotle University of Thessaloniki, Thessaloniki, Greece

T. Maleska Faculty of Civil Engineering and Architecture, Opole University of Technology, Opole, Poland

Grigorios Manoukas Department of Civil Engineering, Aristotle University of Thessaloniki, Thessaloniki, Greece

Dragoș Marcu Popp & Asociații Inginerie Geotehnică S.R.L., Bucharest, Romania

Edoardo M. Marino Department of Civil Engineering and Architecture, University of Catania, Catania, Italy

M. Y. Minch Faculty of Civil Engineering, Wrocław University of Science and Technology, Wrocław, Poland

Roxana Mirițoiu Popp & Asociații Inginerie Geotehnică S.R.L., Bucharest, Romania

Konstantinos Morfidis E.P.P.O.-I.T.S.A.K., Thessaloniki, Greece

Yuya Nishigaki International Division, Tokyu Co., Ltd., Tokyo, Japan

T. Nishimura Future Robotics Technology Center, Chiba Institute of Technology, Narashino-shi, Chiba, Japan

Gholamreza Nouri Department of Civil Engineering, Faculty of Engineering, Kharazmi University, Tehran, Iran

M. Orlando DICEA – Department of Civil and Environmental Engineering, University of Florence, Florence, Italy

Tsutomu Otsuka Structure Maintenance Division, Tokyo Metro Co., Ltd., Tokyo, Japan

F. Pachla Institute of Structural Mechanics, Cracow University of Technology, Cracow, Poland

Shahin Pakzad Department of Civil Engineering, Faculty of Engineering, Kharazmi University, Tehran, Iran

Horațiu Popa Geotechnical and Foundations Department, Technical University of Civil Engineering Bucharest, Bucharest, Romania

N. Postolov Faculty of Civil Engineering, University "Ss. Cyril and Methodius", Skopje, Republic of Macedonia

Juan C. Reyes Universidad de los Andes, Bogotá, Colombia

P. P. Rossi Department of Civil Engineering and Architecture, University of Catania, Catania, Italy

Kota Sasaki Structure Maintenance Division, Tokyo Metro Co., Ltd., Tokyo, Japan

Armando Sierra Universidad de los Andes, Bogotá, Colombia

Barbara Sołtysik Faculty of Civil and Environmental Engineering, Gdansk University of Technology, Gdansk, Poland

Marco Tanganelli Department of Architecture (DiDA), University of Florence, Florence, Italy

T. Tatara Institute of Structural Mechanics, Cracow University of Technology, Cracow, Poland

K. Todorov Faculty of Civil Engineering, University "Ss. Cyril and Methodius", Skopje, Republic of Macedonia

Ryuta Tsunoda Structure Maintenance Division, Tokyo Metro Co., Ltd., Tokyo, Japan

Stefania Viti Department of Architecture (DiDA), University of Florence, Florence, Italy

R. Volcev Faculty of Civil Engineering, University "Ss. Cyril and Methodius", Skopje, Republic of Macedonia

Philip J. Wilkinson Stratagroup, Consulting Engineers Ltd, Hastings, New Zealand

Morteza Tahmasebi Yamchelou Department of Civil Engineering, Faculty of Engineering, Kharazmi University, Tehran, Iran

Mohammad Reza Yosefzaei Department of Civil Engineering, Faculty of Engineering, Kharazmi University, Tehran, Iran

T. Yoshida Future Robotics Technology Center, Chiba Institute of Technology, Narashino-shi, Chiba, Japan

Dan Zamfirescu Technical University of Bucharest, Bucharest, Romania

Zbigniew Zembaty Faculty of Civil Engineering and Architecture, Opole University of Technology, Opole, Poland

Part I
Seismic Load, Ground Motion, Rocking Excitations

Chapter 1
Effect of Soil Compliance on Seismic Response of Slender Towers Under Rocking Excitations

P. Bonkowski, Z. Zembaty, and M. Y. Minch

Abstract Parametric analysis of the effect of soil compliance on the response of a slender tower to combined horizontal and rotational (rocking) of induced ground motion excitations is carried out. In contrast to earlier analyses which use theoretical decomposition of seismic wave field to obtain rotations, this study is using 6 degree of freedom ground motion records to carry on time history integration leading to structural response accounting for "true" phase interactions between horizontal and rocking excitations. The analysis leads to conclusion that the more flexible the sub-soil the smaller the response, however this effect is better pronounced for rotational than horizontal excitations. It was also demonstrated that regardless of the soil compliance, rotational component can have either increasing or decreasing influence on the structural internal forces. In the analysed cases of two sets of excitations such different results were obtained. Clearly more credible rotational-horizontal records of seismic strong ground motion are needed.

Keywords Rotational component · Time history analysis · Seismic analysis · Slender tower · Chimney

1.1 Introduction

Each point on the ground surface can be subjected not only to three, translational components of seismic ground motion but also to three rotations about these axes (Fig. 1.1). However due to lack of measuring techniques and appropriate seismological models these components were omitted or disregarded (see e.g. footnote of

P. Bonkowski (✉) · Z. Zembaty
Faculty of Civil Engineering and Architecture, Opole University of Technology, Opole, Poland
e-mail: p.bonkowski@po.opole.pl; z.zembaty@po.opole.pl

M. Y. Minch
Faculty of Civil Engineering, Wrocław University of Science and Technology, Wrocław, Poland
e-mail: m.minch@pwr.edu.pl

D. Köber et al. (eds.), *Seismic Behaviour and Design of Irregular and Complex Civil Structures III*, Geotechnical, Geological and Earthquake Engineering 48,
https://doi.org/10.1007/978-3-030-33532-8_1

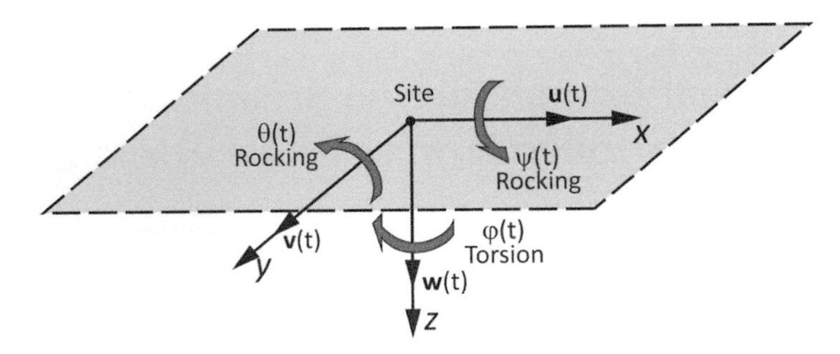

Fig. 1.1 Six components of earthquake ground motion on the ground surface

Richter to his famous monograph (Richter 1958). There are however good reasons to believe that the rotational components exist, particularly in close proximities from the epicentres of earthquakes (Trifunac 1982; Zembaty 2009). Recently the geophysical community accepts the importance of rotational seismic ground motion component and the measuring techniques matured enough and first 6 degree of freedom (6-dof) seismic signals are acquired (e.g. (Takeo 1998; Zembaty et al. 2017)). Very recently the 6-dof seismic records from induced seismic ground motions of Modified Mercalli Intensity (MMI) IV (Zembaty et al. 2017) were applied to compute seismic response of a slender reinforced concrete chimney (Bońkowski et al. 2018). A substantial contribution of rocking component in the overall structural response was demonstrated. The numerical analysis was carried out with the assumption that the slender structure was fixed in the sub-soil. In what follows the numerical analysis of paper (Bońkowski et al. 2018) is extended for compliant soils i.e. those with shear waves velocities from 150 to 800 m/s which roughly corresponds with Eurocode 8 (EC-8) local soil conditions described as types 'A' to 'D'. The analysed structure is an industrial chimney, 160 m high under seismic horizontal and rocking (rotation about horizontal axis) excitations (Fig. 1.2). This analysis complements earlier studies of rotational seismic effects on slender towers where EC-8 seismic code was applied (Zembaty et al. 2016).

1.2 Ground Motion Input

This analysis utilizes ground motion records from monitoring of induced seismicity acquired in the Upper Silesian Coal Basin (south of Poland). The strongest records reached peak ground acceleration PGA = 0.4–0.8 m/s^2 and peak ground velocity (PGV) of about 1–2 cm/s. The recorded quakes reached intensity of Modified Mercalli IV. Detailed information about the monitoring program can be found in the paper by Zembaty et al. (Zembaty et al. 2017).

The earlier analyses of the rotational effects on structures with their inputs taken from analytical and numerical seismic wave decompositions (e.g. (Zembaty 2009; Basu et al. 2013; Falamarz-Sheikhabadi and Ghafory-Ashtiany 2015) applied the

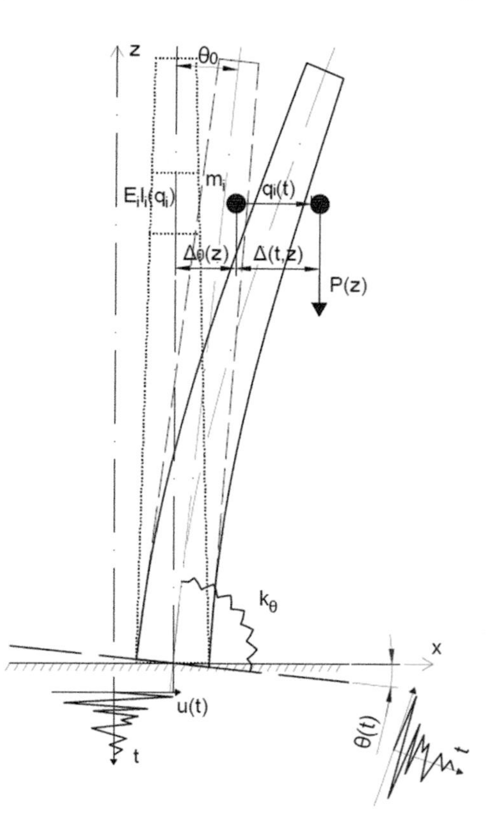

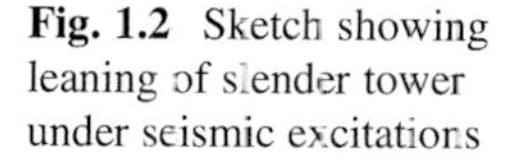
Fig. 1.2 Sketch showing leaning of slender tower under seismic excitations

horizontal and rotational components as acting independently while in reality there are good reasons to assume that they are strongly correlated. In particular such the approach even disregards actual signs of the seismic records playing important role in phase interaction of the response to rocking and horizontal excitations (see e.g. Bońkowski et al. 2018). Thus, applying direct, seismic 6-dof records made it possible to take into account proper phase interaction of these two components which was not the case of publications (Zembaty 2009; Basu et al. 2013; Falamarz-Sheikhabadi and Ghafory-Ashtiany 2015). This is important, as both horizontal and rocking components can substantially magnify or reduce the total seismic response of the slender tower.

1.3 Numerical Analyses

The horizontal and rocking seismic actions have been applied to the 160 m high industrial, reinforced concrete chimney using SAP2000 software (Computers and Structures, Inc 2017). The basic geometry of the structure is shown in Fig. 1.3. More information about the structure can be found in the monograph (Ciesielski 1966) and

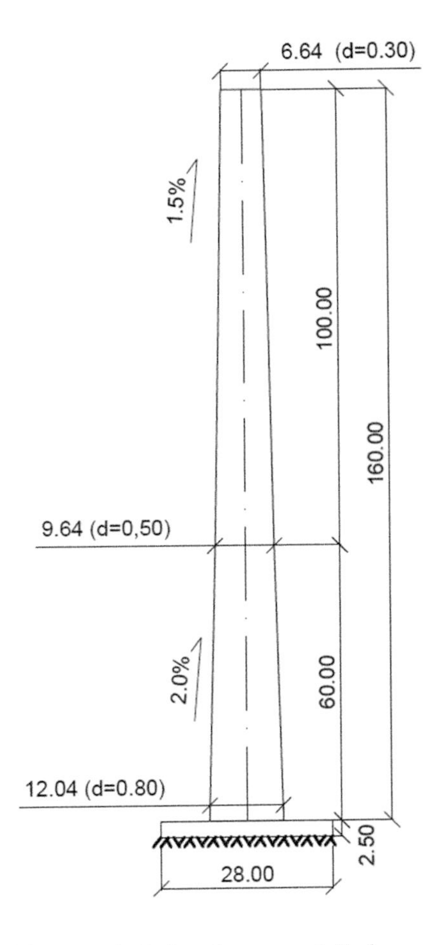

Fig. 1.3 Sketch of the analysed chimney

paper by Zembaty (Zembaty 1987). The chimney shaft has been modelled as a cantilever beam divided into 17 finite elements. The structure is connected with the base by a pin and rotational spring. Masonry lining and insulation have been modelled as additional lumped masses concentrated in the finite element nodes.

To carry out the time history analyses of this paper, translational-rotational records pairs have been chosen from the data set of 6-dof records (see Zembaty et al. 2017), namely translation along North-South direction interacting with the simultaneous rocking about East-West axis *(R E-W)vs(T N-S)* and translation along East-West direction interacting with the rocking about North-South axis *(R N-S)vs(T E-W)*. These records were chosen from the event described in (Zembaty et al. 2017) as IMI 20151212_043336.

To analyse the effect of a soil compliance a familiar eq. (1) was applied (e.g. Castellani 1983) to calculate rotational spring stiffness. Assuming circular footing resting on an elastic half-space:

Table 1.1 Stiffness used in the analysis and first three modal periods

Ground type	K_θ [kNm/rad]	T_1 [s]	T_2 [s]	T_3 [s]	T_4 [s]
Rigid	∞	2.80	0.688	0.289	0.164
A	1.46E+10	2.81	0.691	0.290	0.164
B	6.24E+09	2.82	0.695	0.292	0.164
C	1.32E+09	2.91	0.719	0.301	0.167
D	3.95E+08	3.15	0.776	0.318	0.173

$$K_\theta = \frac{8GR^3}{3(1-\nu)} \tag{1.1}$$

where R = circular footing radius; v = soil Poisson ratio; G = Soil shear modulus given by formula:

$$G = \rho v_s^2 \tag{1.2}$$

in which ρ = Soil unit mass; v_s = shear wave velocity of velocity of the top of the 30 m of ground profile. Shear wave velocity has been taken from table A.1 in Eurocode 8 (CEN 2005). The circular footing radius equals to 12 m. Calculated rotational spring stiffness and first three modal periods have been given in the Table 1.1.

In Fig. 1.4 (a) and (c) bending moments are presented along the height of the chimney for different soil compliance, in (b) and (d) respective rotational ground motion contribution to the overall bending moment has been presented. The minus sign in the contribution means that the rotational ground motion had positive (reducing the total) effect on the internal forces.

It can be seen that rotational ground motion can have either positive or negative effect on the internal forces in the structure depending on the applied direction. This was also noticed for other ground motions not presented in this paper.

The total effect and contribution of the rotational ground motion takes maximum for structure rigidly fixed in the base. When soil compliance increases (natural periods also increase) seismic action effects from both – translational and rotational ground motion decrease. For rotational seismic excitations the decrease is quicker, as bigger is the contribution of higher natural frequencies to its motion. For the analysed structure there is neither a difference in rotational ground motion contribution nor in the total bending moments in the chimney shaft for grounds up to B type.

1.4 Conclusions

Parametric analysis of the effect of soil compliance on the contribution of rotational ground motion to the total bending moments has been performed.

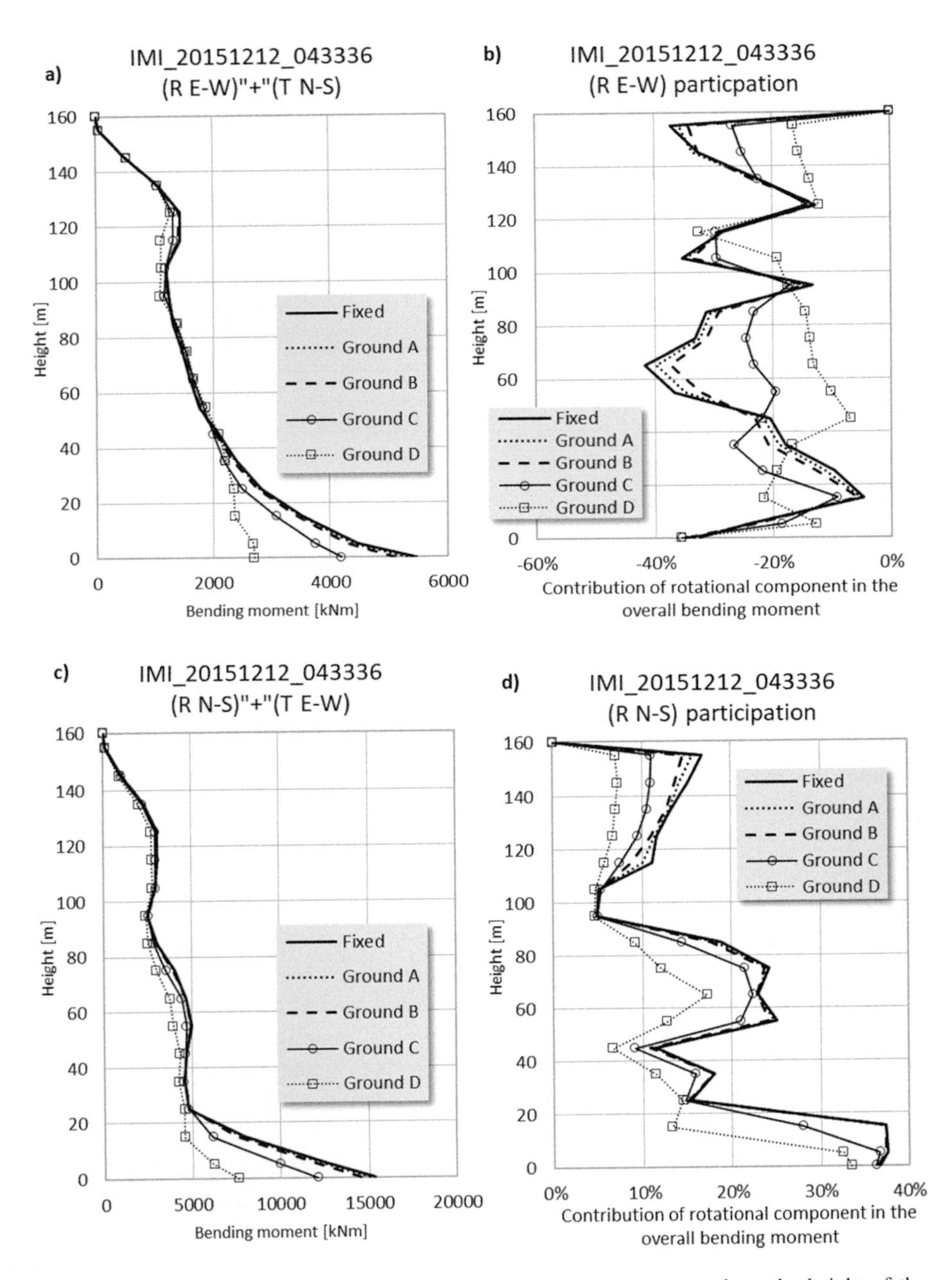

Fig. 1.4 Influence of the soil compliance on the total bending moment along the height of the structure (**a** and **c**) and contribution of rotational component to the total bending moment (**b** and **d**)

Unlike the earlier research applying approximate formulas (e.g. Falamarz-Sheikhabadi and Ghafory-Ashtiany 2012, 2015) or simplified response spectrum approach (e.g. Takeo 1998), this analysis used true 6-dof recorded seismic ground motions. The parametric analysis has an introductory character. Anyway, it demonstrated that soil compliance reduces the total internal forces though this effect is stronger for rotational than for horizontal excitations. The applied ground motion is of small intensity (MMI=IV). Still more records from stronger quakes and different sites are needed to conclude on the overall site effect on slender tower vibrations under combined rocking and horizontal seismic excitations.

Acknowledgments This research was partially supported by statutory fund of Polish Ministry of Science and Higher Education NBS 15/17 and by grant for young researchers on the Faculty of Civil Engineering and Architecture of Opole University of Technology DS – MN/31/WBiA/17.

References

Basu D, Whittaker AS, Constantinou MC (2013) Extracting rotational components of earthquake ground motion using data recorded at multiple stations. Earthq Eng Struct Dyn 42:451–468. https://doi.org/10.1002/eqe.2233

Bońkowski PA, Zembaty Z, Minch MY (2018) Time history response analysis of a slender tower under translational-rocking seismic excitations. Eng Struct 155:387–393. https://doi.org/10.1016/j.engstruct.2017.11.042

Castellani A (1983) Interazione terreno-struttura. In: Costruzioni in zona sismica. Masson Italia Editori, Milano

CEN (2005) EN 1998-6:2005 Eurocode 8: design of structures for earthquake resistance – part 6: towers, Masts and chimneys

Ciesielski R (1966) Budownictwo betonowe. Tom XIII – Zbiorniki, zasobniki, silosy, kominy i maszty. Arkady, Warszawa

Computers and Structures, Inc (2017) SAP2000. Walnut Creek, California, USA

Falamarz-Sheikhabadi MR, Ghafory-Ashtiany M (2012) Approximate formulas for rotational effects in earthquake engineering. J Seismol 16:815–827. https://doi.org/10.1007/s10950-012-9273-z

Falamarz-Sheikhabadi MR, Ghafory-Ashtiany M (2015) Rotational components in structural loading. Soil Dyn Earthq Eng 75:220–233. https://doi.org/10.1016/j.soildyn.2015.04.012

Richter CF (1958) Elementary seismology. W. H. Freeman, San Francisco

Takeo M (1998) Ground rotational motions recorded in near-source region of earthquakes. Geophys Res Lett 25:789–792. https://doi.org/10.1029/98GL00511

Trifunac MD (1982) A note on rotational components of earthquake motions on ground surface for incident body waves. Int J Soil Dyn Earthq Eng 1:11–19. https://doi.org/10.1016/0261-7277(82)90009-2

Zembaty Z (1987) On the reliability of tower-shaped structures under seismic excitations. Earthq Eng Struct Dyn 15:761–775

Zembaty Z (2009) Tutorial on surface rotations from wave passage effects: stochastic spectral approach. Bull Seismol Soc Am 99:1040–1049. https://doi.org/10.1785/0120080102

Zembaty Z, Rossi A, Spagnoli A (2016) Estimation of rotational ground motion effects on the bell tower of Parma Cathedral. In: Seismic behaviour and design of irregular and complex civil structures II. Springer, Cham, pp 35–48

Zembaty Z, Mutke G, Nawrocki D, Bobra P (2017) Rotational ground-motion records from induced seismic events. Seismol Res Lett 88:13–22. https://doi.org/10.1785/0220160131

Chapter 2
Evaluation of Foundation Input Motions Based on Kinematic Interaction Models

Tomasz Falborski

Abstract The present study was designed to demonstrate the importance of base-slab averaging and embedment effects on the foundation-level input motions due to earthquake excitations. Evaluation of foundation-level input motions based on the most commonly adopted kinematic interaction models are presented. In order to conduct this investigation, original records of horizontal accelerations for two case-study buildings were utilized. Computed foundation-level input motions, in both NS and EW directions, were compared to the actual acceleration-time histories recorded at the foundation levels. The results clearly indicate that incorporating base-slab averaging and embedment effects in seismic analyses can modify the dynamic excitation imposed at the foundation level, and, as a consequence, lead to more accurate structural response due to earthquake ground motions.

Keywords Soil-structure interaction · Foundation input motions · Kinematic interaction effects · Seismic response · Irregular structures

2.1 Introduction

Seismic performance of a building structure subjected to strong earthquake excitations may be affected by many different factors (see, for example, Falborski and Jankowski 2013, 2016, 2017), structural pounding (see, for example, Jankowski and Mahmoud 2015, 2016; Naderpour et al. 2016; Sołtysik et al. 2016, 2017), damage level (see, for example, Ebrahimian et al. 2017) etc. Among these factors the interaction between the structure foundation and the underlying soil is indentified as one of the most significant contributors (see, for example, Gazetas 1991; Wolf 1985; Mylonakis and Gazetas 2000; Stewart et al. 1999a, b; Veletsos and Prasad 1989). The dynamic response of a building strongly depends on the ground motions

T. Falborski (✉)
Faculty of Civil and Environmental Engineering, Gdansk University of Technology, Gdansk, Poland
e-mail: tomfalbo@pg.gda.pl

D. Köber et al. (eds.), *Seismic Behaviour and Design of Irregular and Complex Civil Structures III*, Geotechnical, Geological and Earthquake Engineering 48,
https://doi.org/10.1007/978-3-030-33532-8_2

transmitted to the structure, which tend to differ from those recorded in the free field (see, for example, guidelines prepared by National Earthquake Hazards Reduction Program 2012; Abrahamson et al. 1991; Kim and Stewart 2003; Mikami et al. 2008; Stewart and Tileylioglu 2007; Veletsos et al. 1997). The ground motion imposed at the foundation level is reduced due to base-slab averaging (i.e. averaging of variable ground motions across the foundation slab), and embedment effects, which are associated with the reduction of the ground motion that tends to occur with depth in a soil deposit. Both base-slab averaging and embedment effects can be visualized as a filter to the high-frequency (short period) components of the free-field ground motion. Utilizing the free-field motion (usually denoted as u_g), instead of the foundation-level motion (also referred to as the foundation input motion or u_{FIM}), may result in misrepresentation of the actual building response, especially in sesimic analyses of complex and irregular structures.

There are a few theoretical approaches for predicting the relationship between the foundation-input motion and the free-field ground motion. All these mathematical procedures can lead to significant departures from the fixed-base results and, eventually, more accurate and realistic evaluation of the probable structural response to a seismic excitation.

To illustrate the significance of the kinematic interaction effects, the evaluation of foundation input motion using the most frequently adopted procedures (presented in both world literature and building design codes and standards) are presented. Computed foundation input motions, in both NS and EW directions, were compared to the actual acceleration-time histories recorded at the foundation levels.

2.2 Kinematic Interaction Procedures

There are a few theoretical procedures for evaluating the foundation input motions, according to which kinematic interaction effects can be represented in terms of ratios between the response spectrum ordinates for u_{FIM} ($S_{a\ -\ FIM}$) and u_g ($S_{a\ -\ g}$). Therefore, acceleration histories representing the foundation input motions can be predicted by reversing Fourier transform S_a multiplied by base-slab averaging alone or base-slab averaging and embedment factors.

The first procedure utilized in the present study in the one presented by Applied Technology Council (2005), according to which the transfer functions can be estimated as follows:

$$RRS = \frac{S_{a-FIM}(f)}{S_{a-g}(f)} \tag{2.1}$$

$$RRS = RRS_{bsa} \cdot RRS_{emb} \tag{2.2}$$

$$RRS_{bsa} = \begin{cases} 1 - \frac{1}{14100}\left(\frac{B_e}{T}\right)^{1.2} \text{ for } T > 0.2 \text{ sec } (f > 5 \text{ Hz}) \\ 1 - \frac{5}{14100} B_e^{1.2} \text{ for } T \leq 0.2 \text{ sec } (f \leq 5 \text{ Hz}) \end{cases} \quad (2.3)$$

$$RRS_{emb} = \begin{cases} \cos\left(\frac{2\pi \cdot D}{T \cdot n \cdot V_s}\right) \text{ for } T > 0.2 \text{ sec } (f > 5 \text{ Hz}) \\ \cos\left(\frac{10\pi \cdot D}{n \cdot V_s}\right) \text{ for } T \leq 0.2 \text{ sec } (f \leq 5 \text{ Hz}) \end{cases} \quad (2.4)$$

where:

$B_e = \sqrt{ab}$ effective foundation size, where a and b are the full footprint dimensions of the building foundation (in ft),
D foundation embedment (in ft),
V_s shear wave velocity for the site soil conditions (in ft/s),
n shear wave velocity reduction factor for the expected Peak Ground Acceleration (PGA).

It can be noticed that a limiting frequency of 5 Hz was assumed, as it has been recognized to be the most appropriate value for earthquake ground motions. Accordingly, this method may not be fully accurate for predicting the foundation-input motions resulting from mining tremors (which has recently become an issue of major concern of both professional and academic communities in Poland), as their fundamental excitation frequencies usually vary from 5 Hz to 10 Hz (see, for example, Zembaty 2004; Kuźniar and Tatara 2015, 2017).

The second approach employed in the present investigation was proposed by Mylonakis (2006). Transfer functions for evaluating the foundation-input motions are defined as follows:

$$H_u = \frac{S_{a-FIM}(f)}{S_{a-g}(f)} \quad (2.5)$$

$$H_u = H_{u-bsa} \cdot H_{u-emb} \quad (2.6)$$

$$H_{u-bsa} = \begin{cases} \dfrac{\sin\left(a_0\left(\frac{V_s}{V_{app}}\right)\right)}{a_0\left(\frac{V_s}{V_{app}}\right)} \text{ for } a_0 \leq \frac{\pi}{2}\frac{V_s}{V_{app}} \\ \frac{2}{\pi} \text{ for } a_0 > \frac{\pi}{2}\frac{V_s}{V_{app}} \end{cases} \quad (2.7)$$

$$H_{u-emb} = \begin{cases} \cos\left(\frac{D \cdot \omega}{V_s}\right) \text{ for } \frac{D \cdot \omega}{V_s} \leq 1.1 \\ 0.45 \text{ for } \frac{D \cdot \omega}{V_s} > 1.1 \end{cases} \tag{2.8}$$

$$a_0 = \frac{\omega}{V_s} \frac{B_e}{2} \tag{2.9}$$

where:

V_{app} apparent horizontal velocity; for a typical soil site an assumption of $V_{app}/V_s = 10$ can be made.

2.3 Case-Study Buildings

The present study was conducted for two irregular case-study buildings (both located in California, United States) with seismic instrumentation and available recordings provided by the Center for Engineering Strong Motions Data (CESMD), which was established by the United States Geological Survey (USGS) and the California Geological Survey (CGS).

The first case-study structure is a 14-storey reinforced-concrete building located in Los Angeles, CA. The structure has one basement level and measures 45.34 m (148 ft 9 in) tall from the ground surface to the roof. The height of the basement level is 2.75 m (9 ft). The height of the first and second stories is 4.2 m (13 ft 9 in) and 2.75 m (9 ft), respectively, whereas all other above-grade stories are 3.2 m (10 ft 6 in). The building is rectangular in plan, measuring 15.54 m (51 ft) wide by 66.14 m (217 ft) long. It was designed in 1925, and instrumented in 1976 (CSMIP Station No. 24236). A total of 12 accelerometers are located at 4 levels (basement, 8th floor, 12th floor, and main roof). Moreover, three free-field accelerometers are located in the vicinity of the site (CSMIP Station No. 24303), for which the shear wave velocity is 316 m/s (1036.7 ft/s).

The second building is a 14-storey reinforced-concrete structure erected in Santa Rosa, CA. The building has no subterranean levels and measures 37.8 m (124 ft) tall from the ground surface to the roof. The height of the first storey is 3.45 m (11 ft 4 in), and all other above-grade stories are 2.64 m (8 ft 8 in). The gross plan dimensions of the building are 24.38 m (80 ft) wide by 26.21 m long (86 ft), although the typical floor is irregular in plan. The building was designed in 1970, and instrumented in 1985 (CSMIP Station No. 68486). A total of 16 accelerometers are located at 5 levels (ground floor, 5th floor, 9th floor, 12th floor, and roof). Additionally, three free-field accelerometers are located in the vicinity of the site (CSMIP Station No. 68491), for which the shear wave velocity is 363 m/s (1190.9 ft/s).

2.4 Numerical Analysis

In order to evaluate foundation-level input motions using the procedures briefly presented in Sect. 2.2, two different ground motions were utilized in this investigation:

(a) Magnitude 6.4 Northridge Earthquake of 17 January 1994 for the Los Angeles building located 23 km (37 mi) from the epicentre,
(b) Magnitude 6.0 South Napa earthquake of 24 August 2014 for the Santa Rosa building located 42.1 km (67.75 mi) from the epicentre.

Using the actual free-field motions recorded by the seismic stations located in close vicinity to the case-study buildings presented in Sect. 2.3, the foundation-input motions, in both NS and EW directions, were evaluated and compared to the original records at the basement levels. Both recorded and computed peak lateral accelerations are briefly summarized in Table 2.1. Acceleration-time histories are presented in Figs. 2.1, 2.2, 2.3, 2.4, 2.5, 2.6, 2.7, 2.8, 2.9, 2.10, 2.11, 2.12, 2.13, 2.14, 2.15, and 2.16.

Table 2.1 Comparison of peak lateral accelerations

	Peak lateral acceleration in NS direction (×g)				Peak lateral acceleration in NS direction (×g)			
Building, earthquake	u_g	u_{FIM}	u_{FIM} (RRS)	u_{FIM} (H_u)	u_g	u_{FIM}	u_{FIM} (RRS)	u_{FIM} (H_u)
Los Angeles building, 1994 Northridge earthquake	0.389	0.278	0.306	0.297	0.231	0.207	0.205	0.200
Santa Rosa building, 2014 South Napa earthquake	0.049	0.044	0.042	0.043	0.047	0.036	0.036	0.037

where:
u_g recorded free-field motion,
u_{FIM} recorded foundation-level input motion,
u_{FIM} (RRS) foundation-level input motion predicted (computed) with the RRS procedure,
u_{FIM} (H_u) foundation-level input motion predicted (computed) with the H_u procedure.

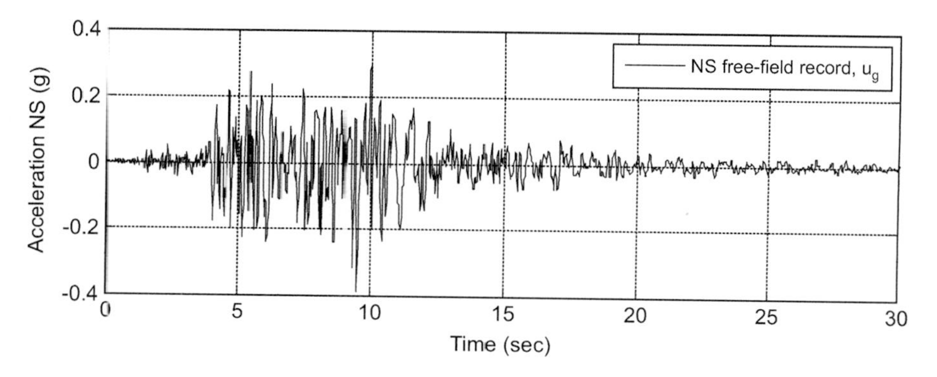

Fig. 2.1 Free-field acceleration-time history record in NS direction (Los Angeles building, Northridge earthquake)

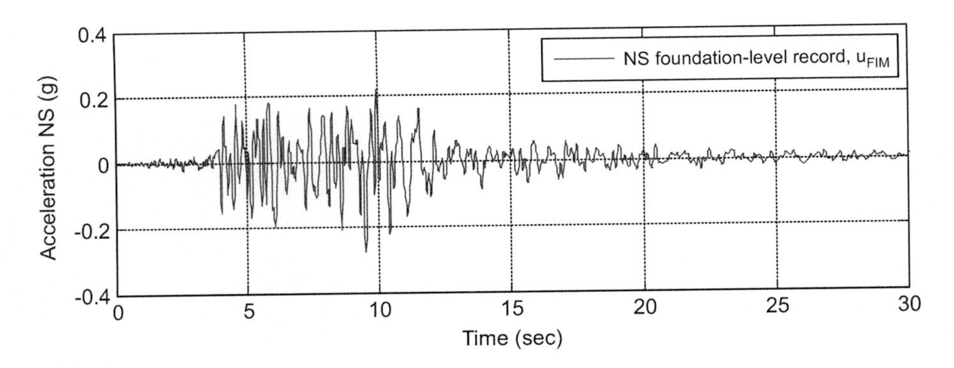

Fig. 2.2 Foundation-level acceleration-time history record in NS direction (Los Angeles building, Northridge earthquake)

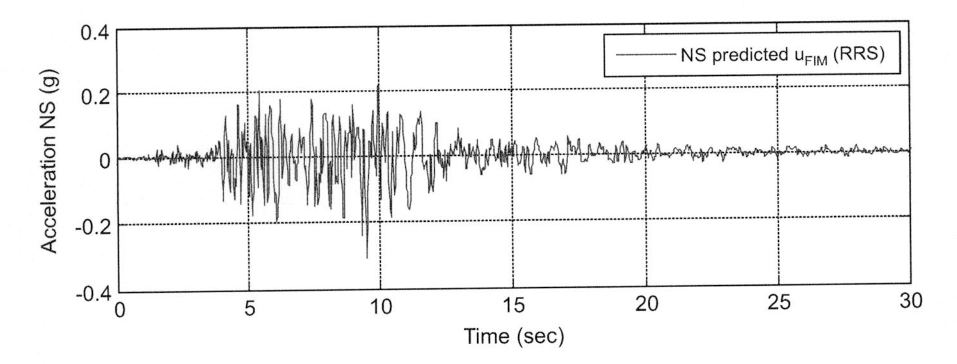

Fig. 2.3 Predicted foundation-level acceleration-time history record in NS direction – RRS (Los Angeles building, Northridge earthquake)

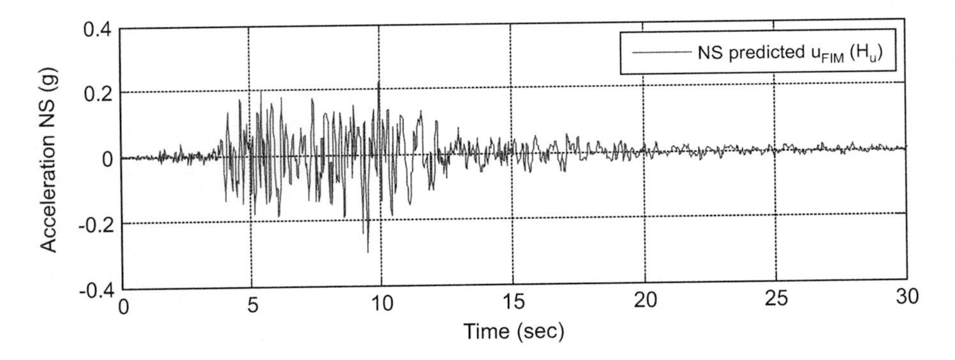

Fig. 2.4 Predicted foundation-level acceleration-time history record in NS direction – H_u (Los Angeles building, Northridge earthquake)

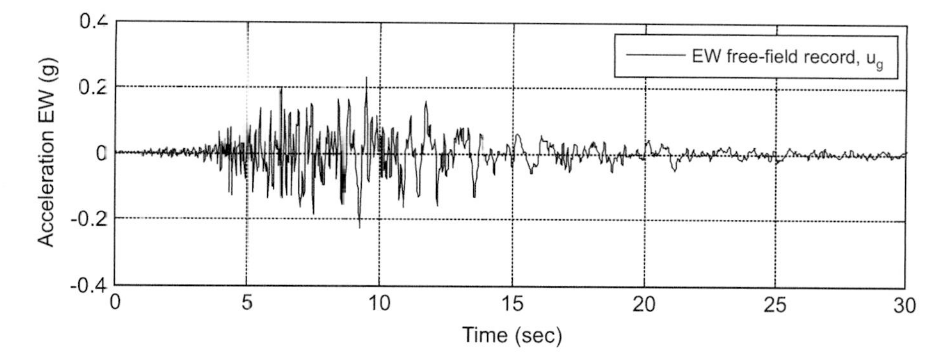

Fig. 2.5 Free-field acceleration-time history record in EW direction (Los Angeles building, Northridge earthquake)

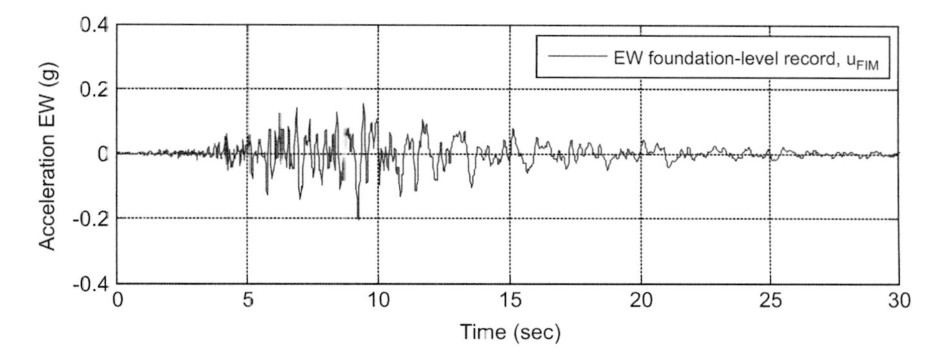

Fig. 2.6 Foundation-level acceleration-time history record in EW direction (Los Angeles building, Northridge earthquake)

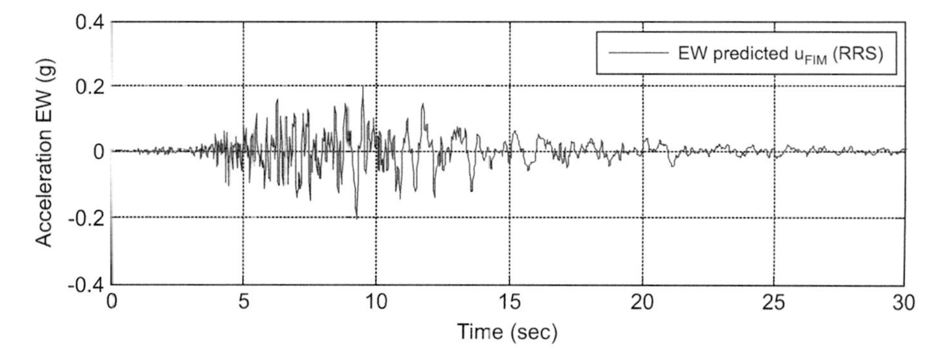

Fig. 2.7 Predicted foundation-level acceleration-time history record in EW direction – RRS (Los Angeles building, Northridge earthquake)

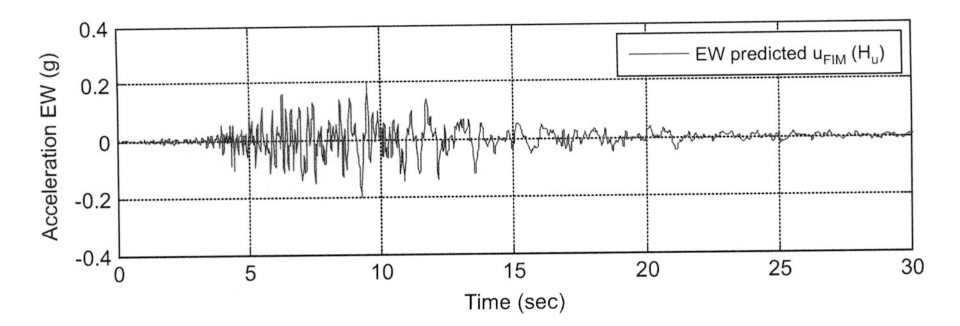

Fig. 2.8 Predicted foundation-level acceleration-time history record in EW direction – H_u (Los Angeles building, Northridge earthquake)

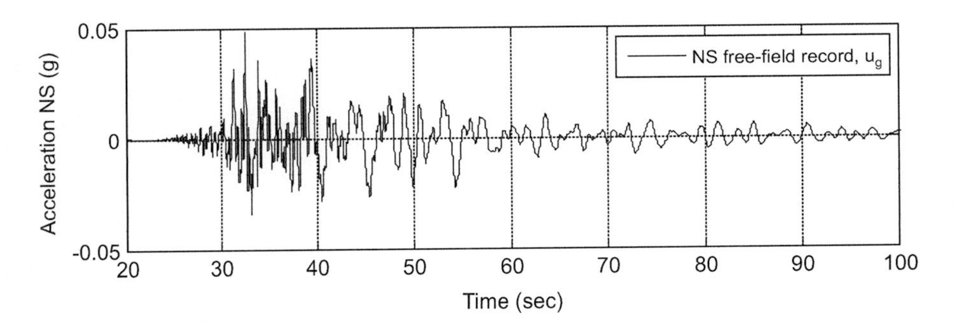

Fig. 2.9 Free-field acceleration-time history record in NS direction (Santa Rosa building, South Napa earthquake)

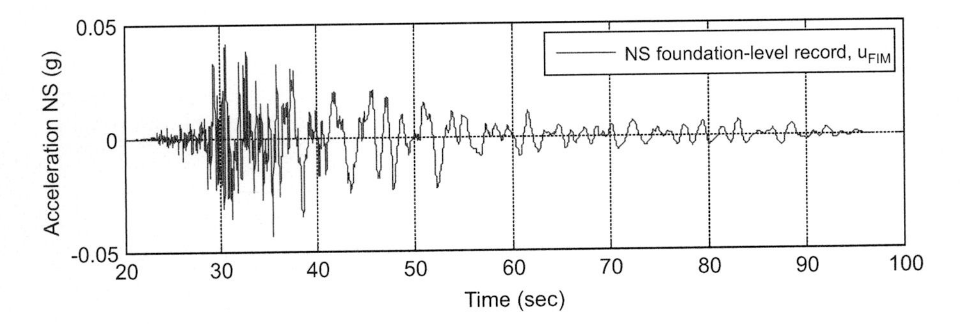

Fig. 2.10 Foundation-level acceleration-time history record in NS direction (Santa Rosa building, South Napa earthquake)

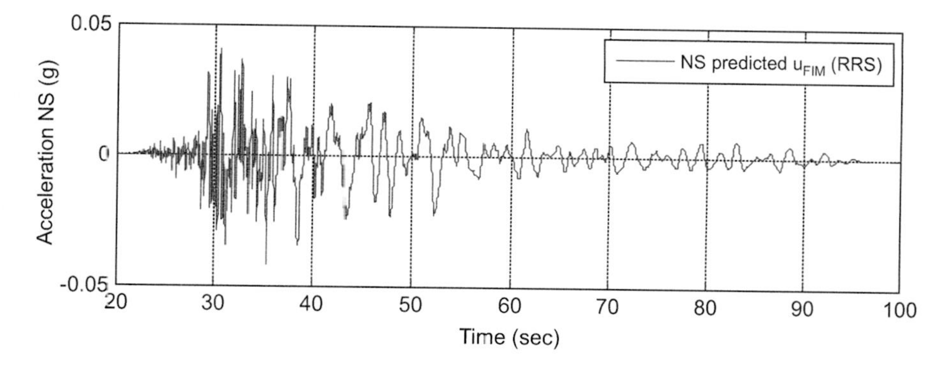

Fig. 2.11 Predicted foundation-level acceleration-time history record in NS direction – RRS (Santa Rosa building, South Napa earthquake)

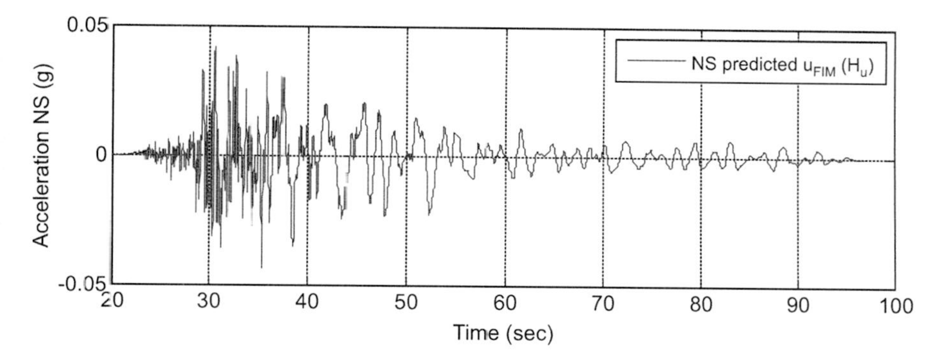

Fig. 2.12 Predicted foundation-level acceleration-time history record in NS direction – H_u (Santa Rosa building, South Napa earthquake)

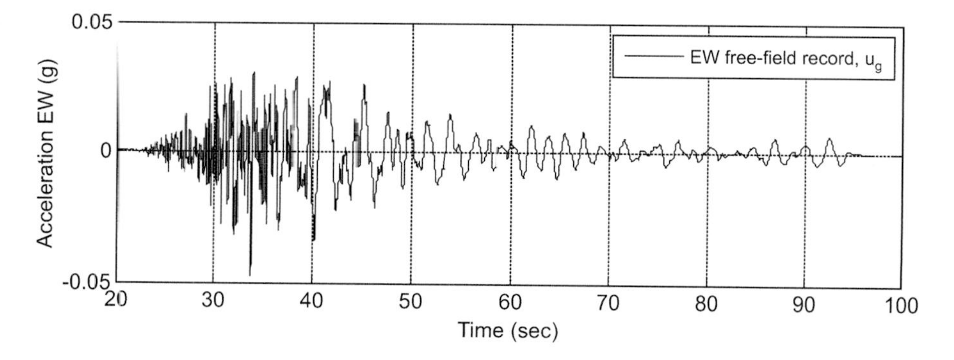

Fig. 2.13 Free-field acceleration-time history record in EW direction (Santa Rosa building, South Napa earthquake)

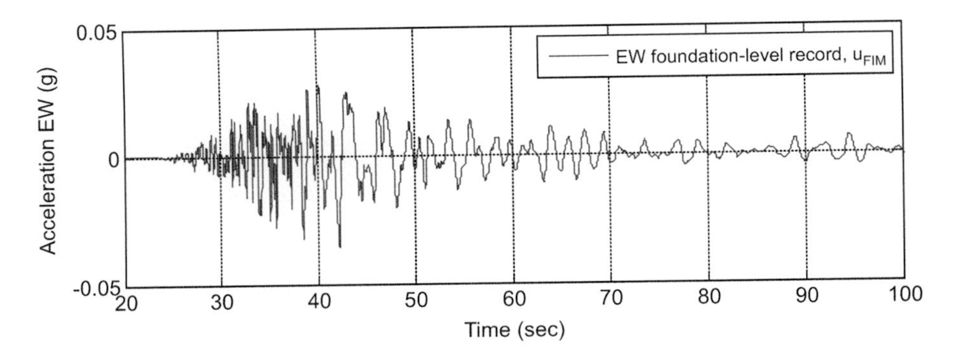

Fig. 2.14 Foundation-level acceleration-time history record in EW direction (Santa Rosa building, South Napa earthquake)

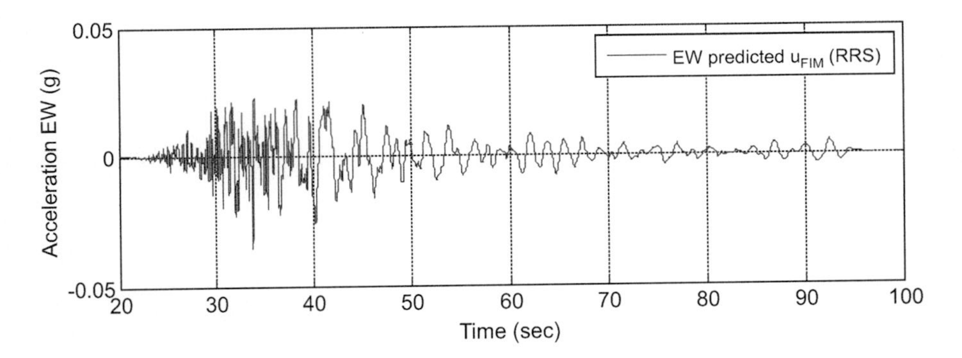

Fig. 2.15 Predicted foundation-level acceleration-time history record in EW direction – RRS (Santa Rosa building, South Napa earthquake)

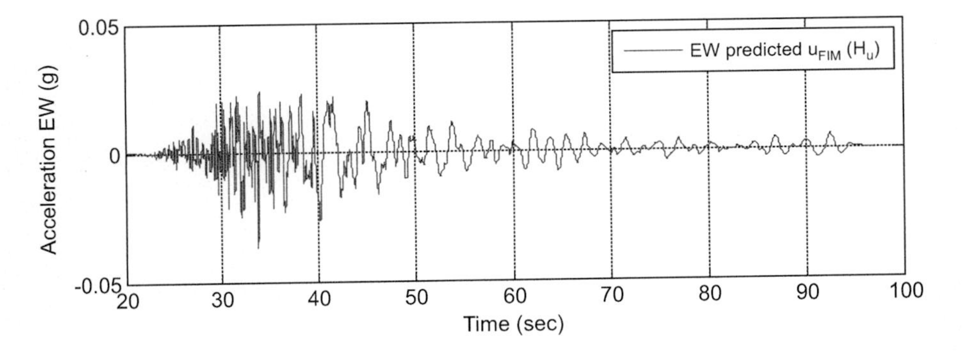

Fig. 2.16 Predicted foundation-level acceleration-time history record in EW direction – H_u (Santa Rosa building, South Napa earthquake)

2.5 Conclusions

The present investigation was carried out to illustrate the importance of base-slab averaging and embedment effects on the foundation-level input motions due to earthquake excitations. Foundation-level input motions were evaluated with the most commonly adopted kinematic interaction procedures. The study was conducted for two case-study buildings with seismic instrumentation and available recordings. Computed foundation-level input motions, in both NS and EW directions, were compared to the original acceleration-time histories recorded at the foundation levels. Specifically, following conclusions can be drawn from the present study:

(a) Close inspection of Table 2.1 shows that the peak lateral accelerations of the originally recorded foundation-level input motions, u_{FIM}, are lower of the order of 10% (Santa Rosa building, 2014 South Napa earthquake, NS direction) to 29% (Los Angeles Building, 1994 Northridge earthquake, NS direction) when compared to those recorded at seismic stations, u_g. The greater reduction in peak lateral accelerations for the Los Angeles building is most likely due to embedment effects, as the Santa Rosa building has no subterranean levels.
(b) Similar values of the predicted peak lateral accelerations with the RRS and H_u procedures to those recorded at the foundation levels confirm the effectiveness of the kinematic interaction procedures for implementing base-slab averaging and embedment effects in seismic analyses.

The results clearly indicate that base-slab averaging and embedment effects in seismic analyses can modify the earthquake-induced motions imposed at foundation level, which may significantly alter structural response during seismic events.

Acknowledgments The author acknowledge accessing strong-motion data through the Center for Engineering Strong Motion Data (CESMD). The networks or agencies providing the data used in this report are the California Strong Motion Instrumentation Program (CSMIP) and the USGS National Strong Motion Project (NSMP).

References

Abrahamson NA, Schneider JF, Stepp JC (1991) Empirical spatial coherency functions for application to soil-structure interaction analyses. Earthquake Spectra 7(1):1–27

Applied Technology Council ATC (2005) Improvement of nonlinear static seismic analysis procedures, FEMA440 report. Redwood City, CA

Ebrahimian M, Todorovska MI, Falborski T (2017) Wave method for structural health monitoring: testing using full-scale shake table experiment data. J Struct Eng 143(4). https://doi.org/10.1061/(ASCE)ST.1943-541X.0001712

Falborski T, Jankowski R (2013) Polymeric bearings – a new base isolation system to reduce structural damage during earthquakes. Key Eng Mater 569–570:143–150

Falborski T, Jankowski R (2016) Behaviour of asymmetric structure with base isolation made of polymeric bearings. In: Zembaty Z, De Stefano M (eds) Geotechnical, geological and

earthquake engineering 40: seismic behaviour and design of irregular and complex civil structures II. Springer, Cham, pp 333–341
Falborski T, Jankowski R (2017) Experimental study on effectiveness of a prototype seismic isolation system made of polymeric bearings. Appl Sci 7(8):808. https://doi.org/10.3390/app8030400
Gazetas G (1991) Foundation vibrations. In: Fang H-Y (ed) Foundation engineering handbook. Springer, Boston
Jankowski R, Mahmoud S (2015) Earthquake-induced structural pounding. Springer, New York
Jankowski R, Mahmoud S (2016) Linking of adjacent three-storey buildings for mitigation of structural pounding during earthquakes. Bull Earthq Eng 14:3075–3097
Kim S, Stewart JP (2003) Kinematic soil-structure interaction from strong motion recordings. J Geotech Geoenviron 129(4):323–335
Kuźniar K, Tatara T (2015) Zastosowanie przybliżonych modeli SSI w przypadku wstrząsów górniczych (application of approximate SSI models in case of mining tremors). Przegląd Górniczy 10:25–30
Kuźniar K, Tatara T (2017) Simple models for determination of the differences of ground and building foundation response spectra in LGC region. Tech Trans 1:65–77
Mikami A, Stewart JP, Kamiyama M (2008) Effects of time series analysis protocols on transfer functions calculated from earthquake accelerograms. Soil Dyn Earthq Eng 28(9):695–706
Mylonakis G, Gazetas G (2000) Seismic soil-structure interaction: beneficial or detrimental? J Earthq Eng 4:277–301
Mylonakis G, Nikolaou S, Gazetas G (2006) Footings under seismic loading: analysis and design issues with emphasis on bridge foundations. Soil Dyn Earthq Eng 26(9):824–853
Naderpour H, Barros RC, Khatami SM, Jankowski R (2016) Numerical study on pounding between two adjacent buildings under earthquake excitation. Shock Vib 2016:1504783. https://doi.org/10.1155/2016/1504783
National Earthquake Hazards Reduction Program NEHRP (2012) Soil-structure interaction for building structures
Sołtysik B, Falborski T, Jankowski R (2016) Investigation on damage-involved structural response of colliding steel structures during ground motions. Key Eng Mater 713:26–29
Sołtysik B, Falborski T, Jankowski R (2017) Preventing of earthquake-induced pounding between steel structures by using polymer elements – experimental study. Procedia Eng 199:278–283
Stewart JP, Tileylioglu S (2007) Input ground motions for tall buildings with subterranean levels. Struct Design Tall Spec Build 16(5):543–557
Stewart JP, Fenves GL, Seed RB (1999a) Seismic soil-structure interaction in buildings I: analytical methods. J Geotech Geoenviron 125(1):26–37
Stewart JP, Fenves GL, Seed RB (1999b) Seismic soil-structure interaction in buildings II: empirical findings. J Geotech Geoenviron 125(1):38–48
Veletsos AS, Prasad AM (1989) Seismic interaction of structures and soils: stochastic approach. J Struct Eng 115(4):935–956
Veletsos AS, Prasad AM, Wu WH (1997) Transfer functions for rigid rectangular foundations. Earthq Eng Struct Dyn 26(1):5–17
Wolf JP (1985) Dynamic soil-structure interaction. Prentice-Hall, Upper Saddle River
Zembaty Z (2004) Rockburst induced ground motion – a comparative study. Soil Dyn Earthq Eng 24(1):11–23

Chapter 3
Transverse and Longitudinal Seismic Effects on Soil-Steel Bridges

T. Maleska, P. Bonkowski, D. Beben, and Z. Zembaty

Abstract Soil-steel bridges and culverts, typically ranging from 3 to 25 m, can be used as an effective alternative for short-span bridges. They can meet the design and safety requirements of traditional bridges but at lower costs and with shorter erection time. For these reasons, soil-steel bridges are more and more often used in road and railway projects in many parts of the world. The purpose of present analysis is more advanced. Respective FEM models of a large soil steel bridge were prepared and eigen problem solved applying SAP 2000 and DIANA programs. Next the dynamic response was computed using time history response analysis with El Centro 1940 record. Two basic cases of seismic loads were analysed, i.e. "XZ" and "YZ" (seismic excitations were induced simultaneously at directions: transversal (X) and vertical (Z), and longitudinal (Y) and vertical (Z)). The totally different way of response of the large soil-steel bridges along and perpendicular to them generates key irregularity effect to study in their seismic response. Soil-steel interaction is also considered using special interface elements. Displacements and bending moments of the bridge are analysed in detail.

Keywords Soil-steel bridge · Culvert · Seismic analysis · Dynamic analysis

3.1 Introduction

The soil-steel bridge structures that are usually built as small and short span transportation objects are attractive due to architectonic values as well as technological advantages resulting in their low costs (Janusz and Madaj 2009). The positive features of soil-steel bridges are as follows: a definitely low construction cost, a significantly shorter construction period and low maintenance expenses. In addition, these bridges naturally blend in the surroundings thank to frequently used facades in

T. Maleska (✉) · P. Bonkowski · D. Beben · Z. Zembaty
Faculty of Civil Engineering and Architecture, Opole University of Technology, Opole, Poland
e-mail: t.maleska@po.opole.pl; p.bonkowski@po.opole.pl; d.beben@po.opole.pl; z.zembaty@po.opole.pl

D. Köber et al. (eds.), *Seismic Behaviour and Design of Irregular and Complex Civil Structures III*, Geotechnical, Geological and Earthquake Engineering 48,
https://doi.org/10.1007/978-3-030-33532-8_3

form of gabion walls. Moreover, these bridge structures are distinguished by a lot of positive exploitation features in comparison to the conventional small reinforced concrete (RC) bridges: (i) the construction efficiency, (ii) the minimal amount of the steel material in relation to the span, (iii) the consistency of the roadway foundation on the transportation route (lack of expansion joints). The latter is especially important in winter period.

The soil-steel bridges can be exposed to dynamic loads that may come from the natural forces (e.g. earthquakes) or traffic loads from railway and road transport. Numerical analysis of soil-steel bridge with a RC relieving slab under the static loads was presented by Beben and Stryczek (Beben and Stryczek 2016). A Finite Element Method computer program was used for this purpose. A positive effect of the relieving slab was confirmed. The main goal of that study was to identify and explain the static loads effects.

Ogawa and Koike (Ogawa and Koike 2001) analysed the influence of seismic activity on the example of several types of pipes (gas pipelines) surrounded by the soil. The slip coefficient between the steel plate and the backfill was defined. Moreover, a simplified method for assessing the plastic deformations of the pipelines affected by earthquakes was proposed. Davis and Bardet (2000) analysed the existing steel pipes surrounded by the soil after earthquakes. Katona (2010) analysed the box and circular RC culverts influenced by seismic activity and on the diversified static load.

The main aim of this paper is a determination of seismic behaviour of a soil-steel bridge under three component ground motion excitations. Particularly the effects of two component horizontal excitations along and perpendicular to the bridge are studied. The combined, net effects of these two seismic loads are not clear for this type of bridges and constitute the spatial seismic irregularity of the structure with respect to the seismic load. To clarify this issue of irregularity, dynamic numerical analyses are carried out. Two finite element programs (DIANA and SAP2000) were applied. Firstly, two FE models of the bridge were prepared to solve respective eigen problem. Next, the benchmark, El Centro 1940 records were used for two cases of seismic loads: with excitations along and vertical to the bridge and perpendicular to the bridge structure. Both numerical analyses of the along and transversal seismic effects were carried out with the vertical El Centro excitation component included in the analysis.

3.2 Short Description of the Analysed Bridge

The subject of analysis is a soil-steel bridge called "ecoduct" (built to pass animals over express roads). Its effective span equals 17.67 m while its vertical height is equal to 6.05 m (Fig. 3.1). The analysed structure consists of a single-span steel shell structure and the compacted backfill. Basic dimensions of the ecoduct structure are presented in Fig. 3.2. The shell is made of corrugated steel plate of thickness $t = 7$ mm which has the corrugation depth of $a = 0.14$ m with pitch $b = 0.38$ m

Fig. 3.1 Side view of soil-steel bridge

(detail in Fig. 3.2). The structure's width is 40.39 m at the top and 53.83 m at the bottom. In the plan the object is situated perpendicularly in relation to the national road. The individual sheets of corrugated plate are connected together using high strength bolts $\phi = 20$ mm using a torque moment of 350–400 Nm. The steel shell structure was supported, by means of steel, uneven-armed channel sections, on two RC strip foundations with width of 4.0 m and length of 57.83 m made of concrete C25/30.

Both ends of the shell were secured and stabilized by creating a reinforced concrete collar with 0.40 × 0.60 m dimensions. The load-bearing structure was designed as a shell consisting of corrugated steel plates backfilled with 0.20–0.30 m thick layer (permeable soil with 10–32 mm grading, compacted to $I_D = 95\%$ according to the Proctor Normal Density scale) for the backfill being in direct contact with the shell. When it comes to backfill in other places, it was compacted to $I_D = 97\%$ which allowed arranging of the ground layers with planting bushes and small trees. The backfilling process was conducted symmetrically on both sides of the shell structure. The differences in the height of backfill on both sides of the shell did not exceed 2–3 layers, i.e. 0.40–0.60 m. Special vibratory plates were used to compact the soil during construction. The soil cover thickness over the steel crown shell equals $h_C = 1.80$ m, which allows planting various kinds of vegetation. Such height of the backfill over the shell results also in increased damping of the entire structure, which may reduce noise coming from the traffic passing below the ecoduct and help to mitigate vibrations of the bridge.

3.3 Numerical Model

3.3.1 General Remarks

SAP2000 and DIANA computer programs using finite element method (FEM) were applied in the further numerical analyses of the soil-steel bridge. The bridge was

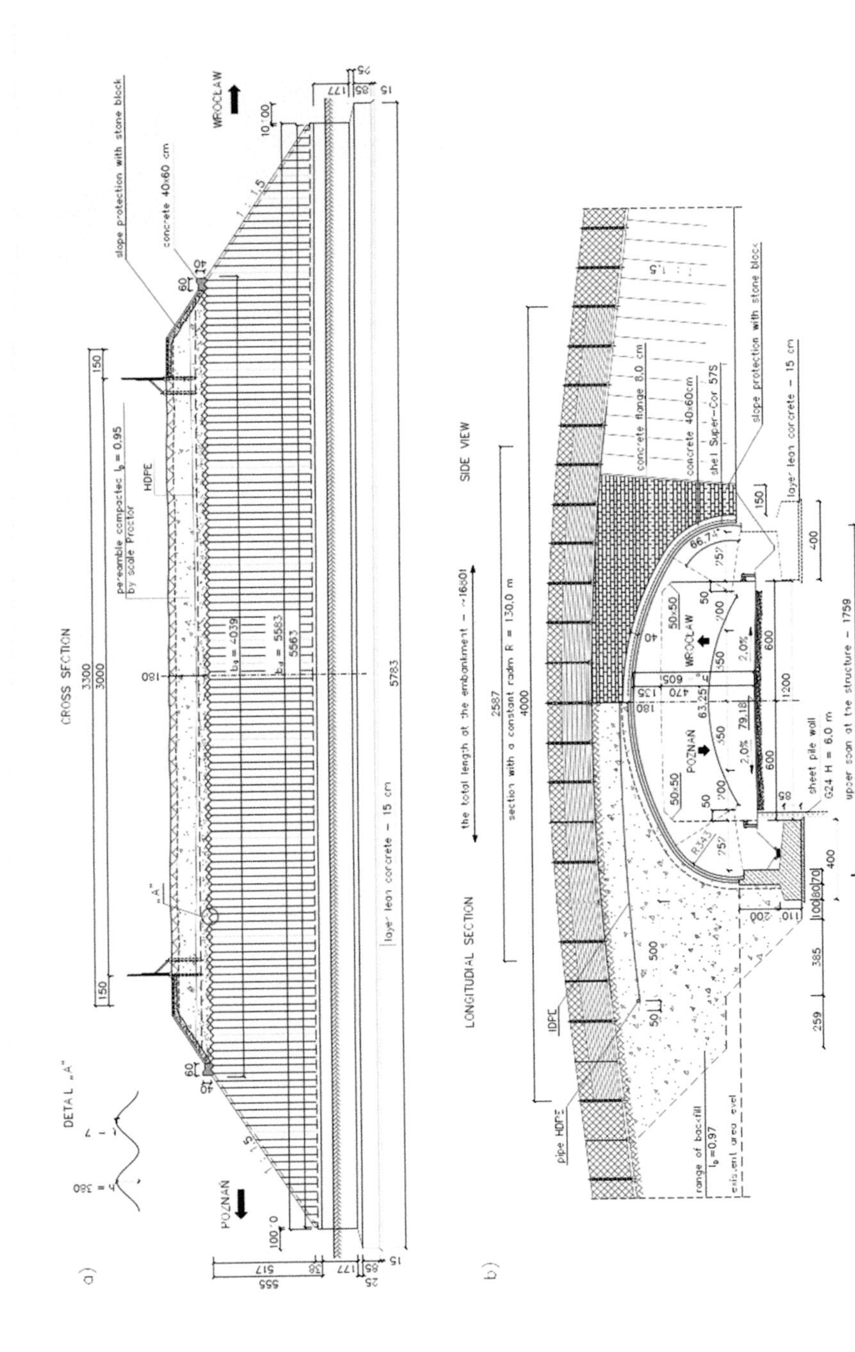

Fig. 3.2 Soil-steel bridge: **(a)** longitudinal section, **(b)** cross section

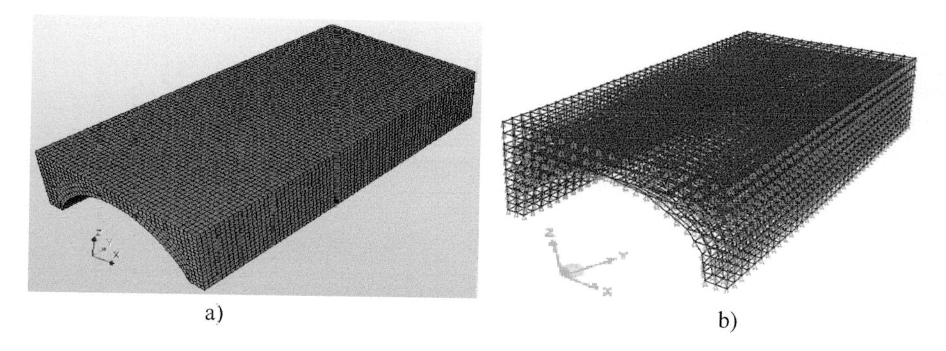

Fig. 3.3 Numerical models created in: (**a**) DIANA, (**b**) SAP2000

modelled as a 3D structure with 3D shell elements for the shell structure, and solid elements for the backfill. Figure 3.3 presents the applied numerical models in both programs. Duncan-Chang nonlinear, elastic hyperbolic and isotropic models were used for the backfill modelling in DIANA (DIANA FEA 2017) and SAP2000 (Introductory Tutorial for SAP 2000 2011) programs, respectively. In the calculation models it was assumed that: (i) the actual dimension of the bridge span equals 17.67 m, (ii) width in the upper part of the shell is equal to 40.39 m, (iii) the structure height is assumed to equal to 6.05 m and (iv) the thickness of the soil over the crown equals 1.8 m.

3.3.2 *Material Characteristics*

For the numerical analyses following materials were applied:

1. Steel plate for shell elements (Q24IF), elastic-plastic model with density of 7850 kg/m^3, Young model of 205 GPa, Poisson ratio of 0.3, the yield strength of steel of 235 MPa, plate thickness of 0.007 m, cross section area of 8.867mm^2/mm, and moment of inertia of 21897.45 mm^4/mm.
2. Backfill in the DIANA program was modelled using solid elements (C3DR8) with Duncan-Chang nonlinear elastic hyperbolic model with material density of 2050 kg/m^3, Young's modulus of 100 MPa, Poisson ration of 0.2, angle of internal friction of 39°, dilation angle of 5°, cohesion of 3 kPa, unloading-reloading stiffness $E_{ur} = 1000$ N/m^2, failure ratio $R_f = 0.7$, reference pressure $P_{ref} = 101{,}350$ N/m^2, exponent for backbone curve $n = 1.1$, exponent for unloading reloading curve $m = 0.25$, minimum tangential stiffness of backbone curve $E_{t,min} = 1200$ N/m^3, and minimum compressive stress 350 N/m^2.

   For the SAP2000 program the elastic-plastic isotropic model was adopted to model the backfill with material density of 2050 kg/m^3, Young's modulus of 100 MPa, and Poisson ratio of 0.2.

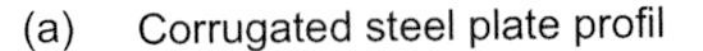
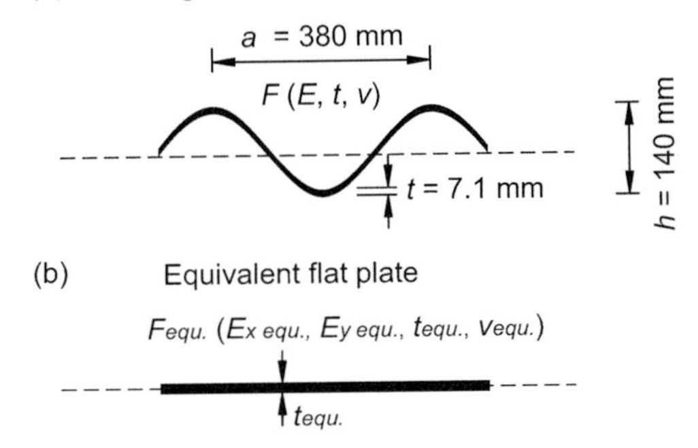

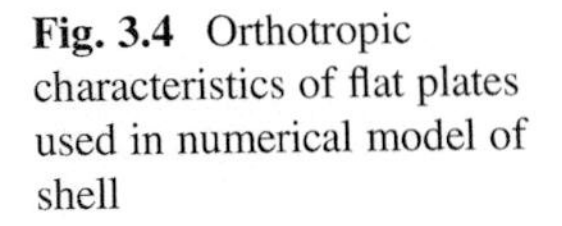
Fig. 3.4 Orthotropic characteristics of flat plates used in numerical model of shell

3. Boundary conditions were assumed as hinged supports on all boundary walls of the soil and support of the shell walls for x, y, z directions.
4. Connections between the backfill and steel shell structure were made by using special interface of the DIANA program which is called “automatic interface” by applying “Coulomb friction” function (Ucci et al. 2011) with rigidity of 100,000 kN/m^3, angle of internal friction 39°, dilation angle of 5°, cohesion of 3 kPa. Additionally the second model with fixed connections was also created in DIANA. In the SAP2000 program only fixed connections were used.
5. Quadratic finite elements with dimension of 0.5×0.5 m for flat shell and backfill in DIANA and SAP2000 programs were applied.

The corrugated steel plate of the shell structure was modelled in DIANA program using shell elements with variable material characteristic plates (Machelski 2008). An orthotropic shell (Fig. 3.4) can be used to calculate the equivalent parameters such as:

- equivalent thickness of plate:

$$t_{equ.} = \sqrt{12(1 - \nu^2)\frac{I}{A}}, \tag{3.1}$$

where I is moment of inertia, A is cross-sectional area, ν is Poisson ratio ($\nu = \nu_x = 0.3$),

- equivalent elastic modulus of material (Young modulus) in circumferential direction of shell:

$$E_{x\ equ.} = E\frac{A}{a\ t_{equ.}}, \tag{3.2}$$

where a is so called pitch of corrugation,

- equivalent elastic shear modulus:

$$E_{y\ equ.} = E\left(\frac{t}{t_{equ.}}\right)^3, \tag{3.3}$$

- equivalent Poisson ratio:

$$\nu_{equ.} = \nu\frac{E_{y\ equ.}}{E_{x\ equ.}}. \tag{3.4}$$

For SAP2000, the orthotropic steel shell is described using scale factors in accordance with (Introductory Tutorial for SAP 2000 2011):

- the axial stiffness:

$$S_a = A \cdot E_x, \tag{3.5}$$

where A is cross-sectional area, and E_x is modulus of elasticity,
- the shear stiffness:

$$S_{Sy} = As_y \cdot g_{xy}, \tag{3.6}$$

$$S_{S_z} = As_z \cdot g_{xy}, \tag{3.7}$$

where As_y, As_z are shear area in X and Y directions, g_{xy} is shear modulus,
- the torsional stiffness:

$$S_T = j \cdot g_{xy}, \tag{3.8}$$

where j is torsional constant,
- the bending stiffness:

$$S_{Bz} = I_{zz} \cdot E_x, \tag{3.9}$$

$$S_{By} = I_{yy} \cdot E_x, \tag{3.10}$$

where I_{zz}, I_{yy} are moment of inertia in Z and Y directions,
- the section mass:

$$M_S = A \cdot m + mpl, \tag{3.11}$$

where m is mass and mpl is mass per unit length,
- the section weight:

$$W_S = A \cdot w + wpl, \tag{3.12}$$

where w is weight and wpl is weight per unit length.

Table 3.1 Natural frequencies *f* [Hz] of soil-steel bridge

Mode	SAP2000 fixed connection	DIANA fixed connection	DIANA with interface
	[Hz]	[Hz]	[Hz]
1	5.272	4.636	4.508
2	5.864	5.401	5.227
3	6.769	5.682	5.433
4	7.833	6.287	6.002
5	8.049	6.582	6.302
6	8.643	7.279	6.913
7	9.003	7.761	7.404
8	9.352	8.030	7.656
9	10.201	8.461	8.069
10	10.249	8.542	8.076
11	10.346	8.733	8.359
12	10.563	9.176	8.748
(...)			
24	13.369	11.772	11.355
(...)			
48	17.712	14.710	14.010
(...)			
62	20.022	16.969	16.180
(...)			
96	23.406	20.205	19.328

3.4 Results of Modal and Seismic Analyses

Modal, linear and non-linear seismic analyses with time history response analyses were carried out for the analysed soil-steel bridge using the El Centro 1940 record (Chmielewski and Zembaty 1998). Table 3.1 shows the results of natural frequencies for three considered cases of numerical models: (i) in SAP2000 with the fixed connection, (ii) in DIANA with the fixed connection and (iii) in DIANA with the interface elements.

Figures 3.5.1 and 3.5.2 present the selected first four natural modes of soil-steel bridge. Results presented in Table 3.1 demonstrate that application of interface elements in DIANA program gives lower natural frequencies. It can be seen that these values are close to each other, which is characteristic for any shell structures. It can be observed that the soil models significant affect the natural frequencies of the bridge as modelled by SAP2000 and DIANA using the assumption of fixed connection between steel and soil.

The calculations of natural frequencies were performed for 96 modes of response, both in SAP2000 and DIANA. For the linear analysis, the calculations were

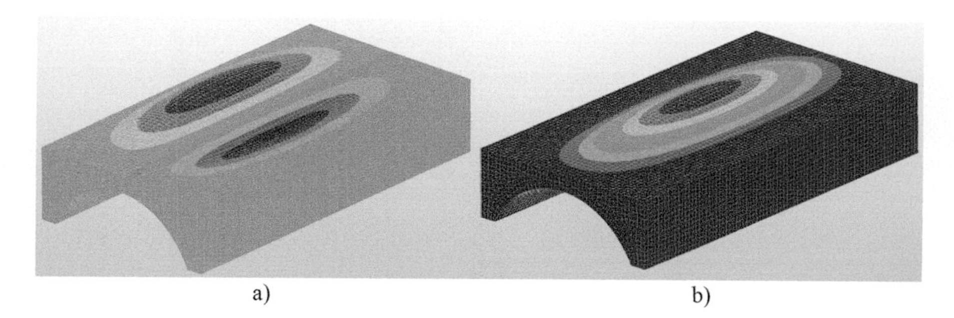

Fig. 3.5.1 Natural modes: (**a**) 1 ($f_1 = 5.989$ Hz), (**b**) 2 ($f_2 = 6.303$ Hz) computed using DIANA program

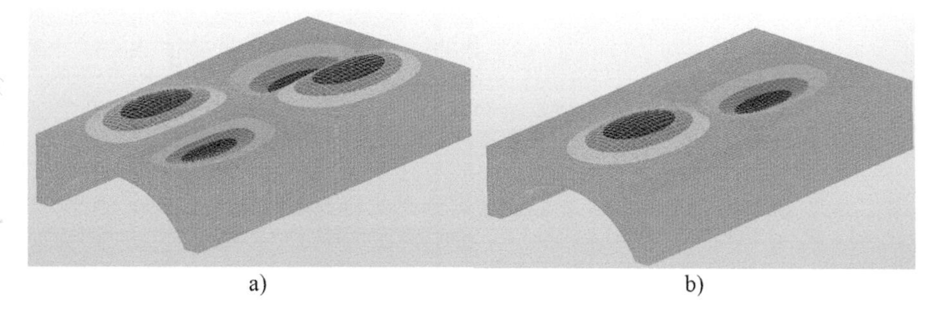

Fig. 3.5.2 Natural modes: (**a**) 3 ($f_3 = 6.612$ Hz), (**b**) 4 ($f_4 = 6.764$ Hz) computed using DIANA program

conducted for 62 modes of natural frequencies (in SAP2000 program). The calculations were stopped when natural frequency exceeded 20 Hz. It should be emphasized that the contribution of natural modes with frequencies exceeding 20 Hz in total seismic response is very low. The non-linear analysis was carried out only for 24 modes (in the DIANA program). This is because the difference of the responses accounting for 12 and 24 modes was very small, namely about 4% for the displacements.

In the next step, a linear (SAP2000) and non-linear (DIANA) analyses of the soil-steel bridge were made using the time history analysis and the El Centro record. Two numerical models were applied in DIANA program, the first model with interface elements and second one with fixed connection. The fixed connection was used also for model prepared in the SAP2000 program. Two cases of the seismic loads were applied, i.e. “XZ” and “YZ” according to Fig. 3.3. It means that seismic excitations were induced simultaneously at directions: transversal (X) and vertical (Z), and longitudinal (Y) and vertical (Z).

It should be underlined that the direction “X” is perpendicular (transversal effects) to the bridge, the direction “Y” is parallel (longitudinal effects) and the direction “Z” is vertical to the bridge. Results of time history of analysed soil-steel bridge are shown in the Tables 3.2 and 3.3 presents the results of the linear analysis

Table 3.2 Result of analysis of soil-steel bridge in SAP2000

	Transversal displacements (X)		Longitudinal displacements (Y)		Vertical displacements (Z)		Bending moments M	
	[mm]	[%]	[mm]	[%]	[mm]	[%]	[kNm/m]	[%]
After 12 mode								
lng & 'z' exc	–	–	2.2	–	5.4	–	10.96	–
trv & 'z' exc	2.5	–	–	–	5.4	–	–	–
After 24 mode								
lng & 'z' exc	–	–	2.2	0	5.6	4	13.67	25
trv & 'z' exc	2.5	0	–	–	5.6	4	–	–
After 48 mode								
lng & 'z' exc	–	–	2.1	5	5.6	4	13.59	24
trv & 'z' exc	2.4	4	–	–	5.6	4	–	–
After 62 mode								
lng & 'z' exc	–	–	2.1	5	5.6	4	13.73	25
trv & 'z' exc	2.4	4	–	–	5.6	4	–	–

Note: lng & 'z' exc = excitation longitudinal (Y) and vertical (Z) to the bridge, trv & 'z' exc = excitation transversal (X) and vertical (Z) to the bridge, [%] = differences between 12 and 24, 48, 62 modes

Table 3.3 Result of analysis of soil-steel bridge in DIANA

	Transversal displacements (X)		Longitudinal displacements (Y)		Vertical displacements (Z)		Bending moments M	
	[mm]	[%]	[mm]	[%]	[mm]	[%]	[kNm/m]	[%]
Fixed connection								
After 12 mode								
lng & 'z' exc	–	–	6.2	–	18.5	–	9.45	–
trv & 'z' exc	8.7	–	–	–	18.5	–	–	–
After 24 mode								
lng & 'z' exc	–	–	6.2	0	18.5	0	9.45	0
trv & 'z' exc	8.6	2	–	–	18.5	0	–	–
With interface								
After 12 mode								
lng & 'z' exc	–	–	5.8	–	26.7		11.29	0
trv & 'z' exc	13.5	–	–	–	26.7	–	–	–
After 24 mode								
lng & 'z' exc	–	–	5.8	0	26.7	0	11.29	0
trv & 'z' exc	13.5	0	–	–	26.7	0	–	–

Note: lng & 'z' exc = excitation longitudinal (Y) and vertical (Z) to the bridge, trv & 'z' exc = excitation transversal (X) and vertical (Z) to the bridge, [%] = differences between 12 and 24 modes

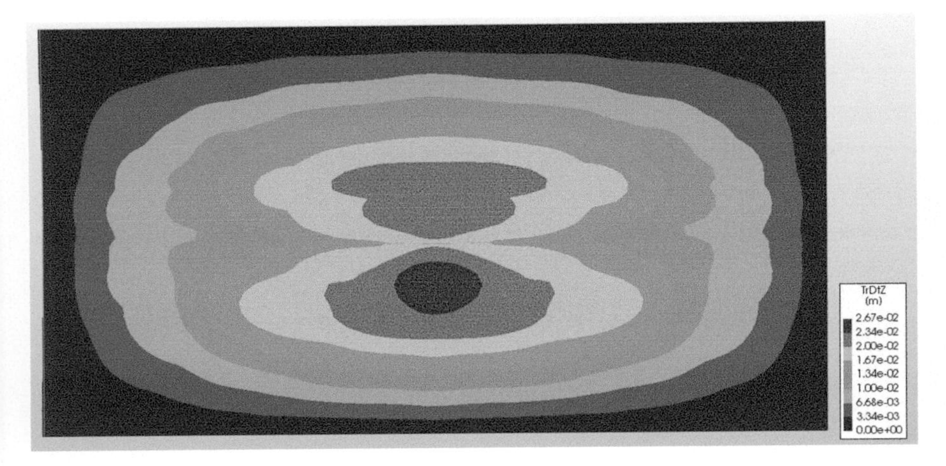

Fig. 3.6 Maximal displacements of the Y direction for model with interface received from DIANA

in SAP2000, while Table 3.3 shows the results of the non-linear analysis in DIANA program. In SAP2000 program, the largest vertical displacements were obtained after 24 modes of natural frequency. In this case, the maximum displacement was 5.6 mm. For 12 modes, the displacements were lower by 4% and amounted to 5.4 mm. The obtained values were the same in both "X" and "Y" directions. In the DIANA program (numerical model with the interface), the maximum vertical displacement was 26.7 mm (Fig. 3.6). This value was obtained for both 12 and 24 modes of natural frequencies. In the case of the numerical model with the fixed connection (in DIANA program), a smaller displacement of 31% was obtained by comparing to the model with the interface, i.e. 18.5 mm. The maximum values occurred in the shell crown.

The maximum bending moments were achieved in the SAP2000 and they amounted to 13.73 kNm/m after 62 modes of natural frequencies. The difference of bending moments between 24 and 48 modes of the response is a few percent. However, the difference between the models with 12 and 24 modes of response does not exceed 25%. In DIANA program, the maximum bending moment was achieved in the numerical model with the interface. The value was equalled 11.29 kNm/m (Fig. 3.7). In the numerical model without of interface, the maximum bending moment was 9.45 kNm/m. Difference between bending moments obtained from both numerical models equal to 16%. The maximum bending moments occurred at the shell crown. It should be noted that the bending moments in all numerical models were analysed only in the longitudinal direction to the bridge.

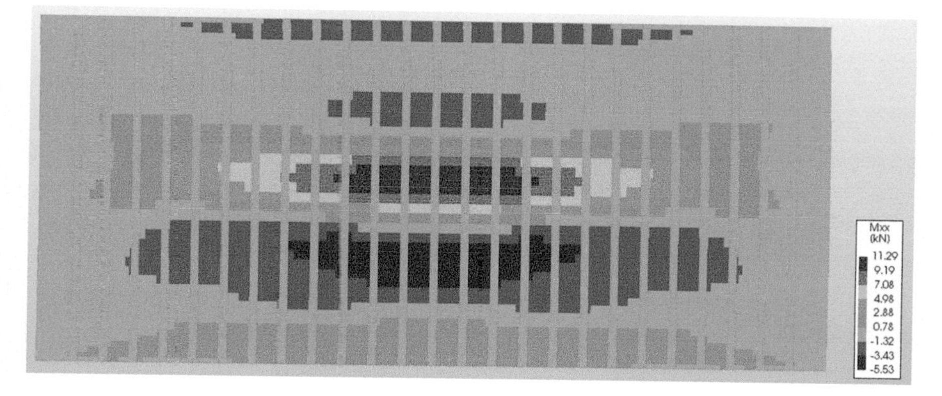

Fig. 3.7 Maximum bending moment in shell structure of the Y direction for model with interface received from DIANA program

3.5 Conclusion

As a result of modal and seismic analyses of soil-steel bridge in the DIANA and SAP2000 programs the following conclusions can be drawn:

1. Seismic study of soil-steel bridge with use of the 12 modes of natural frequencies is enough for analysis of short-to-medium spans of this type of structures. Differences of maximum displacements between the considered numerical models (SAP2000 and DIANA) do not exceed 4% in comparison to the model with 24 modes of natural response. Similar differences occurred both in transversal and longitudinal directions.
2. The obtained results of seismic analysis in the DIANA numerical program confirmed that the application of special interface elements between backfill and steel shell is important. Using the models with and without the interface elements introduces substantial differences in displacements and bending moments. In the case of vertical displacements the presence of interface introduces up to 44% difference in comparison to the model without the interface. However, in the case of bending moments, the difference between both models was smaller and amounted to 19%. It means that the use of interface elements gives more realistic displacements and bending moments.
3. The time history analyses of soil-steel bridge demonstrates quite high differences of obtained values for the linear analyses (SAP 2000) and the non-linear ones (DIANA). For obvious reasons, the non-linear response differs substantially from the linear one in case of soil-steel structures which is the case of culvert and soil-steel bridges. These differences can be seen for both cases of applied excitation loads, i.e. for the "XZ" direction (perpendicular and vertical to the bridge), and the "YZ" direction (parallel and vertical to the bridge). It confirms the irregular behaviour of the soil-steel bridges. These differences can be mostly explained from the way of modelling the structure in SAP2000:

- flat shell with the orthotropic properties using scale factors,
- the soil isotropic model,
- simplified contact elements (fixed connection) to reflect the interaction between various elements (particularly interaction between steel shell and backfill).

4. It can also be concluded that in the future research special attention should be paid to modelling the steel shell and backfill interaction.

References

Beben D, Stryczek A (2016) Numerical analysis of corrugated steel plate bridge with reinforced concrete relieving slab. J Civ Eng Manag 22(5):585–596

Chmielewski T, Zembaty Z (1998) Fundamentals of structural dynamics (in Polish: Podstawy Dynamiki Budowli). Arkady, Warsaw

Davis CA, Bardet JP (2000) Responses of buried corrugated metal pipes to earthquakes. J Geotech Geoenviron Eng ASCE 126:28–39

DIANA FEA (2017) Available online from URL: www.dianafea.com/

Introductory Tutorial for SAP 2000 (2011) https://www.tutorialspoint.com/sap/

Janusz L, Madaj A (2009) Engineering structures from corrugated plates: design and construction. Transport and Communication Publishers, Warsaw

Katona MG (2010) Seismic design and analysis of buried culverts and structures. J Pipeline Syst Eng Pract ASCE 1(3):111–119

Machelski CZ (2008) Modeling of soil-Shell bridge structures. The Lower Silesian Educational Publishers, Wroclaw

Ogawa Y, Koike T (2001) Structural design of buried pipelines for severe earthquakes. Soil Dyn Earthq Eng 21:199–209

Ucci M, Camata E, Spacone E, Lilliu G, Manie J, Schreppers G (2011) Nonlinear soil-structure interaction of a curved bridge on the Italian Tollway A25. Proceedings of 8th international conference on structural dynamics. EURODYN, Leuven, pp 621–628

Part II
Seismic Analysis and Design of Irregular Structures

Chapter 4
Deformation Based Seismic Design of Generally Irregular 3D RC Frame Buildings for Minimized Total Steel Volume

Oren Lavan and Philip J. Wilkinson

Abstract This chapter presents a methodology for designing irregular 3D RC frame buildings by minimizing the total steel volume while satisfying inter-story drifts and material strains limits. Indirectly, this process leads to reduction on the base shear and over-turning moments. The methodology relies only on analysis tools, without any need for knowledge or tools of structural optimization. Although iterative, the methodology requires only a few iterations for convergence. This makes the methodology very attractive for practical use.

Keywords Irregular frame buildings · Setback structures · Structural optimization · Seismic design

4.1 Introduction

Irregular buildings are a common site in many cities around the world. It is their irregularity, however, that often increases their seismic vulnerability (Stathopoulos and Anagnostopoulos 2005; Kyrkos and Anagnostopoulos 2011). It also makes building behaviour harder to predict using simplified analysis methods (Moehle and Alarcon 1986; Valmundsson and Nau 1997; Anagnostopoulos et al. 2015). Thus, complex models and advanced analyses are often used for prediction of their behaviour. Nonetheless, even with advanced analysis types available for verification, procedures (especially simple ones), for design of irregular buildings, are still needed. Ideally, such design methods would lead to economic designs that minimize cost as much as possible, while maintaining desired performance.

O. Lavan (✉)
Technion – Israel Institute of Technology, Haifa, Israel
e-mail: lavan@tx.technion.ac il

P. J. Wilkinson
Stratagroup, Consulting Engineers Ltd, Hastings, New Zealand
e-mail: philip@sgl.nz

D. Köber et al. (eds.), *Seismic Behaviour and Design of Irregular and Complex Civil Structures III*, Geotechnical, Geological and Earthquake Engineering 48,
https://doi.org/10.1007/978-3-030-33532-8_4

This chapter presents a simple iterative approach for the seismic design of 3D irregular RC frame buildings, as proposed by the authors (Lavan and Wilkinson 2016). It involves two stages within each iteration. In the first stage, as analysis is performed on the current design. In the second stage, strengths and stiffnesses are re-designed according to the analysis results and a pre-defined recurrence relation. This procedure is carried out until convergence, which is obtained within a few iterations. The approach targets a design that is as economic as possible while satisfying limits on inter-story drifts and strains (or ductilities). As the approach is transparent and only makes use of analysis tools, without requiring any optimization tools or knowledge, it is more likely to be utilized by the practicing community.

4.2 Problem Statement

The design of RC frame structures requires the dimensions of beams and columns and the amount of steel in these elements. Usually, beams' and columns' dimensions are predetermined based on architectural considerations, gravity design and preliminary seismic design. Deviations from these dimensions is usually not desired. If a feasible design can be reached with these dimensions, the amount of steel in each element is the remaining design variable to be determined by the engineer. Otherwise, some minor changes in elements' dimensions are required. Thus, the main design variables adopted in this research are the amounts of steel in each element, while, in some cases, the dimensions of some elements could be slightly modified.

With the dimensions of the elements set, the amounts of steel strongly affect the construction cost. Therefore, the total steel volume is to be minimized. The amounts of steel control the strength and the stiffness of the element. With lower structural strength, lower base shear and over-turning moments are also expected to reduce. This reduces foundation costs. In turn, lower strength also leads to lower total accelerations (see e.g. Lavan 2015). This results in a reduction in damage to acceleration sensitive non-structural components. Thus, the total steel volume (or sum of moment capacities in all seismic elements) is to be minimized.

While minimum steel volume is desired, a lower steel volume reduces structural strength and stiffness. Clearly larger inter-story drifts and elements' ductility demands (or material strains) are expected. Therefore, it is important to limit those to allowable values. Here, peak inter-story drifts at each story periphery are limited to allowable values. In addition, limits are assigned to allowable beam and column base ductility demands. Different limits are set for the various seismicity levels considered.

4.3 Proposed Methodology

The proposed methodology relies on an optimality criterion of the Fully-Stressed-Design (FSD) type. The earliest FSD type optimality criterion was proposed by Cilley (1900). Cilley also proved it to lead to a formal optimum for the design of the truss of minimum weight under stress constraints. It was shown that, for this problem, in the optimal design all bars reach the allowable stress in at least one loading condition. Accompanied to the FSD optimality criterion is the Analysis/Redesign (AR) algorithm where the engineer performs an analysis on a given design and modifies the cross sections of the bars accordingly. Here, if the stress in the bar is lower than the allowable, the engineer would decrease the bar cross section. Contrarily, if the stress in the bar is larger than the allowable, the engineer would increase the bar cross section. Convergence is usually obtained within a few iterations.

Characteristics of the FSD type have also been identified in optimal designs of seismic retrofitting of buildings using viscous dampers (Lavan and Levy 2006, 2010). The total added damping was to be minimised while inter-story drifts were constrained to allowable values. These optimality criteria were further accompanied by an AR type optimal design scheme (Lavan and Levy 2005, 2009; Levy and Lavan 2006). Such design schemes were also proposed in other problems based on intuitive FSD type criteria (Lavan 2015; Lavan and Daniel 2013; Daniel and Lavan 2015). Comparison of the obtained designs using these methods with those obtained using formal optimization tools (Daniel and Lavan 2014; Lavan and Dargush 2009) revealed that, if not optimal, they were at least near optimal. An AR type algorithm has also been adopted for the seismic design of plane RC frames (Hajirasouliha et al. 2012). In their problem, minimized total steel volume was targeted while limits were assigned to deformations over the height of the building.

In view of the above experience, and the strong relation between the inter-story drift of a given story and the strength and stiffness of the same story, it is expected that an optimized design for the problem stated in the previous section would possess the following intuitive optimality criterion:

For 3D framed structures, the optimal design will attain a flexural strength in beams of peripheral frames, larger than the minimum allowed, only in floors for which the performance measure has reached the allowable.

The performance measure is taken here as the maximum of the normalized (by its allowable value) inter-story drift of the peripheral frame in the story below the floor, and the maximum of the normalized (by their allowable values) ductility demands of all beams in that bent. For the first floor, the maximum of the normalized (by their allowable values) ductility demands of all column bases of that peripheral frame is also considered.

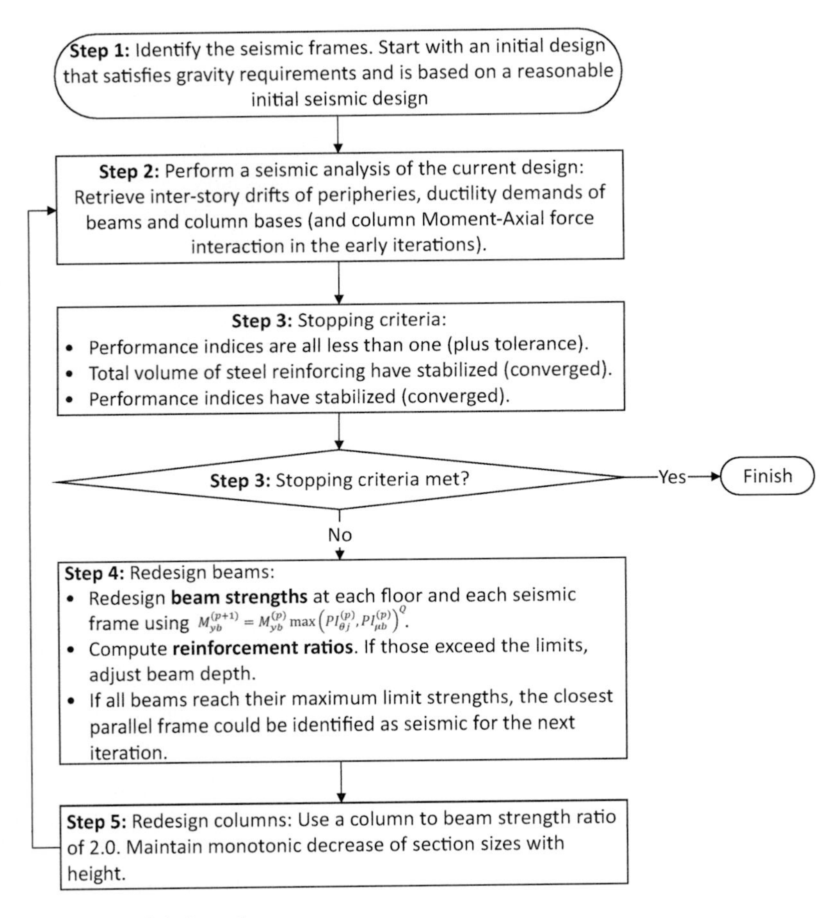

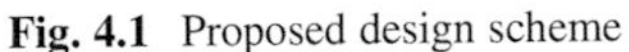
Fig. 4.1 Proposed design scheme

In view of the discussion above, only peripheral frames are considered as lateral load resisting systems. Inner frames are considered as gravity frames and their beams and columns are designed for gravity loads only. Gravity frames are detailed for ductilities congruent with the expected peak deformations. If the use of peripheral frames only does not lead to a feasible design, inner frames are gradually added as lateral load resisting systems one grid at a time from the periphery in.

The proposed scheme is described in Fig. 4.1. Where $M_{yb}^{(p+1)}$ and $M_{yb}^{(p)}$ are the nominal flexural yield strengths of seismic beam b for iterations $p + 1$ and p respectively. $PI_{\theta j}^{(p)}$ and $PI_{\mu b}^{(p)}$ are the parameters $PI_{\theta j}$ and $PI_{\mu b}$ defined in section "Optimization problem" computed for iteration p, respectively. For more details, the reader is referred to (Lavan and Wilkinson 2016).

4.4 Example

The proposed procedure is implemented in this section on the three-story irregular RC frame structure shown in Fig. 4.2. In this figure, the final component dimensions and relative flexural strengths (darker is stronger) are shown. Note that although some beams are larger than adjacent columns, the columns' strength is larger (darker in the figure). The design was performed for two limit states: a serviceability limit state and a life safety limit state. For the serviceability limit state, column bases and beams were assigned with an allowable of ductility of 5 while inter-story drifts were limited to 1% of the story height. For the life safety limit state the allowable ductility was set to 7.3 while 2% inter-story drifts were allowed.

Five sets of two horizontal components of ground motions were selected from the LA10/50 set (Somerville et al. 1997) and were modified using the program SeismoMatch (Seismosoft 2010) to match the design response spectrum. The analysis of the building was performed using Ruaumoko (Carr 2006).

Figure 4.3 presents the Maximum Constraint Violation (MCV) and the Objective Function (OF) value as a function of the iteration count. As can be seen, 14 iterations were required for full convergence.

The behaviour of the final design in terms of inter-story drifts and ductility demands are presented in Fig. 4.4.

As can be seen, most stories of most peripheral frames reached the allowable inter-story drifts in at least one limit state. The others were at their minimum strength. Ductility demands of the beams and column bases were well within their allowable limits. Some limited plasticity in the columns above their bases was apparent. This is very encouraging in such a highly irregular structure.

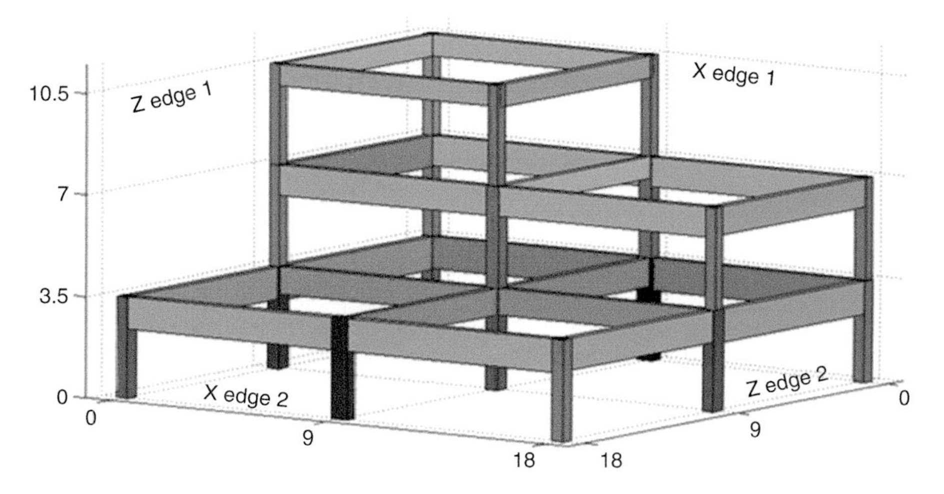

Fig. 4.2 3 story generally irregular RC frame structure: Final component dimensions are shown while relative flexural strengths is indicated by colour (darker is stronger)

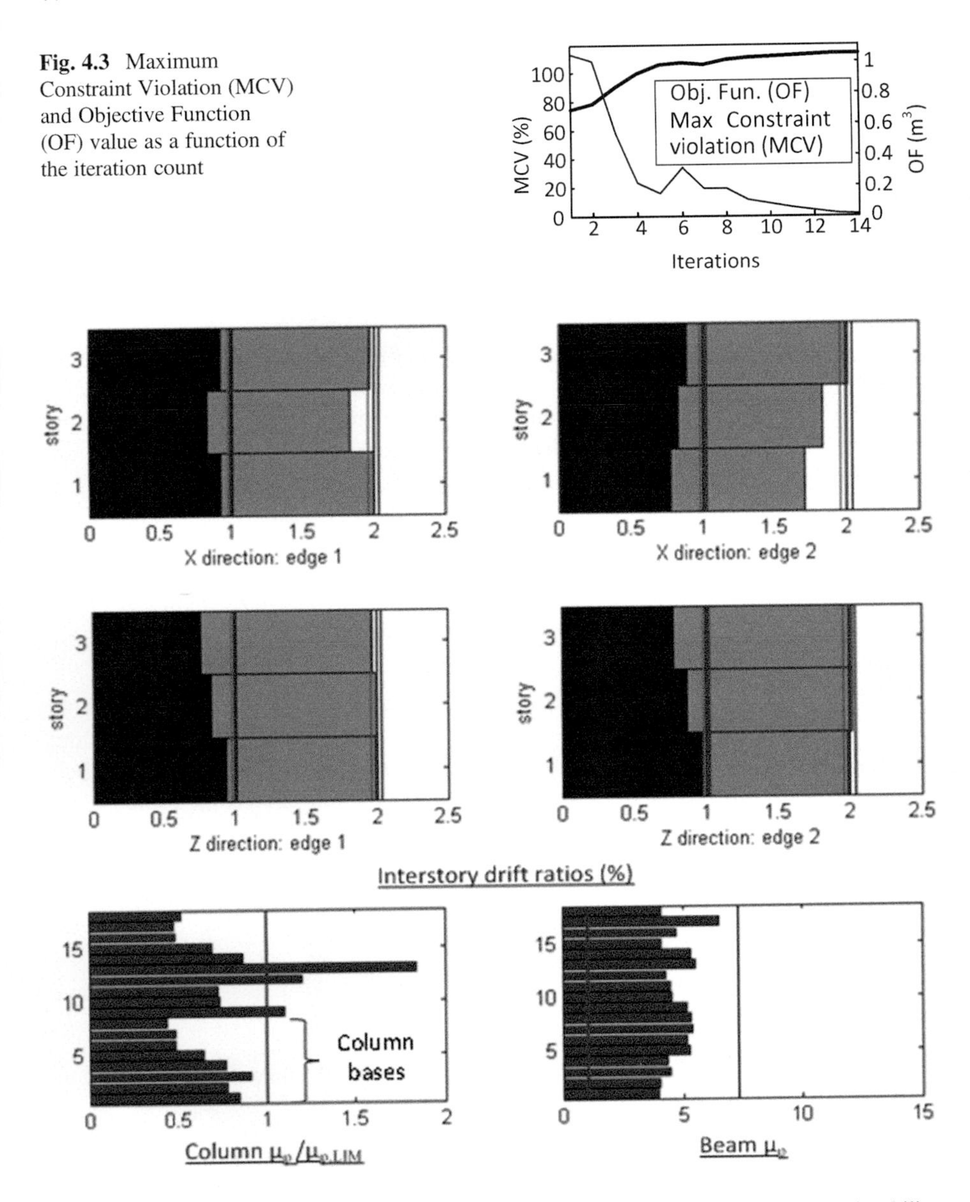

Fig. 4.3 Maximum Constraint Violation (MCV) and Objective Function (OF) value as a function of the iteration count

Fig. 4.4 Analysis-redesign design verification results (the black drift bars show the serviceability mean peak demands and the grey bars the ULS demands)

4.5 Conclusions

A methodology of the analysis/redesign type, for the efficient seismic design of 3D generally irregular frame structures, was presented. The methodology aims at minimizing the total moment capacity of all members, within inter-story drift and ductility demand constraints. As the methodology is very intuitive, and requires analysis tools only, it is intended for practical use in design offices.

The methodology was applied to a 3 story setback structure to demonstrate its utility. The final design was achieved with 14 iterations, which is manageable even though nonlinear time-history analyses are adopted. This further highlights the applicability of the method to design offices. As intended, the final designs attained flexural strengths in beams larger than the minimum gauge, only in floors for which the performance measure reached the allowable. Thus, strength was assigned only where required.

References

Anagnostopoulos SA, Kyrkos MT, Stathopoulos KG (2015) Earthquake induced torsion in buildings: critical review and state of the art. Earthq Struct 8(2):305–377

Carr AJ (2006) RUAUMOKO – program for inelastic dynamic analysis. Department of Civil Engineering, University of Canterbury, Christchurch

Cilley FH (1900) The exact design of statically indeterminate frameworks, an exposition of its possibility but futility. Trans ASCE 43:353–407

Daniel Y, Lavan O (2014) Gradient based optimal seismic retrofitting of 3D irregular buildings using multiple tuned mass dampers. Comput Struct 139:84–97

Daniel Y, Lavan O (2015) Optimality criteria based seismic design of multiple tuned-mass-dampers for the control of 3D irregular buildings. Earthq Struct 8(1):77–100

Hajirasouliha I, Asadi P, Pilakoutas K (2012) An efficient performance-based seismic design method for reinforced concrete frames. Earthq Eng Struct Dyn 41(4):663–679

Kyrkos MT, Anagnostopoulos SA (2011) An assessment of code designed, torsionally stiff, asymmetric steel buildings under strong earthquake excitations. Earthq Struct 2(2):109–126

Lavan O (2015) A methodology for the integrated seismic design of nonlinear buildings with supplemental damping. Struct Control Health Monit 22(3):484–499

Lavan O, Daniel Y (2013) Full resources utilization seismic design of irregular structures using multiple tuned mass dampers. Struct Multidiscip Optim 48(3):517–532

Lavan O, Dargush GF (2009) Multi-objective optimal seismic retrofitting of structures. J Earthq Eng 13(6):758–790

Lavan O, Levy R (2005) Optimal seismic retrofit of irregular 3D framed structures using supplemental viscous dampers. Proceedings of the 4th European workshop on the seismic behaviour of irregular and complex structures. Paper no. 52, Thessalonica, Greece

Lavan O, Levy R (2006) Optimal peripheral drift control of 3D irregular framed structures using supplemental viscous dampers. J Earthq Eng 10(6):903–923

Lavan O, Levy R (2009) Simple iterative use of Lyapunov's solution for the linear optimal seismic design of passive devices in framed buildings. J Earthq Eng 13(5):650–666

Lavan O, Levy R (2010) Performance based optimal seismic retrofitting of yielding plane frames using added viscous damping. Earthq Struct 1(3):307–326

Lavan O, Wilkinson PJ (2016) Efficient seismic design of 3D asymmetric and setback RC frame buildings for drift and strain limitation. J Struct Eng ASCE 143(4):04016205

Levy R, Lavan O (2006) Fully stressed design of passive controllers in framed structures for seismic loadings. J Struct Multidiscip Optim 32(6):485–498

Moehle JP, Alarcon LF (1986) Modeling and analysis methods for vertically irregular buildings. Proceedings of the 8th European conference on earthquake engineering, 3, pp 6.1/57–64, Lisbon, Portugal

Seismosoft (2010) SeismoMatch (version 1.3.0. Build 101): a computer program for spectrum matching of accelerograms. Available at: http://www.seismosoft.com/
Somerville P, Smith N, Punyamurthula S, Sun J (1997) Development of ground motion time histories for phase 2 of the FEMA/SAC steel project. Report No. SAC/BD-97/04
Stathopoulos KG, Anagnostopoulos SA (2005) Inelastic torsion of multistorey buildings under earthquake excitations. Earthq Eng Struct Dyn 34(12):1449–1465
Valmundsson VE, Nau JM (1997) Seismic response of building frames with vertical structural irregularities. J Struct Eng ASCE 123(12):30–41

Chapter 5
Fast Nonlinear Response History Analysis: An Application to Irregular Building Structures

Juan C. Reyes, William A. Avila, and Armando Sierra

Abstract The design or evaluation of high complexity structures requires non-traditional methods to estimate their seismic performance. Although structural engineering still relies on nonlinear static analysis for estimating seismic demands, the nonlinear response history analysis (RHA) is being now increasingly used for design-check and performance evaluation. In this approach, the engineering demand parameters are determined by performing a series of nonlinear RHAs using an ensemble of ground motion records that represent the site's seismic hazard conditions. This type of analysis is computationally demanding and time consuming when applied to three-dimensional computer models subjected to multi-axial excitations. A procedure is presented to reduce the processing time of nonlinear RHA by minimizing the impact on the response of the structure while maximizing the time savings; it has three steps: (i) trimming a segment of the ground-motion record at the beginning; (ii) trimming a segment at the end; (iii) identifying an appropriate time-step to reduce the number of steps required to properly characterize the signal. This procedure has been tested through a parametric study by considering multiple structural periods and response modification factors for multi-story irregular buildings. Results show that the processing time can be reduced without compromising accuracy in estimates of peak EDPs.

Keywords Nonlinear reponse history analysis · Reducing time-steps · Trimming seismic records

5.1 Introduction

Nonlinear response history analysis (RHA) is a useful tool to predict the behaviour of structures subjected to seismic forces. In particular, when a performance-based design is executed based on three-dimensional (3D) RHAs, pairs of seven records

J. C. Reyes (✉) · W. A. Avila · A. Sierra
Universidad de los Andes, Bogotá, Colombia
e-mail: jureyes@uniandes.edu.co;
wa.avila10@uniandes.edu.co; a.sierra1457@uniandes.edu.co

D. Köber et al. (eds.), *Seismic Behaviour and Design of Irregular and Complex Civil Structures III*, Geotechnical, Geological and Earthquake Engineering 48,
https://doi.org/10.1007/978-3-030-33532-8_5

are often used per Chapter 16 of ASCE/SEI 7-10 (American Society of Civil Engineers 2010) according to both California Building Code (International Code Council 2013) and International Building Code (International Code Council 2015). However, in the present, the ASCE/SEI 7-16 (American Society of Civil Engineers 2016) demands eleven pairs of ground motions.

Nonlinear RHAs of 3D computer models of tall buildings or complex structures (e.g., dams and bridges) with significant degrees of freedom can be computationally challenging and time demanding. Even more, when records have high sampling rates, greater to 200 per-second, or are of long duration, the analysis time can be even more critical. Parametric studies or in incremental dynamic analyses (Vamvatsikos and Cornell 2002) of structures subjected to a series of nonlinear RHAs, are clear day-to-day examples of the previous problematic.

With the goal of obtaining a highly efficient RHA without significant error, this research offers a new approach consisting in appropriately trimming the beginning and end of the input record, and then downsampling the remaining record while preserving significant frequency characteristics of the original record, including its S-phase (Reyes et al. 2017). Its main advantage is to obtain a highly efficient RHA managing an acceptable error. The parameter to identify the leading and trailing signals to be trimmed is the maximum roof displacement; this value may be calculated by implementing the uncoupled modal response history analysis UMRHA (Chopra 2007, Chopra and Goel 2004; Reyes et al. 2011a, b, 2015). Arias intensity parameter (Arias 1970) and yield base shear may also be chosen; however, the previously proposed is better as it represents the characteristics of both the ground motion and structural response. In order to illustrate the efficiency of the method, this document describes the results of several 3D-idealized models that correspond to buildings of 3, 6, 9, 12, 15, 18, 21, 24, 27, 30, and 33-stories. Results demonstrated that the method is capable of controlling errors in estimating the peak roof displacement as an engineering demand parameter (EDP), and permitted to compare the EDP from the trimmed and down-sampled records against those from the original records.

5.2 Ground Motions Selected

Two seismic scenarios were considered in this study; for each scenario, eleven ground motions were spectrally matched to the corresponding design spectrum. Selected sites were LA, CA and San Jose, CA, which represent zones of high seismic hazard. Records were selected based on hazard deaggregation analyses and spectral shape. Magnitude ranges were 6.3–8.3 for both cities. Distance ranges were 4.5–65 km and 5–20 km for LA and San Jose sites, respectively. The seismicity of the sites included both near-fault and far-field crustal events.

Chapter 16 of the ASCE/SEI 7-16 allows spectrum-matched ground motions to be used in response history analyses of buildings. As the objective of this investigation is to assess a design problem, time domain spectrum matching was conducted on the original records. RspMatch2005 (a non-commercial computer code),

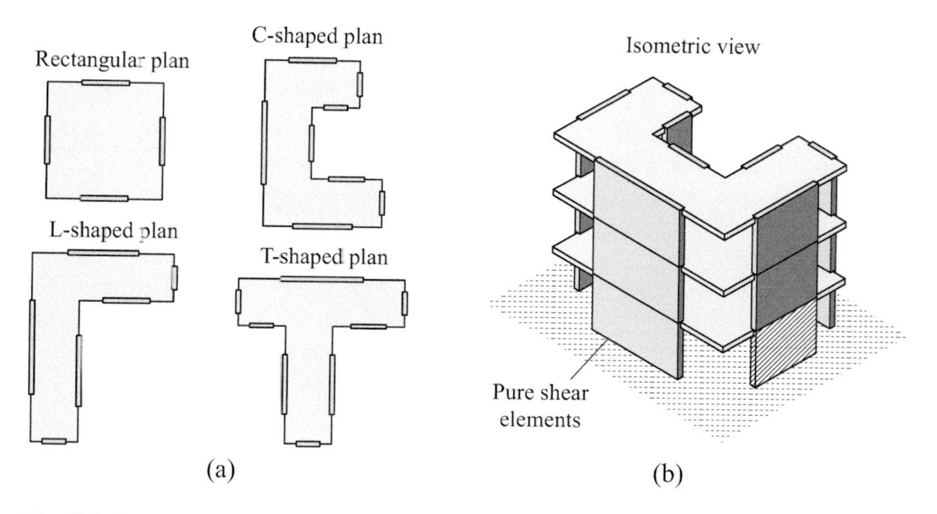

Fig. 5.1 Schematic plan view of reinforced concrete idealized structures used for analysis. (**a**) Floor plans. (**b**) Isometric view of a typical shear building model

recommended as one of the appropriate spectral matching techniques in ATC-82 report, was utilized for this purpose. For near-fault records, it was guaranteed that their velocity pulses were preserved after implementing spectrum matching.

5.3 Structural Systems

Buildings of 3, 6, 9, 12, 15, 18, 21, 24, 27, 30, and 33-stories, with the four different plans shown in Fig. 5.1a, were selected for the analysis phase. The three-dimensional (3D) idealized structure was defined as a shear model containing several vertical elements in each orthogonal horizontal direction (Fig. 5.1b), defined by the trilinear constitutive model shown in Fig. 5.2. The structural systems have a constant initial stiffness k_1 over its height. The latter was adjusted to achieve a prescribed fundamental period T_1 estimated per equations 8–7 in Chapter 12 of ASCE/SEI 7-10 (American Society of Civil Engineers 2010). Considering all the parametric values, a total of 132 structures were able to be characterized.

5.4 Methodology

The proposed method consists in a three-step process; each will be subsequently detailly explained. The maximum roof displacement of an equivalent SDF system is used as the EDP to identify the leading and trailing segments of the signal to be trimmed. The following procedure is taken from Reyes et al. (2017).

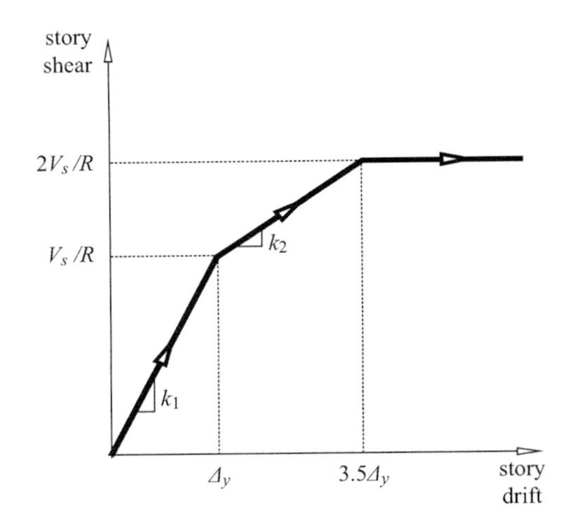

Fig. 5.2 Trilinear constitutive model for vertical elements

5.4.1 Trimming Leading Weak Signal

The leading weak signal that includes the pre-event interval starts from the beginning of the record to the last zero crossing before the roof displacement (u_r) reaches an initial target roof displacement (u_{ri}) defined as (Reyes et al. 2017):

$$u_{ri} = f_i \times \max\left(|u_r(t)|\right) \tag{5.1}$$

where f_i is the displacement modification factor for the leading signal, and | | is the absolute value operator. u_r is computed by implementing UMRHA (Chopra and Goel 2004). Identification of the leading weak signal is illustrated in Fig. 5.3, where the top panel shows ground-motion acceleration record and the bottom panel displays u_r.

5.4.2 Trimming Leading Weak Signal

The trailing weak signal starts from a time instant when the roof displacement (u_r) reaches a final target roof displacement (u_{rf}) and ends at the termination of the record. u_{rf} is defined as (Reyes et al. 2017):

$$u_{rf} = f_f \times \max\left(|u_r(t)|\right) \tag{5.2}$$

where f_f is the displacement modification factor for trailing signal. Identification of the trailing weak signal is illustrated in Fig. 5.4.

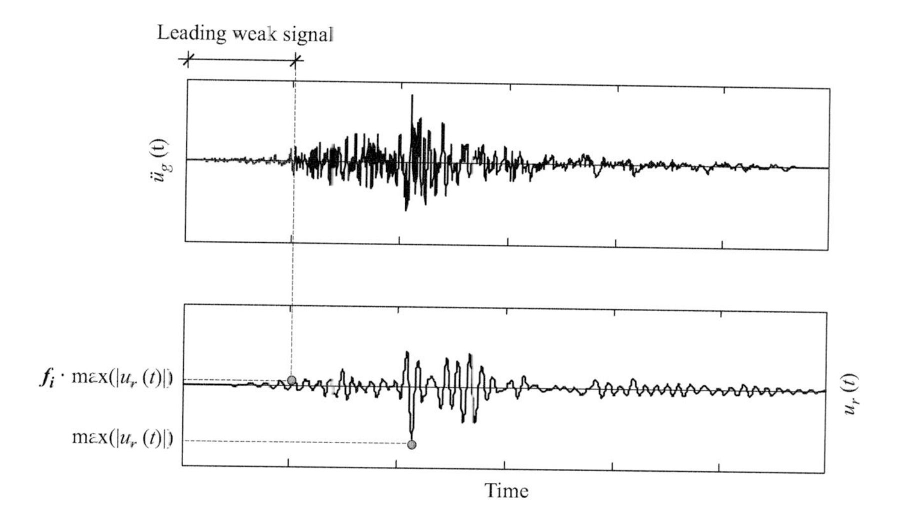

Fig. 5.3 Trimming leading weak signal from the acceleration record (top panel) using modified peak roof displacement, estimated from equivalent SDF system, as a proxy (bottom panel). Maximum roof displacement and its modified value by f_i are marked with yellow circles

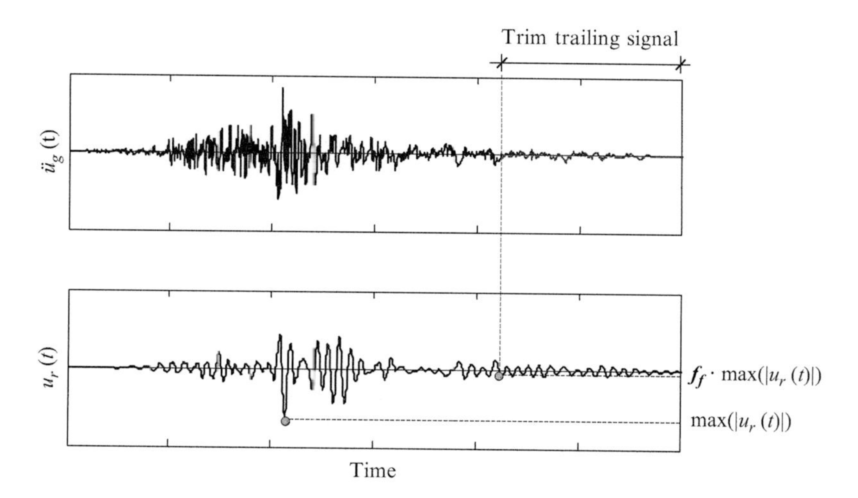

Fig. 5.4 Trimming trailing weak signal from the acceleration record (top panel) using modified peak roof displacement, estimated from equivalent SDF system, as a proxy (bottom panel). Maximum roof displacement and its modified value by f_f are marked with yellow circles

5.4.3 Downsampling

The downsampling is performed following as (Zhong and Zareian 2014):

- Transform the roof displacement time series from time domain to frequency domain;

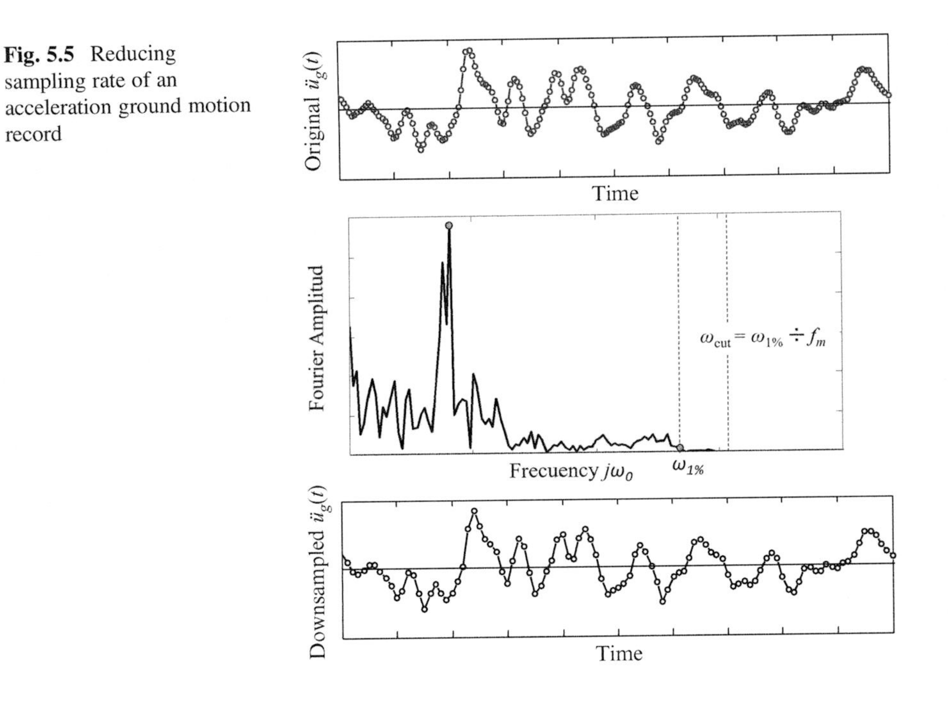

Fig. 5.5 Reducing sampling rate of an acceleration ground motion record

- Identify the largest frequency ($\omega_{1\%}$) associated with an amplitude at least of 1% of the peak response;
- Apply a low-pass filter to the trimmed record with cutoff frequency, $\omega_{cut} = \omega_{1\%} \div f_m$, where f_m is a factor that modifies the usable frequency range.
- Modify time step Δt as a multiple of 0.005 s less than or equal to π/ω_{cut} to eliminate aliasing.
- Resample the filtered record by picking every mth sample, where m represents the new sampling rate.

An example is illustrated in Fig. 5.5 showing the correspondence between the original and downsampled acceleration waveforms; the Fourier spectrum is also included for clarity.

5.5 Results

In order to optimize the processing time while keeping the discrepancies of the estimates of structural response in acceptable limits, a complete study was conducted. It consisted in a parametric research, using two sets of eleven modified records for 132 different structural systems, with various plan shapes, design strengths (R values) and fundamental periods. Additionally, the beginning and end trimming segments, as well as the downsampling (f_i, f_f and f_m) were ranged from 0% to 50% by an interval of 2.5%.

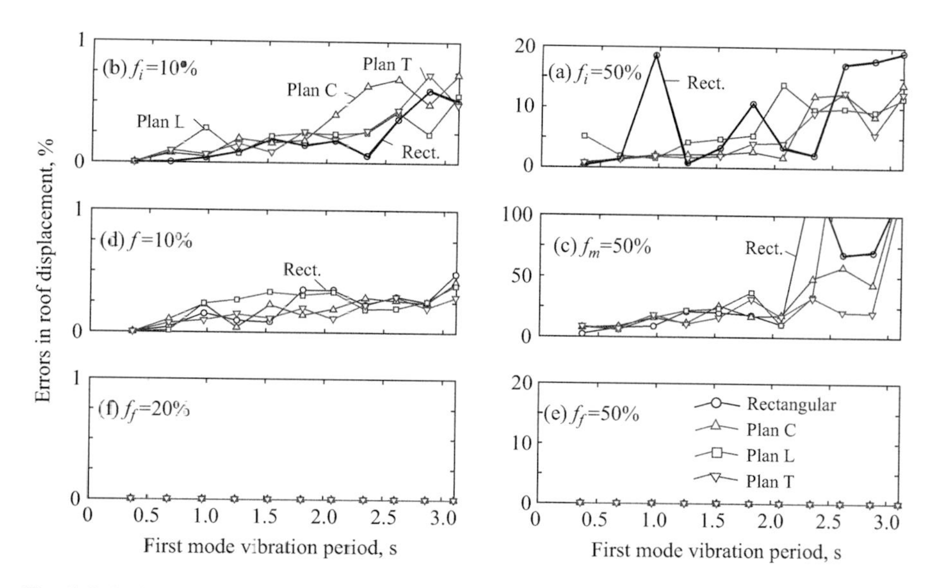

Fig. 5.6 Relative error in peak roof displacements by implementing independently the modifications presented in sections 4.1 to 4.3 for two values of the parameters f_i, f_f and f_m. Each subplot shows the results for various fundamental periods and contain series representing the obtained outcome for each structural plan using $R = 8$

Figure 5.6 presents the relative error in median peak roof displacements by implementing independently the modifications for the eleven records of the LA scenario. Each subplot shows the results for various fundamental periods and contain series representing the obtained outcome for each structural plan using $R = 8$. As expected, the error in peak roof displacement is larger as the value of the parameters f_i, f_f and f_m increases because it leads to a larger segment of the record to be cut or to a larger time step (compare Fig. 5.6a–f). In general, the error increases for longer periods compared to those for shorter ones. If the parameters f_i, f_f and f_m are equal to 10%, 10%, 20%, respectively, the error in roof displacements is less than 1%. For this condition, the unsymmetric-plan buildings with plans C and T seem to be more critical. In Fig. 5.6e, f, the error in the estimation of the peak roof displacement is zero because the trimming of weak trailing signal occurs after the time instant when the peak value of roof displacement is reached. Based on these results, appropriate values of the parameters f_i, f_f and f_m may be 10%, 10%, 20%, respectively.

Figure 5.7 demonstrates the combined effects of proposed values of f_i, f_f and f_m on the error in median peak roof displacement estimates, and the contribution of these modification factors on the time steps saved. The results presented in Fig. 5.7 are for the LA scenario and two R values. The error is within 4% for almost all cases, and the average reduction in time steps is 70% as compared to the original records. The linear structures shows larger errors in roof displacements than the nonlinear ones, being more critical the rectangular plans. The savings in time steps are similar for the four plans considered. For structures with fundamental periods less than 1 s, the time step reduction is mainly due to the trimming leading and trailing signals. For

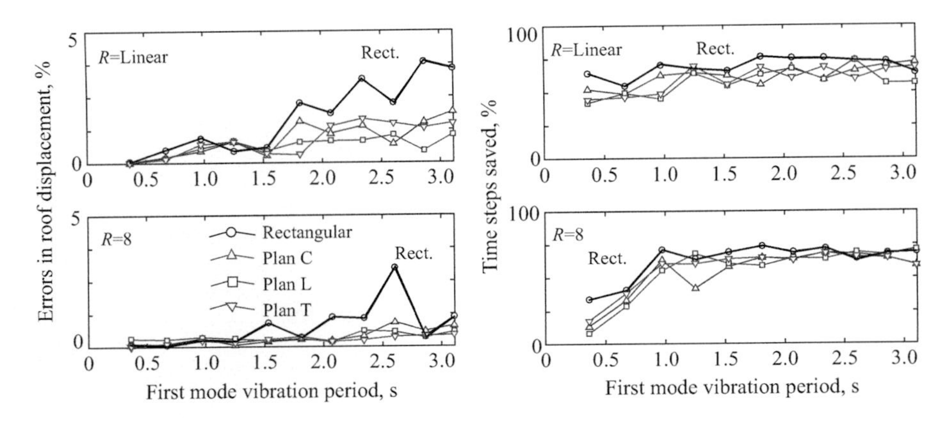

Fig. 5.7 Relative error in peak roof displacements and time steps saved by implementing simultaneously the modifications presented in Sects. 4.1, 4.2, and 4.3 for $f_i = 10\%$, $f_f = 10\%$ and $f_m = 20\%$

structures with the longer fundamental periods, the reduction is mainly due to the downsampling process of the record. This is the benefit of using three different criteria when modifying the ground motion records.

5.6 Conclusions

A practical method where records are modified by trimming leading and trailing weak signals and reducing sampling rates is proposed in this research. This option permits to achieve fast nonlinear response history analysis (RHA) of complex structures or multiple degrees of freedom systems. The method is shown to be successful in limiting the error in roof displacement estimates in most cases because it ensures that the S-phase is preserved in the trimmed record. The goodness of the approximation was measured in terms of the ability to represent the response of unsymmetric-plan MDF systems, from linear to nonlinear, subjected to two sets of eleven ground-motion records. Additional testing results from various EDPs such as floor accelerations, floor velocities, and story-shear still require further validation.

Acknowledgments The authors would like to acknowledge the financial support of Universidad de los Andes, Bogota, Colombia and Christiam Angel due to his support in the final editing tasks.

References

American Society of Civil Engineers (2010) Minimum design loads for buildings and other structures [ASCE/SEI 7-10]. American Society of Civil Engineers, Reston

American Society of Civil Engineers (2016) Minimum design loads for buildings and other structures [ASCE/SEI 7-16]. American Society of Civil Engineers, Reston

Applied Technology Council (2005) Improvement of nonlinear static seismic analysis procedures. Rep. No. FEMA-440, Washington, DC

Arias A (1970) A measure of earthquake intensity. In: Hansen RJ (ed) Seismic design for nuclear power plants. MIT Press, Cambridge, MA, pp 438–483

Chopra AK (2007) Dynamics of structures: theory and applications to earthquake engineering, 4th edn. Prentice-Hall, Upper Saddle River

Chopra AK, Goel RK (2004) A modal pushover analysis procedure to estimate seismic demands for unsymmetric-plan buildings. Earthq Eng Struct Dyn 33:903–927

International Code Council (2013) California building code. Whittier, CA

International Code Council (2015) International building code. Whittier, CA

Reyes JC, Chopra AK (2011a) Three-dimensional modal pushover analysis of buildings subjected to two components of ground motion, including its evaluation for tall buildings. Earthq Eng Struct Dyn 40:789–806

Reyes JC, Chopra AK (2011b) Evaluation of three-dimensional modal pushover analysis for unsymmetric-plan buildings subjected to two components of ground motion. Earthq Eng Struct Dyn 40:1475–1494

Reyes JC, Riaño AC, Kalkan E (2015) Extending modal pushover-based scaling procedure for nonlinear response history analysis of multi-story unsymmetric-plan buildings. Eng Struct 88:125–137

Reyes JC, Kalkan E, Sierra A (2017) Fast nonlinear response history analysis. In: 16th world conference on earthquake engineering, January 9–13, Santiago, Chile

Vamvatsikos D, Cornell CA (2002) Incremental dynamic analysis. Earthq Eng Struct Dyn 31 (3):491–514

Zhong P, Zareian F (2014) Method of speeding up buildings time history analysis by using appropriate downsampled integration time step. In: 10th U.S. national conference on earthquake engineering, July 21–25, Anchorage, Alaska

Chapter 6
Seismic Behaviour of an Irregular Old RC Dual-System Building in Lisbon

Claudia Caruso, Rita Bento, and Edoardo M. Marino

Abstract In this paper, the seismic performance of a reinforced concrete dual-system building with vertical irregularities, built in the 60s in Lisbon, is addressed. The seismic behaviour and the torsional effects of the building are investigated by means of nonlinear static (pushover) analyses and extended N2 method. Then the results are compared with the nonlinear dynamic *Time-History* analysis, the latter considered as the reference solution. A three-dimensional numerical model of the case-study building is developed to account for torsion in the building. The evaluation of the seismic vulnerability of structure is based on performance-based assessment procedures and on the structural safety requirements proposed in Part 3 of Eurocode 8. The main targets of this study are (i) to detect and quantify the main deficiency of this typology of old existing RC buildings, which were not designed to resist the forces induced by torsional vibrations; (ii) to propose suitable retrofitting intervention for this category of buildings.

Keywords Reinforced concrete buildings · Seismic assessment · Dual-system · Extended N2 method

6.1 Introduction

In many European cities, a high percentage of reinforced concrete (RC) dual-system buildings were designed between the 60s and the 80s according to older seismic codes. These buildings typology present characteristics such as smooth reinforcing bars, open ground floors without infills and eccentric RC core walls which determine

C. Caruso · R. Bento (✉)
CERIS, Instituto Superior Técnico, Universidade de Lisboa, Lisbon, Portugal
e-mail: claudia.caruso@tecnico.ulisboa.pt; rita.bento@tecnico.ulisboa.pt

E. M. Marino
Department of Civil Engineering and Architecture, University of Catania, Catania, Italy
e-mail: emarino@dica.unict.it

D. Köber et al. (eds.), *Seismic Behaviour and Design of Irregular and Complex Civil Structures III*, Geotechnical, Geological and Earthquake Engineering 48,
https://doi.org/10.1007/978-3-030-33532-8_6

their higher seismic vulnerability. Also, the seismic response of old RC buildings may be significantly influenced by torsional effects, which can lead to an increase of damage throughout the structure and the consequential need of structural intervention.

In this study, a case study old RC frame-wall building is considered. A three-dimensional (3-D) model is developed to take into account the torsional behaviour, using the open-source software OpenSees (McKenna et al. 2000). A significant effort was devoted to the development of a numerical model that is able to account for the main features of old RC frame-wall structures such as: masonry infills, first storey irregularity, low dissipative behaviour of structural members due to inadequate reinforcement detailing, smooth reinforcing bars. The seismic assessment is performed with an attention on torsional effects. The original N2 (Basic N2) method (Fajfar 1999) and its extension (Extended N2) (Kreslin ab Fajfar 2012) are applied to the case study structure and the results compared with the ones got with nonlinear dynamic *Time-History* analysis (TH).

This work intends to be a contribution on the assessment of the seismic vulnerabilities of this typology of buildings, and a contribution for the development of appropriate retrofitting strategies. A retrofitting intervention based on FRP wrapping is proposed focusing on the main objective to increase the local deformation capacity of the vulnerable structural elements and thus the global deformation capacity of the structure.

This study starts with a brief characterization of the case study building (Sect. 6.2). Afterwards, in Sect. 6.3, a summary review of the assessment methodology and the Extended N2 method is introduced. Then, in Sect. 6.4, the seismic performance of the RC building case study is evaluated and the results of a local strengthening solution are analysed and discussed.

6.2 Building Structure

A case study is chosen, as belonging to the typology of RC dual-system residential buildings in Lisbon designed in the 60s (Fig. 6.1). The building features 8 storeys in elevation. The ground storey is slightly higher, 3.60 meters, all the other storeys are

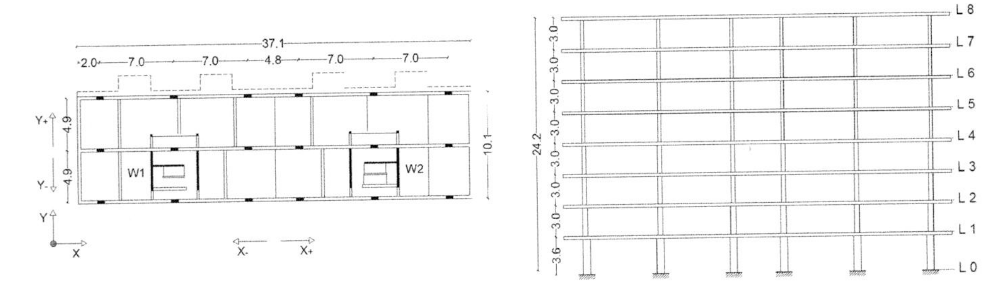

Fig. 6.1 Case study building: structural plan (*left*); cross-section (*right*). Dimensions in [m]

Table 6.1 Floor masses and moments of inertia of mass

	Floor masses [ton]	Moments of inertia of mass [ton.m^2]
Bottom storey	443	54,341
	0.099 m	
Upper storeys	456	53,700
	0.096 m	

3.00 meters high. It presents infills walls at all storeys except at the ground storey which is an open storey.

The building, designed and built before the introduction of modern seismic codes and the modern principles of ductility and energy dissipation capacity, presents some inadequate reinforcement detailing, e.g. (i) insufficient longitudinal and transverse reinforcement in RC columns and walls, which may lead to brittle shear failure, (ii) vertical discontinuity due to reduction in stiffness and size of columns and walls at each storey, typical for building designed for gravity loads only (iii) lack of confined boundary elements in RC walls, which, according to Part 1 of Eurocode8 (EC8-1) (EC8-Part 1: Eurocode 8 2005), are necessary to ensure a ductile behaviour (EC8-1), (iv) smooth reinforcing bars.

Table 6.1 shows the floor masses at the bottom and upper storey and the moment of inertia of mass.

The structure is symmetric in the Y direction and relatively asymmetric in the X direction. Nevertheless, a mass eccentricity was assumed equal to 5% of the plan dimensions in each of the horizontal directions. This mass eccentricity can be considered as an accidental eccentricity, as defined in EC8-1. As a preliminary evaluation of the torsional response of the building, the ratio of the translational period to the rotational period has been evaluated. In fact, typically, values smaller than one denote the structure as torsionally flexible (Anagnostopoulos et al. 2015; Fajfar et al. 2005). For the case study building, these values are equal to 0.96 and 0.99 for the X and Y direction, respectively, showing that the structure is prone exhibit torsional effects.

6.2.1 Numerical Modelling

A 3-D model of the building is developed in OpenSees. Force-based nonlinear beam column elements are used for the structural element and a fibre discretization of the sections is adopted. The effects of confinement on the concrete section core and cover are defined using the Popovics model (Popovics 1973). The constitutive law of the reinforcing steel bars makes use of the Giuffre-Menegotto-Pinto model (Menegotto and Pinto 1973). As for the material properties, a concrete compressive strength of 28 MPa is considered and a steel yield strength of 235 MPa.

Initial stiffness and mass proportional damping is considered for the definition of the TH analysis (Bhatt and Bento 2014), with a 5% damping in the first and fifth mode, being the latter the mode in which almost all sum participation mass is mobilized. For the definition of the elastic response spectrum, necessary for the application of the Extended N2 method, i.e. the N2 method and the linear dynamic response spectrum analyses, a 5% viscous damping is considered.

The presence of infill walls is taken into account by means of two diagonal struts carrying only compressive load (Celarec et al. 2012; Dolšek and Fajfar 2008). A simplified method is assumed to take into account the presence of smooth reinforcing bars and the increased deformability due to strain penetration effects at the base of the vertical structural element. The method, developed in a previous work (Caruso et al. 2018a), consists in reducing the Young Modulus and maximum strength of the steel rebars based on to the rebars' properties and the reinforcement embedment length. In this study, based on the characteristic of the reinforcing bars of the RC walls, the Young Modulus was reduced by 50% and the maximum strength by 30%.

More information on the numerical modelling and characteristics of the buildings analysed may be found in (Caruso et al. 2018b).

6.3 Seismic Safety Assessment

6.3.1 Methodology

The seismic safety assessment of the building is performed by comparing for each vertical element of the structure (columns and RC walls) the seismic demand, in terms of deformations and shear forces, with the capacity, as proposed in the Part 3 of Eurocode 8 (EC8-3).

To assess the seismic performance of the building, and analyse how the torsional behaviour is unfavourable for the assessment of the structural seismic response, the Extended N2 method is applied. The Extended N2 method combines the results of a pushover analysis with those of a Linear Dynamic Response Spectrum Analysis (LDRS), thus accounting for higher mode effects in both plan (torsional effects) and elevation. Correction factors are defined and applied to the results of the pushover analysis to take into account the higher mode effects. The full description of the procedure is described in (Kreslin and Fajfar 2012).

For the pushover analyses lateral forces uniform and modal proportional load patterns are adopted. These forces are applied in the X and Y directions and in positive and negative senses (Fig. 6.1), resulting in four different pushover analyses for each type of load pattern. The target displacements are defined through the N2 method (Fajfar 2000), according to the nonlinear static procedure prescribed in Part 3 of Eurocode 8 (EC8-3) (EC8-Part 3: Eurocode 8 2010). The seismic action is defined by means of the National Annex of EC8-1 elastic response spectrum for soil type B, with a PGA of 0.153 g (return period of 475 years) for earthquake type 1, which represent the critical seismic action for this case study.

A LDRS analysis of the 3-D model is performed. The Complete Quadratic Combination (CQC) rule is used to combine the different modal responses and the results from the two horizontal directions (X and Y) are combined by the Square-Root-of-Sum-of-Squares (SRSS) rule. The sum of the effective modal mass for all the modes considered in each direction is at least 95% of the total mass.

The results of the TH analysis are considered as a benchmark to which the results of the Basic N2 method (which does not account for torsional effects) and the results of the Extended N2 are compared.

To perform the TH analysis, 30 real ground motion records are selected using the methodology developed in (Araújo et al. 2016). Each record has two horizontal components, X and Y, and is scaled to match the elastic spectrum corresponding to the Significant Damage (SD) limit state.

The analyses are performed on a model with accidental eccentricity, i.e. the centre of mass (CM) does not coincide with the geometric centre of the deck (i.e. an accidental eccentricity is considered).

6.4 Analyses of Results

6.4.1 Evaluation of the Target Displacement

Pushover curves as determined by the modal load pattern are showed in Fig. 6.2. The figure shows base shear (V) versus top displacement (d) at the CM for the two sign of loading, positive (+) and negative (−) (see Fig. 6.1) and for the X (Fig. 6.2a) and Y direction (Fig. 6.2b). It is worth noting that the two curves in the X direction are coincident.

The mean result of the TH analyses, used as benchmark to evaluate the applicability of the BasicN2 method, are also shown Fig. 6.2. In this figure, V_{max} corresponds to the mean response of the TH considering the maximum top floor

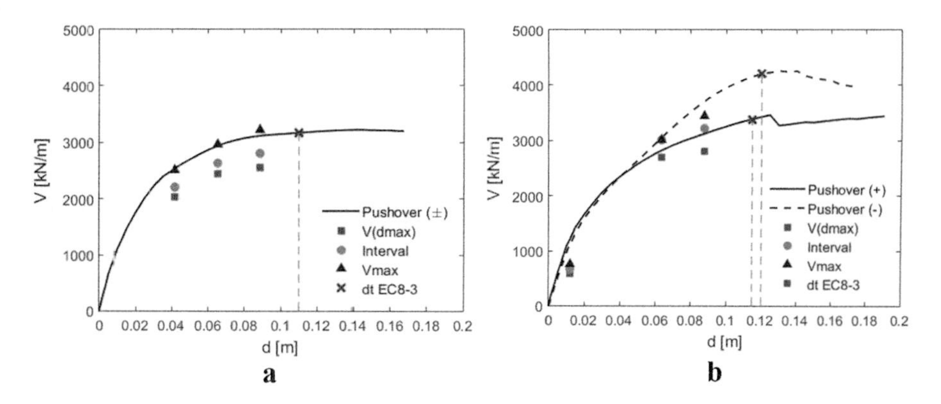

Fig. 6.2 (**a**) Pushover curves in X direction and (**b**) in Y directions

Table 6.2 Target displacements as obtained by the N2 method and TH analysis for SD limit state

	N2 method		
	Modal	Uniform	TH analysis
X+	0.108 m	0.099 m	0.089 m
X−	0.108 m	0.099 m	0.088 m
Y+	0.115 m	0.090 m	0.084 m
Y−	0.120 m	0.096 m	0.088 m

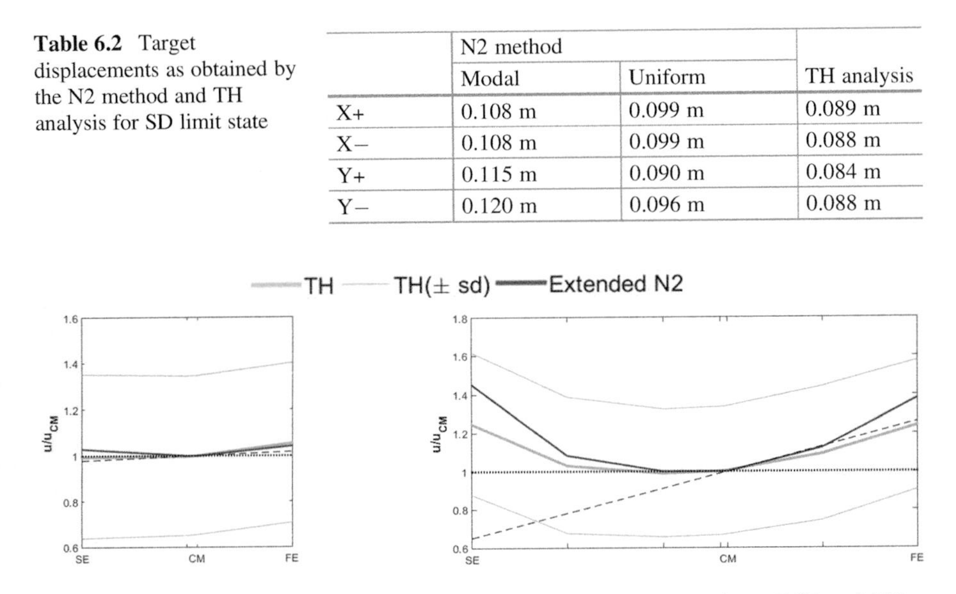

Fig. 6.3 (**a**) Top displacements in plan for X direction and (**b**) for Y direction at PGA = 0.153 g

displacement versus the maximum base shear; $V(d_{max})$ stands for the maximum top floor displacement versus the base shear obtained at the same instant; "Interval" corresponds to the maximum top floor displacement versus the maximum base shear obtained in an interval of time ranging from 1 s before and after the instant correspondent to maximum top floor displacement.

In Fig. 6.2 the values of the target displacement as obtained with the N2 method and for the SD state and the seismic action set by EC8-1 for Lisbon are marked with a red cross. It is evident that the results of the two methods (TH and N2) are quite similar and are summarised in Table 6.2.

The effects of torsion and the effects of higher mode effects on the building response are evaluated in terms of drift, in plan and in elevation, and in terms of shear demand. In order to evaluate the torsional response, the displacement at the centre of mass, at the *flexible* edge (FE) and at the *stiff* edge (SE) of the building are observed, being the FE edge close to the CM.

The distribution of displacement in plan are obtained performing the LDRS and the TH analyses. These displacements are normalized by the top displacement at the mass centre (u_{CM}).

The curve of the in-plan distribution of top displacement determined by the results of LDRS represents the effect of the torsional component as projected by the Extended N2 (through the application of correction factors). In Fig. 6.3 the normalized top displacements u/u_{CM} are presented for the X direction (Fig. 6.3a) and the Y direction (6.3b) at PGA = 0.153 g. It is evident that the results as obtained by the Extended N2 method are very similar to those obtained by the TH analysis, even though the former are slightly conservative.

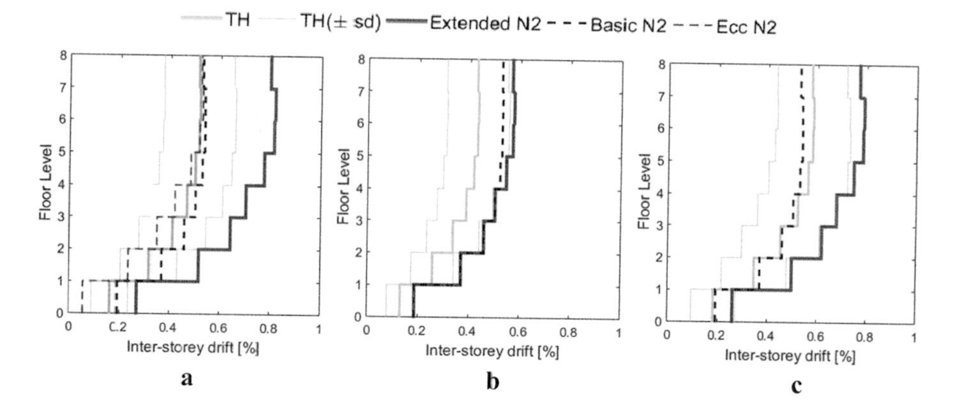

Fig. 6.4 (**a**) Storey-drifts at the SE, (**b**) at the CM and **c** at the FE for the Y direction

The results delivered by the Basic N2 (dashed line in the figure) are not capable of capturing the torsional behaviour of the structure and they totally deviate from the result of the TH.

The interstorey drift ratios are plotted for the Y direction in Fig. 6.4, at the target displacement correspondent to the SD limit state. In the figures, from a to c, the interstorey drifts ratios along the height are showed for the SE, the CM and the FE, respectively. The results denoted as "Basic N2" correspond to the torsionally balanced building and those denoted as Extended N2 are obtained considering the accidental eccentricity. In the X direction, the effects of torsion are very small, as it is evident also from Fig. 6.3a.

The effects of torsion are evident for the higher interstorey drifts provided by the Extended N2 in both FE and SE. Even though the results are quite conservative, it is worth noting that this is also due to higher target displacement obtained with the N2 method (see Table 6.2). Furthermore, it is showed that applying the original N2 method (i.e. without adopting correction factors) on the model which consider the eccentricity in plan (dashed red line in Fig. 6.4a) would lead to underestimate the drifts, which is an unsafe condition.

6.4.2 *Shear Demand and Capacity*

The results of a previous study (Celarec et al. 2012) indicated that the most severe failure mode of the building corresponds to the shear failure (brittle failure mechanism) of the RC walls in both X and Y directions, while the columns have a reasonable flexural behaviour, developing a stable flexural response.

In this study, the shear demand is determined multiplying the results of the pushover analyses with the correction factors. Figure 6.5 shows the shear demand in wall W1 at the target displacement, for the X direction (Fig. 6.5a) and for the Y direction (6.5b).

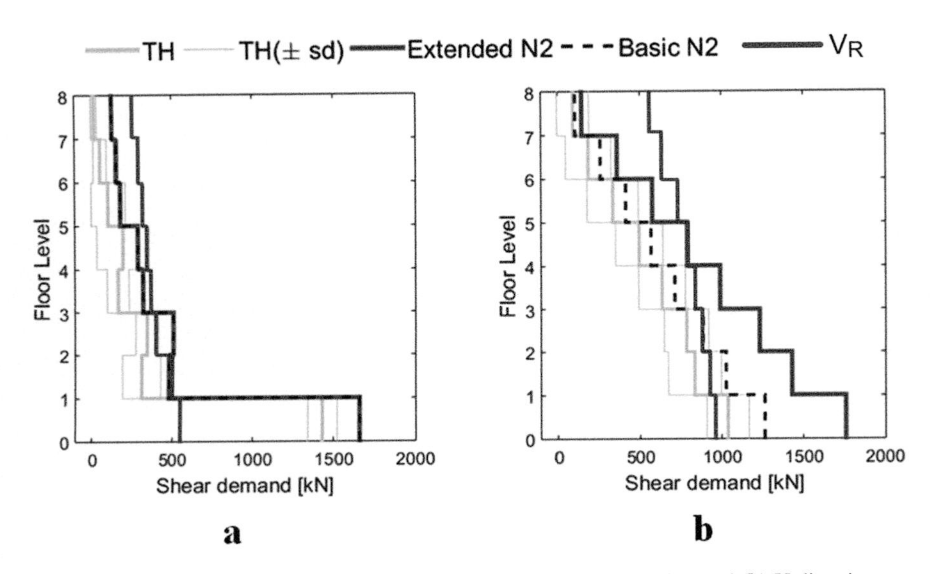

Fig. 6.5 Shear demand and capacity (V_R) for walls W1 in (**a**) X direction and (**b**) Y direction

For the X direction, the Extended and Basic N2 methods deliver almost coincident results, as the correction factors in this direction are very small. Furthermore, the shear demand prediction is conservative when compared to that obtained by the TH analysis.

For the Y direction, the Extended N2 method leads to a conservative estimation of the shear demand when compared to the Basic N2 and the TH analyses. Based on these results it is possible to conclude that the torsional effects aggravate the structural building behaviour, therefore it is important to consider them in seismic assessment procedures.

The shear resistance of RC walls, V_R, is calculated by means of shear resistance to web crushing, $V_{Rd,max}$, and shear resistance as controlled by the stirrups, $V_{Rd,s}$, and verified according to the minimum of the two values.

It is evident from the comparison of the shear demand and capacity (Fig. 6.5) that the walls suffer from brittle shear failure. As stated before, this result is due to the very low amount of transversal reinforcement of the RC walls. Therefore, in the next section a local strengthening solution is addressed to increase the shear capacity of the walls, involving the use of Fibre Reinforced Polymers (FRP).

6.4.3 Retrofitting Strategy

Externally bonded FRPs are used in seismic retrofitting in order to enhance or improve: (i) the deformation capacity of flexural plastic hinges, (ii) deficient lap splices, (iii) shear resistance. To improve the shear capacity of brittle components,

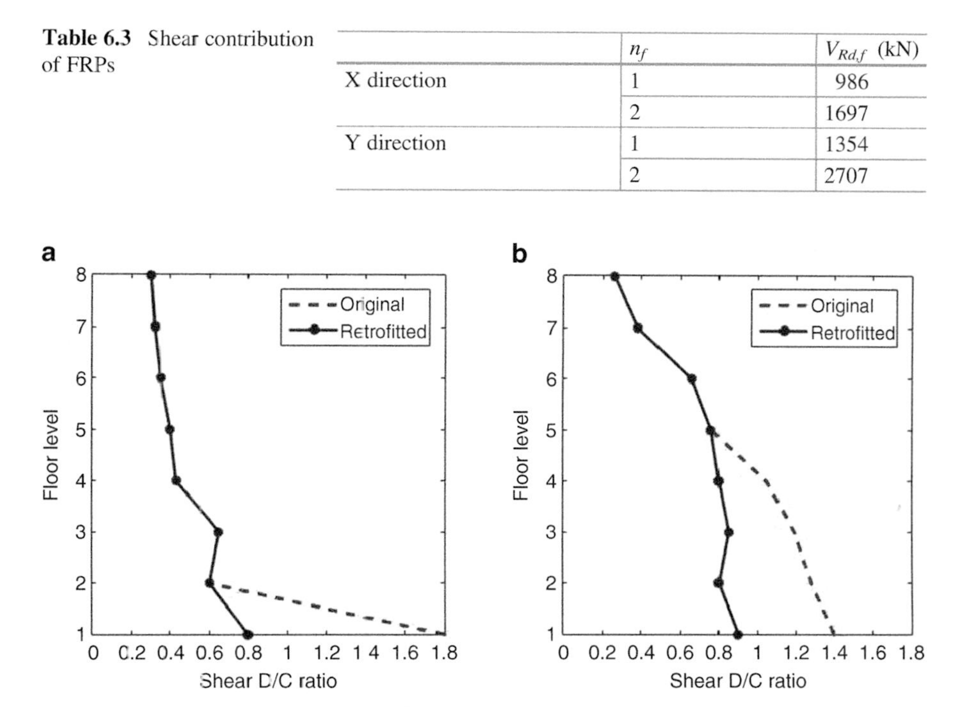

Table 6.3 Shear contribution of FRPs

	n_f	$V_{Rd,f}$ (kN)
X direction	1	986
	2	1697
Y direction	1	1354
	2	2707

Fig. 6.6 Shear D/C ratio for RC walls in (**a**) X direction and (**b**) Y direction

the FRP overlay should be applied with the fibres mainly in the direction in which enhancement of shear strength is pursued. Unlike beams, columns and walls are subjected to a constant shear force within each storey. Hence, if shear strengthening is needed, it should be uniform throughout the height of the vertical element in a storey. Moreover, as the shear demand alternates between opposite values, the main direction of the FRP should be horizontal (Fardis 2009).

The total shear capacity, as controlled by the stirrups and the FRP, is evaluated as the sum of the contribution from the existing concrete member and the contribution from the FRP (Caruso et al. 2018b). The FRP contribution to the shear capacity for full wrapping with FRP may be calculated with Equation A.22 of EC8-3. In Table 6.3 the value of the FRP shear contribution, $V_{Rd,f}$ are reported for different number of layers, n_f.

The shear demand to capacity ratio (D/C) in the RC walls of the Original and Retrofitted schemes at the SD limit state are plotted in Fig. 6.6. All shear D/C ratios fall below unity, indicating positive effects of brace retrofitting on reducing the demand on the shear walls (Fig. 6.6). It is noted that the partial retrofitting at the ground storey results in a decrease in wall shear at the lower storey, but it does not have a negative influence on the upper storeys of the building, where the D/C ratio is unchanged, although shear walls may experience higher demand from the effects of higher modes of vibrations.

6.5 Final Remarks

In this study, the seismic performance assessment of an old RC dual-system building was performed. In particular, the assessment was performed by means of the original (Basic) N2 method and by means of the Extended N2 method, which is capable, by application of correction factors, to account for the higher mode effects in both plan (torsional behaviour) and elevation.

To assess to which extend torsion has an influence on the results obtained by nonlinear static analyses (pushover), the results were compared with the results obtained performing nonlinear dynamic time-history analysis. It was showed that the Extended N2 method leads to a more realistic estimation of the torsional behaviour and, even though the results are slightly conservative, it is essential to capture a general worsening of the seismic structural performance due to the torsion behaviour. Then, a local method of retrofitting was used, involving the partial strengthening at the open ground storey with FRPs-wrapping of single elements (individual RC walls).

As final a final remark, it is suggested that the torsional effects should always be considered for the seismic assessment of old RC buildings, even in symmetric structure, in which an accidental eccentricity is considered (as supported by EC8).

References

Anagnostopoulos SA, Kyrkos MT, Stathopoulos K (2015) Earthquake induced torsion in buildings: critical review and state of the art. Earthq Struct 8(2):305–377

Araújo M, Macedo L, Marques M, Castro JM (2016) Code-based record selection methods for seismic performance assessment of buildings. Earthq Eng Struct Dyn 45:129–148

Bhatt C, Bento R (2014) The extended adaptive capacity spectrum method for the seismic assessment of plan-asymmetric buildings. Earthquake Spectra 30(2):683–703

Caruso C, Bento R, Sousa R, Correia AA (2018a) Modelling strain penetration effects in RC walls with smooth steel bars. Mag Concr Res. https://doi.org/10.1680/jmacr.18.00052

Caruso C, Bento R, Marino EM, Castro JM (2018b) Relevance of torsional effects on the seismic assessment of an old RC frame-wall building in Lisbon. J Build Eng 19:459–471

Celarec D, Ricci P, Dolšek M (2012) The sensitivity of seismic response parameters to the uncertain modelling variables of masonry-infilled reinforced concrete frames. Eng Struct 35:165–177. https://doi.org/10.1016/j.engstruct.2011.11.007

Dolšek M, Fajfar P (2008) The effect of masonry infills on the seismic response of a four storey reinforced concrete frame-a probabilistic assessment. Eng Struct 30(11):3186–3192. https://doi.org/10.1016/j.engstruct.2008.04.031

EC8-Part 1: Eurocode 8 (2005) Design provisions for earthquake resistance of structures. Part 1-1: General rules – seismic actions and general requirements for structures. ENV 1998-1, CEN: Brussels

EC8-Part 3: Eurocode 8 (2010) Design of structures for earthquake resistance. Part 3: Assessment and retrofitting of buildings. ENV 1998-3, CEN: Brussels

Fajfar P (1999) A capacity spectrum method based on inelastic demand spectra. Earthq Eng Struct Dyn 28:979–993

Fajfar P (2000) A nonlinear analysis method for performance based seismic design. Earthquake Spectra 16(3):573–592

Fajfar P, Marusic D, Perus I (2005) Torsional effects in the pushover-based seismic analysis of buildings. J Earthq Eng 6:831–854

Fardis MN (2009) Seismic design, assessment and retrofitting of concrete buildings. Springer, Dordrecht. Geotechnical, Geolcgical, and Earthquake Engineering

Kreslin M, Fajfar P (2012) The extended N2 method considering higher mode effects in both plan and elevation. Bull Earthq Eng 10:695–715

McKenna F, Fenves GL, Scott HM (2000) Open system for earthquake engineering simulation (OpenSEES), Pacific Earthquake Engineering Research Center, University of California, CA

Menegotto M, Pinto PE (1973) Method of analysis for cyclically loaded RC plane frames including changes in geometry and non-elastic behaviour of elements under combined normal force and bending. IABSE Symposium on resistance and ultimate deformability of structures acted on by well-defined repeated loads–Final report

Popovics S (1973) A numerical approach to the complete stress–strain curve of concrete. Cem Concr Res 3(5):583–599

Chapter 7
A Database for Assisted Assessment of Torsional Response of In-Plan Irregular Buildings

F. Barbagallo, M. Bosco, A. Ghersi, E. M. Marino, and P. P. Rossi

Abstract Nonlinear static analysis is currently the most popular method of analysis for the prediction of the seismic response of buildings. However, when dealing with in-plan irregular buildings, the nonlinear static analysis does not allow a proper assessment of the torsional response. In particular, the torsional response modifies the displacement demand of the two sides of the deck with respect to that of the corresponding torsionally balanced system: the displacement demand of the flexible side increases, while that of the stiff side decreases or increases depending on the features of the building. In this paper, the ratio of the maximum displacements of the asymmetric system to the maximum displacement of the corresponding planar system is determined for a large set of single storey systems. The results of this investigation are used to populate a database for the assisted assessment of buildings. This database can be interrogated by means of parameters characterizing the building to be assessed and provides the user with the amplification/deamplification of displacement demand of the two sides of the building with respect to that of the corresponding planar system. The database is conceived in such a way to include new cases without modifying its structure. This allows the database to be easily expanded when new parameters will be explored and new results will be available.

Keywords Irregular buildings · Existing buildings · 3D structural systems · Pushover analysis · Seismic assessment

7.1 Introduction

Nonlinear static analysis is nowadays the most popular method of analysis among structural engineers to assess the seismic behaviour of existing structures. Indeed, this method is simply enough to be handled by professionals and, at the same time,

F. Barbagallo · M. Bosco · A. Ghersi · E. M. Marino (✉) · P. P. Rossi
Department of Civil Engineering and Architecture, University of Catania, Catania, Italy
e-mail: fbarbaga@dica.unict.it; mbosco@dica.unict.it; aghersi@dica.unict.it; emarino@dica.unict.it; prossi@dica.unict.it

D. Köber et al. (eds.), *Seismic Behaviour and Design of Irregular and Complex Civil Structures III*, Geotechnical, Geological and Earthquake Engineering 48,
https://doi.org/10.1007/978-3-030-33532-8_7

allows them to compare the displacements and plastic deformations corresponding to the achievement of a given limit state to values caused by ground motions of assigned intensity levels.

Two of the most important nonlinear static methods existing in literature are the N2 method proposed by Fajfar et al. (Fajfar 1999) and the Capacity Spectrum Method (CSM) proposed by Freeman (Freeman 1998). Variants of these methods are implemented in seismic codes. For example, N2 method is adopted in Eurocode 8 (Eurocode 8 2004) and CSM is adopted in ATC40 (ATC, Applied Technology Council 1996) and in FEMA440 (ATC 2005).

The attractive features of this method of analysis led researchers to spend efforts to verify its effectiveness in the prediction of the seismic response of buildings. Because of the lack of experimental data on the seismic response of actual structures, the effectiveness of nonlinear static methods is generally studied by comparing the seismic response predicted by this method of analysis to that obtained by nonlinear dynamic analysis, which is widely believed the most accurate method of analysis.

Studies carried out in the last years have pointed out that the nonlinear static analysis provides accurate results when the seismic response of the building is mainly affected by a single mode of vibration. Consequently, reasonably accurate results are obtained for low-rise symmetric (Giorgi and Scotta 2013) or very torsionally-stiff buildings (Bento et al. 2010; De Stefano et al. 2013; Fajfar et al. 2005).

Several studies have been devoted to improve the effectiveness of nonlinear static methods in predicting the torsional response of asymmetric buildings (Chopra and Goel 2004; Reyes and Chopra 2011; Fujii 2011; Kreslin and Fajfar 2012; Bosco et al. 2012, 2015; Fujii 2014; Bhatt and Bento 2014; Georgoussis 2015). Many of these studies focus on single-story systems (Palermo et al. 2013, 2017) because it is widely believed that the amplification/deamplification of the displacement demand of in-plan irregular buildings can be approximately simulated by equivalent single-story systems as long as the elastic and inelastic properties of these systems are properly defined. These studies also enumerate the parameters that control the torsional component of the response of asymmetric buildings.

Out of the proposed methods, the extended N2 method (Fajfar et al. 2005; Kreslin and Fajfar 2012) stands out. This method was formulated on the basis of the two following assumption:

1. the elastic response of an asymmetric building, normalised with respect to that of the corresponding torsionally balanced system, provides a conservative estimation of the effect of the torsional component of the response;
2. the deamplification of the seismic response of an asymmetric building in the elastic range of behaviour is larger than that observed when the structure experiences inelastic response.

Based on the first assumption, Fajfar et al. (Fajfar et al. 2005; Kreslin and Fajfar 2012) suggest to evaluate the floor (inelastic) displacement demand of asymmetric buildings by multiplying the displacement demand of the centre of mass, determined by means of the version of N2 method formulated for planar systems (Fajfar 1999), by modification coefficients calculated by an (elastic) modal analysis. The N2 method for planar systems can be applied to the asymmetric structure or to the corresponding torsionally balanced system. Based on the second assumption, the modification coefficients should be taken not lower than one to avoid overestimating the beneficial deamplification of the response caused by the torsional component. The extended N2 method is very easy to apply, however, it is sometimes overly conservative (Bosco et al. 2013).

This paper aims at providing modification coefficients to adjust the results of the original N2 method. However, differently from those proposed in (Fajfar et al. 2005), the modification coefficients are based on the nonlinear response of single storey systems.

A parametric analysis has been conducted on a large set of asymmetric single-storey systems to quantify the effects of the torsional response. The analysed systems are generated by varying the parameters that mostly affect the seismic response of in-plan irregular buildings, i.e. the fundamental period of the corresponding torsionally balanced system, the rigidity eccentricity, the ratio of the torsional to lateral frequencies of the corresponding torsionally balanced system, the strength eccentricity and the ratio of the elastic strength demand to the actual strength of the system.

The in-plan distribution of the displacements demanded by the earthquake excitation is determined by nonlinear dynamic analysis for each asymmetric system and for the corresponding planar system (i.e. the system with restrained deck rotations). The ratio of the maximum displacements of the asymmetric system to the maximum displacement of the corresponding planar system quantifies the modification (amplification or deamplification) of displacement demand caused by the in-plan irregularity. Finally, the results of this investigation are used to populate a database for assisted assessment of buildings.

7.2 Population of the Database

Two sets of single-storey systems rectangular in plan are investigated (Bosco et al. 2017). These systems are characterized by a rigid deck, whose dimensions are equal to $B = 12.5$ m and $L = 29.5$ m. The mass m of each system is equal to 1416 kNs^2/m and is assumed as lumped on the deck. The radius of gyration of mass r_m with respect to the centre of mass CM is equal to 0.312 L. The resisting elements are symmetric with respect to the geometric centre of the deck G. The investigated systems are torsionally restrained and the resisting elements disposed along the x-axis contribute to the 20% of the total torsional stiffness about the centre of rigidity CR. Such a value is typical of rectangular-in-plan structures. The main differences between the structures of the two sets concern the features of the resisting members. In fact, the structures belonging to the first set are representative of multi-storey buildings with braced frames or shear walls. The deck of these systems is sustained by four uni-directional resisting elements disposed in the x-direction and eight uni-directional resisting elements along the y-direction (Fig. 7.1a). The

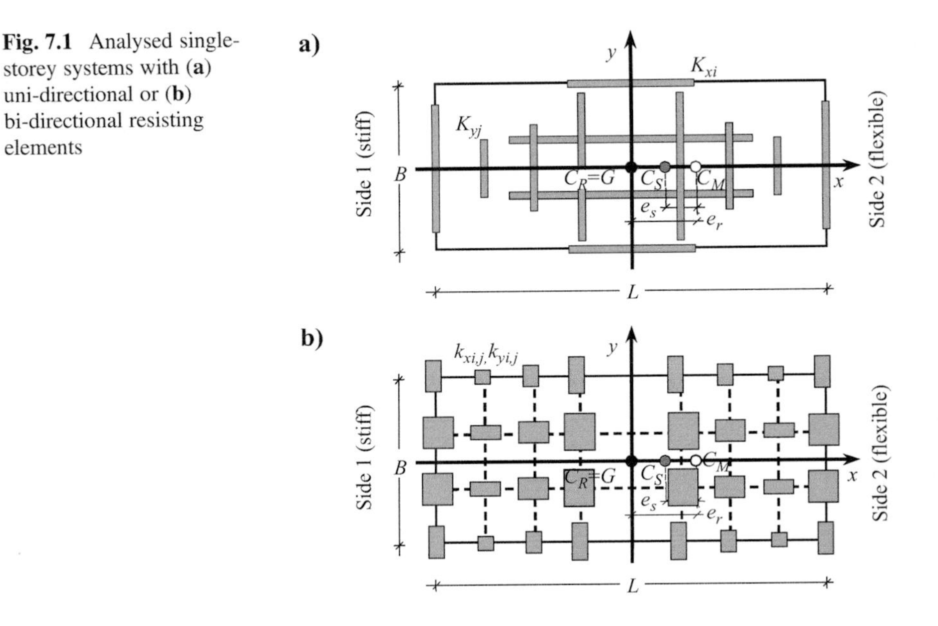

Fig. 7.1 Analysed single-storey systems with (**a**) uni-directional or (**b**) bi-directional resisting elements

uni-directional elements provide lateral stiffness and strength only in their plane and their force-displacement relationship is typically a bilinear curve. On the contrary, the systems belonging to the second suite are representative of multi-storey framed buildings. In these systems, 32 bi-directional resisting elements sustain the deck (Fig. 7.1b) and they supply lateral stiffness and strength in any horizontal direction. In this case, an elliptical yield domain (De Stefano and Pintucchi 2010) is adopted to consider the interaction between resisting forces in the nonlinear range of behaviour. These resisting elements are located in correspondence of the intersection of the bending planes of the resisting elements belonging to the systems of the first set.

All the considered asymmetric systems are mass-eccentric and differ in the parameters that affect the rotational response of the structure:

- the rigidity eccentricity e_r (distance between the centre of rigidity CR and CM);
- the fundamental period of vibration;
- the ratio Ω_θ of the torsional to lateral frequencies of the corresponding torsionally balanced system (obtained by shifting CM into CR);
- the ratio R_μ of the elastic strength demand to the actual strength of the system;
- the strength eccentricity e_s (distance between CR and the centre of strength CS).

In order to create the rigidity eccentricity, the centre of mass CM is shifted along the x-axis. Systems with e_r ranging from $-0.15\ L$ to 0.0 in steps of 0.025 L are generated.

The translational periods of vibration of the torsionally balanced system are supposed to be equal in the x- and y-directions. Values of T_x and T_y in the range from 0.4 s to 1.8 s in steps of 0.2 s are considered.

The in-plan distribution of the lateral stiffness of the resisting elements is determined according to the procedure described in (Ghersi and Rossi 2000) to obtain prefixed values of Ω_θ in the range from 0.60 to 1.40 in steps of 0.05.

The global lateral strength V_u of the planar system is equal along the x- and y-direction and is assigned so as to have prefixed values of the ratio R_μ ranging from 2.0 (systems with moderate inelastic behaviour) to 6.0 (systems with severe inelastic behaviour) in steps of 1.0. The total lateral strength is distributed between the resisting elements to obtain systems with a strength eccentricity e_s ranging from $-0.10\ L$ to $0.10\ L$ in steps of $0.025\ L$.

For the time being, only systems standing on soft soil (soil C according to Eurocode 8) are considered. However, the database is conceived in such a way to include new cases without modifying its structure. This will make possible to easily expand the database when new parameters will be explored and new results will be available.

7.3 Accelerograms

A suite of ten pairs of accelerograms is used to simulate the earthquake excitation. The program SIMQKE (SIMQKE User Manual 1976) is used to generate the accelerograms in compliance with the compatibility conditions of Eurocode 8. The response spectrum for soil type C, design ground acceleration (on soil type A) of 0.35 g and viscous damping ratio of 5% is assumed as target. Each accelerogram is modulated according to a *compound* intensity function characterised by total duration, rise time and duration of the stationary part of the accelerogram equal to thirty, four and seven seconds, respectively. Furthermore, the exponential factors IPOW and ALFA0 that define the rising and decaying phases of the accelerogram are set equal to 2 and 0.25.

The duration of the stationary part has been selected to match the energy content of a previously selected set of 20 natural ground motions representative of seismic events on soft soil and characterized by magnitude in the range from 5.8 to 7.2 and by epicentral distance up to 30 km. The generated accelerograms are characterised by an average value of the Arias Intensity equal to 0.31 m/s and by a significant duration equal to 7.62 s. Further details about the input energy (intended as the maximum input energy during the seismic event) of the ground motions and about their equivalent number of cycles can be found in reference (Amara et al. 2014).

Several studies have pointed out that the significant duration of a ground motion is related to the earthquake magnitude, to the epicentral distance and to the peak ground acceleration (Raghunandan and Abbie 2013). For this reason, in the future the database will be expanded by considering the seismic response of single-storey systems subjected to sets of ground motions with different effective duration.

7.4 Methodology

The seismic response of each building that populates the database has been determined by nonlinear dynamic analysis. The Newmark integration method with parameters β and γ equal to 0.25 and 0.5 (which yields the constant average acceleration method) has been used to evaluate the dynamic response. The time step is set equal to 0.001 s.

A first dynamic analysis has been performed with reference to the asymmetric systems. Particular attention is payed to the displacement demand along the y-axis of the points on the stiff and flexible side of the deck. Specifically, the mean value of the maximum displacements obtained for the 10 pairs of ground motions is considered.

A second analysis has been carried out on the corresponding "planar system", i.e. on the system whose deck rotation has been restrained. Finally, the ratio of the maximum displacement of the asymmetric system to the maximum displacement of the corresponding planar system is determined to quantify how much the torsional response modifies the displacement demand of the stiff and flexible sides of the deck with respect to that of the corresponding torsionally balanced system.

7.5 Effects of the Considered Parameters on the Modification of the Displacement Demand

The modification of the displacement demand of the stiff and flexible sides of the building is determined for each of the considered single-storey systems. First, the obtained results are plotted by contour maps for systems with assigned period, R_μ and Ω_θ as a function of the stiffness and strength eccentricity. Figure 7.2 shows that, for torsionally flexible systems with bi-directional resisting elements, the modification coefficients lead to an amplification of the displacement demand of the stiff side of the systems.

The highest amplifications are obtained for systems with high stiffness eccentricity ($e_r = -0.15\,L$) and characterised by centre of stiffness located between the centre of strength and the centre of mass ($e_s < 0.0\,L$). The maximum amplification of displacement is equal to 1.8 in the case of short period systems ($T_y = 0.4$ s) and $R_\mu = 2$, equal to 1.4 in the case of systems with $T_y = 1.0$ s and $R_\mu = 4$. A slightly larger amplification of the displacement demand is obtained on the flexible side (Fig. 7.3) with a maximum amplification of displacement equal to 2.0 for torsionally stiff systems with $T_y = 0.4$ s (Fig. 7.4).

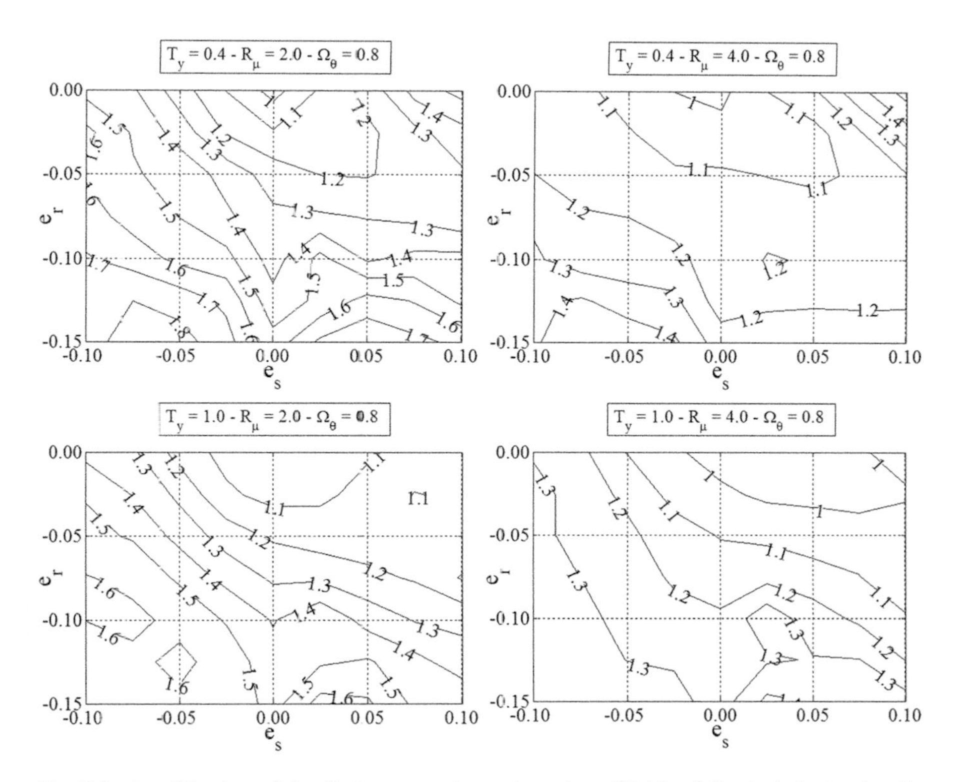

Fig. 7.2 Amplification of the displacement demand on the stiff side of the deck for torsionally flexible systems with bi-directional resisting elements: influence of period and R_μ

7.6 Structure of the Database

The obtained modification coefficients are stored into files organised according to the hierarchical folder tree shown in the screenshot of Fig. 7.5. The first two levels of folder are used to differentiate records referring to systems analysed with accelerograms with different duration of the stationary part and compatible with the spectrum of EC8 for different soil types, respectively. The third folder level is used to distinguish records related to systems with uni- or bi-directional resisting elements. Further, for each considered type of structural system, two other folder levels are branched out to separate records based on the translational period T_y of the structural systems and, for each value of T_y, based on the R_μ factor. Each top-level folder contains a set of 34 files related to structural systems that are homogenous with each other in terms of foundation soil, type of resisting element, translational period and R_μ factor. The basic unit of the database is represented by a pair of files, each one containing a table that reports the modification coefficients of the displacement demand on the stiff or flexible side of the deck for a given value of the ratio Ω_θ depending on the values of normalised rigidity and strength eccentricities. As an example, the table with the modification coefficients related to the stiff side of

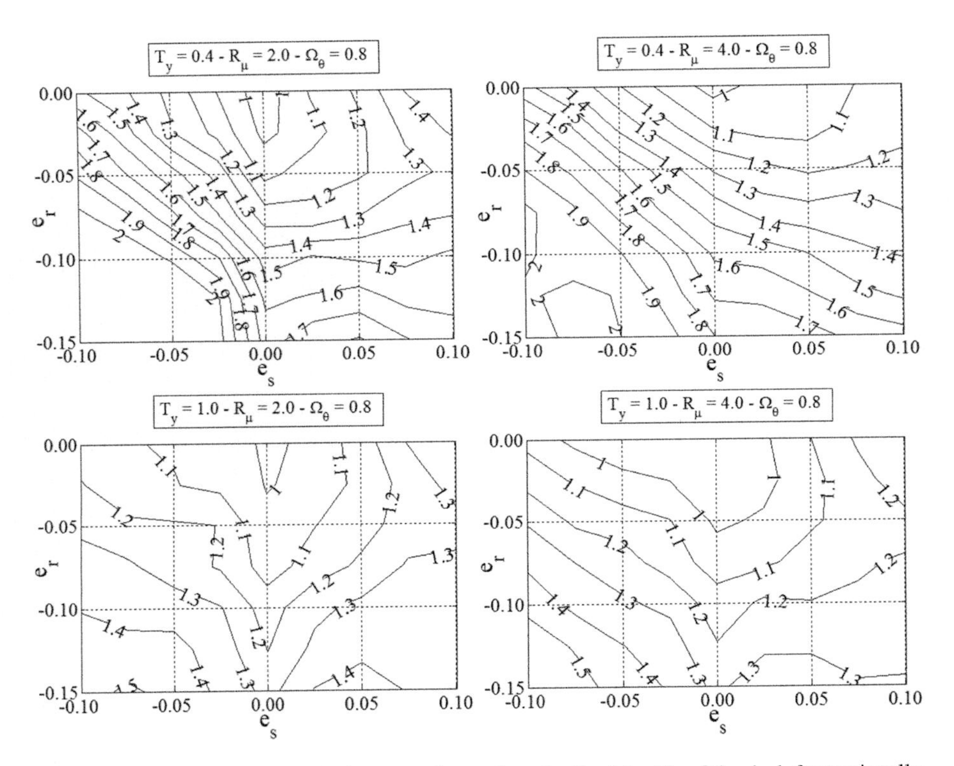

Fig. 7.3 Amplification of the displacement demand on the flexible side of the deck for torsionally flexible systems with bi-directional resisting elements: influence of period and R_μ

systems with $\Omega_\theta = 1.15$ is shown in Fig. 7.6. The row $i = 0$ reports the normalised strength eccentricity (from -0.100 to 0.100) and the column $j = 0$ identifies the normalised rigidity eccentricity (from -0.150 to 0.000). The element T_{ij} of the table is the modification coefficient of a system with rigidity eccentricity $e_r\,(i)$ and strength eccentricity $e_s\,(j)$.

The Database Reader software is provided with an interface that allows the user to formulate the query and interrogate the database. Specifically, the user selects the type of soil (from A to E), the type of system (with uni- or bi-directional resisting elements) and the duration of the stationary part of the accelerograms. Then, the user identifies the point S (T_y, R_μ, Ω_θ, e_r, e_s), which is representative of the structural System under investigation (Fig. 7.7) by assigning the relevant values of T_y, R_μ, Ω_θ, e_r, e_s.

The button "Calculate" starts the Database Reader that selects the neighbourhood about point S by comparing the values of the above-mentioned parameters with the corresponding values of the records of the database. Specifically, all the points P_i representative of records characterised by values of the parameters immediately smaller (T_{y1}, $R_{\mu 1}$, $\Omega_{\theta 1}$, e_{r1} and e_{s1}) or larger (T_{y2}, $R_{\mu 2}$, $\Omega_{\theta 2}$, e_{r2} and e_{s2}) than those of S are selected as schematised in Fig. 7.8.

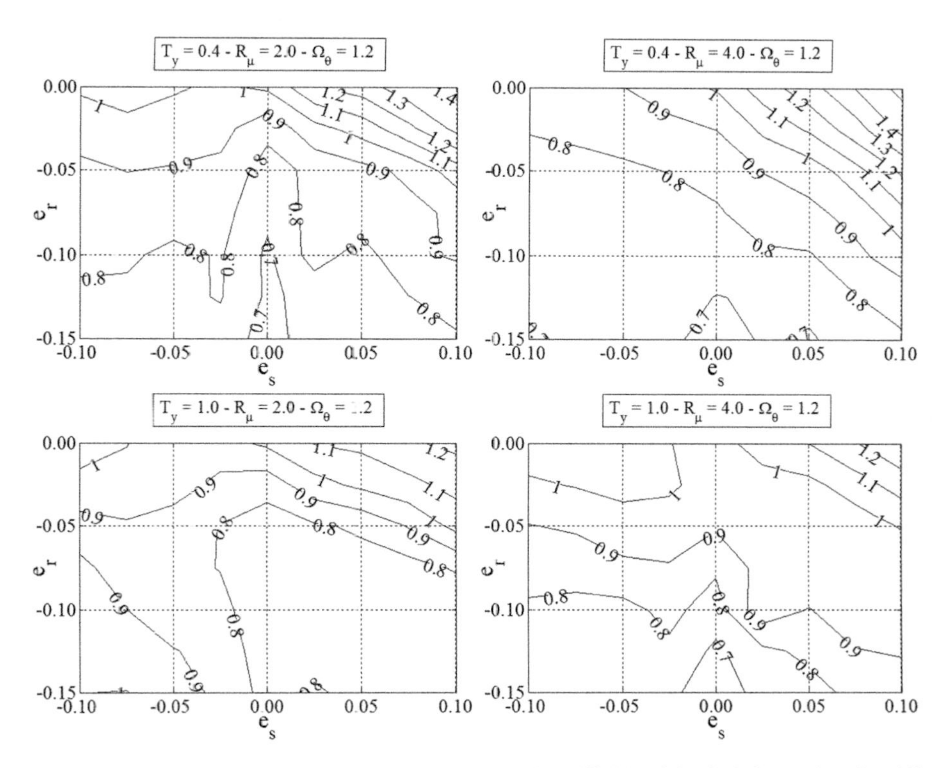

Fig. 7.4 Modification of the displacement demand on the stiff side of the deck for torsionally stiff systems with bi-directional resisting elements: influence of period and R_μ

Totally, the neighbourhood is composed of 32 points. The modification coefficient of the displacement demand of the point S is determined by the "*inverse distance weighting*" method, i.e.

$$AF(\mathrm{S}) = \frac{\sum_{i=1}^{32} w_i AF(\mathrm{S_i})}{\sum_{i=1}^{32} w_i} \tag{7.1}$$

where $AF\ (\mathrm{S_i})$ is the modification coefficient calculated for the system $\mathrm{S_i}$ belonging to the neighbourhood and the weight w_i is calculated as

$$w_i = \frac{1}{d_i^2} \tag{7.2}$$

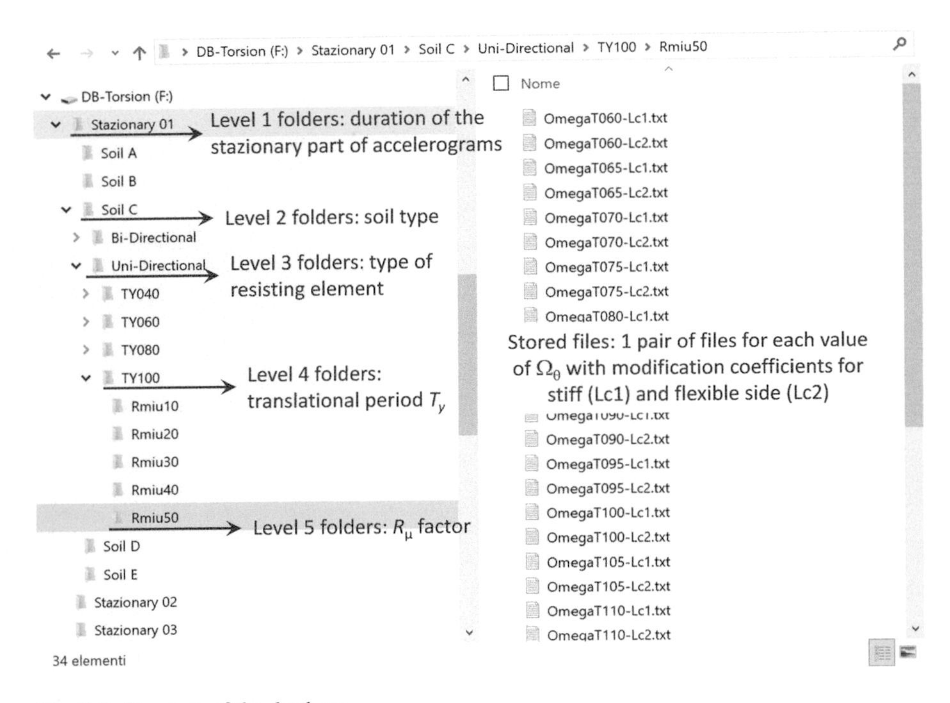

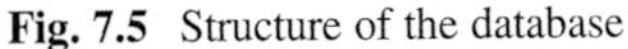

Fig. 7.5 Structure of the database

OmegaT115-Lc1.txt - Blocco note

File Modifica Formato Visualizza ?

	-0.100	-0.075	-0.050	-0.025	0.000	0.025	0.050	0.075	0.100
-0.150	0.7622	0.7857	0.8084	0.8182	0.7813	0.9010	0.9254	0.9486	0.9626
-0.125	0.8180	0.8389	0.8507	0.8908	0.8768	0.9308	0.9695	0.9880	0.9584
-0.100	0.8805	0.8937	0.9200	0.9726	0.9608	1.0623	0.9990	0.9697	0.9255
-0.075	0.9583	0.9505	1.0039	1.0239	1.0132	1.0265	0.9805	0.9358	0.9550
-0.050	1.0057	1.0302	1.0593	1.0320	1.0305	0.9802	0.9289	0.9768	1.0618
-0.025	1.1067	1.0841	1.0638	1.0042	0.9787	0.9747	1.0217	1.1072	1.2206
0.000	1.0752	1.0509	1.0040	0.9842	0.9993	1.0800	1.1457	1.2492	1.3504

Fig. 7.6 Basic unit of the database

$$d_i^2 = \left(T_y - T_{y,i}\right)^2 + \left(R_\mu - R_{\mu,i}\right)^2 + \left(\Omega_\theta - \Omega_{\theta,i}\right)^2 + \left(e_r - e_{r,i}\right)^2 + \left(e_s - e_{s,i}\right)^2 \quad (7.3)$$

In the equation above, d_i represents the distance of the point S representative of the structural system under investigation to the points S_i representative of the systems belonging to the neighbourhood.

DB-Torsion Reader

Directory database:
F:\DataBaseTorsion

Seismic Input
Soil Type C
duration ts 7

Considered Systems
Type of resisting elements
Bi-Directional

Building Features
Period Ty 0.658
er -0.11
es -0.08
Rμ 2.644
Ωθ 1.12

Modification coefficients:
Calculate
Stiff side:
Flexible side:

Data that determine the Level 1, 2, and 3 folders

Data that determine the Level 4 and 5 folders, and the basic units to read

Fig. 7.7 User interface of the database reader software

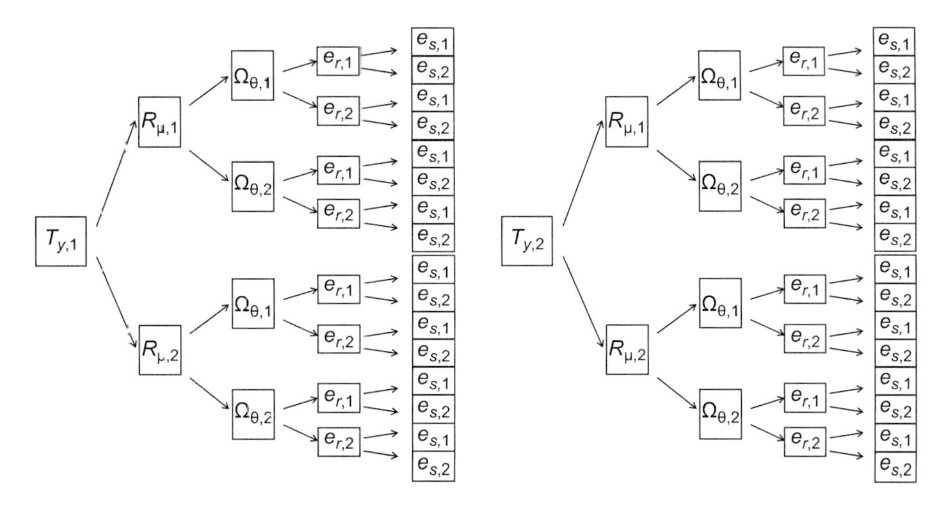

Fig. 7.8 Selection of the neighbourhood

7.7 Conclusions

This paper aims at improving the prediction of the displacement demand given by nonlinear static methods for in-plan irregular buildings. Specifically, modification coefficients are used to adjust the results of the original N2 method, consistently with the *extended N2 method* developed by Fajfar and his team. However, differently from the modification coefficients adopted by the extended N2 method, those derived here are based on the "nonlinear response" of single storey systems generated by varying the parameters that mostly affect the seismic response of in-plan irregular buildings. The modification coefficients are used to populate a database for assisted assessment of torsional response of in-plan irregular buildings. The database provides the modification of the displacement demand produced by torsional effects both on the stiff and flexible side of in-plan irregular buildings. Future studies will be carried out to expand the database. Further, the effectiveness of the modification coefficients obtained by analysing single storey systems will be tested by evaluating the seismic response of multi-storey buildings.

References

Amara F, Bosco M, Marino EM, Rossi PP (2014) An accurate strength amplification factor for the design of SDOF systems with P-Δ effects. Earthq Eng Struct Dyn 43(4):589–611

ATC (2005) Improvement of nonlinear static seismic analysis procedures. FEMA440 report. Redwood City, CA

ATC, Applied Technology Council (1996) Seismic evaluation and retrofit of concrete buildings, vol. 1 and 2, Report No. ATC-40. Redwood City, CA

Bento R, Bhatt C, Pinho R (2010) Verification of nonlinear static procedures for a 3D irregular SPEAR building. Earthq Struct 1(2):177–195

Bhatt C, Bento R (2014) The extended adaptive capacity spectrum method for the seismic assessment of plan-asymmetric buildings. Earthquake Spectra 30(2):683–703

Bosco M, Ghersi A, Marino EM (2012) Corrective eccentricities for assessment by the nonlinear static method of 3D structures subjected to bidirectional ground motions. Earthq Eng Struct Dyn 41:1751–1773

Bosco M, Ghersi A, Marino EM, Rossi PP (2013) Comparison of nonlinear static methods for the assessment of asymmetric buildings. Bull Earthq Eng 11:2287–2308

Bosco M, Ferrara GAF, Ghersi A, Marino EM, Rossi PP (2015) Predicting displacement demand of multi-storey asymmetric buildings by nonlinear static analysis and corrective eccentricities. Eng Struct 99:373–387

Bosco M, Ghersi A, Marino EM, Rossi PP (2017) Generalized corrective eccentricities for nonlinear static analysis of buildings with framed or braced structure. Bull Earthq Eng. https://doi.org/10.1007/s10518-017-0159-x

CEN Eurocode 8 (2004) Design of structures for earthquake resistance. Part 1: General rules, seismic actions and rules for buildings. EN 1998-1:2004. Brussels, Belgium

Chopra AK, Goel RK (2004) A modal pushover analysis procedure to estimate seismic demands for unsimmetric-plan buildings. Earthq Eng Struct Dyn 33:903–927

De Stefano M, Pintucchi B (2010) Predicting torsion-induced lateral displacements for pushover analysis: influence of torsional system characteristics. Earthq Eng Struct Dyn 39:1369–1394

De Stefano M, Tanganelli M, Viti S (2013) On the variability of concrete strength as a source of irregularity in elevation for existing RC buildings: a case study. Bull Earthq Eng 11 (5):1711–1726

Fajfar P (1999) Capacity spectrum method based on inelastic demand spectra. Earthq Eng Struct Dyn 28:979–993

Fajfar P, Marusic D, Peruš I (2005) Torsional effects in the pushover-based seismic analysis of buildings. J Earthq Eng 9:831–854

Freeman SA (1998) The capacity spectrum method as a tool for seismic design. In: Paper presented at the 11th European conference on earthquake engineering, Paris, France

Fujii K (2011) Nonlinear static procedure for multi-story asymmetric frame buildings considering bidirectional excitation. J Earthq Eng 15(2):245–273
Fujii K (2014) Prediction of the largest peak nonlinear seismic response of asymmetric buildings under bi-directional excitation using pushover analyses. Bull Earthq Eng 12(2):909–938
Georgoussis GK (2015) An approximate method for assessing the seismic response of irregular in elevation asymmetric buildings. GeotechGeol Earthq Eng 40:111–122
Ghersi A, Rossi PP (2000) Formulation of design eccentricity to reduce ductility demand in asymmetric buildings. Eng Struct 22:857–871
Giorgi P, Scotta R (2013) Validation and improvement of N1 method for pushover analysis. Soil Dyn Earthq Eng 55:140–147
Kreslin M, Fajfar P (2012) The extended N2 method considering higher mode effects in both plan and elevation. Bull Earthq Eng 10:695–715
Palermo M, Silvestri S, Gasparini G, Trombetti T (2013) Physically-based prediction of the maximum corner displacement magnification of one-storey eccentric systems. Bull Earthq Eng 11(5):1573–1456
Palermo M, Silvestri S, Gasparini G, Trombetti T (2017) A comprehensive study on the seismic response of one-storey asymmetric systems. Bull Earthq Eng 15(4):1497–1517
Raghunandan M, Abbie LB (2013) Effect of ground motion duration on earthquake-induced structural collapse. Struct Saf 41 119–133
Reyes JC, Chopra AK (2011) Three-dimensional modal pushover analysis of buildings subjected to two components of ground motion, including its evaluation for tall buildings. Earthq Eng Struct Dyn 40:789–806
SIMQKE User Manual (1976) NISEE software library. University of California, Berkeley

Chapter 8
Modified Mode-Adaptive Bi-directional Pushover Analysis Considering Higher Mode for Asymmetric Buildings

K. Fujii

Abstract In this paper, the Mode-Adaptive Bi-directional Pushover Analysis (MABPA) previously presented by the author is modified by considering the higher (third) mode response. The modifications proposed in this paper are (a) the peak response of the third mode is estimated from the independent single-degree-of freedom (SDOF) model representing the third mode, and (b) the peak response of each frame is predicted by combining the results from the original MABPA and the third mode response using the square root of the sum of square (SRSS) rule. The modified MABPA is applied to two six-story asymmetric building models with bidirectional setback. The predicted results are compared with time-history analysis results and other simplified procedures; modal pushover analysis (MPA), improved modal pushover analysis (IMPA), a variant of MPA considering bidirectional excitation by (Manoukas G, Athanatopoulou A, Avramidis I, Soil Dyn Earthq Eng 38:88–96, 2012) and (Manoukas G, Avramidis I, Bull Earthq Eng 12(6):2607–2632, 2014), and the original MABPA. The results show that the predicted peak response at "flexible-edge" frame by each simplified procedure agree well with the time-history analysis results. On the contrary, the predicted peak response at the "stiff-edge" frame according to each simplified procedure is different: some of the predicted results, including the results via original MABPA, underestimate the time-history analysis results, while the results via modified MABPA are conservative compare to the time-history analysis results.

Keywords Modified Mode-Adaptive Bi-Directional Pushover Analysis (Modified MABPA) · Asymmetric buildings · Bi-directional excitaion · Pushover analysis

---

K. Fujii (✉)
Department of Architecture, Chiba Institute of Technology, Narashino-shi, Chiba, Japan
e-mail: kenji.fujii@it-chiba.ac.jp

D. Köber et al. (eds.), *Seismic Behaviour and Design of Irregular and Complex Civil Structures III*, Geotechnical, Geological and Earthquake Engineering 48,
https://doi.org/10.1007/978-3-030-33532-8_8

8.1 Introduction

When conducting seismic assessment of an asymmetric building, it is essential to carry out three-dimensional analysis considering all the possible directions of seismic input. For this purpose, the author previously proposed a simplified procedure to predict the largest peak seismic response of an asymmetric building subjected to horizontal bidirectional ground motion acting in an arbitrary angle of incidence (Mode-Adaptive Bi-directional Pushover Analysis, referred to as MABPA) (Fujii 2014). The proposed MABPA has been tested against asymmetric buildings with regular elevation (Fujii 2014) and buildings with bidirectional setbacks (Fujii 2016), and has successfully estimated the largest peak response at a "flexible-edge" frame, while underestimated at a "stiff-side" frame. This is because the original MABPA considers only the response of two modes, while in the response of the underestimated side frame, the third and higher mode responses are significant.

In this paper, the original MABPA is modified by considering the higher (third) mode response. The modified MABPA is applied to two six-story asymmetric building models with bidirectional setback studied in (Fujii 2016). The predicted results are compared with time-history analysis results, other simplified procedures (Chopra and Goel 2004; Reyes and Chopra 2011a, b; Belejo and Bento 2016; Manoukas et al. 2012; Manoukas and Avramidis 2014) and the original MABPA. Note that this paper focuses on comparisons of the accuracy of the modified MABPA to the original MABPA and other simplified procedures for predicting asymmetric buildings with bidirectional setback, and that only the case in which the spectra of the major and minor components of horizontal ground motion are identical is considered. Hence, we include no discussion regarding the influence of the incident direction of seismic input on the response of asymmetric buildings.

8.2 Description of Modified MABPA

The asymmetric building considered in this paper is the N-story building, with $3N$ degrees of freedom ($3N$-DOFs). All the frames of the asymmetric building are oriented in the X or Y directions, which are orthogonal. Let $\boldsymbol{\varphi}_\mathbf{i}$ be the i-th mode vector of an asymmetric building:

$$\boldsymbol{\varphi}_\mathbf{i} = \{\phi_{X1i} \ \cdots \ \phi_{XNi} \ \phi_{Y1i} \ \cdots \ \phi_{YNi} \ \phi_{\Theta 1i} \ \cdots \ \phi_{\Theta Ni}\}^\mathbf{T}. \tag{8.1}$$

In Eq. (8.1), ϕ_{Xji}, ϕ_{Yji} and $\phi_{\Theta ji}$ are the X-, Y-, and rotational components, respectively, of the i-th mode vector at the j-th floor. The tangent of ψ_i, the angle of incidence of the principal axis of the i-th modal response with respect to X-axis, is

$$\tan \psi_i = -\sum_j m_j \phi_{Yji} / \sum_j m_j \phi_{Xji}. \tag{8.2}$$

Let the U-axis be the principal axis of the first modal response, with the V-axis orthogonal to it.

An outline of the original and modified MABPA is summarized in Fig. 8.1. The fundamental assumptions of the modified MABPA are as follows:

1. The spectrum of the horizontal ground motion component acting in *any* angle of incidence is assumed to be identical to the response spectrum of the major component.
2. The building oscillates predominantly in a single mode in each set of orthogonal directions.
3. The principal directions of the first and second modal responses are almost orthogonal.

In this paper, the original MABPA is modified for improving the accuracy of the predicted response at the "stiff-side" frame. Steps 1–5 shown in Fig. 8.1 are the same as the original MABPA (Fujii 2014), while steps 6–8 are the additional steps for consideration of the higher mode. The detail of additional steps is described as follows. Note that the detail of steps 1–5 can be found in the previous study (Fujii 2014).

Let $\boldsymbol{\varphi}_{\mathbf{1ie}}$ be the first mode vector corresponding to the equivalent peak displacement of the first mode, namely, $D_{1U}{}^{*}{}_{\max}$. The second and third mode vectors, $\boldsymbol{\varphi}_{\mathbf{2ie}}$ and $\boldsymbol{\varphi}_{\mathbf{3ie}}$, respectively, are then determined from Eqs. (8.3) and (8.4) in terms of $\boldsymbol{\varphi}_{\mathbf{1ie}}$ and the second and third mode vectors in the elastic range, $\boldsymbol{\varphi}_{\mathbf{2e}}$ and $\boldsymbol{\varphi}_{\mathbf{3e}}$, respectively, considering the orthogonality of the mode vectors.

$$\boldsymbol{\varphi}_{\mathbf{2ie}} = \boldsymbol{\varphi}_{\mathbf{2e}} - \frac{\boldsymbol{\varphi}_{\mathbf{2e}}{}^{\mathrm{T}}\mathbf{M}\boldsymbol{\varphi}_{\mathbf{1ie}}}{\boldsymbol{\varphi}_{\mathbf{1ie}}{}^{\mathrm{T}}\mathbf{M}\boldsymbol{\varphi}_{\mathbf{1ie}}}\boldsymbol{\varphi}_{\mathbf{1ie}}, \tag{8.3}$$

$$\boldsymbol{\varphi}_{\mathbf{3ie}} = \boldsymbol{\varphi}_{\mathbf{3e}} - \frac{\boldsymbol{\varphi}_{\mathbf{3e}}{}^{\mathrm{T}}\mathbf{M}\boldsymbol{\varphi}_{\mathbf{1ie}}}{\boldsymbol{\varphi}_{\mathbf{1ie}}{}^{\mathrm{T}}\mathbf{M}\boldsymbol{\varphi}_{\mathbf{1ie}}}\boldsymbol{\varphi}_{\mathbf{1ie}} - \frac{\boldsymbol{\varphi}_{\mathbf{3e}}{}^{\mathrm{T}}\mathbf{M}\boldsymbol{\varphi}_{\mathbf{2ie}}}{\boldsymbol{\varphi}_{\mathbf{2ie}}{}^{\mathrm{T}}\mathbf{M}\boldsymbol{\varphi}_{\mathbf{2ie}}}\boldsymbol{\varphi}_{\mathbf{2ie}}, \tag{8.4}$$

$$\mathbf{M} = \begin{bmatrix} \mathbf{M_0} & \mathbf{0} & \mathbf{0} \\ \mathbf{0} & \mathbf{M_0} & \mathbf{0} \\ \mathbf{0} & \mathbf{0} & \mathbf{I_0} \end{bmatrix}, \mathbf{M_0} = \begin{bmatrix} m_1 & & 0 \\ & \ddots & \\ 0 & & m_N \end{bmatrix}, \mathbf{I_0} = \begin{bmatrix} I_1 & & 0 \\ & \ddots & \\ 0 & & I_N \end{bmatrix}. \tag{8.5}$$

In Eqs. (8.3), (8.4), and (8.5), $\mathbf{M}$ is the mass matrix; m_j and I_j are the mass and mass moment of inertia of the j-th floor, respectively.

In step 6, pushover analysis of an MDOF model is carried out to obtain the force-displacement relationship representing the third mode response by applying the invariant force distribution $\mathbf{p_3}$, determined as

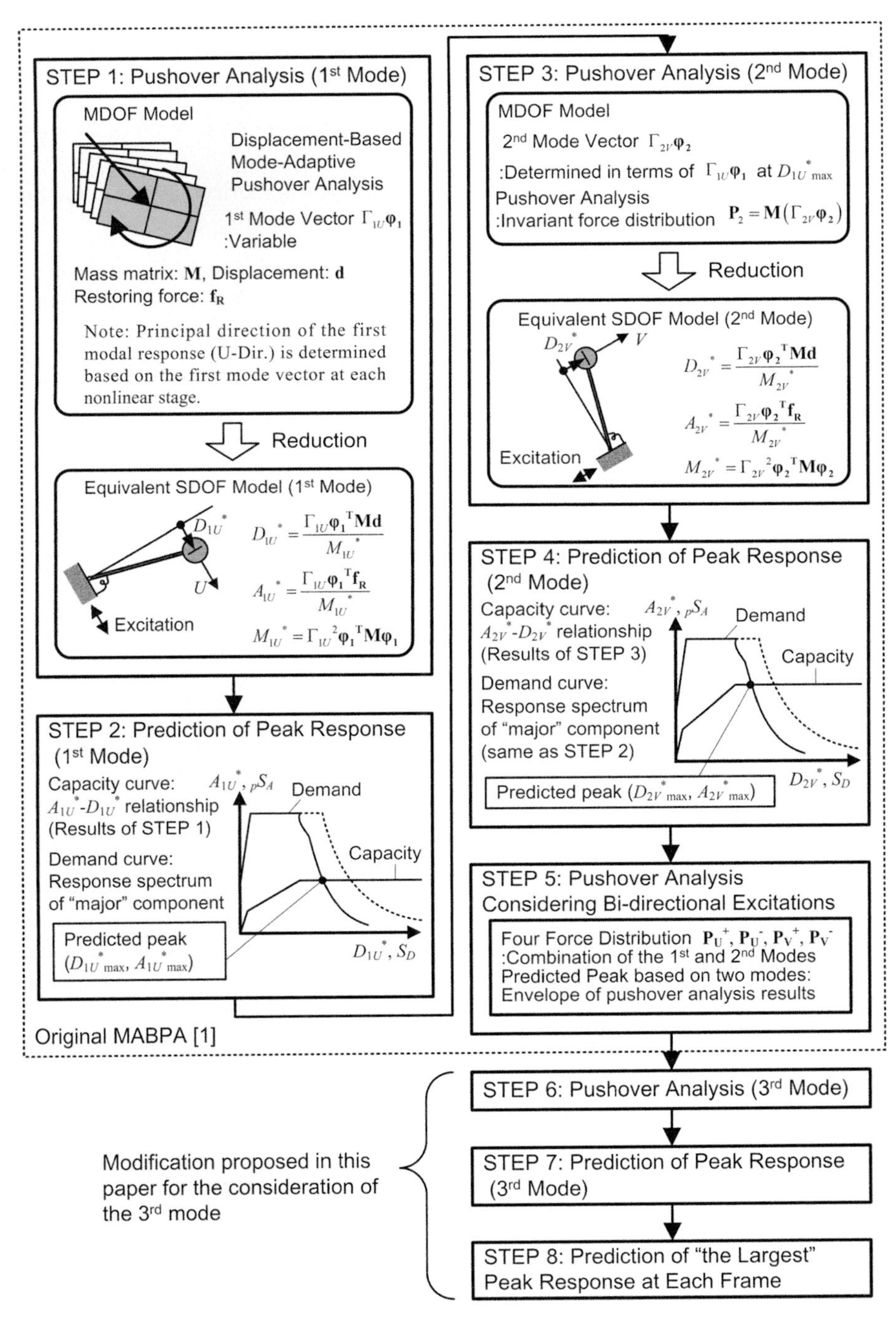

Fig. 8.1 Outline of the modified MABPA

$$\mathbf{p_3} = \mathbf{M}(\Gamma_{3ie}\boldsymbol{\varphi}_{\mathbf{3ie}}). \tag{8.6}$$

In Eq. (8.6), Γ_{3ie} is the modal participation factor of the third mode with respect to the principal direction of the third modal response, which is calculated by Eq. (8.7).

$$\Gamma_{3ie} = \frac{\boldsymbol{\varphi}_{\mathbf{3ie}}{}^{\mathrm{T}}\mathbf{M}\boldsymbol{\alpha}_{\mathbf{3ie}}}{\boldsymbol{\varphi}_{\mathbf{3ie}}{}^{\mathrm{T}}\mathbf{M}\boldsymbol{\varphi}_{\mathbf{3ie}}}, \tag{8.7}$$

$$\boldsymbol{\alpha}_{\mathbf{3ie}} = \{\cos\psi_{3ie}\cdots\cos\psi_{3ie} - \sin\psi_{3ie}\cdots - \sin\psi_{3ie}0\cdots0\}^{\mathrm{T}}, \tag{8.8}$$

where ψ_{3ie} is the angle of incidence of the principal direction of the third modal response with respect to the X-axis corresponding to $D_{1U}{}^*{}_{max}$. The equivalent displacement ${}_nD_3{}^*$ and acceleration ${}_nA_3{}^*$ of the equivalent SDOF model representing the third modal response at each loading step n are determined by

$${}_nD_3{}^* = \frac{\Gamma_{3ie}\boldsymbol{\varphi}_{\mathbf{3ie}}{}^{\mathrm{T}}\mathbf{M}_{\mathbf{n}}\mathbf{d}}{M_{3ie}{}^*}, {}_nA_3{}^* = \frac{\Gamma_{3ie}\boldsymbol{\varphi}_{\mathbf{3ie}}{}^{\mathrm{T}}{}_{\mathbf{n}}\mathbf{f_R}}{M_{3ie}{}^*}, \tag{8.9}$$

$${}_{\mathbf{n}}\mathbf{d} = \{{}_nx_1\cdots{}_nx_N{}_ny_1\cdots{}_ny_N{}_n\theta_1\cdots{}_n\theta_N\}^{\mathrm{T}}, \tag{8.10}$$

$${}_{\mathbf{n}}\mathbf{f_R} = \{{}_nf_{RX1}\cdots{}_nf_{RXN}{}_nf_{RY1}\cdots{}_nf_{RYN}{}_nf_{MZ1}\cdots{}_nf_{MZN}\}^{\mathrm{T}}, \tag{8.11}$$

$$M_{3ie}{}^* = \Gamma_{3ie}{}^2\boldsymbol{\varphi}_{\mathbf{3ie}}{}^{\mathrm{T}}\mathbf{M}\boldsymbol{\varphi}_{\mathbf{3ie}}. \tag{8.12}$$

In Eq. (8.9), ${}_{\mathbf{n}}\mathbf{d}$ and ${}_{\mathbf{n}}\mathbf{f_R}$ are the displacement and restoring force vector at each loading step n, and $M_{3ie}{}^*$ is the equivalent third modal mass with respect to the principal axis of the third modal response.

In step 7, the largest peak equivalent displacement of the third mode, namely, $D_3{}^*{}_{max}$, is obtained using the equivalent linearization technique as in steps 2 and 4. Note that the spectrum used in previous steps is used to predict the third mode response; the spectra of the ground motion component acting in the principal axis of the third modal response is assumed to be the same as that of the major component. In step 8, the largest peak response at each frame is determined. Let r_{kmax12} be the predicted peak response at frame k based on the first and second modal response, which is obtained in step 5, and r_{kmax3} be that based on the third mode. The predicted largest peak response at each frame considering three modes, namely, r_{kmax}, is calculated from

$$r_{k\max} = \sqrt{r_{k\max 12}{}^2 + r_{k\max 3}{}^2}. \tag{8.13}$$

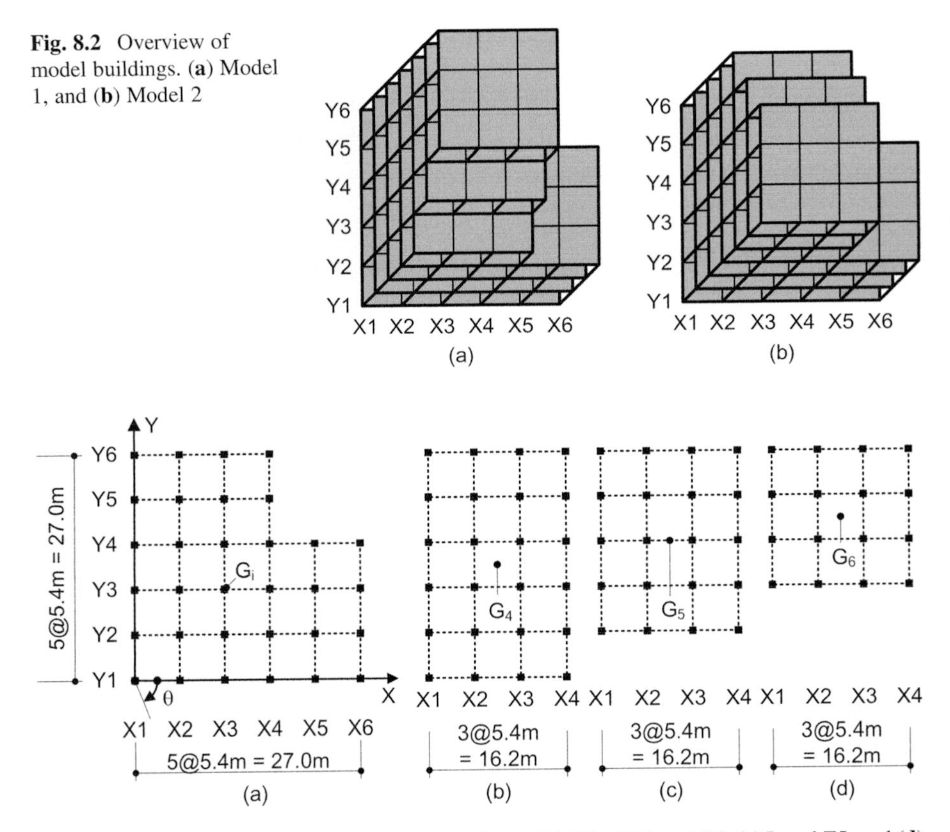

Fig. 8.2 Overview of model buildings. (**a**) Model 1, and (**b**) Model 2

Fig. 8.3 Plan of each floor level for Model 1. (**a**) Level Z0–Z3, (**b**) Level Z4, (**c**) Level Z5, and (**d**) Level Z6

8.3 Buildings and Ground Motion Data

8.3.1 Model Buildings

The building models considered in this study are two six-story reinforced concrete buildings with bidirectional setback, used in a previous study (Fujii 2016). Figure 8.2 shows the overview of two model buildings, namely Models 1 and 2. The plan of each floor level is shown in Fig. 8.3 for Model 1, while the elevations of frames in Models 1 and 2 are shown in Fig. 8.4. Further details of all models can be found in (Fujii 2016).

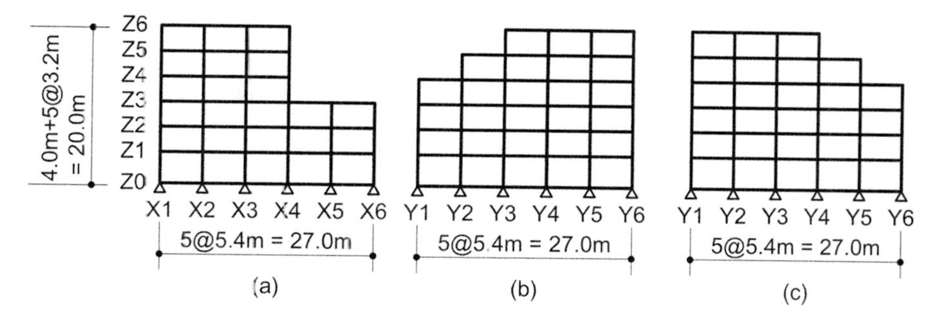

Fig. 8.4 Elevations of frame in Models 1 and 2. (**a**) Frame Y4 (both Models 1 and 2), (**b**) Frame X1 (Model 1), and (**c**) Frame X1 (Model 2)

8.3.2 Ground Motion Data

In this study, the seismic excitation is bidirectional in the X–Y plane, and 10 sets of artificial ground motions are generated. The target elastic spectra of the major and minor components are assumed to be identical in this work. The target elastic spectrum of major and minor components with 5% critical damping – namely ${}_pS_{A\xi}(T,\ 0.05)$ and ${}_pS_{A\zeta}(T,\ 0.05)$, respectively – as determined from the Building Standard Law of Japan (BCJ 2016) for an extremely rare earthquake event considering type-1 soil (rock) is calculated using Eq. (8.14), where T represents the natural period of the SDOF model:

$$
{}_pS_{A\xi}(T,0.05) = {}_pS_{A\zeta}(T,0.05) = \begin{cases} 4.8+45T & \mathrm{m/s^2} & :T \le 0.16\mathrm{s} \\ 12.0 & :0.16\mathrm{s} \le T \le 0.576\mathrm{s} \\ 12.0(0.576/T) & :T > 0.576\mathrm{s} \end{cases}. \tag{8.14}
$$

The phase angle is given by uniform random values, and to consider the time-dependent amplitude of ground motions, the Jenning-type envelope function e (t) proposed by the Building Center of Japan (Otani 2004):

$$
e(t) = \begin{cases} (t/5) & :0\mathrm{s} \le t \le 5\mathrm{s} \\ 1 & :5\mathrm{s} < t \le 35\mathrm{s} \\ \exp\{-0.027(t-35)\} & :35\mathrm{s} < t \le 120\mathrm{s} \end{cases}. \tag{8.15}
$$

Figure 8.5 shows the elastic response spectra of artificial ground motions with 5% critical damping. Note that the artificial ground motions used in this study are generated independently, i.e., there is no correlation between each component.

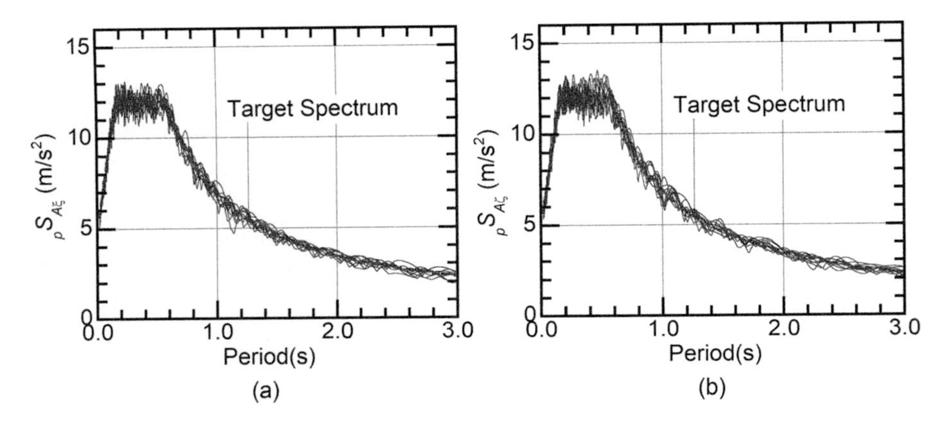

Fig. 8.5 Elastic pseudo acceleration response spectrum. (**a**) "Major" components, and (**b**) "Minor" components

8.3.3 *Analysis Cases*

In this study, the angle of incidence ψ of the major component with respect to the X-axis is varied at intervals of 15° from $(\psi_{1ie} - 90)°$ to $(\psi_{1ie} + 75)°$, where ψ_{1ie} is the angle of incidence of the U-axis corresponding to the predicted peak response D_{1U^*max} of the first mode. Therefore, $10 \times 12 = 120$ cases are considered for the nonlinear time-history analyses of each building model.

The peak response of two building models are also predicted by original MABPA (Fujii 2014), *modal pushover analysis* (MPA) (Chopra and Goel 2004; Reyes and Chopra 2011a, b), *improved modal pushover analysis* (IMPA) (Belejo and Bento 2016), and the procedure presented by Manoukas et al. (2012) and Manoukas and Avramidis (2014), for the comparisons of the accuracy of the predicted peak response of the modified MABPA and other procedures. For the comparisons of the simplified procedures, the following conditions are set:

- The number of modes considered to predict peak response is 3.
- The same equivalent linearization technique is applied to these simplified procedures for predicting the peak response of the equivalent SDOF model. This is done to avoid the discrepancy that occurred from the different technique used to obtain the peak response of the equivalent SDOF model.
- To apply the procedure presented by Manoukas et al. (2012) and Manoukas and Avramidis (2014), the factor for considering bi-directional excitation, κ, is set to be 0.3, according to their recommendation; in Manoukas et al. (Manoukas et al. 2012), $\kappa = 0.3$ is analogous to the 30% combination rule for bidirectional response.

In this study, the 2004 version of MPA (Chopra and Goel 2004) is applied to predict the peak response at each frame under either X- or Y-unidirectional excitation and then these results are combined according to the SRSS rule. This is because

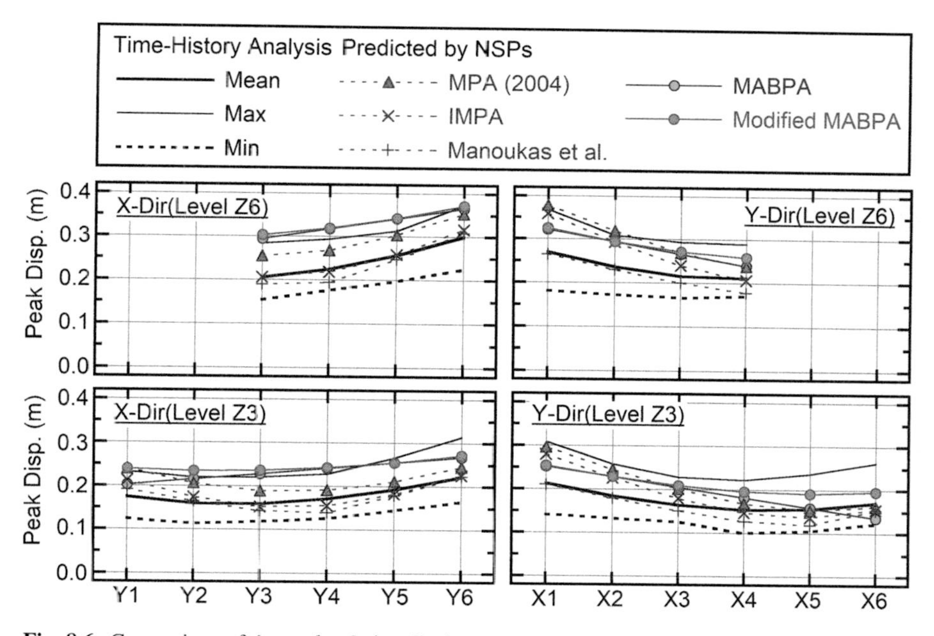

Fig. 8.6 Comparison of the peak relative displacement at each frame of Model 1

it is difficult to directly apply the latest version of the MPA (Reyes and Chopra 2011a, b) to the two considered building models with bidirectional setbacks, where the centers of all floor masses do not lie along the same vertical axis: in such cases, the definition of the story drift at the center of mass, which is required in the latest version of the MPA (Reyes and Chopra 2011a, b), is unclear. Therefore, the procedure applied in this study is referred to as "MPA (2004)", to distinguish it from the latest version (Reyes and Chopra 2011a, b).

8.4 Analysis Results

Figures 8.6 and 8.7 compare the peak relative displacement at each frame as estimated by simplified procedures with that obtained from the time-history analysis results. In these figures, the "max" and the "min" of the time-history analysis are the values obtained as the maximum (or minimum) of 120 time-history analyses cases.

For Model 1 (Fig. 8.6), the peak response predicted by the original MABPA is conservative except for frame X5 in level Z3, while that predicted by the modified MABPA is conservative for all frames. The difference of peak response predicted by the original and modified MABPA is notable at frames Y1, X5 and X6 in level Z3. The peak responses predicted by the other three simplified methods, "MPA (2004)", IMPA and the procedure proposed by Manoukas et al., are close to the mean of the time-history analysis results.

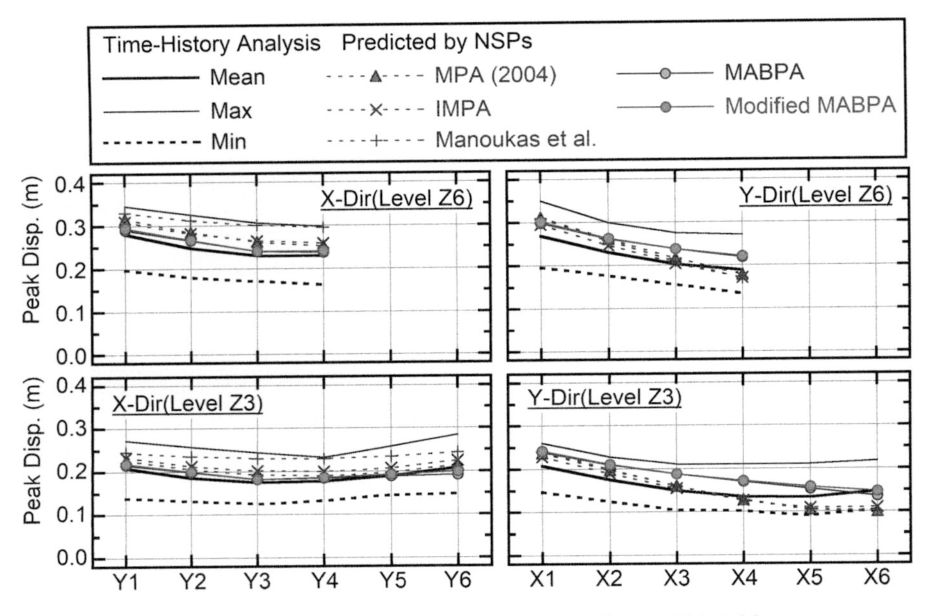

Fig. 8.7 Comparison of the peak relative displacement at each frame of Model 2

For Model 2 (Fig. 8.7), the peak response predicted by the modified MABPA is very close to that predicted by the original MABPA, and both agree very well with the mean of the time-history analysis. The peak responses predicted by other simplified methods are conservative for all frames in the X-direction, while they are unconservative at frames X5 and X6 in the Y-direction.

Figures 8.8 and 8.9 show the peak story drift for Model 1 (Fig. 8.8) and Model 2 (Fig. 8.9), obtained from the time-history analysis results, along with the estimates of the simplified procedures.

In Model 1, the peak response at the "flexible-edge" frames (frame Y6 in the X-direction and frame X1 in the Y-direction) predicted by the original and modified MABPA is slightly larger than their mean obtained from the time-history analysis. The peak responses predicted by the other three simplified procedures are closer to the mean of the time-history analysis results at frame Y6 than those estimated by MABPA. While at frame X1, the peak responses predicted by other procedures are more conservative than those estimated by MABPA. At the "stiff-edge" frame in the Y-direction (frame X6), the original MABPA underestimates the peak response. While the modified MABPA shows improved prediction at frame X6; the predicted peak response by the modified MABPA agrees well with the time-history analysis results. The results predicted by the other three simplified procedures are slightly smaller than the mean of the time-history analysis results.

In Model 2, the original and modified MABPA successfully predict the peak responses; at frames Y1, Y4, Y6 and X6, the estimated peak responses for two procedures agree well with the mean of the time-history analysis results, while at frames X1 and X4, the predicted responses are slightly conservative except for the

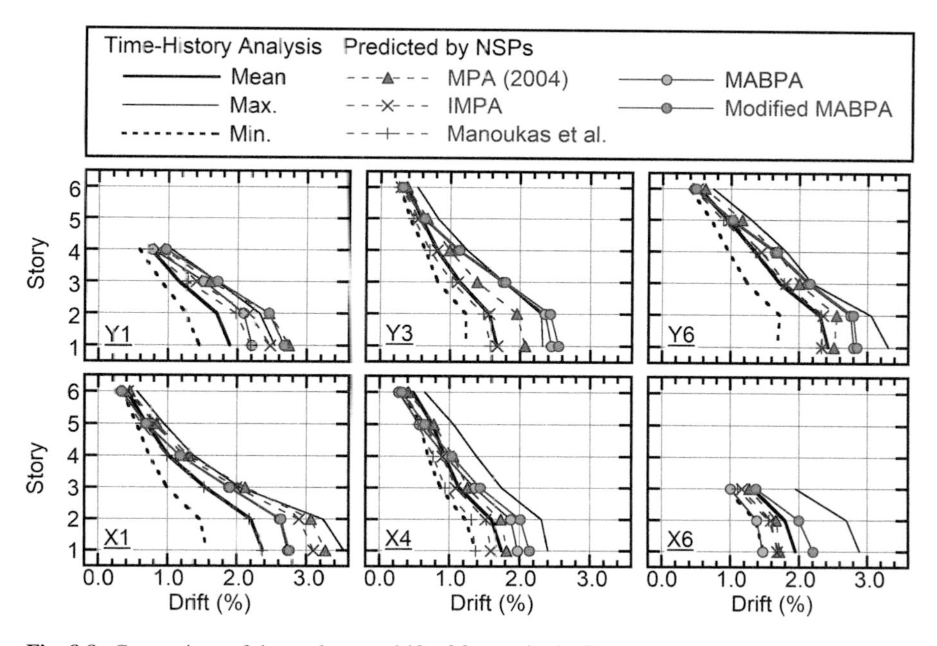

Fig. 8.8 Comparison of the peak story drift of frames in the X and Y directions of Model 1

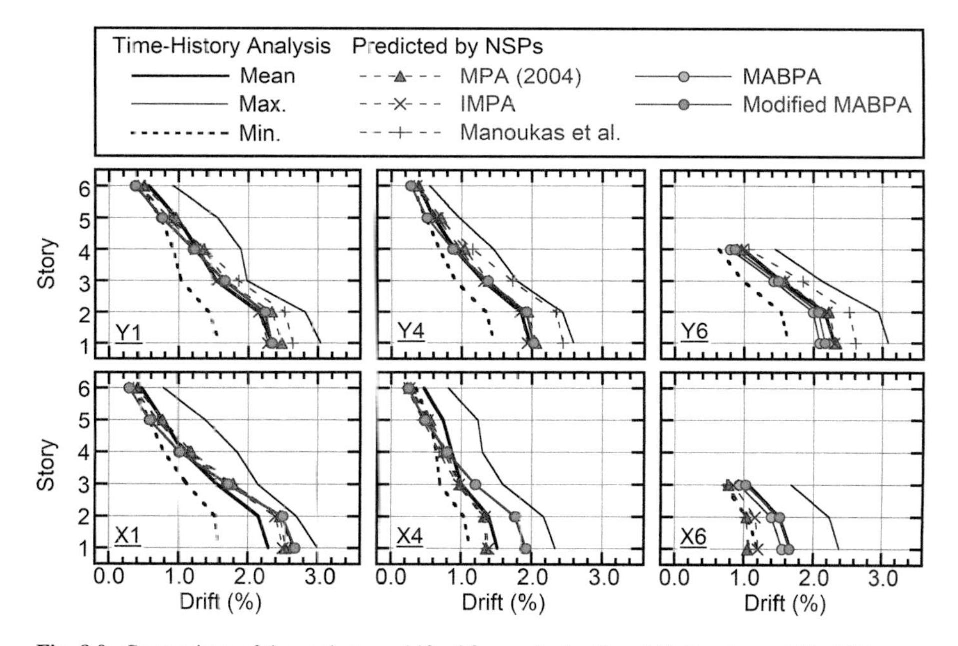

Fig. 8.9 Comparison of the peak story drift of frames in the X and Y directions of Model 2

5th and 6th stories. The predicted peak responses according to the other three procedures agree well with the time-history analysis results except for the "stiff-edge" frame in the Y-direction (frame X6); the three other procedures underestimate the peak response at frame X6.

From the results of the two building models, it may be concluded that the modified MABPA successfully improves the accuracy of peak responses at the "stiff-edge" frames from the original MABPA.

8.5 Discussion and Conclusions

In this paper, the original MABPA is modified by considering the higher (third) mode response. The modifications proposed in this paper are (a) the peak response of the third mode is estimated from the independent single-degree-of freedom (SDOF) model representing the third mode, and (b) the peak response of each frame is predicted by combining the results from the original MABPA and the third mode response using the square root of the sum of square (SRSS) rule. The modified MABPA is applied to two six-story asymmetric building models with bidirectional setback. The predicted results are compared with time-history analysis results and other simplified procedures, "MPA (2004)", IMPA, a procedure proposed by Manoukas et al., and the original MABPA.

The results show that the predicted peak response at the "flexible-edge" frame according to each simplified procedure agrees well with the time-history analysis results. On the contrary, the predicted peak response at the "stiff-edge" frame (frame X6 in both Models 1 and 2) via the simplified procedure differs; some of the predicted results underestimate the time-history analysis results, while the results via modified MABPA are conservative compared to the time-history analysis results.

In conclusion, the modified MABPA successfully improves the prediction of peak response at the "stiff-edge" frame by considering the contribution of the third modal response. Compared with the other simplified procedures discussed in this paper, the modified MABPA may provide more conservative prediction at the "stiff-edge" frame. This is because in the modified MABPA (and the original MABPA), (i) the change of the first mode shape in each loading stage, and (ii) the combination of the first and second modal responses is considered by the envelope of pushover analyses, rather than the SRSS rule. In the "MPA (2004)" and the procedure proposed by Manoukas et al., mode shapes are assumed to remain as those in the elastic range, while the change of the mode shapes is considered in IMPA. However, the combination rule of all modes applied in these three simplified methods (MPA, IMPA and a procedure proposed by Manoukas et al.) is the SRSS (or CQC) rule. The author thinks this may lead the difference from the predicted results via modified MABPA, especially the results in Model 2; as discussed previously (Fujii 2016), the change of mode shapes in the nonlinear range is significant.

Note that in this paper, the *extended N2 method* (Krestin and Fajfar 2012) is not included for the comparisons in this study. This is because in such a building with setback, the definition of the correction factor in the horizontal plane (torsional effect), c_T, for the frames which do not reach the top floor is questionable, since c_T is determined based on the displacement at the top floor. Another simplified procedure, the corrective eccentricity method (Bosco et al. 2012), is also excluded in this study. This is because there is no method (or equations) available to determine the proper corrective eccentricity for buildings with setbacks, where the centers of all floor masses do not lie along the same vertical axis.

References

BCJ (2016) The building standard law of Japan on CD-ROM. The Building Center of Japan, Tokyo
Belejo A, Bento R (2016) Improved modal pushover analysis in seismic assessment of asymmetric plan building under the influence of one and two horizontal components of ground motions. Soil Dyn Earthq Eng 87:1–15. https://doi.org/10.1016/j.soildyn.2016.04.011
Bosco M, Ghersi A, Marino EM (2012) Corrective eccentricities for assessment by nonlinear static method of 3D structures subjected to bidirectional ground motion. Earthq Eng Struct Dyn 41 (13):1751–1773
Chopra AK, Goel RK (2004) A modal pushover analysis procedure to estimate seismic demands for unsymmetric-plan buildings. Earthq Eng Struct Dyn 33(8):903–927
Fujii K (2014) Prediction of the largest peak nonlinear seismic response of asymmetric buildings under bi-directional excitation using pushover analysis. Bull Earthq Eng 12(2):909–938
Fujii K (2016) Assessment of pushover-based method to a building with bidirectional setback. Earthq Struct 11(3):421–443. https://doi.org/10.12989/eas.2016.11.3.421
Krestin M, Fajfar P (2012) The extended N2 method considering higher mode effects in both plan and elevation. Bull Earthq Eng 10(2):695–715
Manoukas G, Avramidis I (2014) Evaluation of a multimode pushover procedure for asymmetric in plan building under biaxial seismic excitation. Bull Earthq Eng 12(6):2607–2632
Manoukas G, Athanatopoulou A, Avramidis I (2012) Multimode pushover analysis for asymmetric buildings under biaxial seismic excitation based on a new concept of the equivalent single degree of freedom system. Soil Dyn Earthq Eng 38:88–96. https://doi.org/10.1016/j.soildyn.2012.01.018
Otani S (2004) Japanese seismic design of high-rise reinforced concrete buildings – an example of performance-based design code and state of practices. In: Proceedings of the 13th world conference on earthquake engineering, Paper No. 5010, Vancouver, Canada, 1–6, August, 2004
Reyes JC, Chopra AK (2011a) Three-dimensional modal pushover analysis of buildings subjected to two components of ground motion, including evaluation for tall buildings. Earthq Eng Struct Dyn 40(7):789–806
Reyes JC, Chopra AK (2011b) Evaluation of three-dimensional modal pushover analysis for unsymmetric-plan buildings subjected to two components of ground motion. Earthq Eng Struct Dyn 40(13):1475–1494

Chapter 9
Structural Irregularities in RC Frame Structures Due to Masonry Enclosure Walls

M. Barnaure

Abstract Masonry infill panels are commonly used in low and mid-rise RC buildings. Under earthquake loadings however, the frame/panel interaction may have significant effects on the structural behaviour of a building. In particular, strong asymmetries due to the presence of masonry infills can lead to a reduction of the building strength under combined vertical and horizontal loading. A four stories frame structure with reinforced concrete members and masonry infills is analysed for horizontal earthquake loading using numerical models. Nonlinear pushover analyses are carried out for a range of alternative configurations. The first configuration refers to the bare frame structure, while the others involve symmetrical and asymmetrical positions for masonry enclosure infill walls interacting with the frames. The infills are modelled as pin joined diagonal frames. Two situations are considered: full infill panels and panels with window openings. The seismic performance of different configurations is analysed. Conclusions are drawn about the influence of structural asymmetries on the capacity of the building. Results show that asymmetric (in plane and/or elevation) masonry enclosure walls can lead to a significant reduction of a building's capacity to withstand earthquake loading. Soft-story configurations are particularly dangerous.

Keywords Diagonal strut · Pushover · Failure criteria · Input energy · Soft-story

9.1 Introduction

Reinforced concrete moment resisting frame structures are common for low and mid-rise buildings. For these buildings, exterior and partition walls are often made with masonry. The influence of these non-structural walls on the seismic behaviour of the buildings is often ignored during design, the panels being only accounted for

M. Barnaure (✉)
Faculty of Civil, Industrial and Agricultural Buildings, Technical University of Civil Engineering, Bucharest, Romania
e-mail: mircea.barnaure@utcb.ro

D. Köber et al. (eds.), *Seismic Behaviour and Design of Irregular and Complex Civil Structures III*, Geotechnical, Geological and Earthquake Engineering 48,
https://doi.org/10.1007/978-3-030-33532-8_9

their mass. Yet masonry infill walls may significantly affect the seismic performance of frame buildings through global (increase in lateral stiffness, stiffness irregularities of the building in plane and/or elevation) and local effects (mainly changing forces and moments in the frame elements) (Barnaure and Stoica 2015).

In many cases, buildings with masonry infills perform very well during earthquake events. Still, there are situations when infills might cause a premature failure of the structural system. The distribution of the infills in plane and elevation strongly influences the global structural behaviour (Cavaleri and Di Trapani 2014). A regular distribution of infills generally leads to an increase in the buildings' stiffness and global capacity (Perrone et al. 2016). On the contrary, irregular distributions are dangerous as they lead to additional torsional effects or soft-storey mechanisms.

It is essential that the role of infill walls is correctly assessed during the design phase. This is no simple task for two reasons. The first one is that numerous factors affect the failure mechanism of masonry infilled RC frames: aspect ratio of the panels, openings, column to beam stiffness ratio, axial loads on columns, type of infill, reinforcing details, etc. (Barnaure and Stoica 2015). Secondly, there is a lack of clear code provisions regarding the modelling of the infills.

This paper explores the influence that structural irregularities – resulting from the presence of masonry infills – can have on the seismic response of RC buildings. A numerical model employing the equivalent strut method is proposed to examine the frame behaviour, with the resulting framework being readily accessible for use by practicing engineers in both the design of new buildings and assessment of existing structures. As the mechanical characteristics of the materials adopted here correspond to typical Romanian buildings, future work could assess the robustness of results to the choice of materials and configurations. Similar models could be used for buildings with different materials or configuration by using modified force-displacement laws for the struts equivalent to the masonry panels.

9.2 Effects of Interaction Between Masonry Infills and RC Frames

9.2.1 Earthquake Observations

Recent earthquake events have shown that the interaction between infill walls and RC frames can be both beneficial and catastrophic. When structural failure occurred, it was generally due to infills creating strong irregularities, either in plane or in elevation, which were not considered during the design phase.

In the 2010 Canterbury earthquake (Kam et al. 2010), RC frames with masonry infill walls behaved very well. Even the older buildings sustained relatively little damage, as the infills increased the stiffness and strength of the bare frames structures.

During the 2009 L'Aquila earthquake, in many cases the strength and stiffness contribution of masonry panels preserved RC buildings from structural damage (Verderame et al. 2011). But irregular distribution of infills (in plane and elevation) was at the same time one of the main causes of the structural collapses observed (Verderame et al. 2011).

According to (Zhao et al. 2009), many reinforced concrete building frames did not perform as intended during the 2008 Wenchuan earthquake. Failure generally occurred in the columns, either through shear failure or through excessive deformation demands at the ground floor level. The main reason for the poor performance was the soft-story mechanisms that occurred due to the presence of ground floor shops facing the street.

Soft story mechanism was also the reason for the structural failure of several multi-storey buildings during the 2015 Gorkha earthquake (Sharma et al. 2016).

In the 2011 Lorca earthquake (Hermanns et al. 2014; De Luca et al. 2014), masonry infills generally had a beneficial contribution to the structural behaviour of the buildings. But the infill panels were also responsible for the structural problems: strong shear forces on columns, soft story mechanisms and torsion moments due to asymmetrical horizontal stiffness distribution. The latter was observed for corner buildings of apartment blocks – several such buildings suffered substantial damage and one even collapsed.

9.2.2 Laboratory Testing

Tests performed on RC frames often show a positive effect of the infills when compared to the bare frame. A few recent studies support this affirmation (Cavaleri and Di Trapani 2014; Basha and Kaushik 2016; Zovkic et al. 2013).

A test on frames designed using current seismic standards infilled with fly ash bricks showed that infilled frames are much stiffer (7–10 times) and stronger (1.6–2.5 times), and also dissipate more energy (1–2.3 times) than the corresponding bare frames (Basha and Kaushik 2016). Similar conclusions regarding the increase in stiffness, strength and energy dissipated were also drawn in a study on RC frames designed according to the Eurocodes infilled with masonry (Zovkic et al. 2013).

A different study, where frames with different infill typologies were tested, showed an increase of 2–4 times in strength on infilled frames (Cavaleri and Di Trapani 2014). The authors of this study concluded, based on their results, that the cyclic behaviour of infilled frames could be predicted with sufficient accuracy by modelling equivalent diagonal struts by means of multilinear plastic link elements (Cavaleri and Di Trapani 2014).

9.3 Numerical Modelling of RC Frame Structures with Infill Panels

The modelling of these structures is an intricate issue because the interaction of the masonry infill panel and the surrounding frame leads to a highly nonlinear inelastic behaviour (Crisafulli et al. 2000). Numerous parameters should be taken into account, such as the mechanical characteristics of masonry; the ultimate strength capacity of panels including the influence of vertical loads; the contact issues and effective contact lengths between frame and infill; the possible failure mechanisms; the presence of openings, etc. (Barnaure and Stoica 2015; Cavaleri et al. 2017).

Over the past decades, many model types have been proposed in the scientific literature, each of them having both advantages and relative shortcomings. These models can be grouped in two classes: micro-models and macro-models (Uva et al. 2012).

Micro-models rely on complex analysis using the finite element method and involve dividing the structure into numerous elements. The RC frame, the panel and their mutual connections are individually modelled and described by constitutive laws. These models can correctly identify the modes of failure and the local effects. Yet they are difficult to implement as they require high computational time and are based on input parameters that are sometimes difficult to assess.

The macro-models resort to simple methods, where a few elements are used to represent the effect of the masonry panel. These simplified methods exhibit advantages in terms of computational simplicity and efficiency, but present several critical aspects that can compromise the reliability of the results (Lima et al. 2014). The most popular method is the equivalent strut, derived from the observation that the load path within the infill mainly follows the diagonal (Mohammad et al. 2016). The masonry panels are replaced by equivalent diagonal struts, which carry loads only in compression. When a single diagonal strut is used, the model cannot correctly predict the evolution of stresses in frame members (Barnaure and Stoica 2015; Crisafulli et al. 2000). For buildings designed without proper shear reinforcement of columns, the brittle behavior that could occur at nodes can only be described by using multi-strut systems (Mohammad et al. 2016; Mohyeddin et al. 2017). Still, single diagonal struts models represent an adequate tool when the analysis is focussed on the overall response of the structure (Barnaure and Stoica 2015; Crisafulli et al. 2000).

9.4 Case Studies

A four stories frame structure with reinforced concrete members and masonry infills is analysed for horizontal earthquake loading through numerical modelling using the ETABS software (ETABS 2015) (Fig. 9.1).

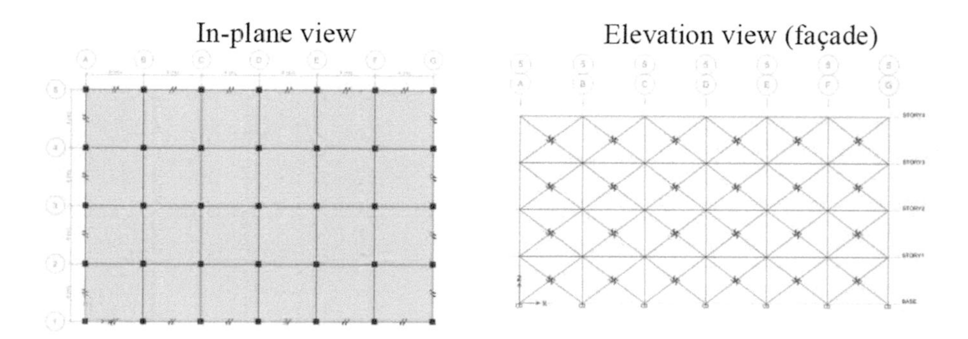

Fig. 9.1 Geometry of the analysed building

The frames are made of C20/25 concrete, reinforced with longitudinal and transversal rebars with $f_{yd} = 300$ MPa. The cross-section of the columns is 40×40 cm, reinforced with 8d20 and 4d25 bars. The beams section is 25×45 cm. Two reinforcement types were considered for the beams. Mid-span, the rebars are 2d16 top and 3d16 bottom, while at its ends the beam has 4d16 top and 3d16 bottom reinforcement. For both columns and beams it was assumed that transverse reinforcement is sufficient as to prevent shear failure. It is our intention to treat this issue in a following study, by using non-centric diagonals and shear hinges in the model, similar to (Mohammad et al. 2016).

The slab is accounted for its mass and in-plane stiffness, but is not considered to contribute to the beams' flexural strength. The masonry panels are modelled as pin-joined diagonals. Two situations are considered: full infill panels and panels with window openings.

Nonlinear pushover analyses are carried out for a range of alternative configurations. The first configuration corresponds to the bare frame structure. Several alternative configurations corresponding to asymmetrical positions for masonry enclosure walls interacting with the frames are considered, as shown in Fig. 9.2.

We are aware that the current trend in the seismic design of structures is to carry out dynamic analysis. Still, pushover analysis can offer reliable information on the structural capacity, while diminishing the risk of high errors due to incorrect parameters (e.g. cyclic laws for the materials) (Cavaleri et al. 2017). As one of the main purposes of this study is to provide a tool for practicing engineers, we chose to use a simplified model, but it is our intention to also perform dynamic analyses in the future.

9.5 Modelling Parameters

Vertical forces are defined on beams and horizontal forces at beam-column joints. An inverted triangular distribution of horizontal forces over the height of the building is assumed.

		In - plane view			
Panel type	Elevation type				
Full	FI	BF (bare frames)	F-4F-FI	F-2X-FI	F-2Y-FI
	SS		F-4F-SS	F-2X-SS	F-2Y-SS
Windows	FI		W-4F-FI	W-2X-FI	W-2Y-FI
	SS		W-4F-SS	W-2X-SS	W-2Y-SS
Panel type	Elevation type				
Full	FI	F-1X-FI	F-1Y-FI	F-XY-FI	
	SS	F-1X-SS	F-1Y-SS	F-XY-SS	
Windows	FI	W-1X-FI	W-1Y-FI	W-XY-FI	
	SS	W-1X-SS	W-1Y-SS	W-XY-SS	
The dotted line represents the façade (s) on which infill panels are considered FI (full infills) configurations have façade panels at all stories SS (soft story) configurations have façade panels on upper stories and bare frames at ground floor level					

Fig. 9.2 Considered case-study configurations

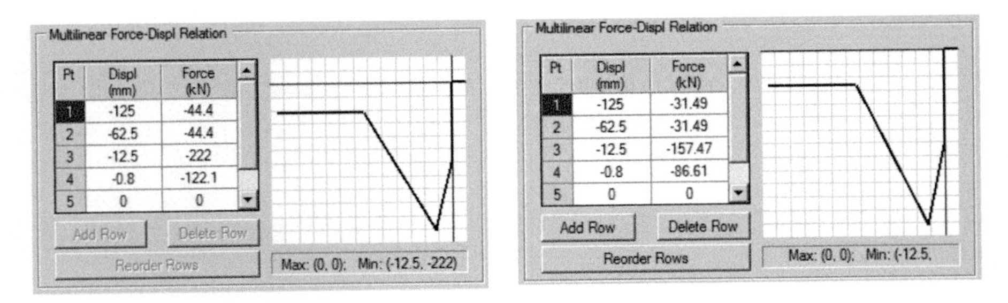

Fig. 9.3 Force-displacement curves for the equivalent diagonal frames: left – full infill panel; right – panel with 150 × 120 window opening

Nonlinear hinges are assigned to the beams and columns (P-M3 hinges for beams and P-M2-M3 for columns). The default hinge properties in the software are used.

The diagonals are link elements with a nonlinear force-displacement curve as shown in Fig. 9.3. The curve for the full panels is based on provisions from the

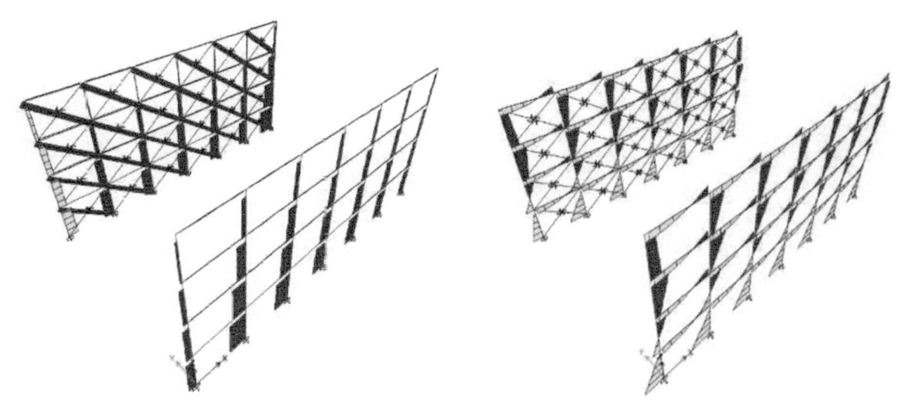

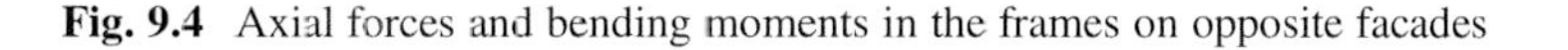

F-1X-FI – Tension/compression in frame members
F-1X-FI – Bending moments in frame members

Fig. 9.4 Axial forces and bending moments in the frames on opposite facades

current Romanian seismic design code (P100-1/2013 2013) and the equations proposed in (Fardis and Panagiotakos 1997). The detailed description of the parameters used for the force-displacement curve as well as comparisons with other types of curves are shown in a previously published paper (Barnaure et al. 2016a). For the panels with window openings, the strength and stiffness can be obtained by multiplying the values for the full panel by a reduction factor, R, as shown in (Al-Chaar 2002). This factor depends on the ratio α between the area of the opening and the area of the panel. The formula is $R_1 = 0.6 \times \alpha^2 - 1.6 \times \alpha + 1$. In this research, a panel size of 360 × 365 cm with a single central window opening of 150 × 120 cm is considered. $\alpha = 0.196$ and, consequently, $R_1 = 0.71$.

9.6 Results and Discussion

The presence of infills can change the force distribution in frame members. Figure 9.4 illustrates the axial and bending moment loads on the opposite facades along X direction for the F1X-FI configuration. It may be concluded that on the infilled façade there is an important modification of axial loads in columns and beams as the façade behaves like a trussed beam. The modification of compression forces is particularly important for the columns at the ends of the façade.

The modification of forces in members determines the modification of the type and position of plastic hinges being formed and a different failure pattern for the structure (Barnaure et al. 2016b). In order to assess the influence on the capacity of the building, the force-displacement curves are traced for all the considered configurations (Figs. 9.5, 9.6, 9.7 and 9.8).

For all the analysed configurations, the presence of masonry infills leads to an increase of the building stiffness. This change in stiffness can lead, depending on the design response spectrum, to higher peak acceleration during earthquakes.

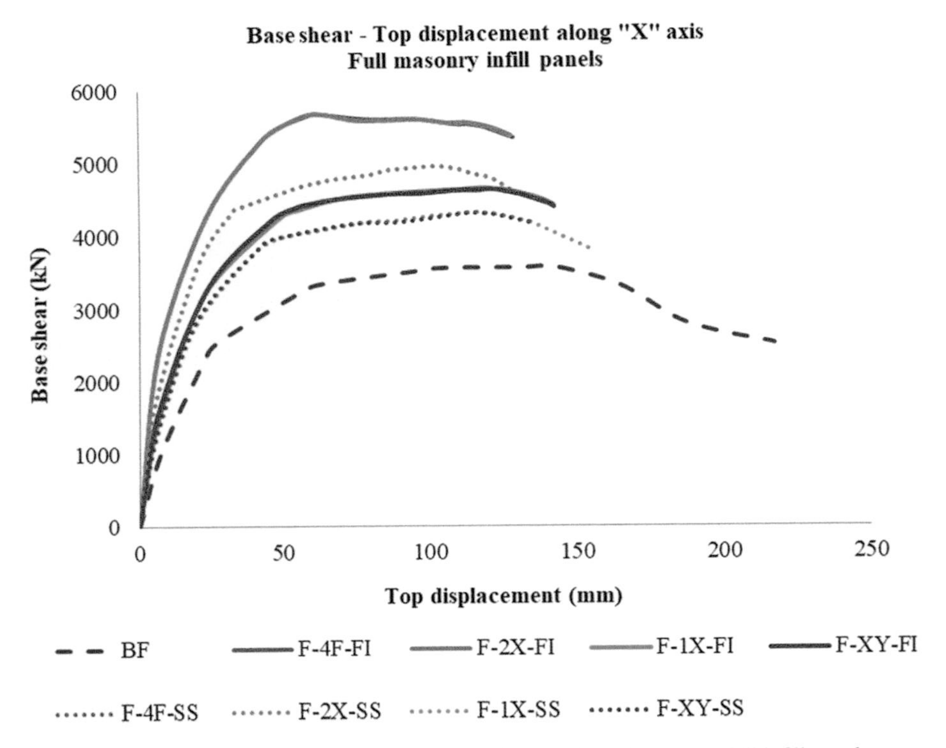

Fig. 9.5 Force – displacement curves for horizontal loads along X axis and full infill panels

The area under the force-displacement curves represents the input energy required to reach collapse. The value of this energy is computed for each considered configuration. The ratios between the input energy for a given configuration and for the corresponding bare frame are shown in Fig. 9.9 for forces along X axis.

By comparing the values for horizontal forces acting along direction X we can reach several conclusions. The first one is that there is no significant influence of the transversal walls, as 4F and 2X results are almost identical, as are 1X and XY. The second conclusion is that, for our considered case study, the regular infill distributions have no significant influence on building strength, while the irregular distributions lead to important diminishments. The values on this strength loss due to the presence of irregularities ranges up to 15% for plane irregularities, up to 20% for soft story irregularities and up to 24% for both in-plane and elevation irregularities.

For horizontal forces acting along direction Y (Fig. 9.10) the same conclusion can be drawn as to the influence of transverse walls. The regular distribution of infills has a positive effect on building strength, with an increase of up to 39% when compared to the bare frame situation. The diminishment due to irregular distribution is less significant than for the X direction, of up to 15%.

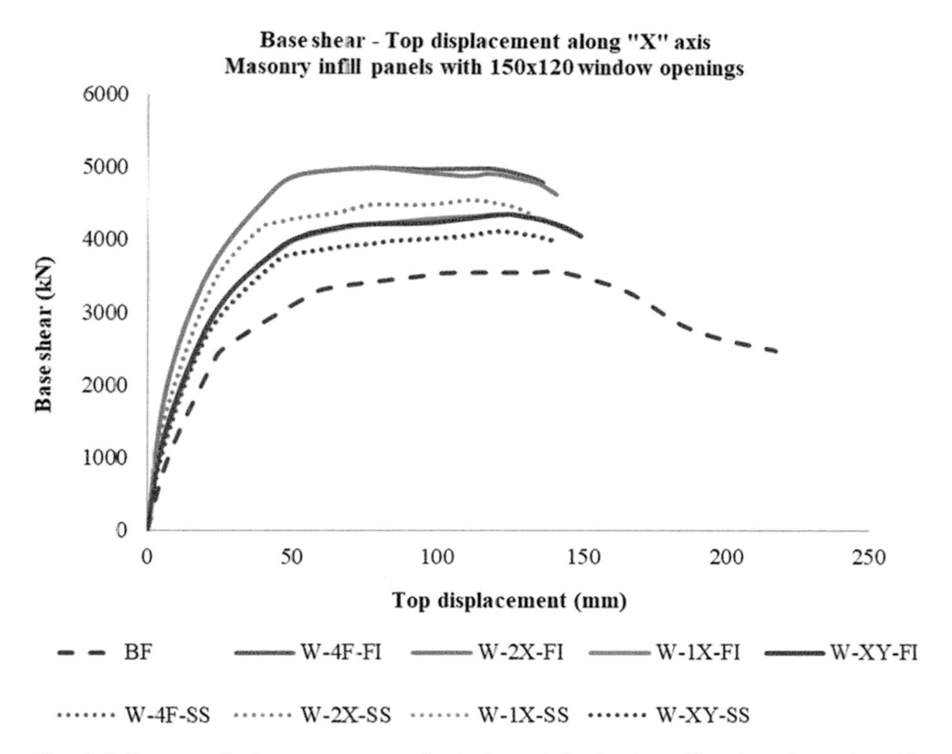

Fig. 9.6 Force – displacement curves for horizontal loads along X axis and panels with 150 × 120 cm window openings

For the considered configurations, the influence of window openings on the behaviour of the building is analysed. As shown in Fig. 9.11, the influence is generally not significant, within only a few percent from the full infill values.

The influence of soft-story configurations is also assessed. With a few exceptions, these configurations lead to an important decrease of capacity for the analysed building, as shown in Fig. 9.12. The observed capacity reduction ranges up to 21%, when compared to the same building with similar infills at all the stories.

9.7 Conclusions

In this study, the seismic response of RC frame buildings with irregularities due to masonry enclosure walls is assessed using numerical models.

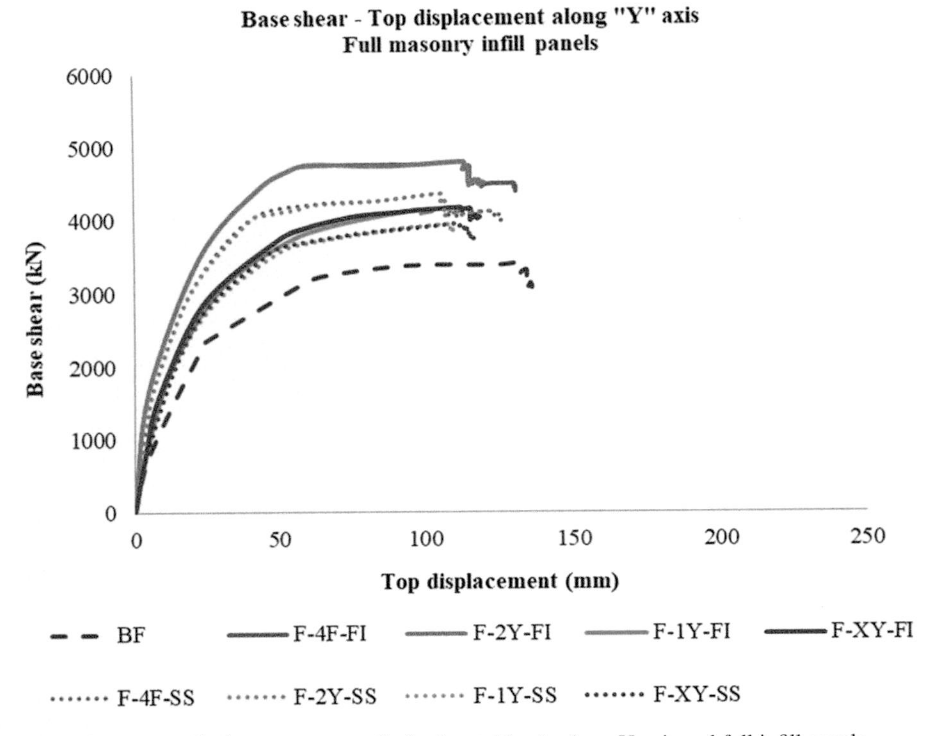

Fig. 9.7 Force – displacement curves for horizontal loads along Y axis and full infill panels

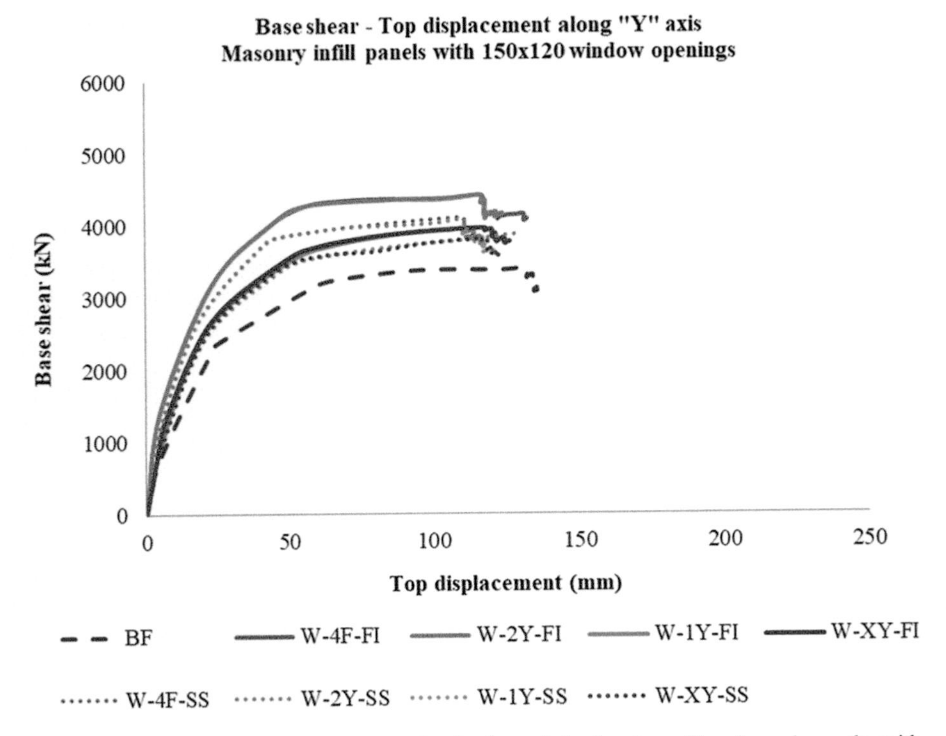

Fig. 9.8 Force – displacement curves for horizontal loads along Y axis and panels with 150 × 120 cm window openings

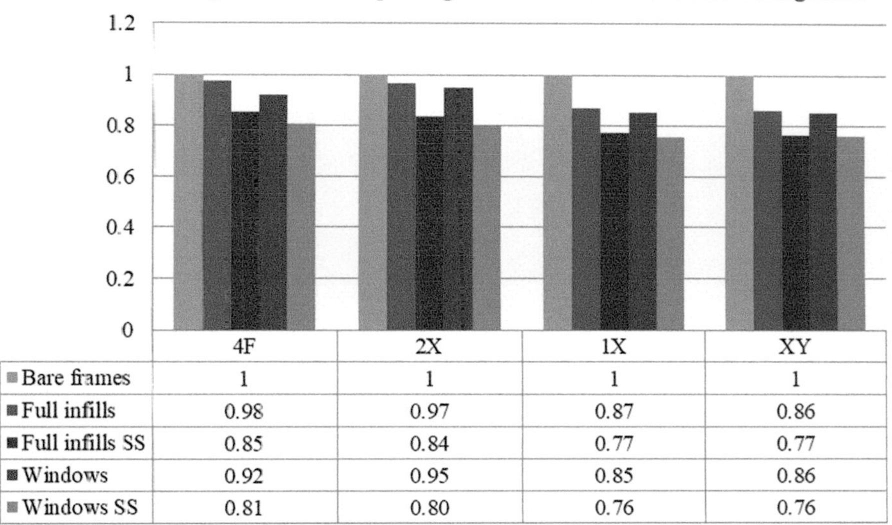

	4F	2X	1X	XY
Bare frames	1	1	1	1
Full infills	0.98	0.97	0.87	0.86
Full infills SS	0.85	0.84	0.77	0.77
Windows	0.92	0.95	0.85	0.86
Windows SS	0.81	0.80	0.76	0.76

Fig. 9.9 Ratios between the input energy that leads to collapse and the input energy that leads to collapse for the corresponding bare frame (forces along X axis)

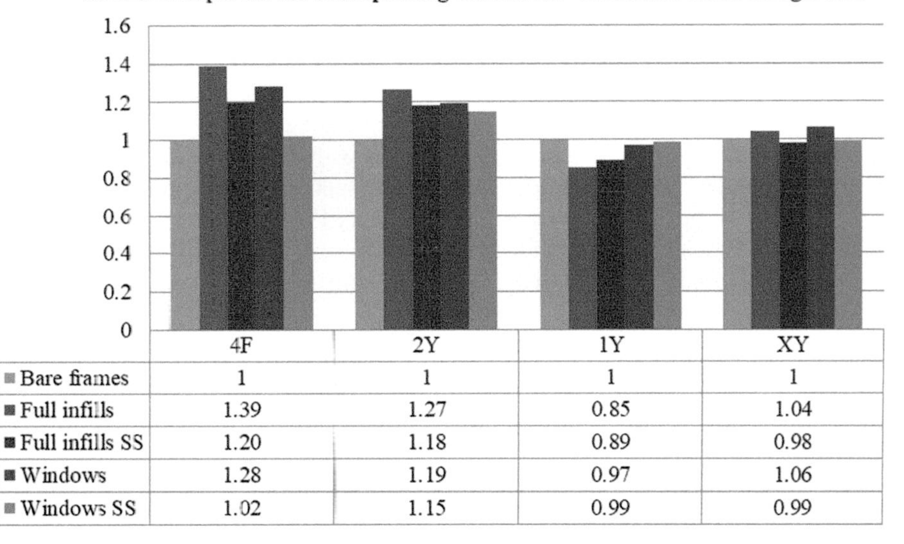

	4F	2Y	1Y	XY
Bare frames	1	1	1	1
Full infills	1.39	1.27	0.85	1.04
Full infills SS	1.20	1.18	0.89	0.98
Windows	1.28	1.19	0.97	1.06
Windows SS	1.02	1.15	0.99	0.99

Fig. 9.10 Ratios between the input energy that leads to collapse and the input energy that leads to collapse for the corresponding bare frame (forces along Y axis)

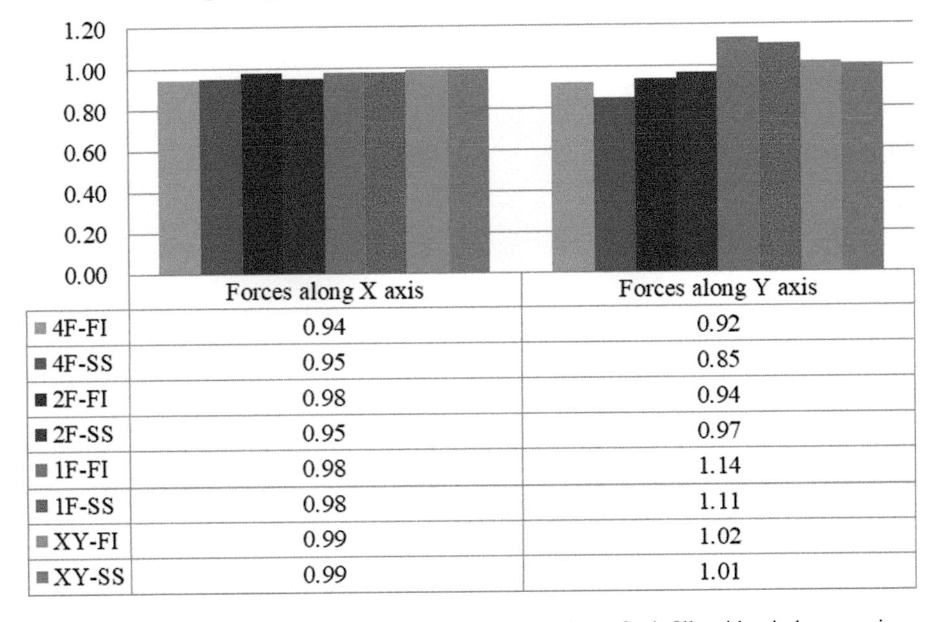

	Forces along X axis	Forces along Y axis
4F-FI	0.94	0.92
4F-SS	0.95	0.85
2F-FI	0.98	0.94
2F-SS	0.95	0.97
1F-FI	0.98	1.14
1F-SS	0.98	1.11
XY-FI	0.99	1.02
XY-SS	0.99	1.01

Fig. 9.11 Ratios between the input energy that leads to collapse for infills with window openings and the input energy for the corresponding full infill

The numerical simulations show that infill masonry panels can significantly modify the structural behaviour of reinforced concrete framed structures. Regular distribution of infills can have a positive effect on building strength when compared to the bare frame situation. On the contrary, irregular panel distributions can lead to important diminishments of building capacity to withstand lateral loading. For the analysed situations, the strength loss due to the presence of irregularities ranges up to 15% for plane irregularities, up to 20% for soft story irregularities and up to 24% for both in-plane and elevation irregularities.

This means that irregular configurations, if not taken into account during the structural design process, can lead to serious damage and even to the partial or full collapse of the buildings during earthquakes.

Even though the proposed model is simple enough to be used by practicing engineers for current design or evaluation, the obtained results are in accordance with building behaviour observed during recent earthquake events.

At the present time, there is a lack of clear code provisions regarding the modelling of RC frames with masonry infills. Single-strut equivalent diagonal models have certain limitations, in particular regarding the evaluation of shear forces that might develop in columns. Still, the proposed model could be used for the design of new buildings, as the columns designed in conformity with current code

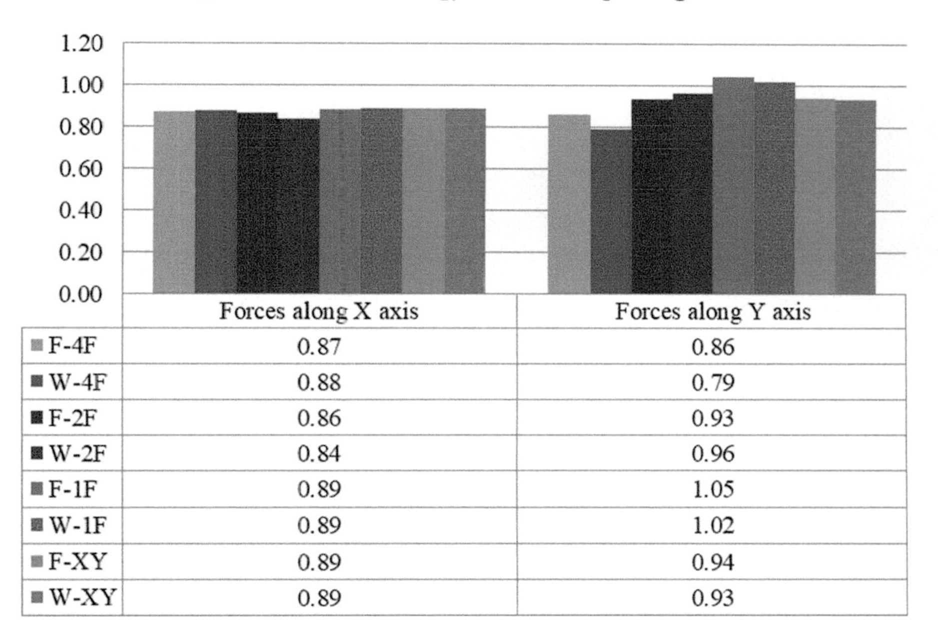

	Forces along X axis	Forces along Y axis
F-4F	0.87	0.86
W-4F	0.88	0.79
F-2F	0.86	0.93
W-2F	0.84	0.96
F-1F	0.89	1.05
W-1F	0.89	1.02
F-XY	0.89	0.94
W-XY	0.89	0.93

Fig. 9.12 Ratios between the input energy that leads to collapse for soft-story configurations and the input energy for the corresponding full infill

provisions are less susceptible to failure due to the additional shear forces resulting from frame/panel interactions. For the assessment of older buildings, more complex models should be used, with multiple non-centric struts, as well as shear hinges.

References

Al-Chaar G (2002) Evaluating strength and stiffness of unreinforced masonry infill structures (ERDC/CERL TR-02-1). Engineer Research and Development Center, Construction Engineering Research Laboratory, Champaign

Barnaure M, Stoica DN (2015) Analysis of masonry infilled RC frame structures under lateral loading. Math Model Civil Eng 11(1):29–39

Barnaure M, Ghita AM, Stoica DN (2016a) Seismic performance of masonry-infilled RC frames. Urban Arhitect Constructii 7(3):229–238

Barnaure M, Ghita AM, Stoica DN (2016b) Influence of the infill panels masonry type on the seismic behaviour of reinforced concrete frame structures. In: The 1940 Vrancea earthquake. Issues, Insights and Lessons Learnt. Springer, pp 319–331

Basha SH, Kaushik HB (2016) Behavior and failure mechanisms of masonry-infilled RC frames (in low-rise buildings) subject to lateral loading. Eng Struct 111:233–245

Cavaleri L, Di Trapani F (2014) Cyclic response of masonry infilled RC frames: experimental results and simplified modeling. Soil Dyn Earthq Eng 65:224–242

Cavaleri L, Di Trapani F, Asteris PG, Sarhosis V (2017) Influence of column shear failure on pushover based assessment of masonry infilled reinforced concrete framed structures: a case study. Soil Dyn Earthq Eng 100:98–112
Crisafulli FJ, Carr AJ, Park R (2000) Analytical modelling of infilled frame structures-a general review. Bull N Z Soc Earthq Eng 33(1):30–47
De Luca F, Verderame GM, Gómez-Martínez F, Pérez-García A (2014) The structural role played by masonry infills on RC building performances after the 2011 Lorca, Spain, earthquake. Bull Earthq Eng 12(5):1999–2026
ETABS (2015) Computer & Structures, Inc. Berkeley
Fardis MN, Panagiotakos TB (1997) Seismic design and response of bare and masonry-infilled reinforced concrete buildings. Part II: Infilled structures. J Earthq Eng 1(3):475–503
Hermanns L, Fraile A, Alarcón E, Álvarez R (2014) Performance of buildings with masonry infill walls during the 2011 Lorca earthquake. Bull Earthq Eng 12(5):1977–1997
Kam WY, Pampanin S, Dhakal RP, Gavin H, Roeder CW (2010) Seismic performance of reinforced concrete buildings in the September 2010 Darfield (Canterbury) earthquakes. Bull N Z Soc Earthq Eng 43(4):340–350
Lima C, De Stefano G, Martinelli E (2014) Seismic response of masonry infilled RC frames: practice-oriented models and open issues. Earthq Struct 6(4):409–436
Mohammad AF, Faggella M, Gigliotti R, Spacone E (2016) Seismic performance of older R/C frame structures accounting for infills-induced shear failure of columns. Eng Struct 122:1–13
Mohyeddin A, Dorji S, Gad EF, Goldsworthy HM (2017) Inherent limitations and alternative to conventional equivalent strut models for masonry infill-frames. Eng Struct 141:666–675
P100-1/2013 (2013) Code for seismic design – Part I – Design prescriptions for buildings [in Romanian]. Ministry of Regional Development and Public Administration, Bucharest
Perrone D, Leone M, Aiello MA (2016) Evaluation of the infill influence on the elastic period of existing RC frames. Eng Struct 123:419–433
Sharma K, Deng L, Noguez CC (2016) Field investigation on the performance of building structures during the April 25, 2015, Gorkha earthquake in Nepal. Eng Struct 121:61–74
Uva G, Raffaele D, Porco F, Fiore A (2012) On the role of equivalent strut models in the seismic assessment of infilled RC buildings. Eng Struct 42:83–94
Verderame GM, De Luca F, Ricci P, Manfredi G (2011) Preliminary analysis of a soft-storey mechanism after the 2009 L'Aquila earthquake. Earthq Eng Struct Dyn 40(8):925–944
Zhao B, Taucer F, Rossetto T (2009) Field investigation on the performance of building structures during the 12 May 2008 Wenchuan earthquake in China. Eng Struct 31(8):1707–1723
Zovkic J, Sigmund V, Guljas I (2013) Cyclic testing of a single bay reinforced concrete frames with various types of masonry infill. Earthq Eng Struct Dyn 42(8):1131–1149

Chapter 10
Influence of the Soil Initial Shear Modulus on the Behaviour of Retaining Walls for Deep Excavations in Bucharest – Case Studies

Alexandra Ene, Oana Carașca, Roxana Mirițoiu, Dragoș Marcu, and Horațiu Popa

Abstract Initial shear modulus (G_0) is an important soil parameter used in geotechnical and earthquake engineering, when modelling geotechnical structures using advanced models for the soil behaviour or when modelling the soil-structure interaction in seismic analysis. Starting from the theory of wave propagation, it is known that this parameter is obtained through *in situ* or laboratory tests, by measurements of the shear wave velocity of the soil. This paper presents an analysis of the initial shear modulus values obtained by seismic tests in depth, carried out during the geotechnical investigations of several projects developed by the authors in Bucharest area. The results obtained from these tests were related to the stratigraphy on site, to relate the values obtained with the geological stratums typical for the city. The further performed analysis was approaching the influence of adopting different values for the initial shear modulus, within the interval obtained from the abovementioned investigations on the behaviour of retaining walls for deep excavations in the urban area of Bucharest. The analysis of the behaviour was performed through Finite Element Method, using the hyperbolic constitutive law with account of the initial shear modulus for the soil. The results obtained in terms of efforts and deformations for different values of initial shear modulus, keeping the rest of the parameters constant, are then presented and discussed. Also, the results are compared to

A. Ene (✉) · R. Mirițoiu · D. Marcu
Popp & Asociații Inginerie Geotehnică S.R.L., Bucharest, Romania
e-mail: alexandra.ene@p-a.ro

O. Carașca
Popp & Asociații Inginerie Geotehnică S.R.L., Bucharest, Romania

Geotechnical and Foundations Department, Technical University of Civil Engineering Bucharest, Bucharest, Romania

H. Popa
Geotechnical and Foundations Department, Technical University of Civil Engineering Bucharest, Bucharest, Romania

D. Köber et al. (eds.), *Seismic Behaviour and Design of Irregular and Complex Civil Structures III*, Geotechnical, Geological and Earthquake Engineering 48,
https://doi.org/10.1007/978-3-030-33532-8_10

measurements of the deformations carried out in inclinometer casings installed in diaphragm walls and in ground extensometers installed in the ground, considering real cases.

Keywords Initial shear modulus · Diaphragm wall · Hyperbolic constitutive law · Inclinometer · Seismic investigations

10.1 Introduction

The soil stiffness and its non-linear dependency on the amplitude of the deformation is a feature which must be considered in the analysis of geotechnical structures for a realistic estimation of the displacements. Advanced constitutive models account for the initial shear modulus, also called "small strain shear modulus". In the same time, this parameter is useful in the analysis of structures in interaction with the foundation ground in case of seismic actions (Benz 2007; Atkinson and Sallfors 1991; Atkinson 2000; Benz et al. 2009; Santos and Correia 2000, 2001; Clayton 2011; NIST GCR 12-917-21 2012).

The advanced design of civil structures considers several approaches for considering the soil-structure interaction (analytical or numerical), dependent on the deformability parameters of the foundation ground.

The initial soil stiffness may be determined either by laboratory or *in situ* tests. The laboratory tests on soil samples must simulate the real loading conditions, such as drainage, stresses etc., in the same time giving punctual, but not overall results about the ground massif. The preferred method is site testing, considering that the disturbance of the soil samples is minimize and because of the better coverage of the investigated soil volume as well. However, the determination of the soil stiffness by *in situ* tests is indirect, based on theoretical and empirical correlations.

For determining the soil stiffness by *in situ* tests, the main procedure is generating shear waves (in depth or at the ground surface) and measuring the velocity of these waves through sensors located at different distances and depths. The most common tests carried out in Romania are: Downhole (ASTM D7400-08 2008), Crosshole (ASTM D4428/D4428M - 07 2007), cone penetration test with seismic probe (sCPT) (Robertson et al. 1989; McGann et al. 2015) and flat dilatometer test with seismic probe (sDMT) (Hryciw 1990).

This paper presents an analysis of the initial shear modulus (soil stiffness in small strain domain) values for the typical strata of Bucharest obtained by seismic tests in depth, carried out during the geotechnical investigations of 13 projects developed in distinct areas of Bucharest. The main investigation method used is the Downhole test, but also some other methods were used and presented (sCPT, sDMT, Crosshole). The values of the initial shear modulus G_0 obtained on these sites were used in analysis of the retaining structures of the deep excavations, to obtain more real estimations of the structures behaviour in terms of stresses and deformations. The results of the design are then compared to measurements in inclinometer casings measuring the horizontal displacements of the retaining walls and in ground extensometer casings measuring the vertical deformations of the foundation ground, carried out throughout the construction stage of several monitored projects.

10.2 Determination of Soil Stiffness Modulus in Small Strain Domain for Projects in Bucharest

It is well known that the soil stiffness modulus in small strain domain (initial shear modulus), G_0, is equal to the square of shear wave velocity – v_S – measured for a specific soil layer, multiplied by the soil density, ρ (Poisson 1831).

As aforementioned, for measuring the values of the propagation velocities for the shear waves (seismic waves) for soils in Bucharest, the results of Downhole (ASTM D7400-08 2008), Crosshole (ASTM D4428/D4428M-07 2007), sCPT (Robertson et al. 1989; McGann et al. 2015) and sDMT (Hryciw 1990) *in situ* tests were used, carried out during the field investigations for the Geotechnical Reports of each site. The preferred and most often used method is the Downhole test, considering that it requires only one borehole, which is previously drilled for geotechnical purposes (soil sampling with determination of geotechnical parameters). Hence, the succession of soil layers and the main characteristics thereof are well known, increasing the accuracy of the data processing and interpretation of the tests.

Typically, the soil stratification specific to Bucharest has the following succession:

1. Old and new fillings found in surface, resulted from various sources and periods of the city development;
2. Upper sandy clayey complex, "Clays of Bucharest" or "Bucharest Loam" comprised of silty-clayey soil deposits and pockets of clayey sands (stiff);
3. Upper sandy complex "Colentina Gravels" comprised of sands and small gravels (medium dense);
4. Intermediate lacustrine complex consisting in general of clays or silty-clays with bounding surfaces (stiff);
5. Intermediate sandy complex, "Mostiștea Sands" consisting of medium and fine sands, sometimes with clayey or sandy inserts (medium dense);
6. Inferior lacustrine complex, consisting of fine clays and sands;
7. Frăteşti layers, the oldest quaternary formations in the area, at relatively high depth (approximately 100–180 m) consisting of sands and gravels with clayey inserts (medium dense to dense).

Within the presented projects, the boreholes carried out for Downhole tests were all 50-m deep, crossing through all the macro-layers until reaching the Inferior lacustrine complex of fine clays and sands (No. 6 above). However, considering the heterogenous character of the soil stratification specific to Bucharest and since the locations of the projects are spread all over the city area, in several Downhole boreholes, the layers were found at different depths and some of the layers were

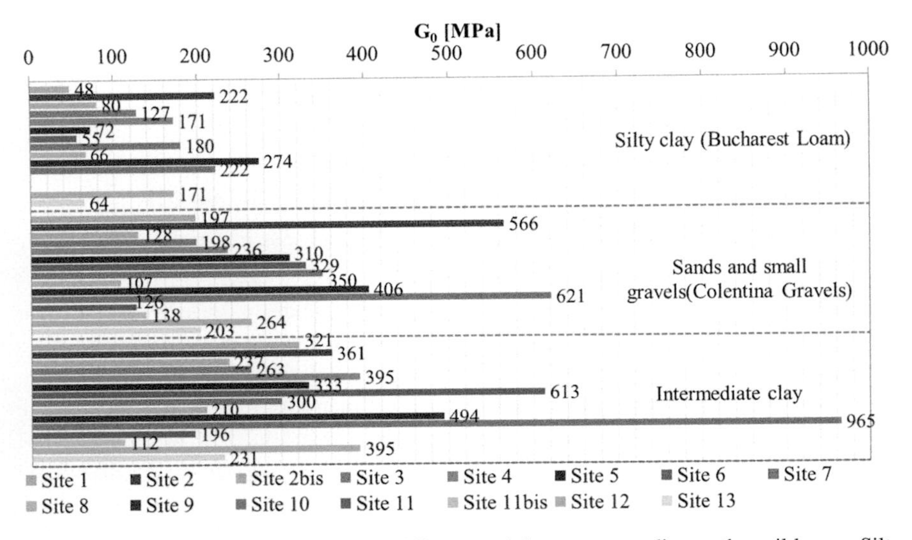

Fig. 10.1 Values obtained for the initial stiffness modulus, corresponding to the soil layers: Silty Clay (Bucharest Loam), Sands and small gravels (Colentina Gravels), Intermediate Clay

missing. Also, the thicknesses of the encountered layers are distinct depending on the location of the project (Popa et al. 2016).

For this paper, the processing was focused on the first three important layers (Bucharest Loam, Colentina Gravels and Intermediate Clays), considering that these are the soil layers that most of the retaining works designed for deep excavation pits are crossing, hence that give the most important soil loads behind the retaining walls. The filling layer was neglected, due to its heterogeneity from one site to another.

For the purpose of the present study, the mean value of the initial shear modulus G_0 and the rest of the geotechnical parameters were used in the calculations (Fig. 10.1).

The processing of the results was carried out for each of the three layers, using all the available values. The charts displayed in Fig. 10.2 show the values of the stiffness modulus in small strain domain corresponding to the middle depth of the soil layer measured from the surface of the ground, on each of the investigated sites. The values are displayed in comparison to the mean value (dashed line) and the standard deviation resulted from the statistical analysis.

10.3 Retaining Walls for Deep Excavations – Case Studies

In order to assess the influence of adopting different values for G_0 initial shear modulus on different retaining wall solutions for deep excavations, 4 sites – case studies – representative for the urban area of Bucharest were considered. The analysis was performed keeping constant the value of the soil resistance parameters

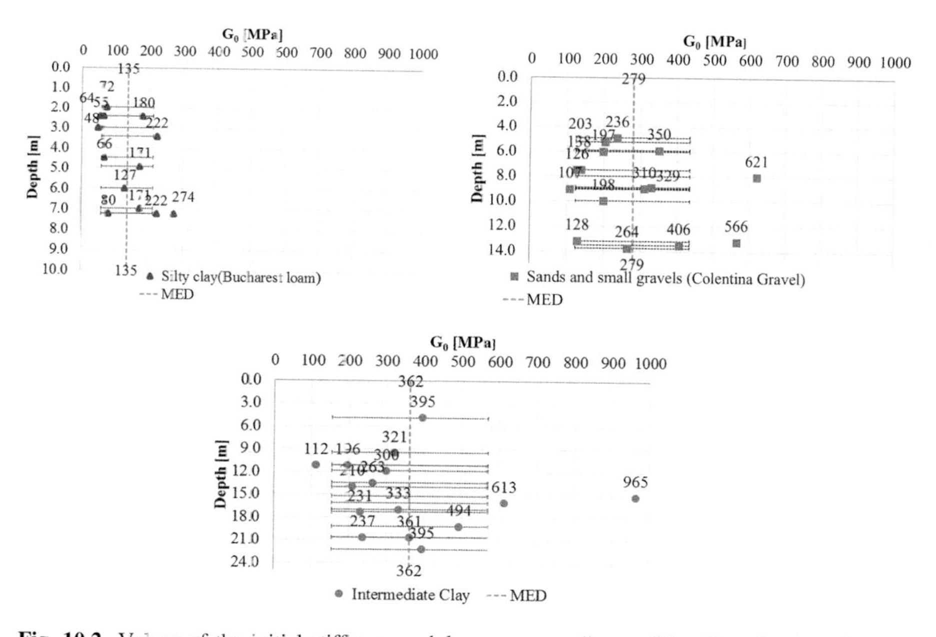

Fig. 10.2 Values of the initial stiffness modulus, corresponding to Silty Clay, Sands and small gravels and Intermediate Clay, respectively

Table 10.1 Numerical parameters of soil models

Case study	Dimensions of the model	Soil cluster mesh	Retaining wall mesh	Interface
1 (Site 9)	35 m deep, 60 m wide	1626 Triangular 15-node elements	25 elements of 5 node line type	103 elements of 5-node line using 4-point Newton-Cotes integration type
2 (Site 11)	33 m deep, 70 m wide	2166 Triangular 15-nodes elements	31 elements of 5 node line type	127 elements of 5-node line using 4-point Newton-Cotes integration type
3 (Site 2)	32 m deep 60 m wide	1776 Triangular 15-nodes elements	28 elements of 5 node line type	115 elements of 5-node line using 4-point Newton-Cotes integration type
4 (Site 11)	46 m deep 65 m wide	8404 Triangula5 15-nodes elements	48 elements of 5 node line type	195 elements of 5-node line using 4-point Newton-Cotes integration type

(for which mean values were considered) and using different values for G_0 initial shear modulus as they were obtained from site investigation.

Calculations were performed using the 2D Finite Element model for plane strain state in Plaxis 2017 software (Plaxis 2D 2017). Stresses and deformations were determined using for soil a nonlinear constitutive model, namely Hardening Soil with small stiffness model (Benz 2007). The boundary conditions of the models consist in fixing their bottom against all directions, their vertical boundaries against horizontal directions and in considering the ground surface free in all directions (Table 10.1).

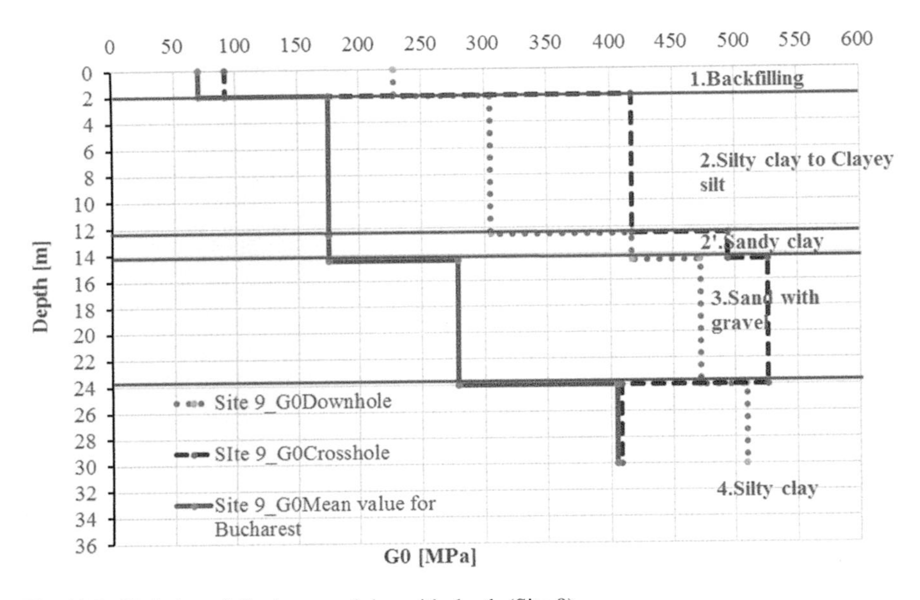

Fig. 10.3 Variation of G_0 shear modulus with depth (Site 9)

10.3.1 Cantilever Retaining Walls

Case Study 1 (Site 9) is represented by a cantilever pile retaining wall (with 16 m length) for a 7.70 m deep excavation from the natural ground level. The lithology of this site consists of about 12 m thick stiff silty clay (the Bucharest loam), followed by a thin sandy clay layer of about 1–2 m thick, about 10 m thick medium dense sand with gravel (Colentina gravels) and then stiff silty clay (Intermediate clays). The ground water level was found 2 m below the excavation level

In Fig. 10.3 are represented the variations for G_0 initial shear modulus obtained from Downhole, Crosshole and a mean value for Bucharest. It can be observed that for both Downhole and mean value the obtained values for G_0 increase with depth, which is expected. Also, the values obtained from Crosshole increase up to 24 m depth where a sudden decrease of its value was recorded for the silty clay layer (layer 4). Although the influence of the 4th layer is not significant for the retaining wall behaviour, seismic evaluation of the site conditions in simplified approaches (considered down to 30 m depth) or even for more advanced soil-structure interaction can lead to less accurate results due to erroneous evaluation of this soil layer stiffness.

As expected, it can be observed that the values obtained for displacement and bending moments decrease when the values of G_0 (for the same layer) increase – especially when comparing site specific determinations to mean values for Bucharest (Fig. 10.4).

The values for displacement resulted about 25% lower (when using the values from Crosshole) and about 30% lower (when using the values from Downhole) with respect to the values obtained using the mean values for Bucharest. The values for

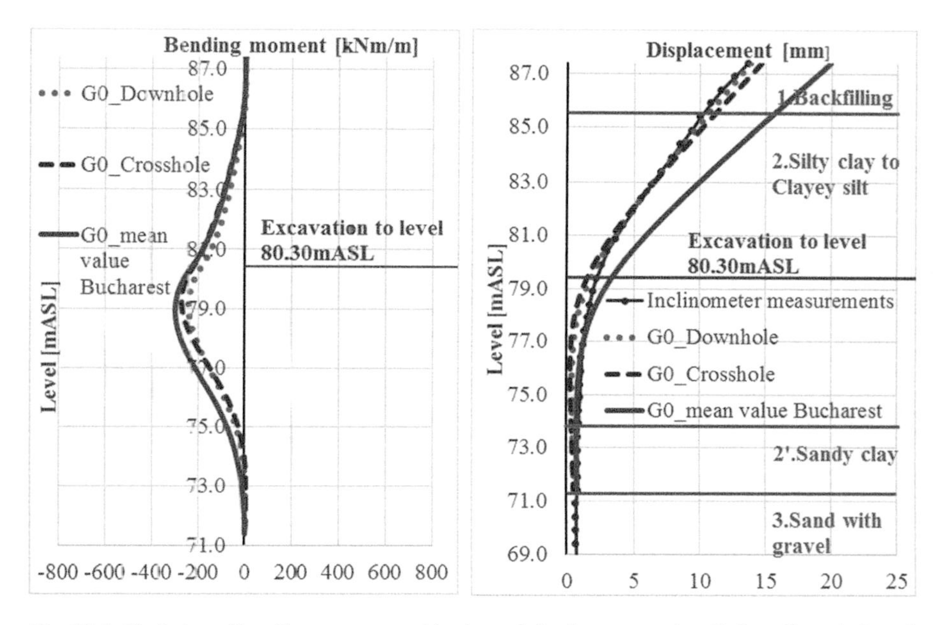

Fig. 10.4 Variation of bending moment and horizontal displacement using G_0 from Downhole and Crosshole tests as well as a mean value for Bucharest (Site 9)

bending moment resulted about 9% lower (when using the values from Crosshole) and about 18% lower (when using the values from Downhole) as compared with the values obtained by using the mean values for Bucharest.

Comparing the horizontal displacements obtained from FEM analysis with those from measurements in inclinometer casing it can be observed that while G_0 from Downhole and Crosshole gives similar results to real measurements, the horizontal displacements obtained with G_0 from mean value for Bucharest resulted about 30% greater than the measured one.

Case Study 2 (Site 11) consists of a deep excavation supported by a cantilever diaphragm wall (with 19 m length) for a 6 m deep excavation from the upper level of the diaphragm wall, while outside the diaphragm wall a preliminary 4 m high sloped excavation was executed. The ground water level was found 3.5 m above the excavation level and the lithology consisted of about 5.5 m thick backfilling, followed by a thin medium dense sand with gravel layer (Colentina gravels) of about 3.5 m thick, about 4.5 m thick layer of stiff silty clay (Intermediate clay), about 2 m thick layer of clayey sand, about 10 m thick stiff clay and then fine sands (Mostistea sands).

Also for this site (Fig. 10.5) it can be observed the dependence with depth of G_0 obtained from Downhole test and mean value for Bucharest, as it was expected. Moreover, the value obtained from sCPT decreases for the sand with gravel (3rd layer) and silty clay layer (4th layer), and since these two layers interact with the retaining wall their influence is significant especially in terms of horizontal displacements (Fig. 10.6).

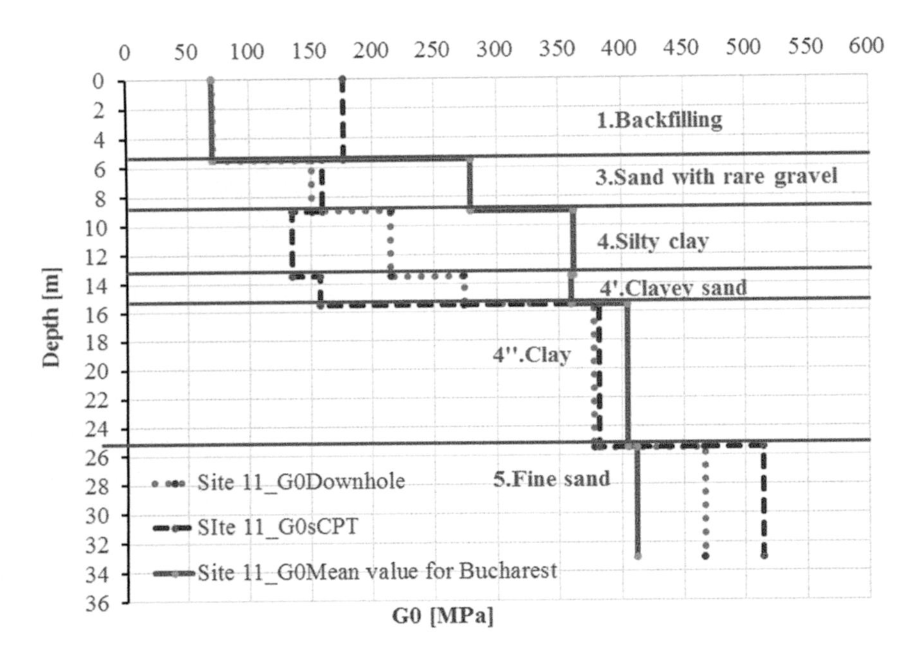

Fig. 10.5 Variation of G_0 shear modulus with depth (Site 11)

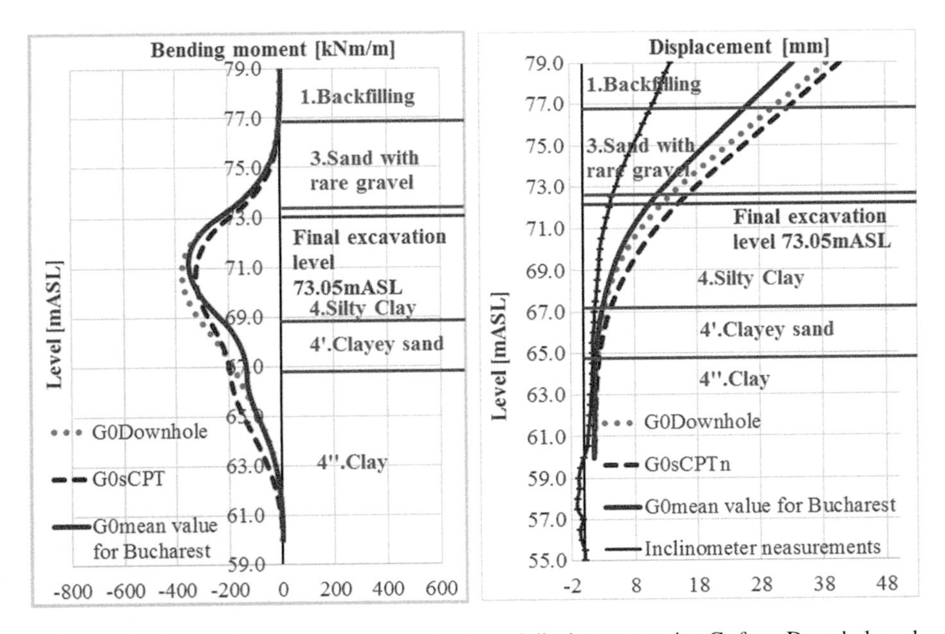

Fig. 10.6 Variation of bending moment and horizontal displacement using G_0 from Downhole and sCPT tests as well as a mean value for Bucharest (Site 11)

As expected, the values obtained for horizontal displacements and bending moments decrease when G_0 value (for the same layer) increases – especially when comparing values obtained from Downhole test and mean values for Bucharest to sCPT values (Fig. 10.6).

The values for displacements resulted 16% greater (when using G_0 from Downhole) and about 25% greater (when using G_0 from sCPT) as compared to the values obtained using the mean values for Bucharest. The difference in bending moments between the three cases are lower than 10%.

Comparing the horizontal displacements obtained from FEM analysis with those from inclinometer measurements it can be observed that the values are about 136% greater (when using the mean values for Bucharest) and about 200% greater (when using values from Downhole and sCPT) compared to those measured on site.

10.3.2 Retaining Walls Supported by One Level of Steel Struts

Case Study 3 (Site 2) consists of a deep excavation supported by inclined struts installed in the cap beam and in the raft executed in the central area. The excavation is 7 m deep, and the retaining diaphragm wall has 19 m length. The ground water level was found 4 m above the excavation level and the lithology consist of approximately 4 m of backfilling, followed by a thin stiff silty clay layer of about 2 m thick (Bucharest loam), about 10 m thick medium dense sand with rare gravel (Colentina gravels), a 6 m thick layer of stiff clay (Intermediate clays), a layer of medium sand of about 3.5 m thick (Mostistea sands), a very thin layer of clay with sand insertion and then fine sands

For this site (Fig. 10.7) it can be observed that the values obtained for G_0 from Downhole are much larger than in the previous cases, as well as, from those obtained from sDMT for this site. Also, the value obtained from Downhole decreases for the silty clay (2nd layer) and clay layers (4th layer).

The G_0 values obtained from both sDMT test (performed only to 16 m depth) and the mean values for Bucharest increase with depth, as it was expected.

The values for displacements obtained with G_0 from sDMT and from mean values for Bucharest are almost similar and with 13% and 17% greater than those obtained using G_0 from Downhole, for the excavation below struts stage and struts removal stage respectively. Also, it can be observed that, for the struts removal stage, the maximum horizontal displacement obtained from FEM analysis is 40% greater (when using the values from Downhole) and about 47% greater (when using the values from sDMT and mean values for Bucharest) with respect to those obtained from inclinometer measurements (Fig. 10.8).

The values for bending moments obtained using G_0 from sDMT and from mean values for Bucharest are almost similar and with 11% greater than those obtained using G_0 from Downhole for the excavation below struts stage. For the struts removal stages the bending moment values are almost similar for all three cases (Fig. 10.9).

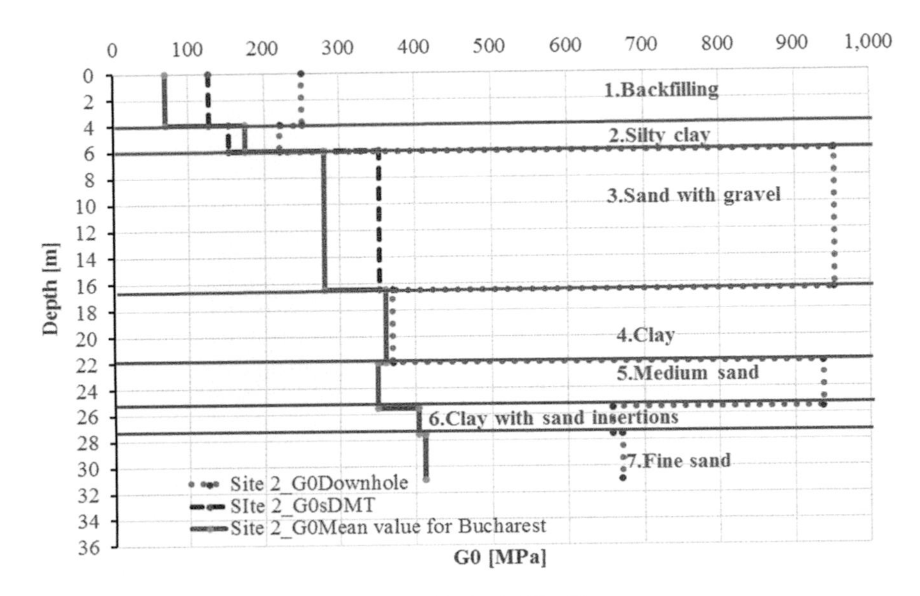

Fig. 10.7 Variation of G_0 shear modulus with depth (Site 2)

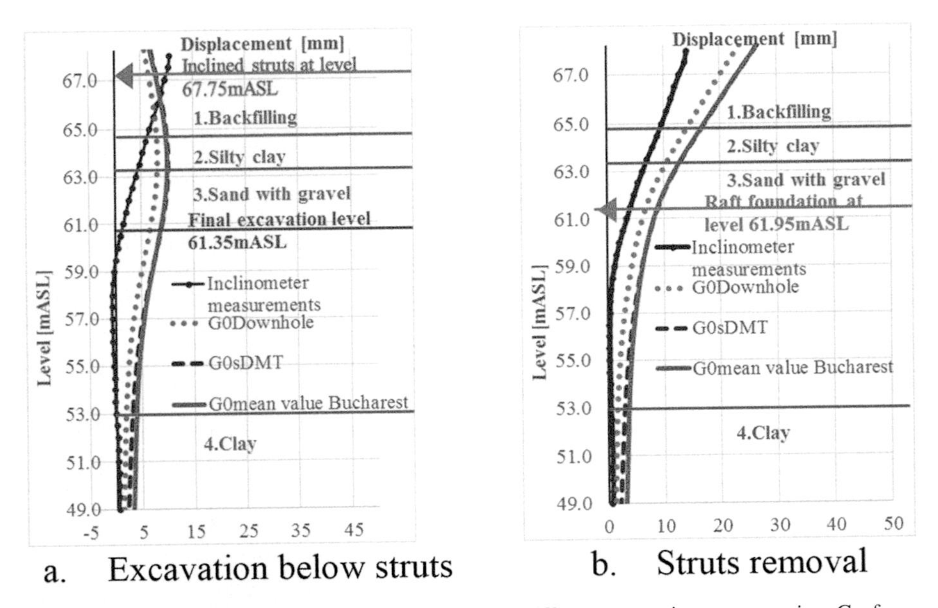

Fig. 10.8 Variation of horizontal displacement for different execution stages using G_0 from Downhole and sDMT tests, as well as, a mean value for Bucharest (Site 2)

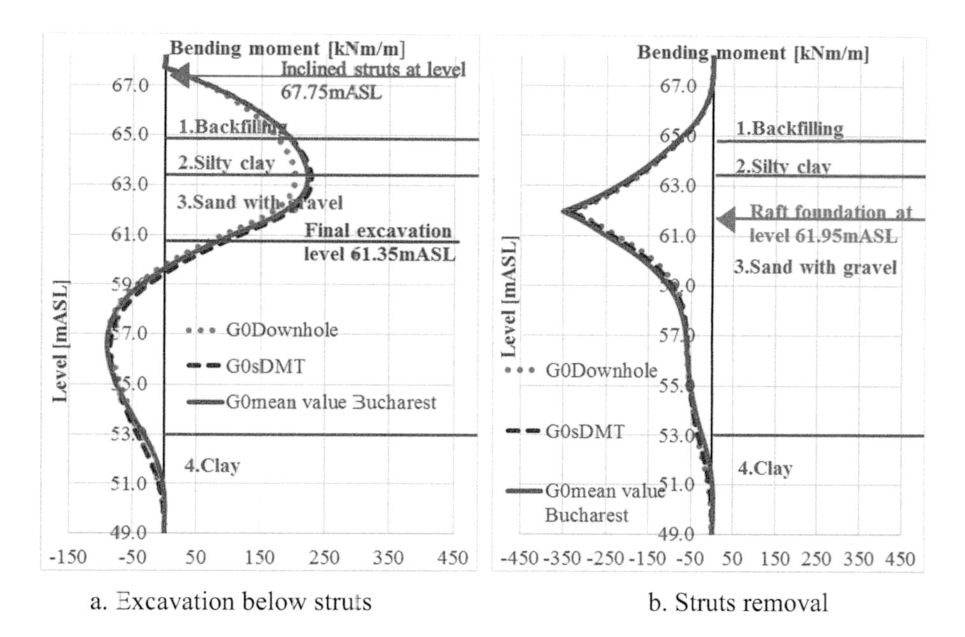

Fig. 10.9 Variation of bending moment for different execution stages using G_0 from Downhole and sDMTtests as well as a mean value for Bucharest (Site 2)

10.3.3 Retaining Walls Supported by Two Levels of Steel Struts

Case Study 4 (Site 1) consists of a deep excavation supported by two levels of horizontal steel struts connected to the diaphragm wall by means of a steel wale. The excavation is approximately 12 m deep, and the retaining wall has 30 m length. The ground water level was found 6.8 m above the excavation level and the lithology consist of approximately 2.5 m of backfilling, followed by a thick sand with gravel layer of about 5 m thick (Colentina gravels), about 5.5 m thick clay and silty sand layer (Intermediate clays), a 2 m thick layer of fine sand (Mostistea sands), a thick layer of clay of about 11.5 m thick and then clay with limestone concretions.

For this site (Fig. 10.10) it can be observed that the value for G_0 from mean values for Bucharest decreases for the 11.5 m thick clay layer (6th layer), situation that is not confirmed by the value obtained from the Downhole test. Due to the thickness of this layer, its influence is significant for the retaining wall behaviour and it is confirmed by the fact that the horizontal displacement in the clay layer resulted 25% lower when using the values from Downhole in comparison with the mean values for Bucharest (Fig. 10.11).

Comparing the horizontal displacements from FEM analysis with those from site measurements it can be observed that the measured displacements resulted 70% lower than those obtained from FEM analysis.

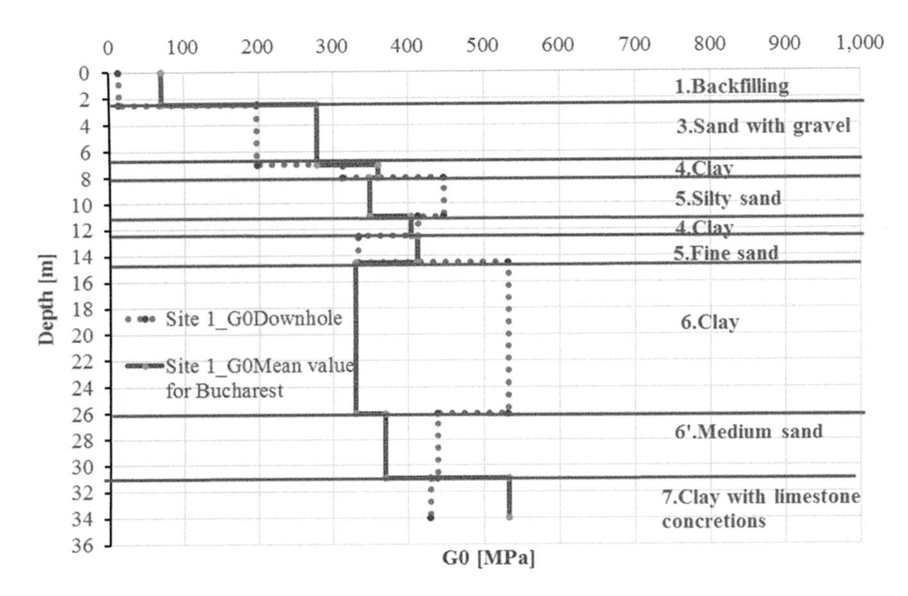

Fig. 10.10 Variation of G_0 shear modulus with depth (Site 1)

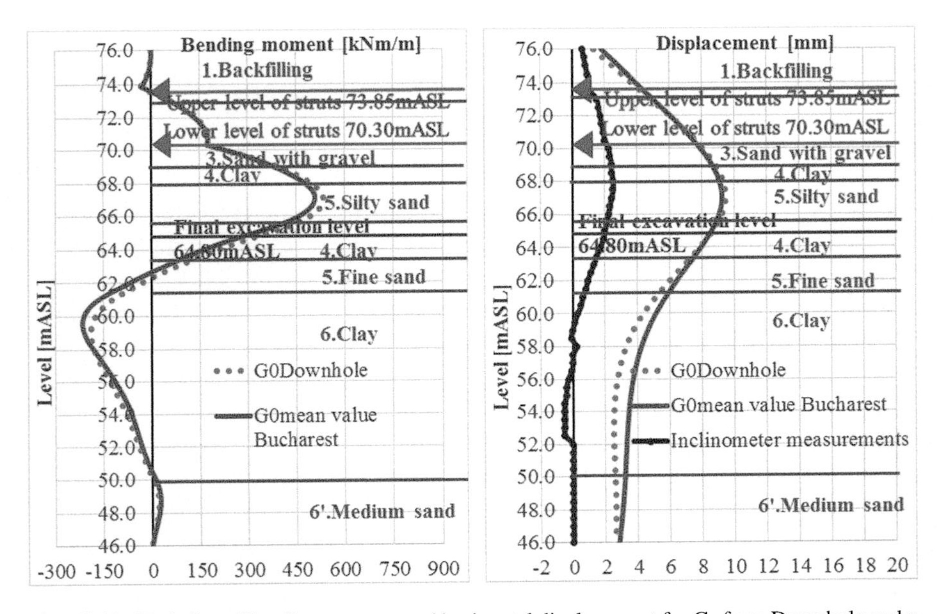

Fig. 10.11 Variation of bending moment and horizontal displacement for G_0 from Downhole and a mean value for Bucharest (Site 1)

Also, the values for bending moments obtained using G_0 from Downhole test and from mean values for Bucharest are similar.

10.4 Predicted Vertical Displacements Compared to Measurements

In addition to the horizontal displacements, the vertical displacements were also monitored by means of ground extensometer casings.

In Fig. 10.12 are presented the measured vertical displacements, as well as, those obtained from FEM analysis performed for Site 2 and Site 11. The vertical displacements obtained with G_0 from Downhole and sDMT are almost similar with the ones obtained with mean values for Bucharest. Although, an exception was registered for calculation performed using G_0 values from sCPT, the vertical displacement obtained being approximately 50% greater than those obtained using Downhole and mean values for Bucharest.

The maximum soil heave measured on site increases with approximately 250% related to the maximum value obtained from FEM analysis (using G_0 initial modulus from Downhole) for Site 2 and with 156% greater related to the maximum value obtained from FEM analysis (using G_0 initial modulus from Downhole) for Site 11.

10.5 Conclusions

The present article presents the variation of G_0 initial shear modulus obtained from various *in situ* test for four sites in Bucharest area, as well as its influence on the retaining wall in terms of horizontal and vertical displacements and bending moments.

Variation of the soil initial modulus even in relatively small area (all within the margins of Bucharest city) is very important, thus precaution is needed in estimating this parameter or comparing with experience. Besides from the differences in soil layers thickness and properties, it can also be intuited that also the variation of the testing procedures and equipment might lead to such discrepancies. It is then necessary to emphasize the importance of more uniform practice, especially since there is not yet a standardized local or international procedure for such field tests.

Variation of initial shear modulus leads to more or less variation of efforts and especially of the displacements of retaining structures depending also on other factors such as: backfilling thickness and variation, adjacent structures or different works, other geotechnical parameters etc. It is obvious that if prediction is performed with significant underestimation, even considering codes safety margins it might

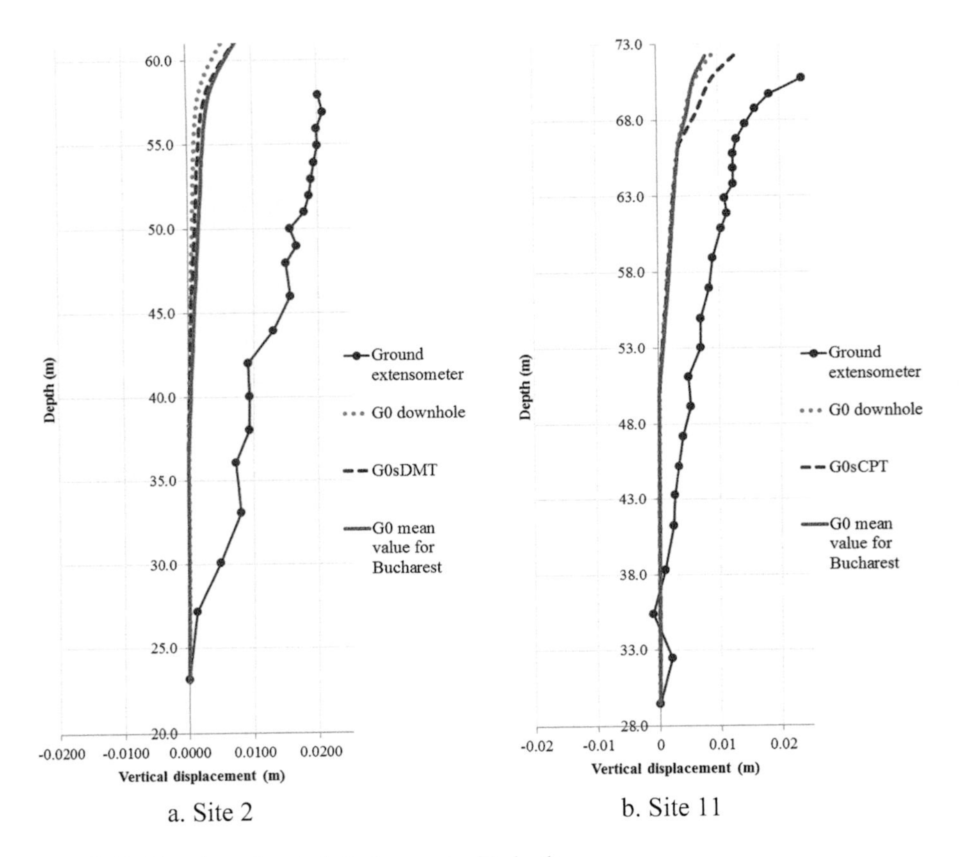

Fig. 10.12 Variation of vertical displacement with depth

lead to hazardous results and, on the other hand, if prediction is significantly overestimated and combined with codes safety margins (characteristic values of geotechnical parameters chosen as cautious estimate or by statistical processing) the design becomes highly costly related to the necessary safety.

For Sites 9 and 11 it can be noticed that the initial shear modulus G_0 obtained from Downhole and mean value for Bucharest increases with depth, while the ones obtained from Crosshole and sCPT decreases for the silty clay layer and sand with gravel layer. For Site 2 the initial shear modulus G_0 obtained from sDMT and mean values for Bucharest increases with depth, while the one obtained from Downhole decreases for the silty clay layer and clay layers respectively. In the case of Site 1, G_0 from mean values for Bucharest decreases for the clay layer, situation that is not confirmed by the value obtained from the Downhole test.

The values for bending moment resulted between 11% and 18% lower (for Site11, Site 9 and Site 2), when using G_0 from Downhole Test as compared to the values obtained with G_0 from mean values for Bucharest. For Site 1 the values for bending moments obtained using G_0 from Downhole test and from mean values for Bucharest are similar.

The values for horizontal displacements resulted between 13% and 30% lower (for Site 1, Site 2 and Site 9) when using G_0 from Downhole Test in comparison with the mean values for Bucharest. For Site 11 the results obtained are different, meaning that when using G_0 from Downhole and sDMT the horizontal displacements are up to 25% greater than those obtained with the mean values for Bucharest.

With the exception of Site 9, where the horizontal displacements obtained from site measurements and those obtained using G_0 from Downhole and Crosshole were similar, for the other sites the horizontal displacements obtained from FEM analysis (using G_0 from Downhole, sCPT or sDMT) resulted between 40% and 200% greater than those obtained from site measurements.

Moreover, as it can be noticed from the presented measurements results, compared with the calculation results, the vertical displacement (heave) is poorly estimated compared with the actual measurements, even when the horizontal displacements obtained from numerical calculations are calibrated with the measured ones (Site 9).

Further research on the influence of variation the initial shear modulus in the design should be considered for the evaluation of seismic site response analysis (site seismic hazard evaluation) – usually performed within the first 30 m of depth or more – or in more advanced soil-structure interaction analysis for seismic design.

References

ASTM D4428/D4428M - 07 (2007) Standard test methods for crosshole seismic testing (ASTM Committee D18 on Soil and Rock)

ASTM D7400-08 (2008) Standard test methods for downhole seismic testing (ASTM Committee D18 on Soil and Rock)

Atkinson JH (2000) Non-linear soil stiffness in routine design. Géotechnique 50(5):487–508

Atkinson JH, Sallfors G (1991) Experimental determination of soil properties. Proceedings of the 10th ECSMGE, vol 3, pp 915–955

Benz T (2007) Small-strain stiffness of soils and its numerical consequences. Dissertation, Institut fur Geotechnik der Universitat Stuttgart, Germany

Benz T, Schwab R, Vermeer P (2009) Small-strain stiffness in geotechnical analyses, Ernst & Sohn Verlag für Architektur und technische Wissenschaften GmbH & Co. KG, Berlin, Bautechnik Special issue 2009 – Geotechnical Engineering

Clayton C (2011) Stiffness at small strain: research and practice. Géotechnique 61:5–37

Hryciw RD (1990) Small strain shear modulus of soil by dilatometer. J Geotech Eng Div ASCE 116:1700–1715

McGann C, Bradley A, Taylor M, Wotherspoon M, Cubrinovski M (2015) Applicability of existing empirical shear wave velocity correlations to seismic cone penetration test data in Christchurch New Zealand. Soil Dyn Earthq Eng 75:76–86

NIST GCR 12-917-21 (2012) Soil-structure interaction for building structures. U.S. Department of Commerce, National Institute of Standards and Technology

Plaxis 2D (2017) Part 1: Reference manual & Part 3: Material models

Poisson SD (1831) Memoir on the propagation of motion in elastic media. Mémoires de l'Académie des Sciences de l'Institut de France 10:549–605

Popa H, Carașca O, Ene A, Marcu D (2016) Analiza rigidității pământului în domeniul deformațiilor mici obținută prin teste in-situ pentru amplasamente din București. In: Proceedings of the XIII Romanian national conference for soil mechanics and geotechnical engineering. Cluj-Napoca, Romania, MEDIAMIRA, pp 399–406

Robertson PK, Campanella RG, Gillespie D, Rice A (1989) Seismic CPT to measure in-situ shear wave velocity. J Geotech Eng 112(8):791–803

Santos A, Correia AG (2000) Shear modulus of soils under cycling loading at small and medium strain level. In: 12th world conference on earthquake engineering, p 530

Santos A, Correia AG (2001) Reference threshold shear strain of soil. Its application to obtain a unique strain-dependent shear modulus curve for soil. In: 15th international conference SMGE, vol 1, pp 267–270, A.A. Balkema, Istanbul

Chapter 11
Seismic Performance of Uneven Double-Box Tunnel Sections for Subway

Tsutomu Otsuka, Kota Sasaki, Shinji Konishi, Yuya Nishigaki, Kouichi Maekawa, and Ryuta Tsunoda

Abstract The center pillars of the common double box tunnel of subways have been investigated and their seismic performance has been verified in practice by taking advantage of the lesson of collapsed Daikai station in 1995. However, these verified box tunnels are either horizontally-arranged or vertically-arranged box culvert, and the seismic performance of the transition between these box culverts have not yet been examined. In this study, the failure mode of these uneven tunnels and its characteristic shape are analyzed, and the seismic performance regarding the safe internal spaces and the effects of cross-section shape is discussed.

Keywords Subway · Cut and cover tunnel · Uneven double box section · Seismic performance · 3D finite element analysis

11.1 Introduction

In Japan, subway tunnels constructed by the cut-and-cover method are typically horizontally-arranged double box tunnels with rectangular cross-sections as shown in Fig. 11.1. The common width is about 10 or 11 m. The cut-and-cover method generally involves the excavation of public roads, installing tunnel frames in place and then covering them.

However, some tunnels had to be built under narrow roads in Tokyo and not to secure an enough width for horizontally-arranged double box tunnel; thus, there are some vertically-arranged double box tunnels with longitudinally long cross sections.

T. Otsuka (✉) · K. Sasaki · S. Konishi · R. Tsunoda
Structure Maintenance Division, Tokyo Metro Co., Ltd., Tokyo, Japan
e-mail: tsu.ootsuka@tokyometro.jp

Y. Nishigaki
International Division, Tokyu Co., Ltd., Tokyo, Japan

K. Maekawa
Department of Civil Engineering, School of Engineering, Tokyo University, Tokyo, Japan

D. Köber et al. (eds.), *Seismic Behaviour and Design of Irregular and Complex Civil Structures III*, Geotechnical, Geological and Earthquake Engineering 48,
https://doi.org/10.1007/978-3-030-33532-8_11

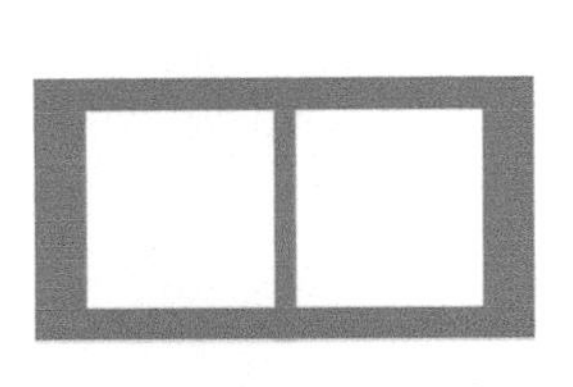

a) Horizontally Oriented Box b) Vertically Oriented Box

Fig. 11.1 Typical double box tunnel with rectangular cross-section

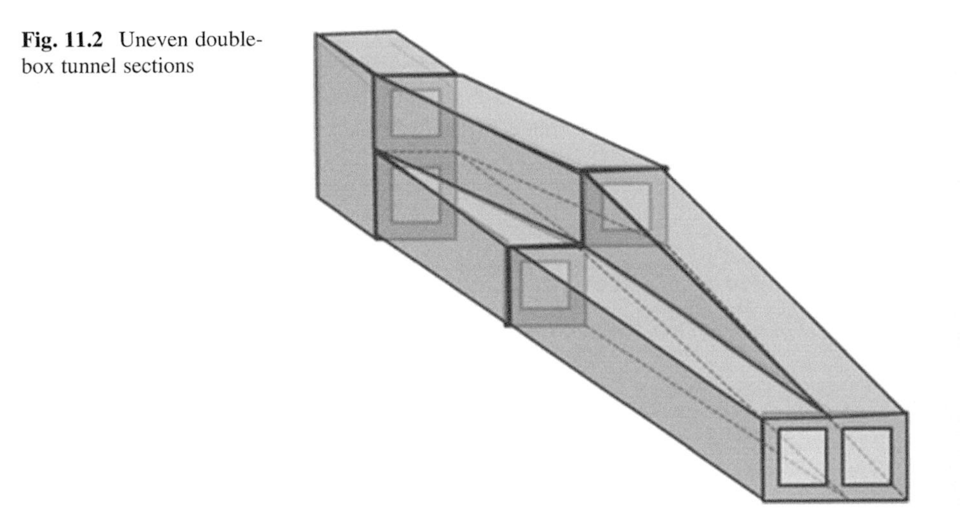

Fig. 11.2 Uneven double-box tunnel sections

To connect these different shape tunnels, there are transition sections as shown in Fig. 11.2. In this paper, the transition is referred as "uneven double-box tunnel sections." In these sections, the cross-section of tunnels gradually shifts in the vertical direction.

Design Standard for Railway Structures and Commentary (Seismic Design)(Railway Technical Research Institute 2012) describes the explanations of the concept and methodology of verifying the seismic performance of common horizontally-arranged double box tunnels (Kawanishi et al. 2012) such as the one shown in Fig. 11.1. However, the standard does not describe any explanations of uneven double-box tunnel sections as shown in Fig. 11.2. Seismic loads acting on such special shaped tunnel sections result in complex deformation modes that could lead to localization of damage or failures that are difficult to repair. While the seismic performance of some special shaped tunnels at Metropolitan Expressway ramps (Hiroshi et al. 2008, 2009; Naoto et al. 2004) have been examined, uneven double-box tunnels for railway have not yet been sufficiently investigated.

Fig. 11.3 Analytical cross-section shape

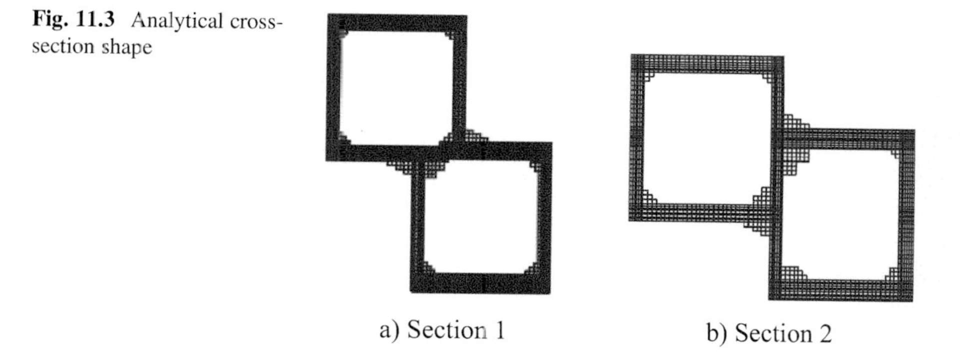

Table 11.1 Specified concrete mix proportion and other input condition for analysis

W/C	Water	Cement	Sand	Gravel
60%	145 kg/m^3	241 kg/m^3	832 kg/m^3	977 kg/m^3

Rebar yield strength: 370 N/mm^2
Soil friction angle: 45°
Soil cohesion: 0 N/mm^2
Initial soil rigidity: 43 N/mm^2

Based on these circumstances, this paper explains the use of nonlinear analysis to investigate failure modes of earthquakes in uneven double-box tunnel sections due to effects of tunnel shapes as fundamental investigation.

11.2 Preliminary Analysis

The simulation analysis has been examined through loading experiments on model tunnel by using the 3D nonlinear analysis (Koichi et al. 2003; Nam et al. 2006; Satoshi et al. 2007) for RC structure systems. This method of analysis makes it possible to track rebar yielding and cracks that result from opening and closing in multiple directions, which is nonlinear behaviour peculiar to RC structures. In this study, the nonlinear static responded displacement method (Kazuhiko 1994; Akira 1992) is adopted to verify a fundamental seismic performance. Figure 11.3 shows mesh generations of tunnel sections with the RC parts. These cross-section shapes were created from the design drawings of an actual subway tunnel.

For the analysis, the 3D thermodynamic responded model DuCOM-COM3 (An and Koichi 1997; Koichi et al. 2009; Okhovat and Maekawa 2009) is adopted. Table 11.1 shows the physical properties to be input for the analysis. Soil is distributed to 1.05 m off the left and right edges of the structure, and elastic matter is distributed further away from the structure. The red part in Fig. 11.4 is the RC part, the blue represents the soil, and the yellow represents the elastic matter as pseudo-elements. Earth covering was set to 3.00 m, around the same amount that actually covers. The soil beneath the RC structure was also set to 1.05 m. The ratio of

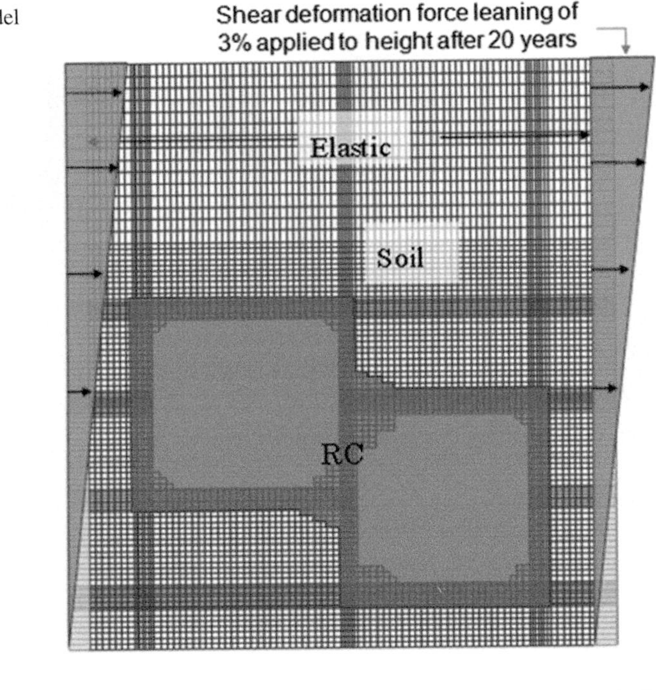

Fig. 11.4 Analytical model (static analysis)

Table 11.2 Environmental condition of the tunnel structure

		First 28 days	After
Inside of the tunnel	RH	99.99%	40%
	Temp.	20 °C	25 °C
Outside of the tunnel	RH	99.99%	99%
	Temp.	20 °C	20 °C

reinforcement of distributing bar was assumed to be a uniform 0.5%. Other reinforcement was installed as the steel ratio of each element based on design drawings.

For relative humidity (RH) and temperature, it is assumed that RH of inside and outside of the tunnel is 99.99% for 28 days after concrete depositing, and set the temperature to 20 °C (Table 11.2). After that, the RH is decreased to 40% and increased the temperature to 25 °C inside of the tunnel, while the RH is decreased to 99% and the temperature is the same for outside of the tunnel.

Given the fact that 20 years have passed since the concrete depositing, 3% shear displacement to the height of model was loaded as an earthquake ground action as shown in Fig. 11.4. This displacement is assumed as ground action due to large earthquake, because of verification of a failure mode. After the structure tilted toward the right, shear displacement was loaded to opposite direction to tilt toward the left, then return to the original position.

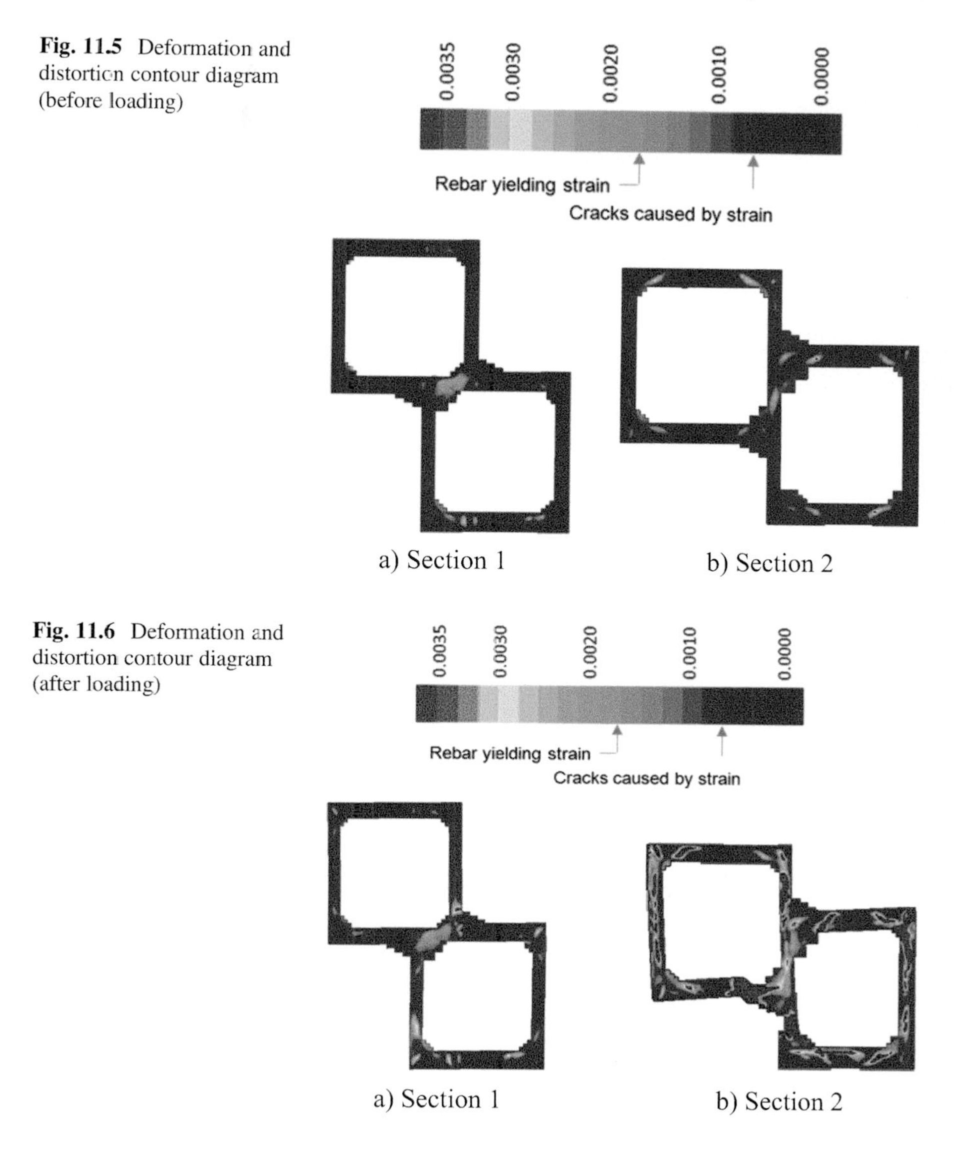

Fig. 11.5 Deformation and distortion contour diagram (before loading)

Fig. 11.6 Deformation and distortion contour diagram (after loading)

11.3 Failure Modes for Uneven Double-Box Tunnel Sections

Figures 11.5 and 11.6 show the results of analysis. These figures show the model deformation (with 2x magnification of deformation) and the maximum principal strain distribution (tension made positive) before and after loading by shear displacement.

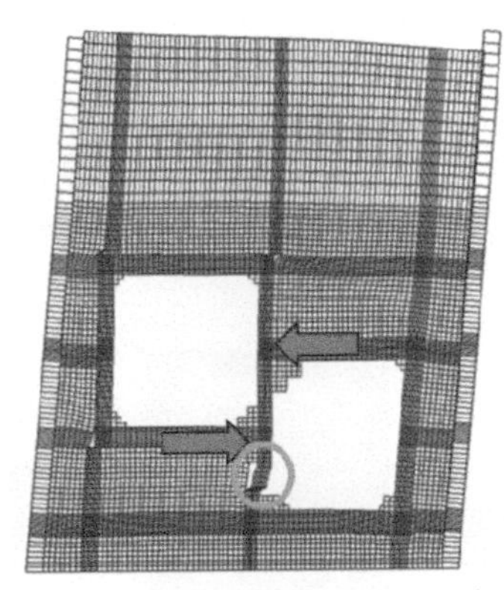

a) Uneven Double-box Tunnel

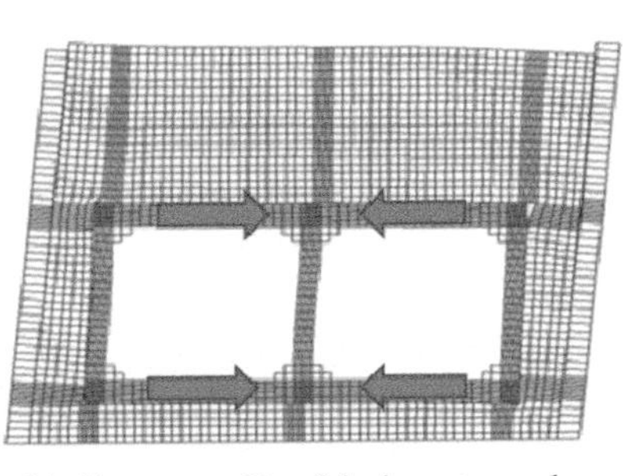

b) Common Double box tunnel

Fig. 11.7 Model deformation diagram

Especially, the large localized strain occurred at the haunch and joint part of each box before loading. These phenomena are the result of thermal stress from the initial concrete installation, cracks under the long term due to dry shrinkage, and concrete creep and delayed failure of concrete due to sustained earth pressure. In Section 1, although the large strain occurred at the joint part of each box before loading, there is almost no change in strain condition at the joint part after loading. On the other hand, Section 2 shows signs of shear failure caused by loading in the lower slab of left-side box and the lower of interior wall.

Considering the above, the different shape of the cross-section result in different condition of internal forces before loading, and that the potential for subsequent failure modes and shear failure also differs.

11.4 Comparison with Common Double Box Tunnel Deformation

Figure 11.7 shows diagrams of model deformation of uneven double-box tunnel and common horizontally-arranged double box tunnel by shear displacement. A common double box tunnel is deformed uniformly as the upper and lower slabs receive compressive forces. On the other hand, an uneven double-box tunnel is disformed ununiformly and the deformation is concentrated on the part of the structure, as the upper and lower slabs push against the interior wall as shown in Fig. 11.7.

As just described, the respective forces of each part of the tunnels are transferred through different pathways for each type of tunnel. Thus, the location and mechanism of failure for uneven double-box tunnels are quite different from common double box tunnel.

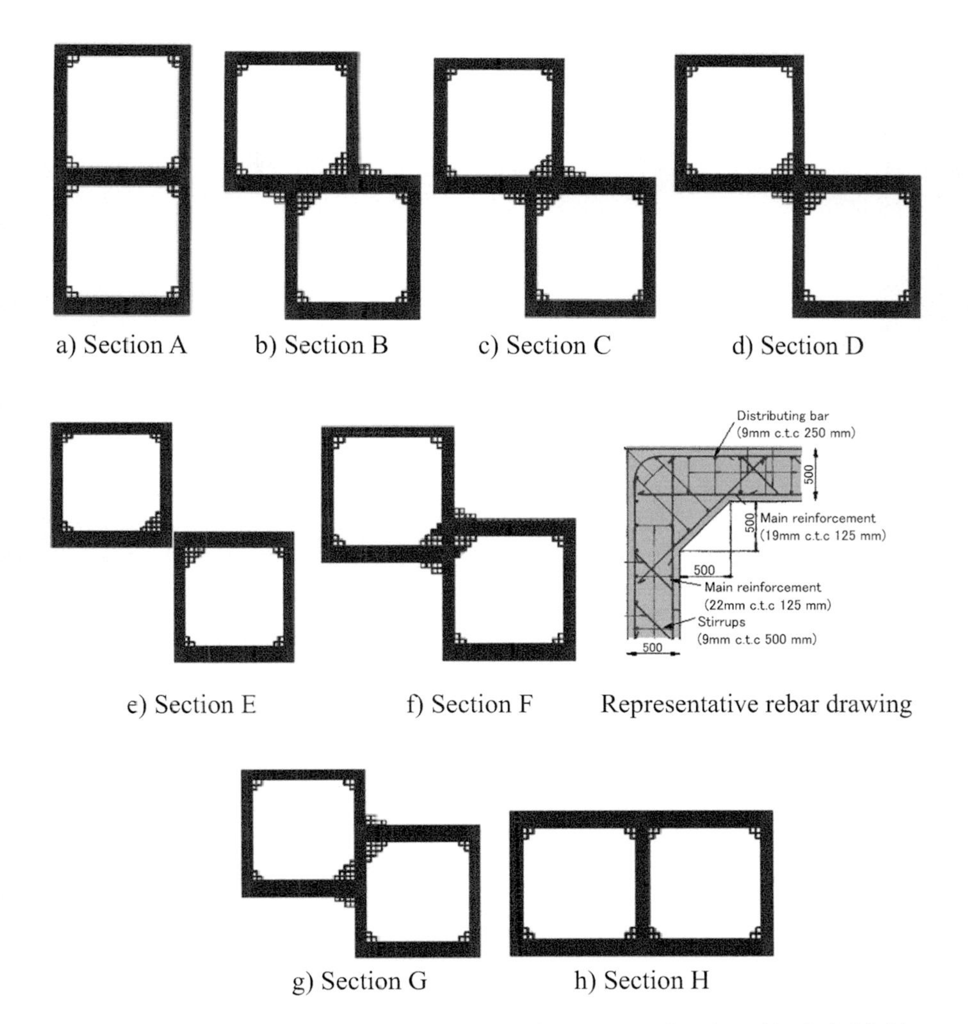

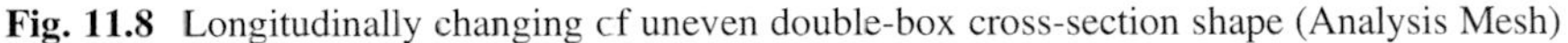

Fig. 11.8 Longitudinally changing of uneven double-box cross-section shape (Analysis Mesh)

11.5 Effects of Uneven Double-Box Tunnel Section Forms

(1) Analytical conditions

The fundamental investigation on the effects of uneven double-box tunnel cross-section shapes against the structural damage due to shear displacement was conducted by a numerical analysis (DuCOM-COM3). The analytic models of uneven double-box tunnel section with different vertical alignments were considered based on the existing subway tunnel sections, and eight different cross-section models were considered. These sections are distinguished with a letter from A to H as shown in Fig. 11.8.

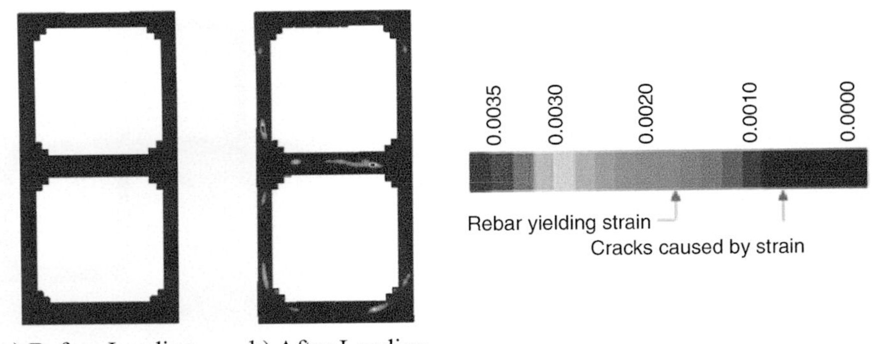

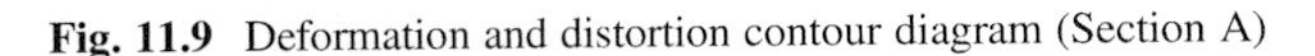

Fig. 11.9 Deformation and distortion contour diagram (Section A)

The distance from the tunnel centre position to the ground was assumed as 8.00 m. However, the earth covering of Section H was too deep to compare with other sections, so that the depth of overburden of Section H was set equal to that of Section G. The other conditions for model layouts were the same as shown in Fig. 11.4. The Environmental conditions were also the same as shown in Table 11.2.

The ratio of reinforcement was assumed to be 0.5% uniformly, and the main reinforcements with 22 mm in diameter were installed based on the rebar arrangement drawing. Moreover, the stirrups with 9 mm in diameter were assumed at an interval of roughly 500 mm.

In this investigation, the elapsed years were set at 5 years from concrete placement to appropriately verify the seismic structural damages, in light of the results of preliminary analysis in which the elapsed years were set at 20 years and the large localized strain occurred at haunches and joints of each box before loading. The deformation of structure became larger as the rigidity of joints became smaller, but localized share failure was considered to be less likely to occur in general. The purpose of this investigation was to verify the capacity to maintain internal space of tunnel.

For earthquake ground behaviour, 3% shear displacement to the height of model was loaded. The loading process was the same as the preliminary analysis.

(2) Results of analysis

Figures 11.9, 11.10, 11.11, 11.12, 11.13, 11.14, 11.15, 11.16 show the results of analysis of eight cross-sections. The figures show the model deformation (with 2x magnification of deformation) and the maximum principal strain distribution (tension made positive) before and after shear displacement was given.

In Section A, the strain was the lowest of all sections for both before and after loading. It was supposed that internal space of tunnel could be maintained.

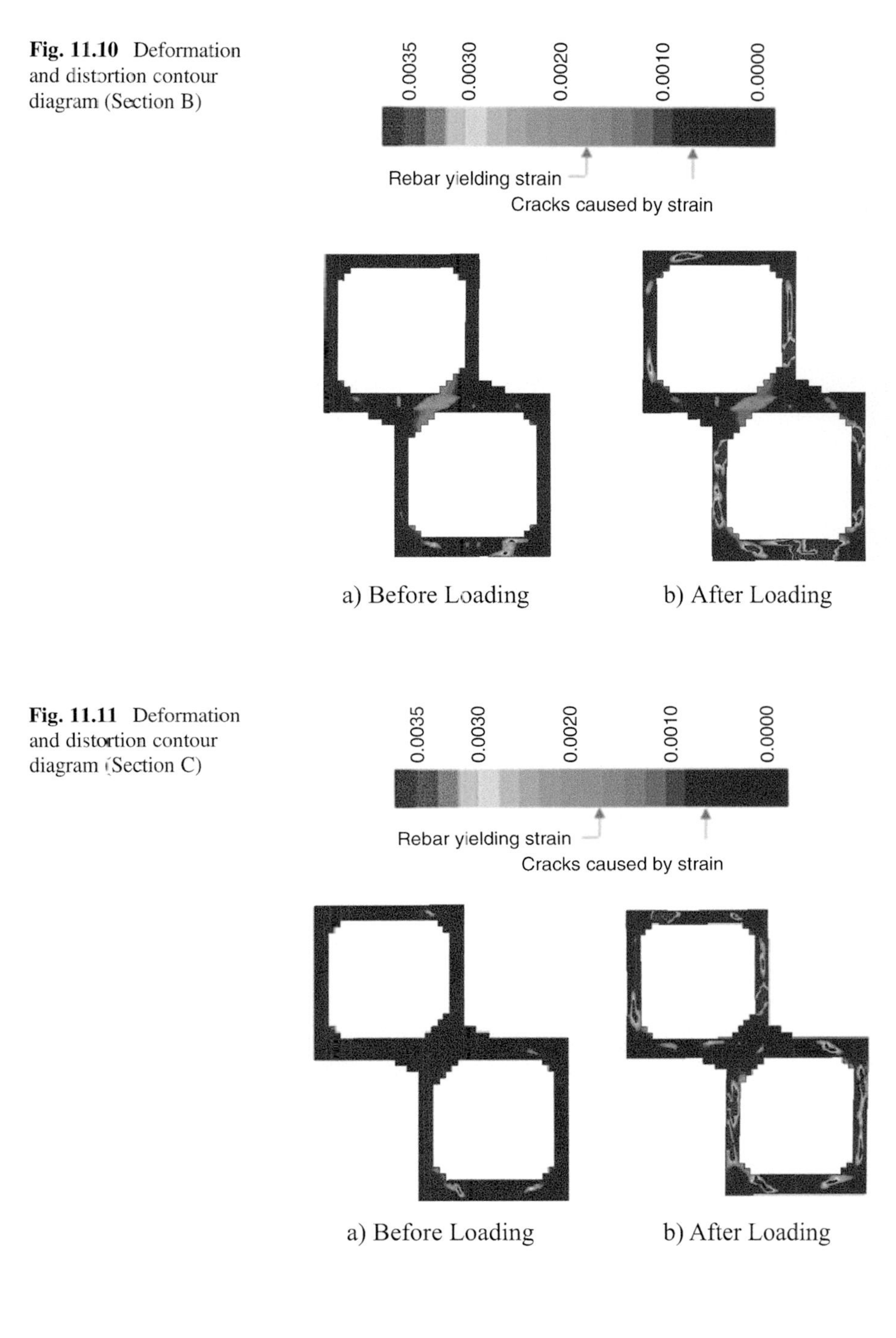

Fig. 11.10 Deformation and distortion contour diagram (Section B)

Fig. 11.11 Deformation and distortion contour diagram (Section C)

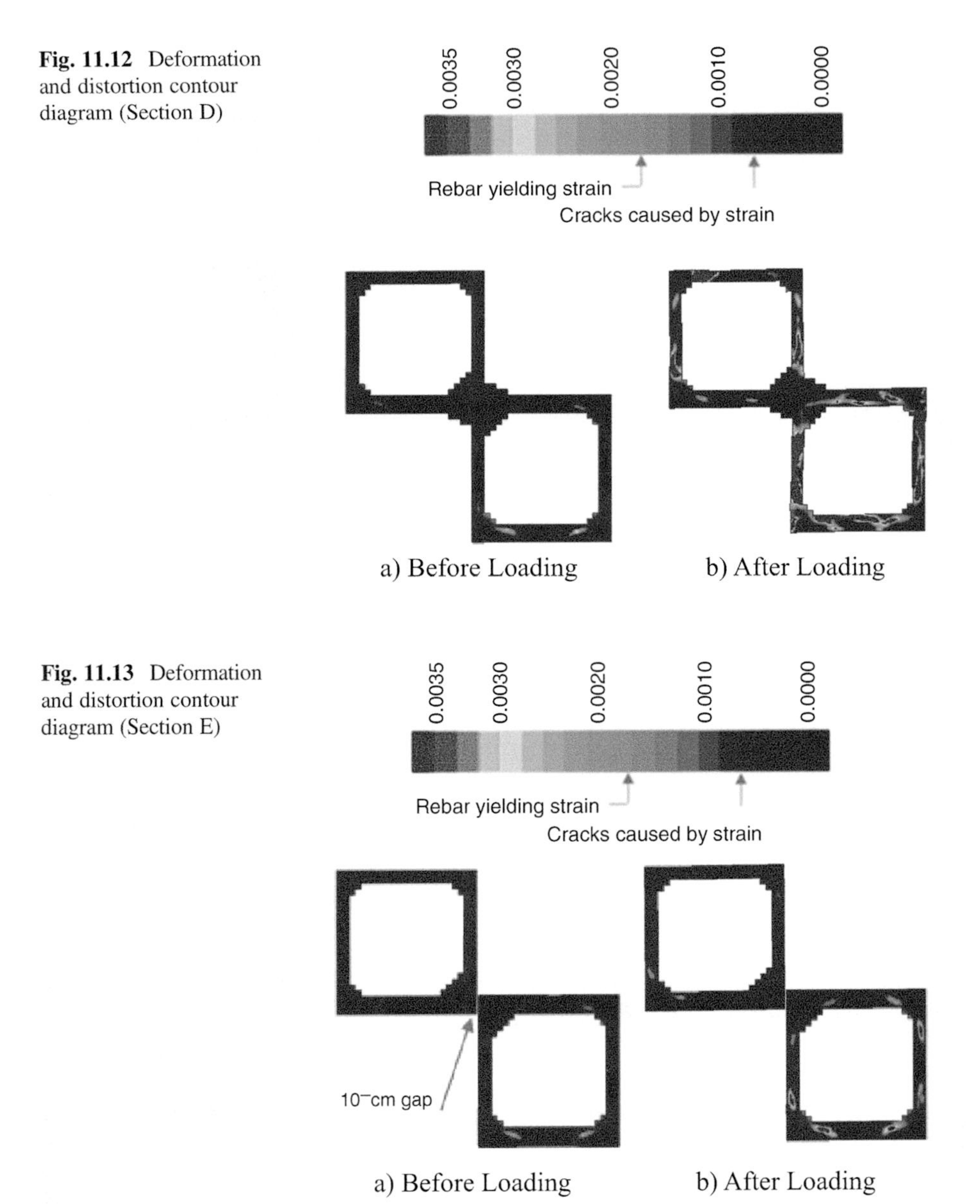

Fig. 11.12 Deformation and distortion contour diagram (Section D)

Fig. 11.13 Deformation and distortion contour diagram (Section E)

In Section B, the strain occurred in which rebar yields at the haunch and joints of two boxes before loading, and Fig. 11.10 shows signs of compression damage of concrete caused by loading in the bottom slab of lower box. However, the strain on the sidewalls was relatively lower and it was a sign of bending failure. Thus, this section can maintain the internal space even if given damages by shear displacement.

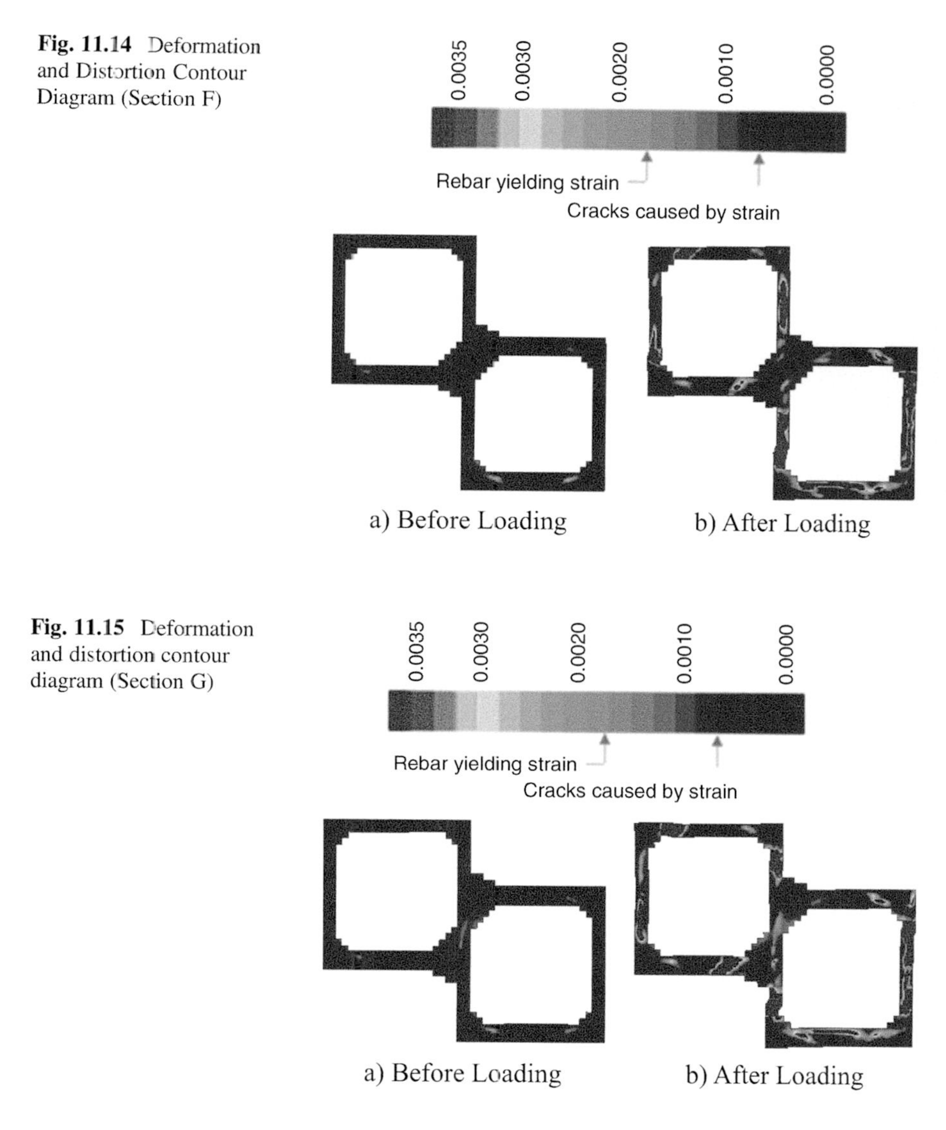

Fig. 11.14 Deformation and Distortion Contour Diagram (Section F)

Fig. 11.15 Deformation and distortion contour diagram (Section G)

In Section C, the relatively large strain occurred at haunch before loading, however, no damage of the structure was observed. Figure 11.11 shows signs of compression damage of concrete caused by loading almost the same as Fig. 11.10, and there were presumed to be bending failure. Thus, this section also can maintain the internal space.

In Section D, large strain hardly occurred at the haunch of lower box before loading. Figure 11.12 shows that compression damage of concrete caused by loading occurred on the sidewalls of lower box. However, there were presumed to be bending failure. Thus, this section also can maintain the internal space.

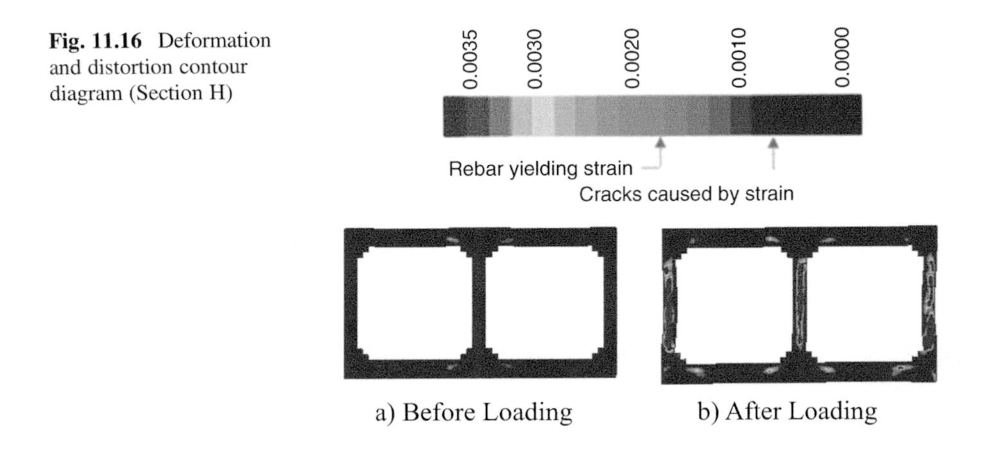

Fig. 11.16 Deformation and distortion contour diagram (Section H)

In Section E, this section was not uneven double-box tunnel, because the two boxes were independent of each other as shown in Fig. 11.13. There were two single-boxes in close proximity (offset distance is only 10 cm). The strain was not large for both before and after loading. It was supposed that there was no damage due to shear displacement.

In Section F, large strain hardly occurred at the haunch of lower slab of right-side box before loading, and no damage of the structure was observed. Figure 11.14 shows that compression damage of concrete caused by loading occurred on the right-side walls of right-side box and interior wall. However, these were assumed to be bending failure. Thus, this section also can maintain the internal space, and the transfer of axial force can be maintained as long as there is no potential for shear failure.

In Section G, large strain hardly occurred at the haunch before loading, and no damage of the structure was observed. However, Fig. 11.15 shows a sign of shear failure caused by loading on the lower part of interior wall. Moreover, compression damage of concrete also occurred at other parts, which were presumed to be bending failure. This section shows the most severe damage in all sections. Thus, it was supposed that the capacity to maintain the internal space would be degraded in this section. In addition, the vertical slipping of double-box tunnel section was presumed to be more severe condition compared with the horizontal slipping one.

In Section H, this was a common horizontally-arranged double-box tunnel. The strain was not large before loading. Figure 11.16 shows that compression damage of concrete caused by loading occurred on the sidewalls and centre pillars. However, these were presumed to be bending failure. Thus, this section also can maintain the internal space.

The angle of shear displacement to be given as the ground action was found to be proportional to the average angle of tunnel shear deformation in most cases. However, in Section D, F and G, the average angle of tunnel shear deformation became larger due to accumulation while loading to tilt toward the left.

11.6 Conclusion

In this study, failure mode of earthquakes in uneven double-box tunnel cross-sections due to the effects of tunnel shapes was examined by comparing responses calculated with non-liner static analysis as fundamental research. Specifically, the results showed that shear displacement of ground up to 3% could potentially cause tunnel sections the loss of capacity to maintain the internal space due to shear failure of RC structures. The results were described as follows:

The large localized strain occurred and different shapes of the cross-sections resulted in different conditions of internal forces before loading.

A horizontally-arranged double-box tunnel deforms uniformly. However, for uneven double-box tunnels, the upper and lower slabs pushed against the interior wall, which likely caused large localized deformation.

The analysis to investigate the effect of uneven double-box tunnel sections showed that compression damage of concrete caused by loading occurred in all sections except Section G. There were presumed to be bending failure and could also maintain the internal space.

The vertical slipping of uneven double-box tunnel section was presumed to be more severe condition against earthquake ground behaviours comparing with the horizontal slipping one. In this study, Section G showed a sign of shear failure caused by loading on the lower part of interior wall.

The experiments to evaluate the results of analysis on uneven double-box tunnel sections were planned based on this fundamental research. Moreover, it was intended that more practical, highly accurate ways to evaluate actual structures was found through continuous investigation.

Acknowledgments For our numerical analysis, the quasi-equilibrium thermodynamic analysis system DuCOM-COM3 was used. It was developed using JSPS Grant-in-Aid for Scientific Research (S) (23226011).

References

Akira T (1992) Research on how seismic loads act in response displacement method. J JSCE 441 (I-18):157–166

An X, Koichi M (1997) Failure analysis of underground RC frame subjected to seismic. In: Proceedings of JSCE, No.571/V-36, pp 251–268

Hiroshi D, Hatsuku T, Tsuyoshi I, Muneo H, Takemine Y, Naoto O, Itami H (2008) Hull 3D seismic response analysis of underground ramp tunnel structure using large-scale numerical computation. In: Proceedings of the 14th world conference on earthquake engineering

Hiroshi D, Yoshihiro T, Naruhiko K, Shogo O, Takemasa Y, Muneo H, Toru K (2009) Experimental study on design method for the connection between steel segmental linings and concrete bodies. J JSCE A 65(3):718–737

Kawanishi T, Kiyono J, Nishiyama S (2012) A study on the accuracy of a static analysis method for cut and cover tunnels. In: Proceedings of the 15th world conference on earthquake engineering

Kazuhiko K (1994) Earthquake-resistant design of underground structures. Kajima Institute Publishing

Koichi M, Hajime O, Pimanmas A (2003) Non-linear mechanics of RC. SPON Press
Koichi M, Tetsuya I, Toshiharu K (2009) Multi-scale modeling of structural concrete. Taylor & Francis
Nam SH, Song HW, Byun KJ, Maekawa K (2006) Seismic analysis of underground reinforced concrete structures considering elasto-plastic interface element with thickness. Eng Struct 28:1122–1131
Naoto O, Hirokoshi K, Takemine Y, Tachibana K, Akiba H (2004) Dynamic behaviour of underground Motoway junction due to large earthquake. Proceedings of the13th world conference on earthquake engineering
Okhovat MR, Maekawa K (2009) Seismic damage control of underground structures associated with reduced stiffness of soil foundation. In: Proceedings of the fifth congress on forensic engineering 362(35)
Railway Technical Research Institute (2012) Design standard for railway structures and commentary (seismic design). Maruzen Publishing
Satoshi T, Koichi M, Kazuhiko K (2007) Three-dimensional cyclic behaviour simulation of RC columns under combined flexural moment and torsion coupled with axial and shear forces. J Adv Concr Technol 5(3):409–421

Chapter 12
Failure Probability of Regular and Irregular RC Frame Structures

M. Kosič, M. Dolšek, and P. Fajfar

Abstract In the paper, the Pushover-based Risk Assessment (PRA) method is summarized and applied to the estimation of the "failure" probability of two reinforced concrete frame buildings. The first building is a modified version of the well-known SPEAR building, designed according to Eurocode 8. Although the building is asymmetrical, the influence of torsion is moderate, so the building can be considered as a representative of regular buildings. The second building has the same structural layout, but has infill walls included only in the upper stories, which induce irregularity along the height of the building. The comparison of the "failure" probabilities obtained for the two examples indicates a lower, although still significant, seismic risk for the regular code-conforming variant of the building (0.75% over the lifetime of the structure). A three to five times larger "failure" probability is obtained for the irregular variant of the building, for which a soft first storey effect is predicted.

Keywords Seismic risk · Failure probability · Pushover analysis · N2 method · Regular and irregular structures · Reinforced concrete frame

12.1 Introduction

Due to large uncertainties involved in the simulation of the seismic response of structures, an important task of the performance-based earthquake engineering is the development of new design and assessment methods that have a probabilistic basis and try to properly take into account the inherent uncertainty. Despite of their theoretical advantages, probabilistic methods have not yet been implemented in structural design practice, with the exception of nuclear power plant structures. In order to facilitate a gradual introduction of probabilistic considerations into practice,

M. Kosič (✉) · M. Dolšek · P. Fajfar
Faculty of Civil and Geodetic Engineering, University of Ljubljana, Ljubljana, Slovenia
e-mail: mirko.kosic@fgg.uni-lj.si; matjaz.dolsek@fgg.uni-lj.si; peter.fajfar@fgg.uni-lj.si

D. Köber et al. (eds.), *Seismic Behaviour and Design of Irregular and Complex Civil Structures III*, Geotechnical, Geological and Earthquake Engineering 48,
https://doi.org/10.1007/978-3-030-33532-8_12

simplified practice-oriented approaches for the determination of seismic risk are needed. One of the methods that allow for practice-oriented estimation of "failure" probability (i.e. the probability of exceeding the near collapse limit state) is the Pushover-based Risk Assessment method (PRA) (Dolšek and Fajfar 2007; Fajfar and Dolšek 2012; Kosič et al. 2017). The PRA method combines the probabilistic seismic assessment method in closed form, developed by Cornell and co-authors (Cornell et al. 2002), upon which the SAC-FEMA guidelines are based (FEMA 350 FEMA 2000), and the pushover-based N2 method (Fajfar 2000). Recently proposed default values for the dispersion of the capacity at "failure" (Kosič et al. 2014, 2016) can be used.

In the paper, the pushover-based risk assessment (PRA) method is summarized and applied for the estimation of the "failure" probability of a regular and irregular variant of a reinforced concrete (RC) frame building.

12.2 Summary of the Pushover-Based Risk Assessment (PRA) Method

The "failure" probability of building structures, i.e. the probability of exceeding the near collapse limit state (NC), which is assumed to be related to a complete economic failure of a structure, can be estimated as (Fajfar and Dolšek 2012; Cornell 1996)

$$P_{NC} = \exp\left[0.5\ k^2 \beta_{NC}^2\right] H(S_{a,NC}) = \exp\left[0.5\ k^2 \beta_{NC}^2\right]\ k_0 S_{a,NC}^{-k}, \qquad (12.1)$$

where $S_{a,\ NC}$ is the median NC limit-state spectral acceleration (shortly called the capacity at "failure") and β_{NC} is the logarithmic standard deviation of $S_{a,\ NC}$ due to record-to-record variability and modelling uncertainty. The parameters k and k_0 are related to the hazard curve which is assumed as linear in the logarithmic domain ($H(S_a) = k_0\ S_a^{-k}$). The capacity at "failure" $S_{a,\ NC}$ is estimated using the pushover-based N2 method (Fajfar 2000), whereas predetermined dispersion values are used for β_{NC}. In the original formulation of the method (Fajfar and Dolšek 2012), the intensity measure used for risk assessment was peak ground acceleration (*PGA*). In this study, the spectral acceleration at the period of the equivalent SDOF model $S_a(T^*)$ is used instead, like in (Kosič et al. 2017). In the following subsections the determination of the parameters in Eq. (12.1) is presented.

12.2.1 *Determination of Seismic Hazard Parameters* k *and* k_0

The procedure used for the estimation of the parameters k and k_0 depends on the available seismological data. Basically, three options are available:

(a) If a hazard curve for $S_a(T^*)$ is available at the location of the building, k and k_0 are estimated by fitting the hazard curve with a linear function in logarithmic domain over a range of intensities, e.g. [0.25 1.25]·$S_{a,\ NC}$ (Dolšek 2012).
(b) If only hazard maps are available for two return periods R_1 and R_2, the following equations are used

$$k = \ln\left(R_1/R_2\right)/\ln\left(S_{a,1}/S_{a,2}\right) \text{ and } k_0 = 1/\left(R_1 \cdot S_{a,1}^{-k}\right) = 1/\left(R_2 \cdot S_{a,2}^{-k}\right), \quad (12.2)$$

where $S_{a,\ 1}$ and $S_{a,\ 2}$ are the values of the intensity measure for the selected return periods.

(c) If only one value of the intensity measure corresponding to a specific return period R ($S_{a,\ R}$) is available, the parameter k has to be assumed, as proposed in (Fajfar and Dolšek 2012), and k_0 is estimated as follows

$$k_0 = 1/\left(R \cdot S_{a,R}^{-k}\right). \quad (12.3)$$

Considering Eq. (12.3), the Eq. (12.1) can be written in the form

$$P_{NC} = \exp\left[0.5\ k^2 \beta_{NC}^2\right] \frac{1}{R} \left(\frac{S_{a,NC}}{S_{a,R}}\right)^{-k}, \quad (12.4)$$

where explicit calculation of k_0 is no longer required. The Eq. (12.4) can be used equivalently to Eq. (12.1) in the case of the procedures (b) and (c).

12.2.2 Estimation of the Capacity at "Failure"

The capacity at "failure" $S_{a,\ NC}$ can be estimated using the N2 method. First, the pushover analysis with an invariant distribution of lateral forces is performed ($P_i = m_i \Phi_i$). The pushover curve is idealized with a bilinear relationship, and the relations of the N2 method (Fajfar 2000) are used for the calculation of the MDOF-SDOF transformation factor Γ and the characteristics of the equivalent SDOF model, i.e. the mass m^*, the yield displacement D_y^*, the yield force F_y^*, the period T^*, and the yield spectral acceleration S_{ay}:

$$\Gamma = \frac{m^*}{\sum m_i \Phi_i^2}, \quad m^* = \sum m_i \Phi_i, \quad D_y^* = \frac{D_y}{\Gamma}, \quad F_y^* = \frac{F_y}{\Gamma}, \quad T^* = 2\pi\sqrt{\frac{m^* D_y^*}{F_y^*}}, \quad S_{ay} = \frac{F_y^*}{m^*}, \tag{12.5}$$

where D_y and F_y are the yield displacement and yield force of the MDOF system, respectively. Φ_i is the component of the assumed shape vector in ith storey (typically the first mode shape), and m_i is the mass of ith storey.

Next, the displacement at "failure" D_{NC} has to be determined. A widely accepted definition of the "failure" (near collapse (NC) limit state) at the level of the structure is still not available. Two possible definitions are: (i) the NC of the structure corresponds to the NC limit state of the most critical vertical element; (ii) the NC of the structure corresponds to 80% strength at the softening branch of the pushover curve. In this paper, the first definition is used. For a building with important influence of higher modes in the plan and/or elevation an extended formulation of the N2 method (Kreslin and Fajfar 2012) can be used for calculation of D_{NC}.

The ductility at "failure" μ_{NC} is calculated as the ratio of the displacement at the "failure" D_{NC} and the yield displacement of the structure D_y. Finally, the capacity at "failure" $S_{a,\,NC}$ is calculated as the product of the yield spectral acceleration S_{ay} and the reduction factor due to ductility R_μ, as follows

$$S_{a,NC} = S_{ay} \cdot R_\mu(\mu_{NC}, T^*); \quad R_\mu = \begin{cases} (\mu_{NC} - 1)\dfrac{T^*}{T_C} + 1 & T^* \leq T_C \\ \mu_{NC} & T^* > T_C \end{cases}, \tag{12.6}$$

where T_C is the characteristic period of the ground motion (T_C in Eurocode 8 CEN 2004).

12.2.3 Dispersion of the Capacity at "Failure"

Extensive studies have been made by the authors in order to determine typical dispersions of the capacity at "failure" for reinforced concrete (RC) building structures using $S_a(T^*)$ as the intensity measure (Kosič et al. 2014, 2016). Based on these studies, in a simplified approach it may be reasonable to assume $\beta_{NC} = 0.5$ as an appropriate estimate for RC building structures. This value takes into account record-to-record and modelling uncertainty.

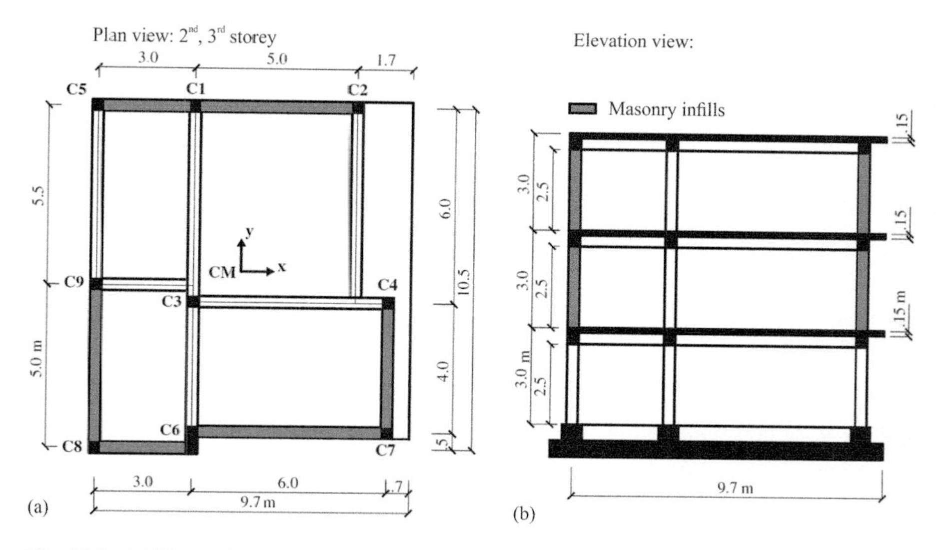

Fig. 12.1 (**a**) Plan and (**b**) elevation view of the code-conforming SPEAR building with schematic presentation of the disposition of masonry infill walls in the case of the infilled variant of the building (infill walls are included only in the 2nd and 3rd storey)

12.3 Case Study: Estimation of the "Failure" Probability for a Regular and Irregular Variant of a RC Frame Building

In this chapter, the PRA method is applied for a practice-oriented estimation of "failure" probability, i.e. the probability of exceeding the near collapse (NC) limit state, of a regular and irregular variant of a reinforced concrete (RC) frame building. The results are used to assess the influence of the irregularity on the seismic performance of the examined building.

12.3.1 The Investigated Structures and Input Parameters

The plan and elevation view of the investigated building are presented in Fig. 12.1. The first variant of the building, i.e. bare frame variant, is a modified version of the well-known SPEAR building, designed according to Eurocode 8 DCH provisions (Rozman and Fajfar 2009). Although the building is asymmetrical, the influence of torsion is moderate, so the building can be considered as a representative of regular buildings. The second variant of the building, i.e. the infilled variant, has the same layout and reinforcement as the first building, but has masonry infill walls included only in the upper stories, which induce irregularity along the height of the building (Fig. 12.1b). The properties of the masonry infill walls are taken after a previous study (Celarec and Dolšek 2013). Infill walls are assumed to be 25 cm thick with a cracking strength and the Young's modulus of 0.36 MPa and 1.50 GPa, respectively.

It is assumed that both variants of building are located in a moderate seismic region, for which the only available seismological data is the design peak ground acceleration on rock $PGA = 0.25\ g$ (475-year return period). The soil characterization according to Eurocode 8 (CEN 2004) is soil type C ($S = 1.15$).

12.3.2 Structural Modelling

The structural models are created using the PBEE toolbox (Dolšek 2010), which allows a rapid generation of nonlinear models for OpenSees (Open system for earthquake engineering simulation (OpenSees) 2017), definition of different type of analyses, and advance post-processing of the results. The modelling of the buildings is performed in accordance with the principles described in previous studies by the authors (Kosič et al. 2014, 2016, 2017). As an exception, the ultimate rotations in, both, the columns and beams at the near collapse (NC) limit state, which corresponds to 80% of the maximum moment in the post-capping region, are estimated using the EC8-3 (CEN 2005) formula for secondary elements ($\gamma_{el} = 1.0$, representing mean estimates). The masonry infill walls are modelled by means of two diagonal struts, which can carry only compression loads. The force-displacement relationship of the diagonal struts is defined according to the procedure used in (Celarec and Dolšek 2013; Celarec et al. 2012). In this study, the influence of potential shear failure due to the effect of masonry infill walls is not taken into account. Extensive description of the structural modelling implemented in PBEE toolbox can be found in (Dolšek 2010).

12.3.3 Pushover Analysis, the Parameters of the Equivalent SDOF Systems and Consideration of the Influence of Higher-Mode Effects

The first step of the PRA method is the pushover analysis for the two variants of the building. The structures are analysed independently in both principal directions (X and Y) and for both signs of seismic loading. The results for the most critical sign of seismic loading are considered (i.e. +X and + Y direction). The obtained pushover curves, plastic mechanisms and corresponding damage in the plastic hinges at near collapse (NC) limit state are presented in Figs. 12.2 and 12.3.

The pushover curves are idealized with a bilinear relationship considering an equal-area principle up to the displacement at maximum force (see Fig. 12.2). The relations of Eq. (12.5) are used for calculation of the MDOF-SDOF transformation factor Γ and the characteristics of the equivalent SDOF models. A summary of the

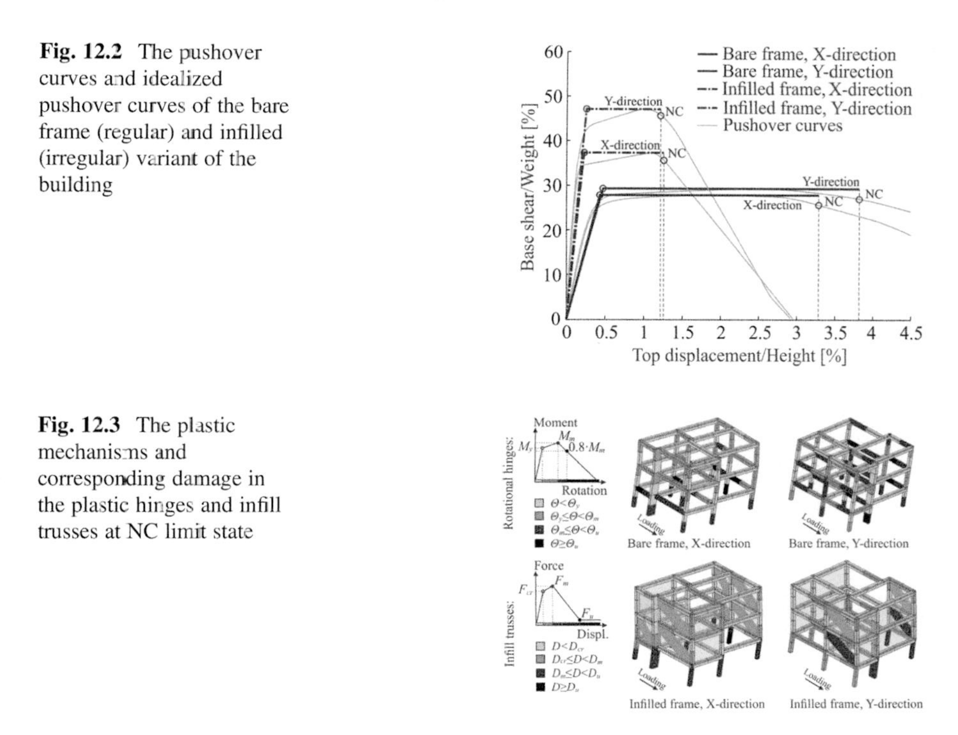

Fig. 12.2 The pushover curves and idealized pushover curves of the bare frame (regular) and infilled (irregular) variant of the building

Fig. 12.3 The plastic mechanisms and corresponding damage in the plastic hinges and infill trusses at NC limit state

Table 12.1 The characteristics of the idealized SDOF systems for the bare frame (regular) and infilled (irregular) variant of the building

Structure:	F_y^* (kN)	D_y^* (m)	D_{NC}^* (m)	m^* (t)	T^* (s)	Γ	F_y/W	S_{ay} (g)
Bare frame, X	627	0.03	0.23	195.8	0.61	1.25	0.28	0.33
Infilled frame, X	998	0.02	0.11	271.8	0.45	1.05	0.37	0.37
Bare frame, Y	651	0.03	0.26	185.8	0.60	1.27	0.29	0.36
Infilled frame, Y	1165	0.02	0.09	247.0	0.40	1.14	0.47	0.48

characteristics of the idealized SDOF systems for the bare frame (regular) and the infilled (irregular) variant of the building is presented in Table 12.1.

Due to asymmetry in the plan of the building, the influence of the higher-mode effects in the plan (the effects of inelastic torsion) is taken into account in the estimation of the displacement at "failure" D_{NC}, whereas the influence of the higher mode effect in the elevation of the building is considered negligible.

According to Extended N2 method (Kreslin and Fajfar 2012), the influence of higher mode effects in the plan of the building is considered by multiplying the displacements by correction factors C_T, which depend on the location in the plan. A linear modal response spectrum analysis with consideration of the Eurocode 8 elastic spectrum (CEN 2004) for soil type C and a peak ground acceleration of 0.29 g (1.15·0.25 g) is performed in addition to the pushover analysis. The correction factors C_T are computed as a ratio between normalized roof displacements

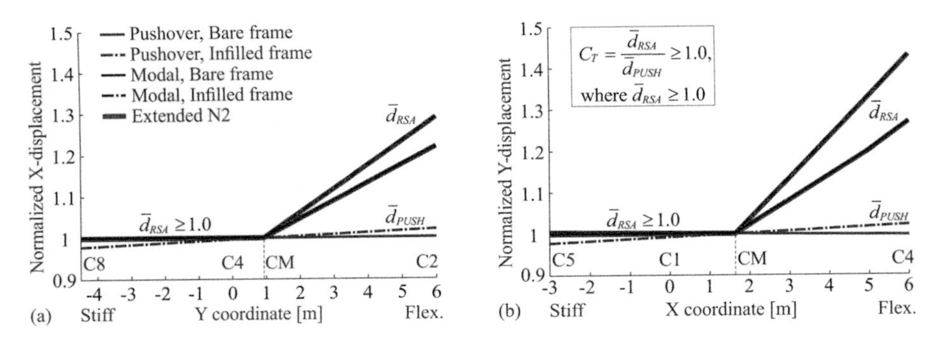

Fig. 12.4 Roof displacements (normalized to top displacement at the centre of mass (CM)) determined by elastic modal analysis and pushover analyses at NC limit state for both variants of the building in (**a**) X and (**b**) Y direction

(normalized to roof displacement at the centre of mass) from the modal response spectrum analysis ($\bar{d}_{RSA}$) and from the pushover analysis ($\bar{d}_{PUSH}$), as presented in Fig. 12.4.

Considering the influence of torsion, the critical elements in X and Y direction are columns C1 and C4, respectively, which first attain the NC limit state (see Figs. 12.1 and 12.3). An exception is the bare frame variant of the building, for which the critical element in Y direction is the column C6. The obtained displacements at "failure" for the equivalent SDOF models are presented in Table 12.1.

12.3.4 Estimation of the Capacity at "Failure" with the N2 Method

The second step of the PRA method is the estimation of the capacity at "failure" with N2 method. For this purpose, the ductility at "failure" of the structures μ_{NC} is computed first as described in Sect. 12.2.2. The ductilities μ_{NC} amount to 7.6 and 8.1 (X and Y direction, respectively) in the case of the bare frame variant of the building, and to 5.6 and 4.8 (X and Y direction, respectively) in the case of the infilled variant of the building. Next, the reduction factor due to energy dissipation capacity R_μ is calculated using Eq. (12.6). In the case of the bare frame variant of the building ($T^* \geq T_C = 0.6$ s for soil type C according to (CEN 2004), the equal displacement rule applies in both directions ($R_\mu = \mu_{NC}$). On the contrary, the calculated reduction factors R_μ in the case for the infilled variant of the building ($T^* < T_C = 0.6$ s) are lower than the corresponding ductilities μ_{NC} (i.e. R_μ equals to 4.4 and 3.5 in X and Y direction, respectively). Finally, the capacity at "failure" $S_{a,NC}$ is calculated as the product of the reduction factor R_μ and the yield spectral acceleration S_{ay} (Eq. 12.6). The computed capacities at "failure" are presented in Table 12.2. Note that the process of calculation of $S_{a,NC}$ can be visualized in the acceleration-displacement (AD) format (see Fig. 12.5).

Table 12.2 Summary of the "failure" probabilities obtained for the bare frame (regular) and infilled (irregular) variant of the building, and of the input data used in computations

Structure:	$S_{a,\ NC}$ (g)	β_{NC}	k_0	k	$H(S_{a,\ NC})$	P_{NC}	P_{NC}^{50} (%)
Bare frame, X	2.48	0.50	$7.44 \cdot 10^{-4}$	3.0	$0.49 \cdot 10^{-4}$	$1.50 \cdot 10^{-4}$	0.75
Infilled frame, X	1.66	0.50	$7.82 \cdot 10^{-4}$	3.0	$1.70 \cdot 10^{-4}$	$5.24 \cdot 10^{-4}$	2.59
Bare frame, Y	2.89	0.50	$7.82 \cdot 10^{-4}$	3.0	$0.32 \cdot 10^{-4}$	$0.99 \cdot 10^{-4}$	0.50
Infilled frame, Y	1.69	0.50	$7.82 \cdot 10^{-4}$	3.0	$1.62 \cdot 10^{-4}$	$4.98 \cdot 10^{-4}$	2.46

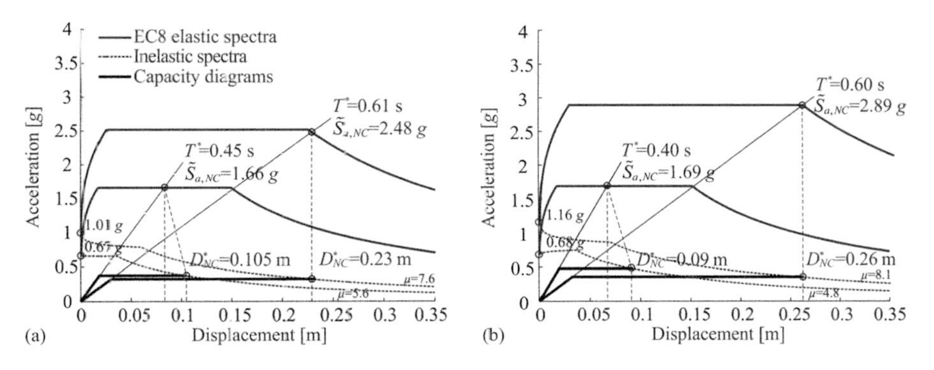

Fig. 12.5 Elastic and inelastic demand spectra corresponding to NC limit state, and capacity diagrams of the equivalent SDOF models in acceleration-displacement (AD) format for both variants of the building in (**a**) X and (**b**) Y direction

12.3.5 *Estimation of "Failure" Probabilities*

The final step of the PRA method is the calculation of the "failure" probabilities P_{NC} using Eq. (12.1). Due to lack of seismic hazard data for the location of the building (only the design PGA_{475} is available), the procedure (c) from Sect. 12.2.1 is used. A typical slope of the seismic hazard curve $k = 3$ is assumed, and the parameter k_0 is calculated using Eq. (12.3) for each building, considering $R = 475$ years and $S_{a,\ 475}$ is calculated from the Eurocode's elastic acceleration spectrum (Type 1, soil type C) for $PGA_{475} = 0.29\ g$ (soil factor $S = 1.15$ is included). Because only PGA_{475} is available, a fixed shape of the ground-motion spectrum has to be assumed for estimation of $S_{a,\ 475}$. For both variants of the building, the dispersion of the capacity at "failure" $\beta_{NC} = 0.50$ is used. The computed "failure" probabilities are presented in Table 12.2. The "failure" probabilities in 50 years are computed as $P_{NC}^{50} = 1 - (1 - P_{NC})^{50}$. Equivalently, the same results are obtained using Eq. (12.4), which does not require the calculation of the parameter k_0.

12.3.6 Comparison of the "Failure" Probabilities Obtained for the Two Variants of the Building and Discussion of Results

In this section, the results obtained for the two variants of the building are compared and used for an assessment of the influence of the irregular distribution of infill walls over the height of the building.

A comparison of the pushover curves, presented in Fig. 12.2, reveals that the irregular distribution of infill walls over the height of the building increases the strength and the stiffness of the structure. However, the deformation capacity of the building is significantly reduced. This is due to the presence of relatively strong and stiff infill walls in the upper storeys, which induce a concentration of damage in the bottom storey, producing a soft storey effect. As a result, the capacity at "failure" $S_{a,NC}$ of the infilled variant of the building is decreased by 30% to 40% compared to the bare-frame variant.

It is interesting to observe that the distribution of infills increases the eccentricity of the building (especially in Y direction). In the case of the infilled variant of the building, larger normalized roof displacements $\bar{x}_{RSA}$ are obtained due to different lengths of infill walls at the opposite sides of the plan.

The "failure" probabilities of the bare frame (regular) variant of the building are in the order of $1{\cdot}10^{-4}$ and $1.5{\cdot}10^{-4}$ per year (about 0.5% and 0.75% in 50 years). The computed "failure" probabilities are in line with the expected values for code-conforming structures obtained from previous studies, i.e. between 0.25% and 1% in 50 years (Fajfar and Dolšek 2012; Kosič et al. 2014, 2016, 2017). A similar value (1% in 50 years) is also assumed in FEMA P-1050-1 (BSSC, Building Seismic Safety Council 2015) as the target probability for the design of buildings of ordinary importance, however, it is based on the actual collapse and not on the near collapse limit state. On the other hand, in the case of the infilled variant of the building, which has an irregular distribution of infills over the height, the estimated "failure" probabilities are three to five times larger (e.g. $5{\cdot}10^{-4}$ per year or 2.5% in 50 years). The difference between the "failure" probabilities of the regular and irregular variant of the building is considerable. For example, it is larger than the ratio between the "failure" probabilities of ordinary (importance class II) and very important structures (importance class IV, $\gamma_i = 1.4$), which can be obtained from Eq. (12.1). Considering that same hazard parameters k_0 and k ($k = 3$), the same overstrength ($S_{a,\ NC,\ IV} / S_{ad,\ IV} \approx S_{a,\ NC,\ II} / S_{ad,\ II}$) and the same dispersion β_{NC} apply for both types of structures, it can be shown than the ratio between the "failure" probabilities of ordinary and very important structures roughly equals to $P_{NC,II}/P_{NC,IV} = (S_{a,NC,IV}/S_{a,NC,II})^k \approx (S_{ad,IV}/S_{ad,II})^k = \gamma_i^k = 2.7$. This indicates that the irregular distribution of infill walls leads to a significant decrease in the seismic performance of the infilled variant of the building. Nevertheless, proper detailing of reinforcement and minimum requirements of the Eurocode 8 (CEN 2004) significantly limited the detrimental influence of the irregular distribution of infills. Despite the formation of a soft-storey mechanism, the "failure" probabilities of the infilled variant are still at least an order of magnitude smaller than the "failure" probabilities of old buildings, designed and built without observing appropriate codes for seismic resistance (e.g. between 5% and 30% in 50 years (Fajfar and Dolšek 2012; Kosič et al. 2014, 2016, 2017). Such an observation can be mainly attributed to much larger deformation capacities of code-conforming structural components (especially columns), which can, despite the concentration of damage in a single storey, still accommodate relatively large drift demands.

12.4 Conclusions

In the paper, the pushover-based risk assessment (PRA) method is summarized and applied for the estimation of the "failure" probability, i.e. the probability of exceeding the near collapse (NC) limit state, of a regular and irregular variant of a RC frame building. The irregular variant of the building has infill walls included only in the upper stories, which induce irregularity along the height of the building. The "failure" probabilities for the two variants of the building are compared and used to assess the influence of the irregular distribution of infill walls along the height of the building.

The comparison of the "failure" probabilities obtained for the two examples indicates a lower, although still significant, seismic risk for the regular code-conforming variant of the building (0.75% over the lifetime of the structure). A three to five times larger "failure" probability is obtained for the irregular variant of the building, for which a soft first storey effect is predicted. It should be noted, however, that the "failure" probability obtained for the irregular variant of the building is still at least an order of magnitude smaller than a typical "failure" probability of old buildings, designed and built without observing appropriate codes for seismic resistance. Such an observation can be mainly attributed to large deformation capacities of code-conforming structural components (especially columns), which can, despite the concentration of damage in a single storey, still accommodate relatively large drift demands.

It is worth noting that the absolute values of the estimated "failure" probability are sensitive to the input data and simplifying assumptions made. The comparisons of "failure" probabilities between structures, however, are more reliable and can provide valuable additional data needed for decision-making. Simple and computationally inexpensive probabilistic methods, such as the PRA method, may facilitate the gradual introduction of probabilistic considerations into the structural engineering practice.

Acknowledgments The results presented in this paper are based on work that has been continuously supported by the Slovenian Research Agency (J2-4180, P2-0185). This support is gratefully acknowledged.

References

BSSC, Building Seismic Safety Council (2015) NEHRP recommended seismic provisions for new buildings and other structures (FEMA P-1050-1). Federal Emergency Management Agency, Washington, DC

Celarec D, Dolšek M (2013) Practice-oriented probabilistic seismic performance assessment of infilled frames with consideration of shear failure of columns. Earthq Eng Struct Dyn 42 (9):1339–1360

Celarec D, Ricci P, Dolšek M (2012) The sensitivity of seismic response parameters to the uncertain modelling variables of masonry-infilled reinforced concrete frames. Eng Struct 35:165–177

CEN, Comité Européen de Normalisation (2004) Eurocode 8: design of structures for earthquake resistance. Part 1: General rules, seismic actions and rules for buildings. EN 1998-1:2004. Brussels, Belgium

CEN, Comité Européen de Normalisation (2005) Eurocode 8: design of structures for earthquake resistance. Part 3: Assessment and retrofitting of buildings. EN 1998–3:2005, Brussels, Belgium

Cornell CA (1996) Calculating building seismic performance reliability: a basis for multi-level design norms. In: Proceedings of the 11th world conference on earthquake engineering, Mexico City, Mexico, 23–28 June 1996

Cornell CA, Jalayar F, Hamburger RO, Foutch DA (2002) Probabilistic basis for 2000 SAC federal emergency management agency steel moment frame guidelines. J Struct Eng ASCE 128 (4):526–533

Dolšek M (2010) Development of computing environment for the seismic performance assessment of reinforced concrete frames by using simplified nonlinear models. Bull Earthq Eng 8 (6):1309–1329

Dolšek M (2012) Simplified method for seismic risk assessment of buildings with consideration of aleatory and epistemic uncertainty. Struct Infrastruct Eng 8(10):939–953

Dolšek M, Fajfar P (2007) Simplified probabilistic seismic performance assessment of plan-asymmetric buildings. Earthq Eng Struct Dyn 36(13):2021–2041

Fajfar P (2000) A nonlinear analysis method for performance-based seismic design. Earthquake Spectra 16(3):573–592

Fajfar P, Dolšek M (2012) A practice-oriented estimation of the failure probability of building structures. Earthq Eng Struct Dyn 41(3):531–547

FEMA (2000) Recommended seismic design criteria for new steel moment frame buildings. Report No. FEMA 350, SAC Joint Venture, Washington, DC

Kosič M, Fajfar P, Dolšek M (2014) Approximate seismic risk assessment of building structures with explicit consideration of uncertainties. Earthq Eng Struct Dyn 43(10):1483–1502

Kosič M, Dolšek M, Fajfar P (2016) Dispersions for pushover-based risk assessment of reinforced concrete frames and cantilever walls. Earthq Eng Struct Dyn 45(13):2163–2183

Kosič M, Dolšek M, Fajfar P (2017) Pushover-based risk assessment method: a practical tool for risk assessment of building structures. In: Proceedings of the 16th world conference on earthquake engineering, Santiago, Chile, 9–13 January 2017

Kreslin M, Fajfar P (2012) The extended N2 method considering higher mode effects in both plan and elevation. Bull Earthq Eng 10:695–715

Open system for earthquake engineering simulation (OpenSees) (2017) Pacific Earthquake Engineering Research Center, University of California, Berkeley, CA. Available from: http://opensees.berkeley.edu

Rozman M, Fajfar P (2009) Seismic response of a RC frame building designed according to old and modern practices. Bull Earthq Eng 7(3):779–799

Chapter 13
Assessment of Nonlinear Static Analyses on Irregular Building Structures

Gabriel Dănilă

Abstract Nonlinear static analysis is many times chosen, in engineering practice, to predict the seismic demands in building structures. Despite its simplicity, the nonlinear static analysis based on invariant load patterns has certain limitations caused by its inability to account for the variation of the dynamic characteristics, of the building structure, resulting from inelastic behavior and the higher modes effect. The paper presents a comparative study on three building structures, for which were made adaptive and nonadaptive nonlinear static analyses and incremental nonlinear time history analyses were performed. The analyzed buildings structures have elevation irregularity, except for the smallest which is regular. The capacity curves, resulting from nonlinear static analyses were compared with the mean capacity curve resulting from incremental nonlinear time-history analyses. The study has shown that the adaptive nonlinear static analysis, with displacements load pattern, caught, with enough accuracy, the behavior of irregular building structures of low and medium height. The nonlinear static analyses may not sufficiently predict the seismic demand for the tallest building structure, having deep elevation irregularity.

Keywords Adaptive non-linear static analyses · Elevation irregular structures · Incremental time-history analyses

13.1 Introduction

The nonlinear static analysis is a simple and efficient alternative to the non-linear time-history analysis. Despite its simplicity, the nonlinear static analysis can provide important information regarding the post-elastic behavior of the building structures (Antoniou and Pinho 2004). The nonlinear static analysis based on invariant load patterns has certain limitations caused by its inability to account for the

G. Dănilă (✉)
"Ion Mincu" University of Architecture and Urbanism, Bucharest, Romania

D. Köber et al. (eds.), *Seismic Behaviour and Design of Irregular and Complex Civil Structures III*, Geotechnical, Geological and Earthquake Engineering 48,
https://doi.org/10.1007/978-3-030-33532-8_13

variation of the dynamic characteristics of the building structure, resulting from inelastic behavior and the higher modes effect (Antoniou et al. 2005; Elnashai 2001). The paper presents a comparative study on three building structures, for which were made adaptive and nonadaptive pushover analyses and incremental nonlinear time history analyses were performed.

13.2 Description of the Analysed Buildings Structures

The study was made on three reinforced concrete buildings structures, with varying number of stories. The first building structure has 4 stories (ground floor and 3 stories), the second one has 9 stories (ground floor and 8 stories) and the third one has 15 stories (ground floor and 14 stories). The location of the analysed buildings structures was considered the Bucharest town, due to the long predominant periods of the seismic movements.

The 1st building structure is a plane frame with 3 spans, of 6 meters and the following stories heights: the ground floor – 4m, the 1st, the 2nd and the 3rd floor – 3,2 m each.

The 2nd building structure is also a plane frame with three spans of 6 m and nine stories. The ground floor is 4 m height and the 1...8 stories are 3.2 m height. At the 4th storey is presented a drawback, resulting an elevation irregularity.

The 3rd building structure is a plan frame with three spans of 6 m and 15 stories. The ground floor has 4 m height and the stories 1...15 have 3.2 m height. At the 4th and 9th stories are presented two drawbacks, resulting in a severe elevation irregularity (Fig. 13.1).

The dynamic properties of the analysed building structures, resulted from the modal analysis are presented in the Table 13.1.

13.3 The Seismic Action

The seismic action is described by six artificial time-histories, compatible with the design spectrum for Bucharest and one recorded time-history, on the INCERC site, corresponding to the N-S component of the March 4, 1977 earthquake. The artificial time-histories were generated by means of the SeismoArtif (SeismoArtif [computer software] 2012) computer program and the recorded time-history was scaled, to the maximum ground acceleration of 0.24 g. Figure 13.2 shows the elastic response spectra, corresponding to the seven time-histories.

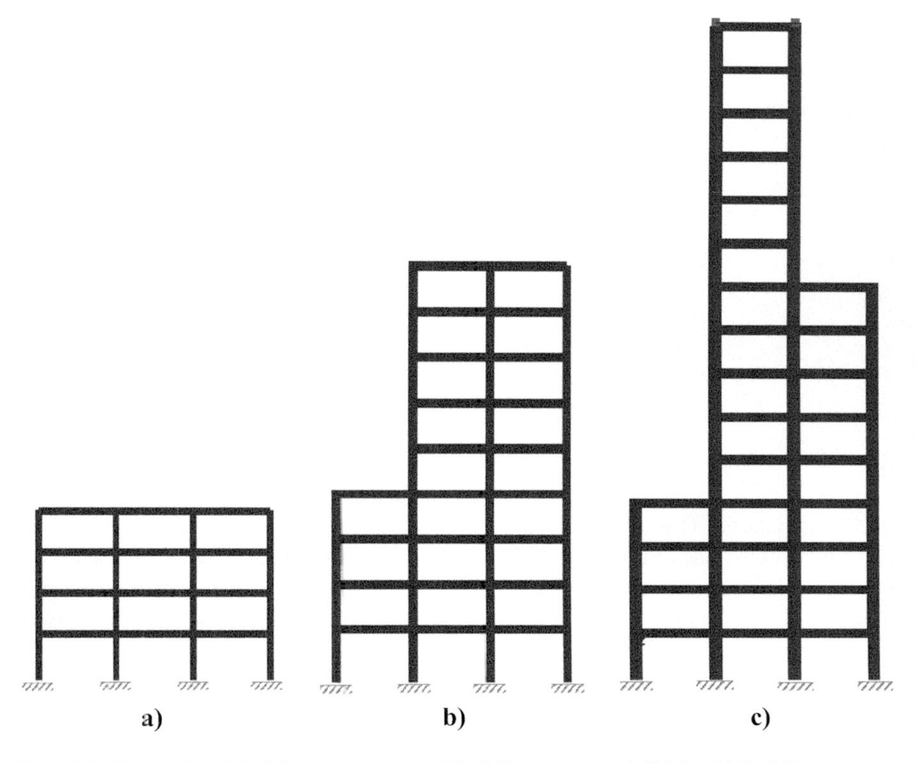

Fig. 13.1 The analysed building structures: (**a**) Building structure 1 (BS1); (**b**) Building structure 2 (BS2); (**c**) Building structure 3 (BS3)

Table 13.1 Dynamic properties of the analysed building structures

Dynamic properties	BS1		BS2		BS3	
The vibration modes	1st	2nd	1st	2nd	1st	2nd
The vibration periods	0.79 s	0.25 s	1.12 s	0.42 s	1.31 s	0.56 s
The mass participation factors	90.09%	7.64%	76.03%	15.0%	65.22%	18.8%

13.4 Non-linear Static Analyses. Comparative Study

13.4.1 Modelling Issues in the Non-linear Static Analysis and in the Incremental Dynamic Analysis

The non-linear static analysis and the incremental dynamic analysis were performed using the SeismoStruct v6.0 (SeismoStruct [computer software] 2012) computer program. The concrete modelling was made using the (Mander et al. 1988) model, with constant confinement. The reinforcement was modelled with a bilinear

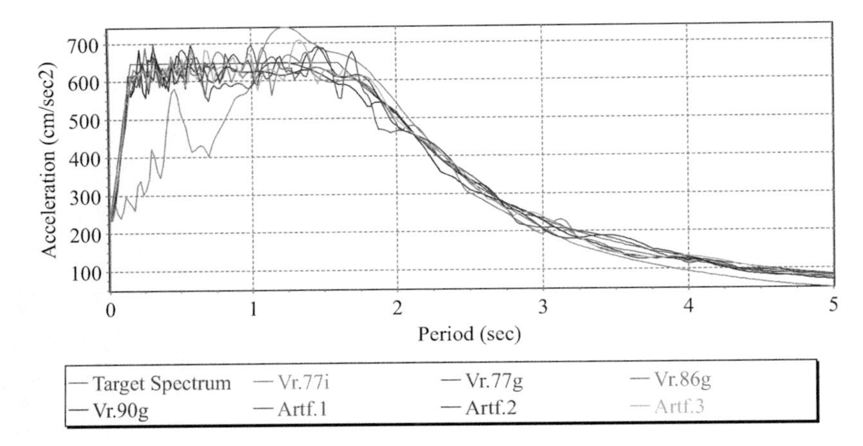

Fig. 13.2 The elastic response spectra (Dănilă et al. 2014)

hardening model. The inelastic behavior of the elements was taken into account using the fiber model (distributed plasticity model). The fiber model considers the element cross-section meshed in fibers, each fiber being a concrete area or a longitudinal reinforcement. Each fiber is defined by a uniaxial stress-strain relationship.

The structural elements of the BS1 and of the BS2 were meshed into 100 fibers per section, with four integration sections per element, considering force-based interpolation functions. For the BS3, the structural elements were meshed into 150 fibers per section, with four integration sections per element, also considering force-based interpolation functions.

13.4.2 The Incremental Dynamic Analyses

The incremental dynamic analyses were performed using the time-histories defined in Sect. 13.3. Thus, seven incremental dynamic analyses were performed for each BS. Figure 13.3 shows the "dynamic capacity curves" of the three analysed building structures.

In the incremental dynamic analyses, the Hilber-Hughes-Taylor integration method was used, with a $\Delta t = 0.01$ s time step and integration parameters $\alpha = -0.001$, $\beta = 0.2505$ and $\gamma = 0.501$. The elastic damping was taken into account with the use of the damping proportional to the tangent stiffness.

The dynamic capacity curves were plotted considering the maximum response of the building structures for different scale factors of the time-histories. There were used seven scale factors in the analyses of BS1 and BS2 and ten scale factors in the analyses of BS3.

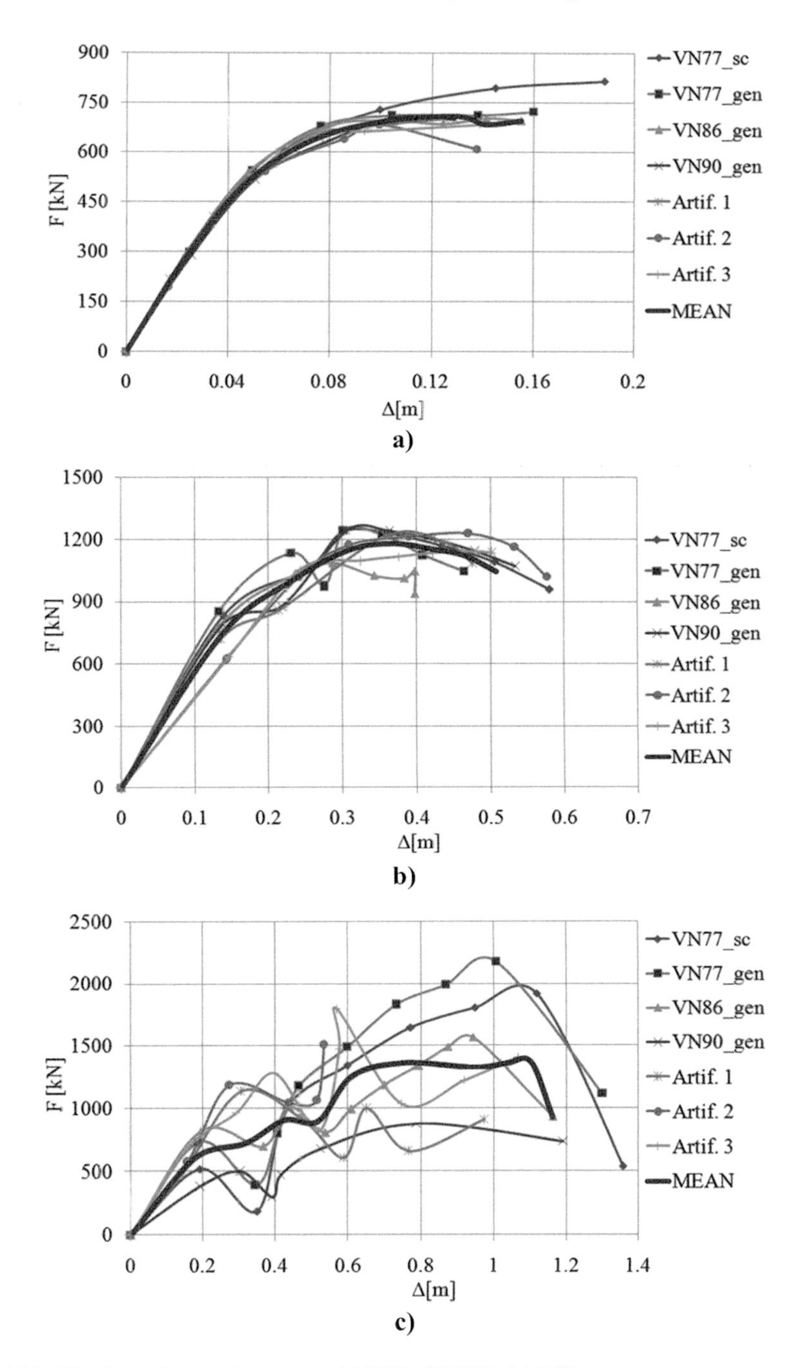

Fig. 13.3 The dynamic capacity curves: (**a**) BS1; (**b**) BS2; (**c**) BS3

13.4.3 The Non-linear Static Analyses

The comparisons between non-linear static analyses were performed in terms of capacity curves. The capacity curves resulting from non-linear static analyses were compared with the mean dynamic capacity curves (M-INDA).

The non-linear static analyses were performed by pushing the building structures in the positive and in the negative x-direction, except for the BS1 where the capacity curves where computed only for positive x-direction. The following types of non-linear static analyses were performed:

- Force-based adaptive pushover (FAP);
- Displacement-based adaptive pushover (DAP);
- Non-adaptive pushover with the force load pattern distributed according to the fundamental eigenmode (MFP);
- Non-adaptive pushover with the force load pattern distributed linearly (LFP);
- Non-adaptive pushover with the force load pattern distributed uniformly (UFP);
- Non-adaptive pushover with the displacement load pattern distributed according to the fundamental eigenmode (MDP);
- Non-adaptive pushover with the displacement load pattern distributed linearly (LDP);
- Non-adaptive pushover with the displacement load pattern distributed uniformly (UDP).

Figure 13.4 shows the capacity curves resulting from non-linear analyses and the mean dynamic capacity curve for BS1.

Due to the fact that BS1 is regular, of low height, the differences between pushover analyses are small. The only difference occurs in the case of UDP and

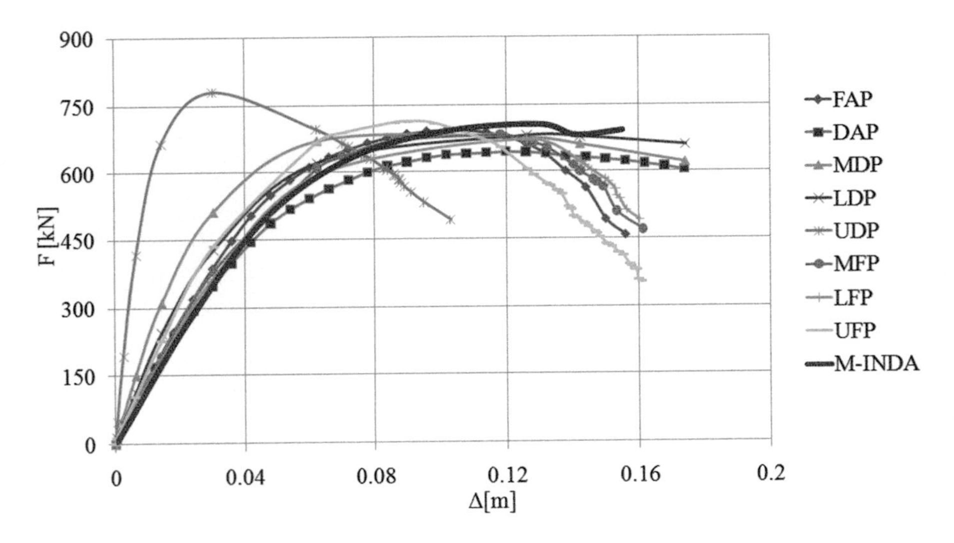

Fig. 13.4 The capacity curves for BS1

MDP, where the capacity curves are diverging from the mean dynamic capacity curve. There are also differences between UFP and M-INDA, knowing the fact that this type of analysis is used to determine the maximum shear forces.

As a result of the analyses carried-out on BS1, it can be stated that there are minor differences between adaptive and non-adaptive pushover analyses for low regular structures. To very high post-elastic displacements the differences between adaptive and non-adaptive pushover analyses increases, due to the changes in the dynamic characteristics of the building structure.

Figure 13.5 shows the capacity curves resulting from non-linear analyses and the mean dynamic capacity curve for BS2. Because BS2 is irregular, pushover analyses on both positive and negative x-direction were performed. The static capacity curves are compared to the mean dynamic capacity curve for both directions.

BS2 is an irregular structure of medium height, for which the contribution of the higher vibration modes to the total building response is significant. In the elastic regime the differences between adaptive and non-adaptive pushover analyses are very small, except for LDP, UDP, MDP and UFP, where the capacity curves are diverging from the mean dynamic capacity curve. In the post-elastic regime there are differences between adaptive and non-adaptive pushover analyses, the maximum accuracy being obtained in the case of DAP, whatever the direction of loading.

Figure 13.6 shows the capacity curves resulting from non-linear analyses and the mean dynamic capacity curve for BS3. Because BS2 has strong elevation irregularity, pushover analyses on both positive and negative x-direction were performed. The static capacity curves are compared to the mean dynamic capacity curve for both directions.

In the elastic domain the differences between adaptive and non-adaptive pushover analyses are very small, except for LDP, UDP, MDP and UFP, where the capacity curves are diverging from the mean dynamic capacity curve. The post-elastic behaviour of BS3 highlights the limitations of adaptive and non-adaptive pushover analyses. None of the static capacity curves approaches the mean dynamic capacity curve on both positive and negative x-direction.

13.5 Conclusions

The study has evaluated the effectiveness of adaptive and non-adaptive pushover analyses for three building structures.

For low and regular building structures, it can be stated that there are minor differences between adaptive and non-adaptive pushover analyses. To very high post-elastic displacements the differences between adaptive and non-adaptive pushover analyses increases, due to the changes in the dynamic characteristics of the building structure.

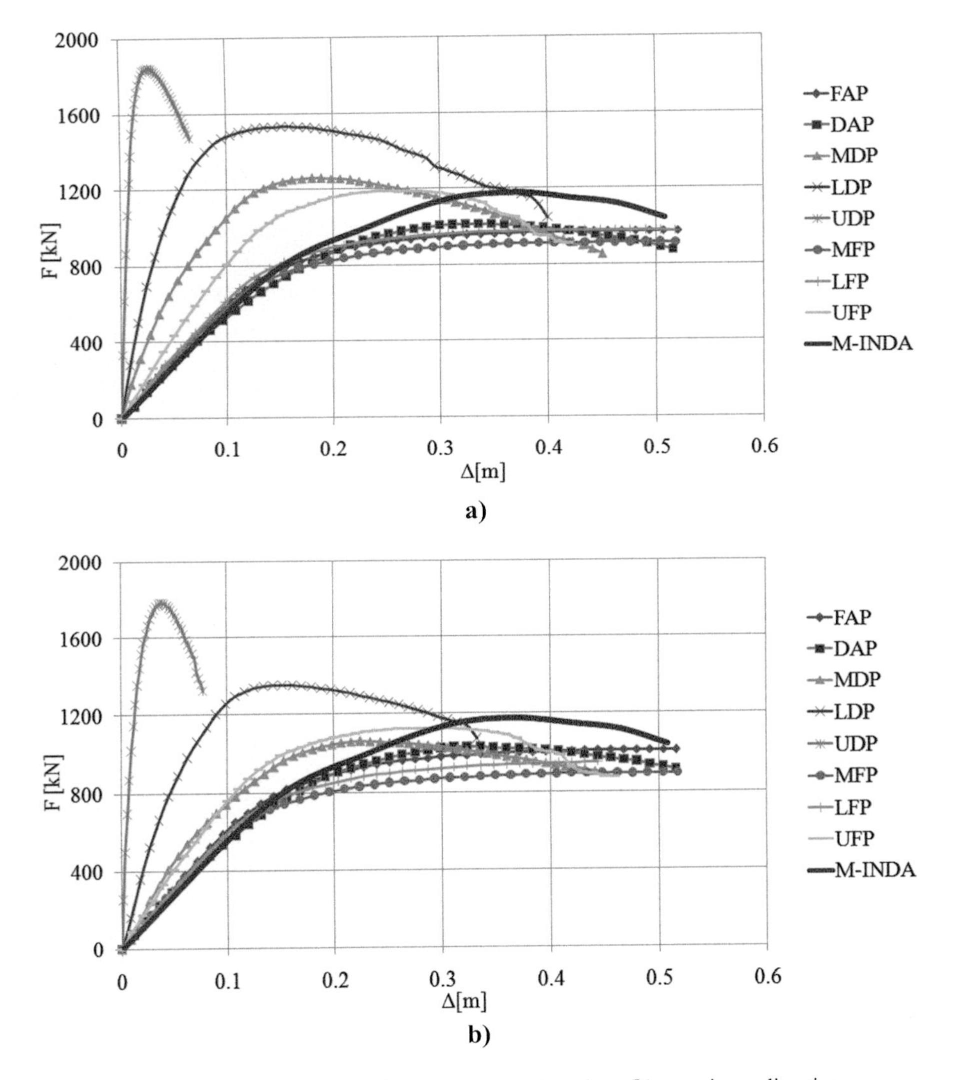

Fig. 13.5 The capacity curves for BS2: (**a**) positive x-direction; (**b**) negative x-direction

For irregular building structures of medium heights, in the elastic regime the differences between adaptive and non-adaptive pushover analyses are very small. In the post-elastic regime there are differences between adaptive and non-adaptive pushover analyses, the maximum accuracy being obtained in the case of displacement-based adaptive pushover.

For tall irregular building structures, in the elastic domain the differences between adaptive and non-adaptive pushover analyses are very small. The post-elastic behaviour highlights the limitations of adaptive and non-adaptive pushover analyses. None of the static capacity curves approaches the mean dynamic capacity curve on both

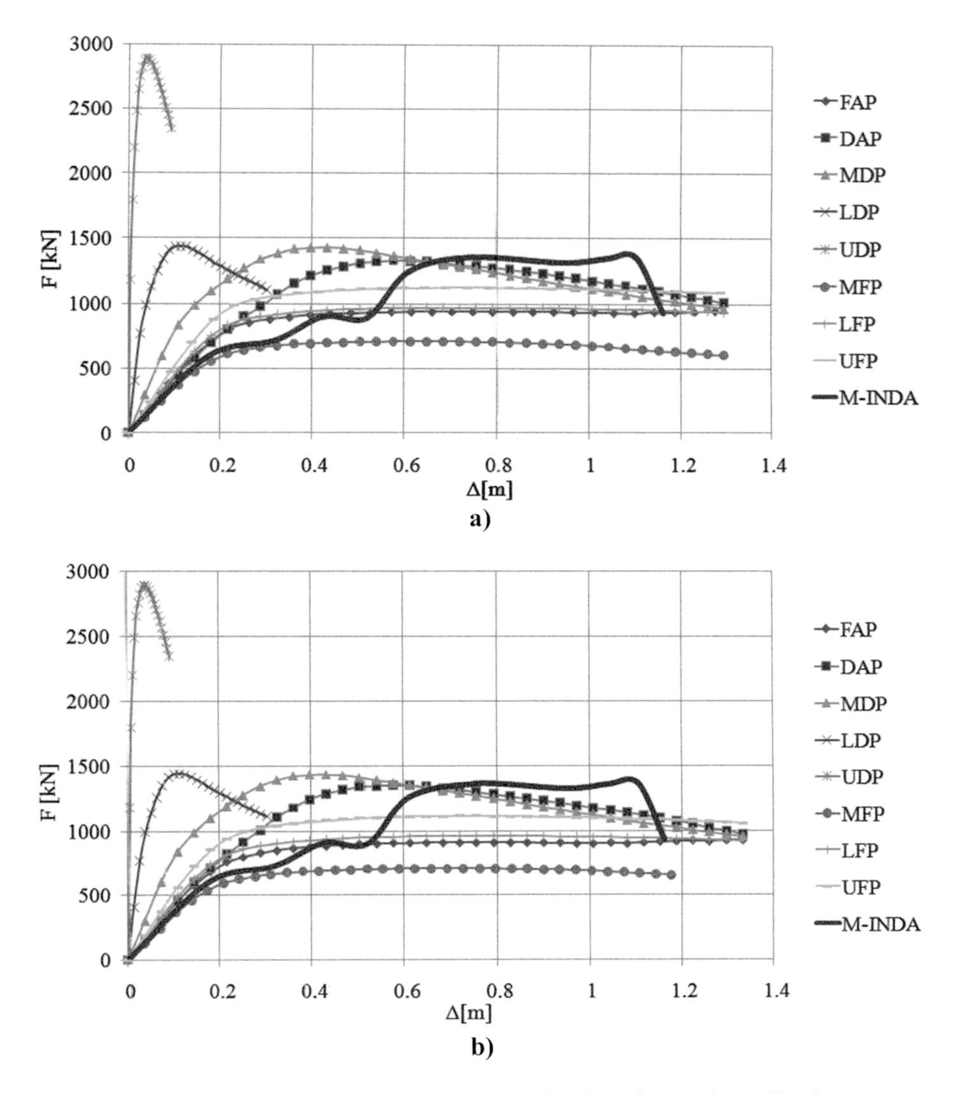

Fig. 13.6 The capacity curves for BS3: (**a**) positive x-direction; (**b**) negative x-direction

positive and negative x-direction. The behaviour of the building structure in post-elastic domain indicates the non-linear dynamic analysis as the only analysis than can provide "accurate" informations about the non-linear behaviour of tall irregular building structures.

Acknowledgments The author would like to thank to SeismoSoft Company for providing free educational licences of SeismoStruct and SeismoArtif computer programs without which this study could not have been achieved.

References

Antoniou S, Pinho R (2004) Advantages and limitations of adaptive and non-adaptive force-based pushover procedures. J Earthq Eng 8(4):497–522

Antoniou S, Rovithakis A, Pinho R (2005) Development and verification of a fully adaptive pushover procedure. In: Proceedings of 12th European conference on earthquake engineering, Ispra

Dănilă G, Petrescu V, Chescă AB (2014) Smart structures subjected to seismic actions. Acta Techn Napoc Civil Eng Architect 57(1)

Elnashai AS (2001) Advance inelastic static (pushover) analysis for earthquake applications. Struct Eng Mech 12(1):51–69

Mander JB, Priestley MJN, Park R (1988) Theoretical stress-strain model for confined concrete. J Struct Eng ASCE 114(8):1804–1826. https://doi.org/10.1061/(ASCE)0733-9445

SeismoArtif [computer software] (2012) Pavia: SeismoSoft srl. Available: http://www.seismosoft.com

SeismoStruct [computer software] (2012) Pavia: SeismoSoft srl. Available: http://www.seismosoft.com

Chapter 14
Seismic Assessment of an Irregular Unreinforced Masonry Building

Gabriel Dănilă and Adrian Iordăchescu

Abstract The paper presents a study on a construction, erected in the 1870–1920 period, composed of two interconnected buildings. The main A building is reserved to living, having a basement, ground floor, one story, penthouse and a loft. The building is approximately rectangular with a dead wall on the North side property limit, being connected with the B building on the South side. Located on the South, the B building is an exterior staircase with a terrace floor and a penthouse, being an unique element as conception in the beginning twentieth century architecture of Bucharest. The construction is not classified as a historical monument, but is placed in the protected area of a historical monument of class B (LMI 2015: B-II-m-B-19768). The construction has several deficiencies of which the most important is the differential settlement of about 25 cm in transversal direction, respectively 10 cm in the longitudinal direction. The irregular shape, because of the link between the two buildings, leading to major torsional effects and the presence of some structural walls, supported directly by the floors, are other deficiencies of the construction. The seismic evaluation was carried out by linear static and dynamic analyses, taking into account the second order effect. There were proposed strengthening and straightening solutions for the construction. For building strengthening seismic isolation method was adopted, reducing drastically the torsion effects and the efforts in the structural elements. Building straightening was achieved by means of compensation presses placed at the basement level.

Keywords Irregular masonry building · Seismic isolation · Differential settlement · Time-history analyses

G. Dănilă (✉) · A. Iordăchescu
"Ion Mincu" University of Architecture and Urbanism, Bucharest, Romania

D. Köber et al. (eds.), *Seismic Behaviour and Design of Irregular and Complex Civil Structures III*, Geotechnical, Geological and Earthquake Engineering 48,
https://doi.org/10.1007/978-3-030-33532-8_14

14.1 Introduction

According to the Romanian legislation in force, the safety of the existing building stock constitutes a national interest action for limiting or avoiding the effects of a potential disaster. In order to solve this problem the building's owners are obliged to adopt measures for their safety.

For the building located on Temişana entrance, no. 7, district 1, Bucharest, it was considered the base performance objective (BPO), consisting of meeting the requirements corresponding to the life safety performance level for the seismic action with the mean recurrence interval of 100 years. The performance objective and the mean recurrence interval of the seismic action were considered according to the provisions of P100-3/2008 (Universitatea Tehnică de Construcţii Bucureşti 2008) seismic code. The established performance objective largely determines the cost and the complexity of the seismic rehabilitation works as well as the benefits that can be gained in terms of safety, degrading reduction and reduction of the operational discontinuity in the case of a major seismic event.

The building located on Temişana entrance, no. 7, district 1, Bucharest was executed before 1920 when there were no seismic norms.

14.2 Building Description

The exact construction date of the building is not certain but is in between 1870–1920, taking into account that the land was bought between 1912 and 1914.

The building is not classified as a historical monument, but it is placed in the protected area of a historical monument of class B (LMI 2015: B-II-m-B-19768) (Figs. 14.1, 14.2, and 14.3).

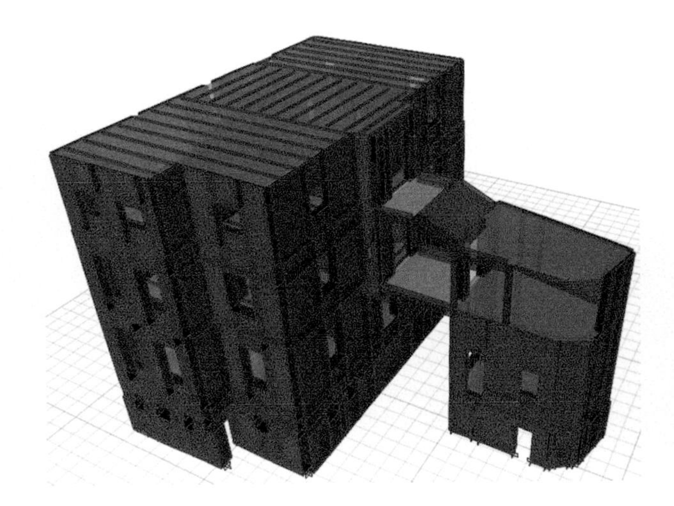

Fig. 14.1 The analysed building

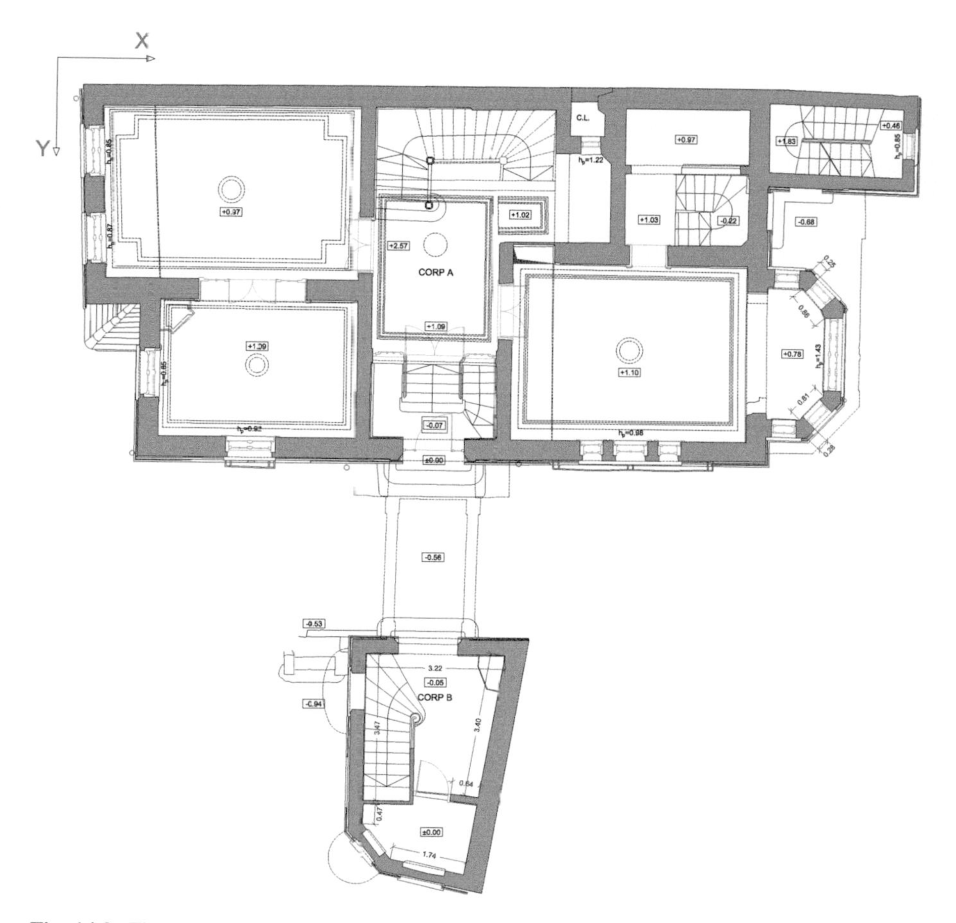

Fig. 14.2 The ground floor plan of the building

The building is composed of two parts. The main building (A) is reserved to living, having a basement, ground floor, one storey, penthouse and a loft. The building is approximately rectangular with a dead wall on the North side property limit, being connected with the B building on the South side.

The A building structure is made of unconfined brick masonry. The basement is covered with a reinforced concrete slab, the walls are made of brick masonry of 55–60 cm and the stairs are made of reinforced concrete. The ground floor and the storey one are covered with wooden floors, the exterior walls have 45–50 cm thickness and the interior walls have 28 cm thickness.

The building B is an annex building, unique element as conception in the beginning of the twentieth century architecture of Bucharest, having the functionality as exterior staircase with terrace floor and penthouse, providing exterior accesses to the main A building. The building structure is made of unconfined brick masonry.

Fig. 14.3 West facade surveying with damages indication

The building analyses were performed using the Etabs v2015 (Etabs [computer software] 2015) computer program. The masonry walls and the coupling beams were modelled with shell elements. The floor beams were modelled with frame elements.

14.3 Structural Configuration Deficiencies and Damage Description

After the earthquakes from November 10, 1940 (Gutenberg-Richter magnitude 7.4), March 4, 1977 (Gutenberg-Richter magnitude 7.2), August 30, 1986 (Gutenberg-Richter magnitude 7.0) and May 31, 1990 (Gutenberg-Richter magnitude 6.7), the building revealed global conformation deficiencies, such as significant global torsion effects, the low quality of the structural materials and the soil-structure interaction which leads to differential settlements of about 25 cm in the transversal direction and 10 cm in the longitudinal direction.

There can be noticed significant damages to the structural walls, generated by the differential settlements and by the seismic actions, which leads to inclined cracking and to the decomposition of the respective masonry walls in the corner areas. It is also noted that some masonry walls have no vertical continuity, being supported by steel beams embeded in the floors, as well as the existence of more recent consolidation elements (steel rods and vertical and horizontal steel elements for stiffening), probably introduced after March 4, 1977 earthquake, being visible on the west facade and on the dead wall.

The building has numerous structural degradings of the exterior walls and a major vertical detachment – of about 4 cm at the top – of the building B from the building A. Also, there are non-structural partitionings, added with time, by the inhabitants and the intervention with inadequate finishes in the annex spaces (sanitary groups). The facade plaster is partial degraded due to the humidity caused by the meteoric or capillarity water.

14.4 The Seismic Action

The seismic action is described by four recorded seismic motions and three artificial accelerograms. Because there were made spatial linear time-history analyses, the recorded accelerograms were used with all three components (two horizontal components and one vertical component). In the case of the artificial accelerograms, the seismic action was applied also in the three directions of the building, complying with the provisions of paragraph 4.5.3.6.2 (4) from the P100-1/2006 (Universitatea Tehnică de Construcții București 2006) seismic code.

For the recorded accelerograms, the horizontal component, with the maximum ground acceleration, was scaled to 0.24 g and the other two components (one horizontal and one vertical) were scaled with the same scaling factor.

The artificial accelerograms were generated by means of SeismoArtif (SeismoArtif [computer software] 2012) computer program, using random processes, to fit the target spectrum – design spectrum from P100-1/2006 (Universitatea Tehnică de Construcții București 2006) seismic code, corresponding to Bucharest city. It were applied the envelope shapes Compound for the artificial accelerogram (1), Saragoni&Hart for the artificial accelerogram (2) and Exponential for the artificial accelerogram (3).

The Table 14.1 shows the main parameters that characterize the seismic actions used in the study.

14.5 The Seismic Assessment

The assessment of an existing building is done to determine possibilities of the building structure to undertake the gravitational and seismic loads, depending on the overall configuration, the dimension of the structural elements, the quality of the used materials, the wear condition and the possible damage caused by accidental or extraordinary demands.

Table 14.1 Parameters that characterise the seismic actions

The accelerogram	PGA [cm/s^2]	PGV [cm/s]	PGD [cm]	The predominant period [s]
INCERC(1977)_NS	235.6	85.4	52.2	1.16
INCERC(1977)_EV	195.9	33.9	17.1	0.78
INCERC(1977)_V	128.1	16.1	16.2	0.16
INCERC(1986)_NS	218.8	35.1	9.6	0.56
INCERC(1986)_EV	235.6	24.9	7	0.5
INCERC(1986)_V	44.6	6.2	2.2	0.7
ISPH(1986)_H1	235.6	38.7	9.9	0.8
ISPH(1986)_H2	208.5	23.6	13.8	0.3
ISPH(1986)_V	108.8	9.4	22.2	0.42
Otopeni (1986)_NS	235.7	27.9	8.3	0.42
Otopeni (1986)_EV	132.54	12.1	7.4	0.26
Otopeni (1986)_V	65.4	4.5	7.6	0.2
Artif. 1	235.9	86.7	69.2	0.32
Artif. 2	235.9	52.4	24.5	1.34
Artif. 3	235.5	74.2	90.5	0.56

Table 14.2 Dynamic properties of the existing building structure

		Modal participating mass ratios [%]					
Mode	Period [s]	UX	UY	UZ	RX	RY	RZ
1	0.369	0	0.63	0	0.39	0	0
2	0.307	0.32	0	0	0	0.15	0.32
3	0.262	0.33	0	0	0	0.20	0.29

For the seismic assessment of the building located on Temişana entrance, no. 7, District 1, Bucharest, there were made linear static and dynamic analyses. The building analyses were performed taking into account the stiffness degradation, expressed by reducing the masonry modulus of elasticity, according to the values provided in CR 6/2006 (Monitorul Oficial al Romaniei, Partea I, Nr. 807 bis/26. IX.2006 2006) design code. The dynamic properties of the building structure are given in Table 14.2, from which can be distinguished the torsion effects on the second and third mode of vibration.

Figure 14.4 presents the displacement response of the existing building structure in both horizontal directions. The response in displacements is highlighted for each seismic action described in the Chap. 4.

Figure 14.5 presents the acceleration response of the existing building structure in both horizontal directions. The response in accelerations is highlighted for each seismic action described in Chap. 4.

Figure 14.6 presents the shear forces of the existing building structure in both horizontal directions.

The values of the relative displacements are below the allowable ones, the maximum one being of 2.51‰ in x – direction and of 2.14‰ in y – direction for the Serviceability Limit State. At the Ultimate Limit State the maximum relative displacements are of 0.50% in x – direction and of 0.43% in y – direction. The

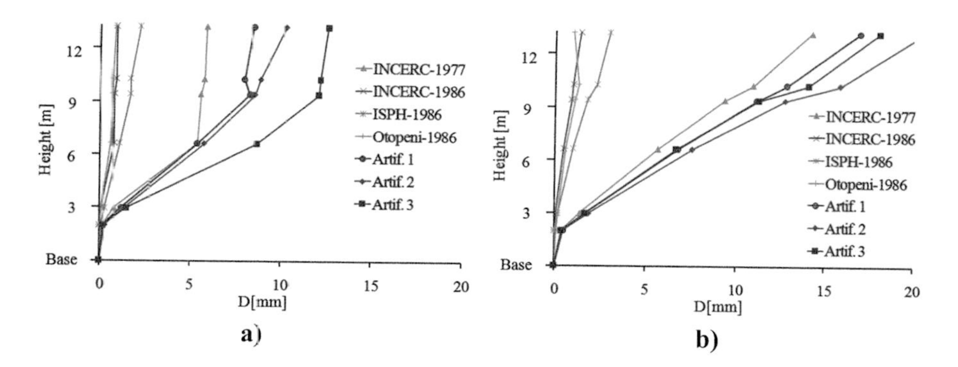

Fig. 14.4 The relative displacements of the building structure: (**a**) x – direction; (**b**) y – direction

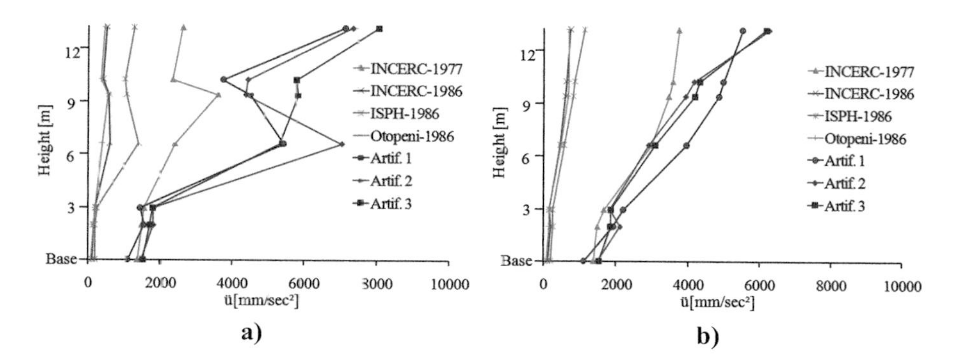

Fig. 14.5 The absolute accelerations of the building structure: (**a**) x – direction; (**b**) y – direction

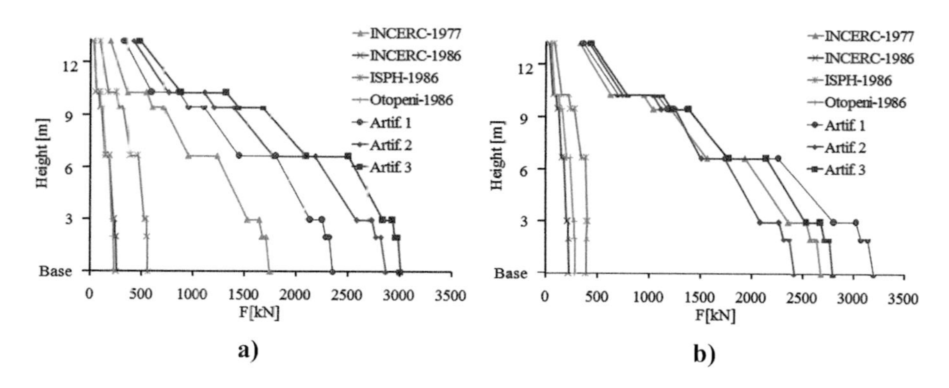

Fig. 14.6 The shear forces on the building structure: (**a**) x – direction; (**b**) y – direction

allowable values of the relative displacements are considered, according to P100-1/2006 (Universitatea Tehnică de Construcții București 2006) seismic code, 5‰ for the Serviceability Limit State and 2.5% for the Ultimate Limit State.

Based on the dimensions and on the axial forces of the relevant structural elements, there was computed the strength capacity for horizontal loads for each relevant structural element, taking into account the smallest value, determined for one of the following:

- Eccentric compression with eccentricity in the plane of the walls;
- Shearing of the horizontal joint;
- Shearing in inclined sections;

The strength check was performend accordind to CR 6/2006 (Monitorul Oficial al Romaniei, Partea I, Nr. 807 bis/26.IX.2006 2006) design code.

The safety index, defined as the ratio between the Capacity and the Demand (Capacity/Demand), was computed for both horizontal directions, leading to the following values:

- For x – direction: Capacity/Demand = 0.66;
- For y – direction: Capacity/Demand = 0.49.

The most vulnerable element is the ground floor masonry wall of building B, which sustains the intermediate staircase slab, placed along the x-direction of the building, due to the important torsional effects and the low axial force, which leads to a Demand/Capacity index of 0.26.

14.6 The Structural Interventions Measures

Based on the seismic assessment there were proposed the intervention measures for retrofitting, elimination of the global torsion effects and straightening of the building.

Table 14.3 provides the coordinates of the mass and rigidity centers, the excentricities between the two centers and the allowable eccentricities – computed

Table 14.3 Mass and rigidity center coordonates and eccentricities

Building level	XCM [m]	YCM [m]	XCR [m]	YCR [m]	EX [m]	EY [m]	EXa [m]	EYa [m]
±0.00	10.08	14.54	10.07	15.08	0.01	0.54	1.98	1.88
+0.96	10.10	14.81	10.11	15.57	0.01	0.76	1.98	1.88
+4.62	10.20	14.36	9.79	16.51	0.41	2.15	1.98	1.88
+7.40	10.52	14.53	9.99	16.82	0.53	2.29	1.98	1.88
+8.23	9.83	15.14	9.94	16.82	0.11	1.68	1.98	0.96
+11.19	9.16	15.42	9.73	16.62	0.57	1.20	1.98	0.96

XCM, YCM – mass center coordinates on X and Y directions
XCR, YCR – rigidity center coordinates on X and Y directions
EX, EY – eccentricities between mass and rigidity center on X and Y directions
EXa, EYa – allowable eccentricities between the mass and rigidity center on X and Y directions

according to CR 6/2006 (Monitorul Oficial al Romaniei, Partea I, Nr. 807 bis/26. IX.2006 2006) design code.

According to CR 6/2006 (Monitorul Oficial al Romaniei, Partea I, Nr. 807 bis/26. IX.2006 2006) design code the allowable eccentricity between the mass center and the rigidity center is 0.1 L (L being the building length perpendicular to the seismic attack). Analysing the values from Table 14.3 it can be observed that the eccentricities between the mass center and the rigidity center along the y-direction exceed the allowable excentricities, except for the ±0.00 and + 0.96 building levels.

For the intervention measure there was adopted the seismic isolation method using 32 high damping rubber bearings. The isolation plan was established at the level of the basement, in the upper part of the basement walls, above the ground level.

The characteristics of the high damping rubber bearings are given below:

$d_{dc} = 0.332$ m – the design displacement corresponding to the design earthquake;
$G = 0.25$ MPa – the shear modulus of the rubber bearing;
$D = 46$ cm – the bearing diameter;
$t_r = 6$ mm – the thickness of the rubber layer;
$n_r = 58$ – the number of the rubber layers;
$t_s = 2$ mm – the thickness of the steel plates;
$d_y = 3.318$ cm – the yield displacement of the rubber bearing;
$k_e = 510.4$ kN/m – the elastic stiffness of the rubber bearing;
$k_p = 80.2$ kN/m – the post-elastic stiffness of the rubber bearing;
$F_y = 16.9$ kN – the yield force of the rubber bearing;
$F_{max} = 40.9$ kN – the maximum force of the rubber bearing, corresponding to the design displacement;

The dynamic properties of the isolated building are given in Table 14.4, from which it can be observed the drastically diminishing of the torsion effects.

Figure 14.7 presents the displacement response of the isolated building structure in both horizontal directions. The response in displacements is highlighted for each seismic action described in Chap. 4. On both x and y directions, the maximum displacements are given by the *Artif. 1* seismic action and the minimum displacements are given by the *INCERC-1986* seismic action.

Figure 14.8 presents the acceleration response of the isolated building structure in both horizontal directions. The response in accelerations is highlighted for each seismic action described in Chap. 4. Along both x and y directions, the maximum accelerations are given by the *Artif. 3* seismic action and the minimum accelerations are given by the *Otopeni-1986* seismic action.

Table 14.4 Dynamic properties of the existing building structure

		Modal participating mass ratios [%]					
Mode	Period [s]	UX	UY	UZ	RX	RY	RZ
1	3.52	0.68	0.25	0	0	0	0.07
2	3.46	0.27	0.73	0	0	0	0
3	2.86	0.06	0.02	0	0	0	0.93

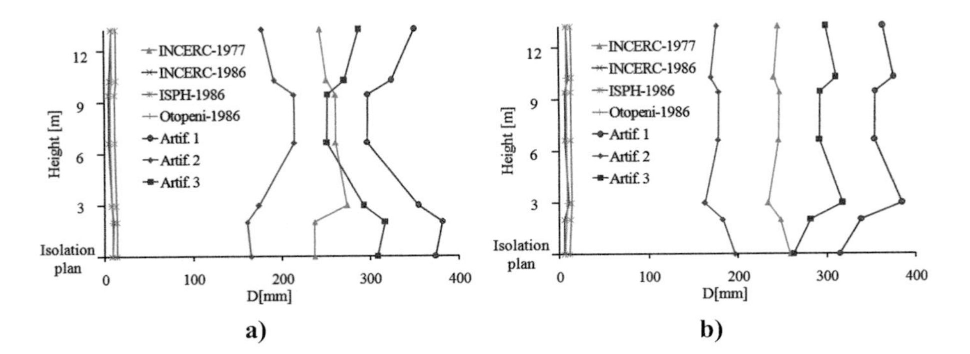

Fig. 14.7 The relative displacements of the isolated building structure: (**a**) x – direction; (**b**) y – direction

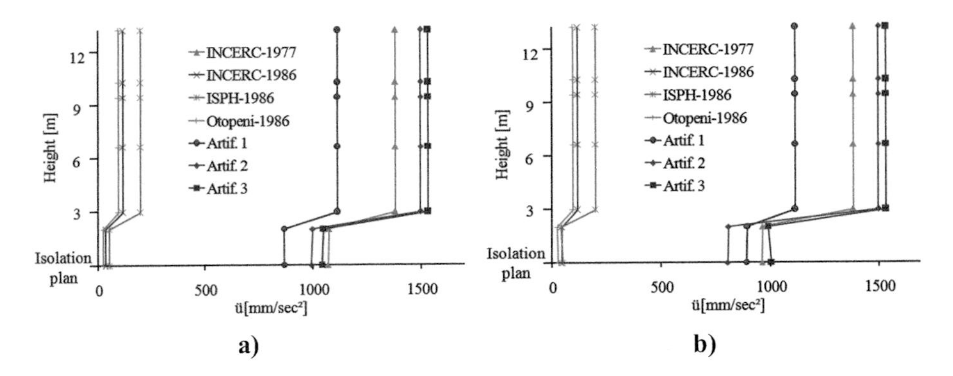

Fig. 14.8 The absolute accelerations of the isolated building structure: (**a**) x – direction; (**b**) y – direction

Figure 14.9 presents the shear forces for the isolated building structure in both horizontal directions. On both x and y directions, the maximum shear forces are given by the *Artif. 3* seismic action and the minimum shear forces are given by the *Otopeni-1986* seismic action.

The structural intervention measures assume the following succession of the main stages:

- It is drawn, on the basement walls, the upper and lower level of the new elements, the Upper Bearing Frame (UBF) and the Lower Bearing Frame (LBF), which will be introduced into the structure;
- It is marked on the basement walls the position of the holes where the steel tables will be mounted in order to achieve the UBF and the LBF.
- The installation of the steel tables into the created holes is done after laying a layer of high strength mortar at the bottom.
- After finishing the preparatory operations on a certain area, one can proceed to mounting the reinforcement and casting the frame.
- Similarly, is done for LBF.

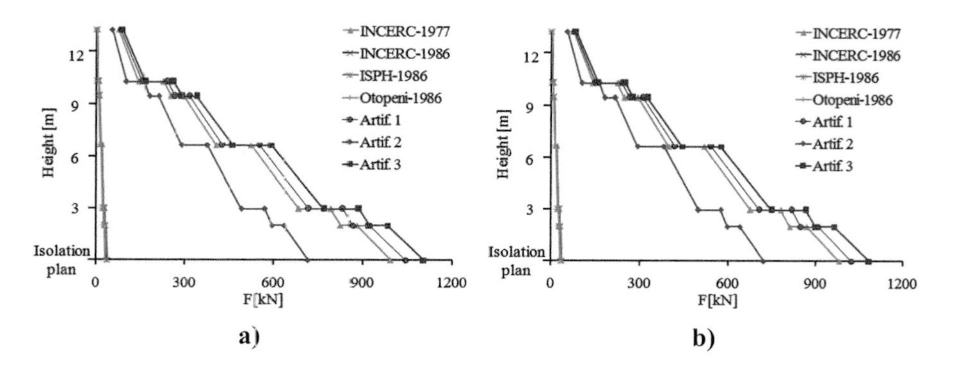

Fig. 14.9 The shear forces on the isolated building structure: (**a**) x – direction; (**b**) y – direction

- After the UBF and the LBF are executed, it is possible to proceed with the installation of the differential pressure compensation presses and then applying the differential loading for straightening the building.
- The mounting of compensating presses is done by dismounting the masonry between the UBF and the LBF.
- The transfer of the building loads to the compensation presses is done only after all the compensating presses have been installed and after the masonry has been dismantled between them.
- After building straightening, seismic bearings can be installed between the UBF and LBF.
- The transfer of the loads on the bearings is made only after all seismic bearings have been assembled.

14.7 Conclusions

The seismic assessment of the building located on Temisana entrance, no. 7, District 1, Bucharest, aimed to highlight the level of antiseismic protection, retrofitting and straightening of the building.

The construction has several deficiencies of which the most important is the differential settlement of about 25 cm in transversal direction, respectively 10 cm in the longitudinal direction. The building has important torsion effects due to the irregular plan shape. The retrofitting was carried out through the seismic isolation method, reducing drastically the torsion effects and the forces in the structural elements. Even if the building is still irregular it behaves like a regular one with insignificant torsional effects.

Figure 14.10 presents, for comparison, the mean displacement response of the isolated and fixed base building structure in both horizontal directions.

Figure 14.11 presents, for comparison, the mean acceleration response of the isolated and fixed base building structure in both horizontal directions.

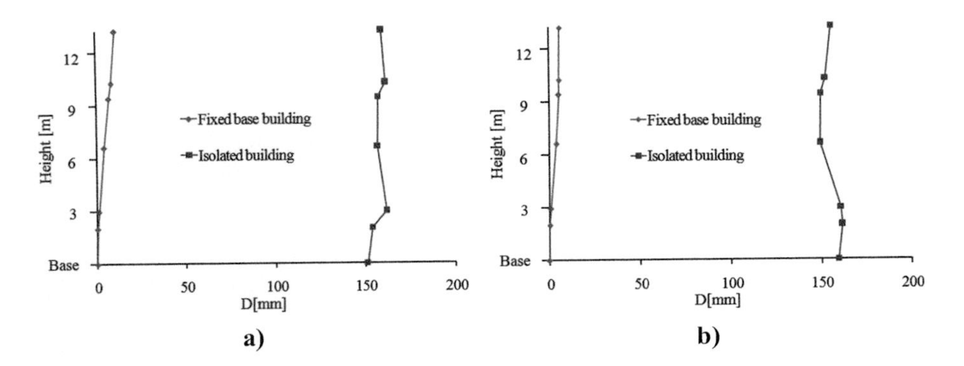

Fig. 14.10 The mean relative displacements of the isolated and fixed base building: (**a**) x – direction; (**b**) y – direction

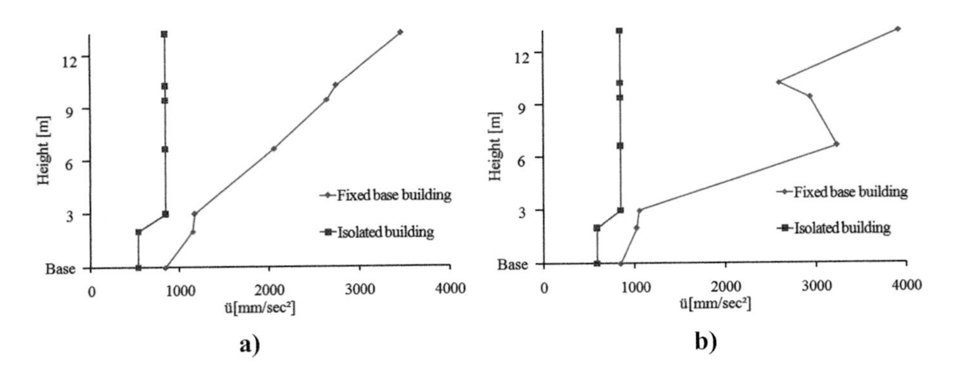

Fig. 14.11 The mean absolute accelerations of the isolated and fixed base building: (**a**) x – direction; (**b**) y – direction

Figure 14.12 presents the mean shear forces on the isolated and fixed base building in both horizontal directions.

The evaluation of the safety index for the isolated building structure, was leading to the following values:

- For x – direction: Capacity/Demand = 1.86;
- For y – direction: Capacity/Demand = 1.27.

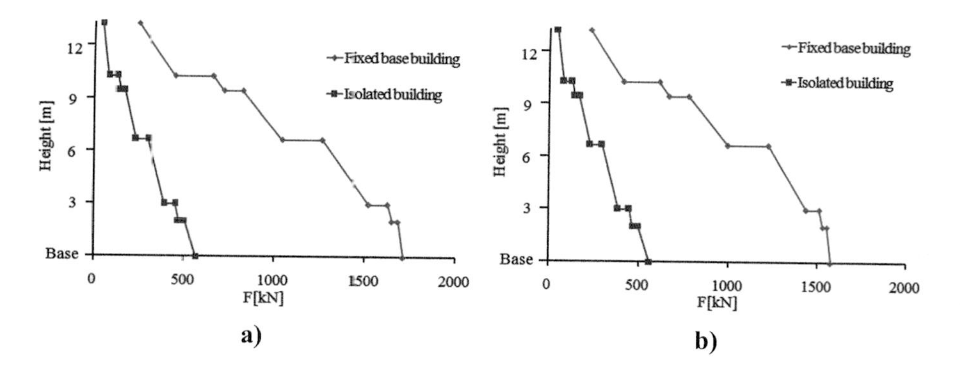

Fig. 14.12 The mean shear forces on the isolated and fixed base building: (a) x – direction; (b) y – direction

References

Etabs [computer software] (2015) Computer & Structures. Inc

Monitorul Oficial al Romaniei, Partea I, Nr. 807 bis/26.IX.2006 (2006) Cod de proiectare pentru structuri din zidărie, indicativ CR 6-2006. București

SeismoArtif [computer software] (2012) Pavia: SeismoSoft srl. Available: http://www.seismosoft.com

Universitatea Tehnică de Construcții București (2006) Cod de proiectare seismică P100: Partea I – Prevederi de proiectare pentru clădiri. P100-1/2006. București

Universitatea Tehnică de Construcții București (2008) Cod de evaluare si proiectare a lucrărilor de consolidare la clădiri existente, vulnerabile seismic. P100-3/2008. București

Chapter 15
Assessment of Global Torsional Sensitivity of Common RC Structural Walls Layout Types

Ionuț Damian, Dietlinde Köber, and Dan Zamfirescu

Abstract The objective is to validate the conclusions of a study about the nonlinear behaviour of the asymmetrical RC structural wall systems that was based on the one storey equivalent model time history response. The main instrument of verification is the time history nonlinear response of the full 3D multistorey model. The time history analysis was conducted using Perform 3D software. Three structural types are selected for this study: type I (strong torsionally restrained, $\Omega_\theta = 1.4$) corresponding to the most common structural wall type structure in Romania; type VI – the central core structure with perimeter frames along one main direction; and type III the central core structure without perimeter frames ($\Omega_\theta = 0.3$). The parameters which were varied during the study are: (a) stiffness to mass center eccentricity, and (b) q –behavior factor. Nine spectrum compatible accelerograms were used. The seismic characteristics were the ones corresponding to Bucharest. The results were judged in function of two response parameters: (1) R – safety factor, the minimum capacity to demand displacement corresponding to a vertical element and (2) R_l – the drift amplification of the floor extremities due to torsion relative to pure translation. The main conclusions to be validated are:

1. If the substantial decrease of the q behaviour factor accounted for torsionally flexible structures has a significant beneficial effect on the safety factor.
2. If the perimeter frames have a beneficial effect for the behaviour of the central core structure and for the suitability of this type of structure for seismic areas.

Keywords Central core structure · Time history analysis · Nonlinear response · Safety factor · Drift amplification

---

I. Damian · D. Köber (✉) · D. Zamfirescu
Technical University of Bucharest, Bucharest, Romania
e-mail: ionut.damian@utcb.ro; dzam@utcb.ro

D. Köber et al. (eds.), *Seismic Behaviour and Design of Irregular and Complex Civil Structures III*, Geotechnical, Geological and Earthquake Engineering 48,
https://doi.org/10.1007/978-3-030-33532-8_15

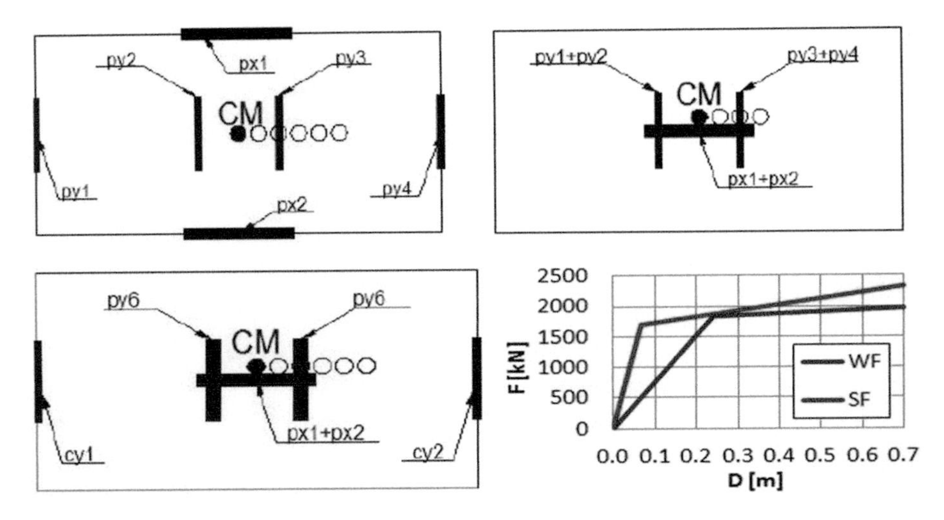

Fig. 15.1 Analyzed structural types: (**a**) torsionally restricted (type I); (**b**) central core structure without perimeter frames (type III); (**c**) central core type with perimeter frames (type VI); (**d**) Weak frames and stiff frames capacity curves

15.1 Introduction

Multi-storey buildings are very common for overpopulated cities. One of the plan layouts that covers, on one hand, architectural and usage needs and respects, on the other hand, structural requirements, is the central core system. Due to the concentration of stiffness and strength in the middle part of the plan layout, central core structures are (depending on the plan dimensions of the core with respect to the overall structural plan dimensions) often torsional sensitive.

The design of such structures is not regulated by nowadays seismic design codes (EN1998-1, P100-1/2013). This paper presents some main conclusions regarding the seismic behaviour of multi-storey plan irregular and torsional sensitive structures, as an attempt to offer guidelines for practical engineers. Three methods of investigation are used: (a) dynamic nonlinear analysis on simplified three degrees of freedom (3DOF) structures (Gutunoi 2014); (b) dynamic nonlinear analysis on multi degree of freedom (MDOF) structural models; (c) simplified SESA method (Köber and Zamfirescu 2009, 2010), suitable for practical design.

Three structural types are investigated (Fig. 15.1). The first one (type I) is a torsionally restricted structure ($\Omega_\theta = 1.4$), corresponding to the most common structural wall type structure in Romania. The second one is a central core structure with perimeter frames (type VI). The third one is a central core structure without perimeter frames (type III, $\Omega_\theta = 0.3$).

Eccentricities of 5%, 10%, 15%, 20% and 25% are considered by shifting the centre of mass from its symmetric position.

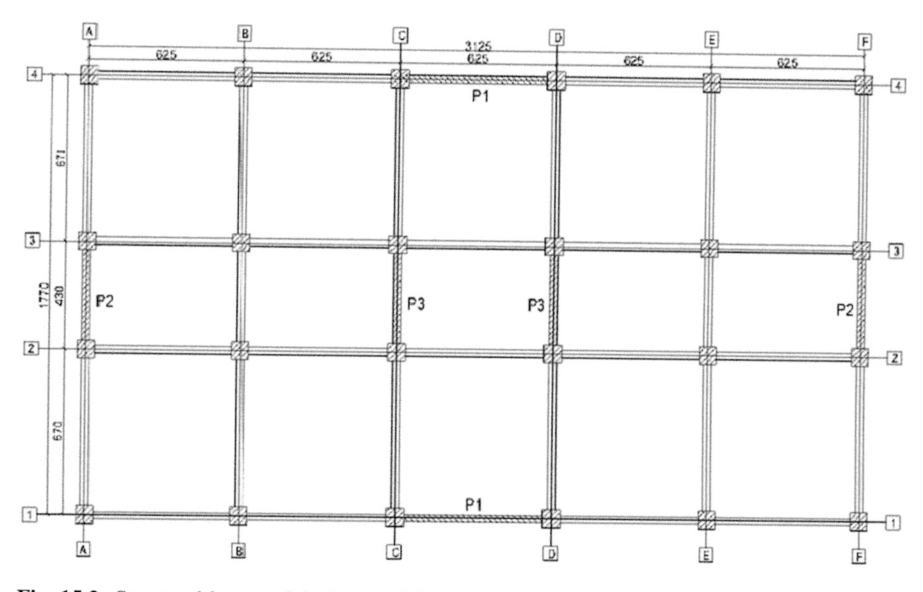

Fig. 15.2 Structural layout of the base building

The seismic input is given by a set of 9 accelerograms, one natural input (corresponding to the Vrancea '77 earthquake, NS registration) and 8 synthetic accelerograms, generated according to the elastic response spectrum from P100-1/2013 (Postelnicu et al. 2012). According to the seismic characteristics of Bucharest, a corner period $T_c = 1.6$ s was considered and a PGA $= 0.24$ g. Unidirectional seismic input along transversal direction of the building was applied.

Rayleigh damping model proportional with mass and elastic stiffness matrix was used. The percentage of critical damping was considered 5% for translational vibration mode along transversal direction and for torsional vibration mode.

15.2 Hypotheses and Parameters Used in the Study

The structural types analyzed are part of an ample study conducted at TUCEB in 2014 (Gutunoi 2014). Type I structure represents the base building (Fig. 15.2) and was designed according to P100-1/2006, the penult version of the Romanian seismic design code. The structure has 10 stories, each having 3 m height.

Each structural type was generated based on equivalent strength plan approach. The bilinear curve of each strength plan was generated based on pushover analysis of the whole building and of the individual axe. However, the force-displacement curve for an axe is not the same in the 3D structure compared to the 2D approach, especially for axes containing frames (B, E, 2 and 3 – Fig. 15.2). Therefore, the concept of stiff frames (SF) and flexible frames (FF) was introduced in order to take into account the different possible behaviour on 3DOF models (Fig. 15.1d).

For the simplified computation (methods (a) and (c) mentioned in Sect. 15.1) the real MDOF structures were reduced to 3DOF models, considering following main hypotheses:

- unlimited floor stiffness for axial forces;
- vertical regular structures were considered;
- fixed connections at base level;
- the deformed shape of the MDOF structure is described by the displacement at one level;
- definition of structural components by bilinear force-displacement curves;
- only centre of mass was shifted, stiffness centre is kept constant.

The correspondence between structural types (Fig. 15.1) and base structure (Fig. 15.2) is:

- *py1* and *py4* represent strength plans corresponding to axe A and half of B, and half of E and F, respectively;
- *py2* and *py3* represent strength plans corresponding to axe C and half of B, respectively half of E and D;
- *px1* and *px2* represent strength plans corresponding to axe 4 and 3, respectively axe 1 and 2;
- *cy1* and *cy2* represent strength plans corresponding to frames on axe B, respectively E.

The behaviour factor q was considered once 4.6 for all the structural types, which is the value used for the design of the base structure and then a reduced value of 3 corresponding to a central core structure for type III and type VI layouts.

The bilinear curves of the strength planes of equivalent 3DOF model are adjusted from the MDOF curves based on basic SDOF-MDOF equivalence method. When changing eccentricity, the strength of each plane is changing accordingly.

The rotation capacity of each element was evaluated according to EN 1998-3 and was converted to horizontal displacement capacity of each strength plan based on pushover analysis.

The influence of the hysteretic model was studied as well, knowing the typical cyclic behaviour of reinforced concrete structures. Three hysteretic models were considered:

- bilinear kinematic hardening (BKH);
- peak oriented model without unloading stiffness degradation (POM);
- peak oriented model with unloading stiffness degradation (POM USD).

Results from dynamic nonlinear analysis are evaluated by the following two parameters:

1. R – safety factor, the minimum capacity to demand displacement corresponding to a vertical element;
2. R_I – the drift amplification due to torsion relative to pure translation.

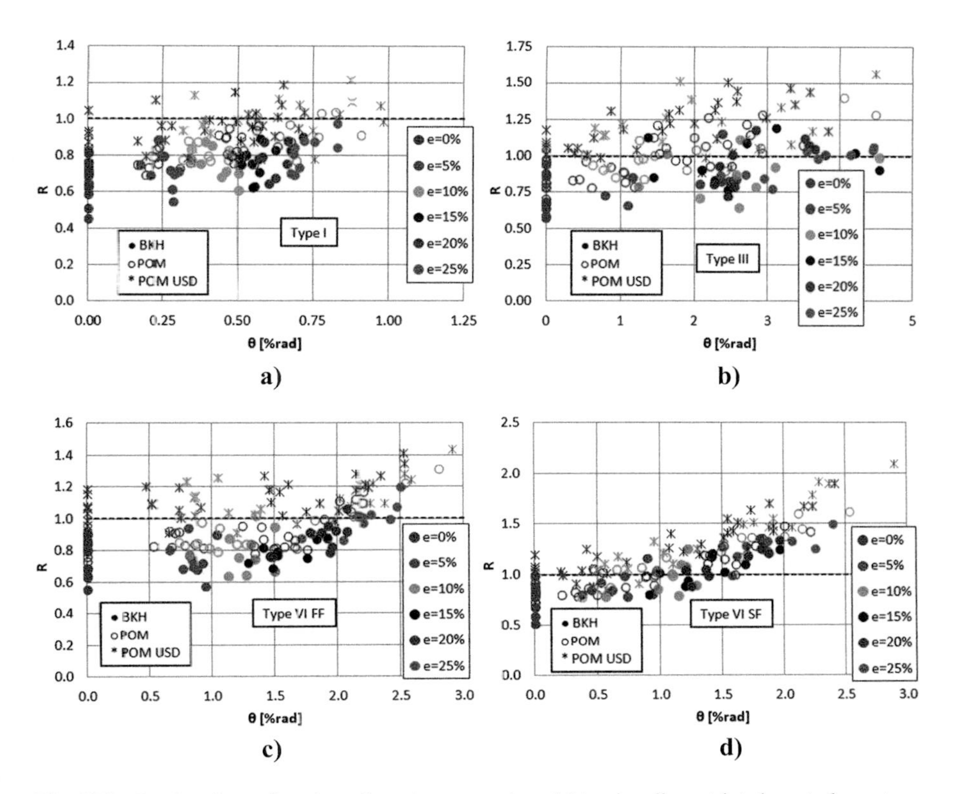

Fig. 15.3 *R* safety factor function of maximum rotation: (**a**) torsionally restricted; central core type (**b**) without perimeter frames; (**c**) with flexible perimeter frames; (**d**) with stiff perimeter frames

In order to check if a praxis suitable investigation method offers accurate results for central core structural layouts, the SESA method was applied and its results were compared to those from dynamic nonlinear analysis on simplified models, in terms of displacement of the centre of mass, structural edge displacement and structural rotation.

15.3 Results from Dynamic Nonlinear Computation on 3DOF Models (Method a)

Figure 15.3 shows results in terms of the safety factor R, with respect to the structural rotation θ, for all analyzed structural types and an unreduced behaviour factor, equal to 4.6.

The R safety factor is defined for the most loaded structural wall (py4, see Fig. 15.1). For the torsionally restricted structure (walls perpendicular to the direction of the seismic input do not yield) the structural response is predictable and

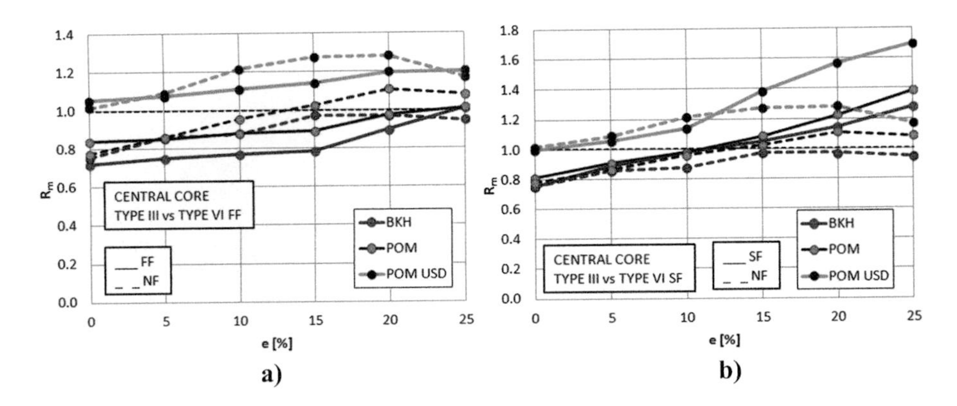

Fig. 15.4 Mean value of R safety factor for central core structure (**a**) no perimeter frame vs. flexible frame; (**b**) no perimeter frame vs. stiff frame

stable, a rise in eccentricity corresponding to a greater rotation. The maximum R value does not match the maximum structural rotation.

For the central core structures the R safety factor is defined with respect to the perimeter frames. The frames were modelled once as stiff frames SF (counting 11.5% of the total structural stiffness) and once as flexible frames FF (counting 3.5% of the total structural stiffness). The contribution of the stiff perimeter frames to the structural response does not change the predominant torsional failure. Frames are fully loaded and experience high displacement amplifications due to torsion. For flexible frames and eccentricities up to 15% failure occurs in the most loaded wall (the nearest to the mass centre), while for greater eccentricity values failure occurs in the perimeter frames parallel to the seismic input.

The central core structure having flexible frames behaves better than the one having stiff frames (Fig. 15.4) because the flexible frames do not yield until maximum demand, and so, they have an important contribution in restraining the torsional structural response.

Figure 15.5 shows results in terms of safety factor R, with respect to the eccentricity, for all analyzed structural types and an unreduced behaviour factor, equal to 4.6. As it was expected, the most favourable behaviour corresponds to the torsionally restricted structure, while the most unfavourable behaviour corresponds to central core structure with stiff perimeter frames.

Figure 15.6 shows results in terms of the drift amplification R_I, with respect to the eccentricity, for all analyzed structures and an unreduced behaviour factor, equal to 4.6. It is important to note that for a torsionally restricted structure, increasing eccentricity produces drift reduction for several accelerograms when using BKH model. This is presented in Fig. 15.6a, where for BKH hysteretic model the mean is applied once to all accelerograms and once for accelerograms with positive amplification factors (+BKH). This happens because BKH hysteretic model dissipates a large amount of seismic energy.

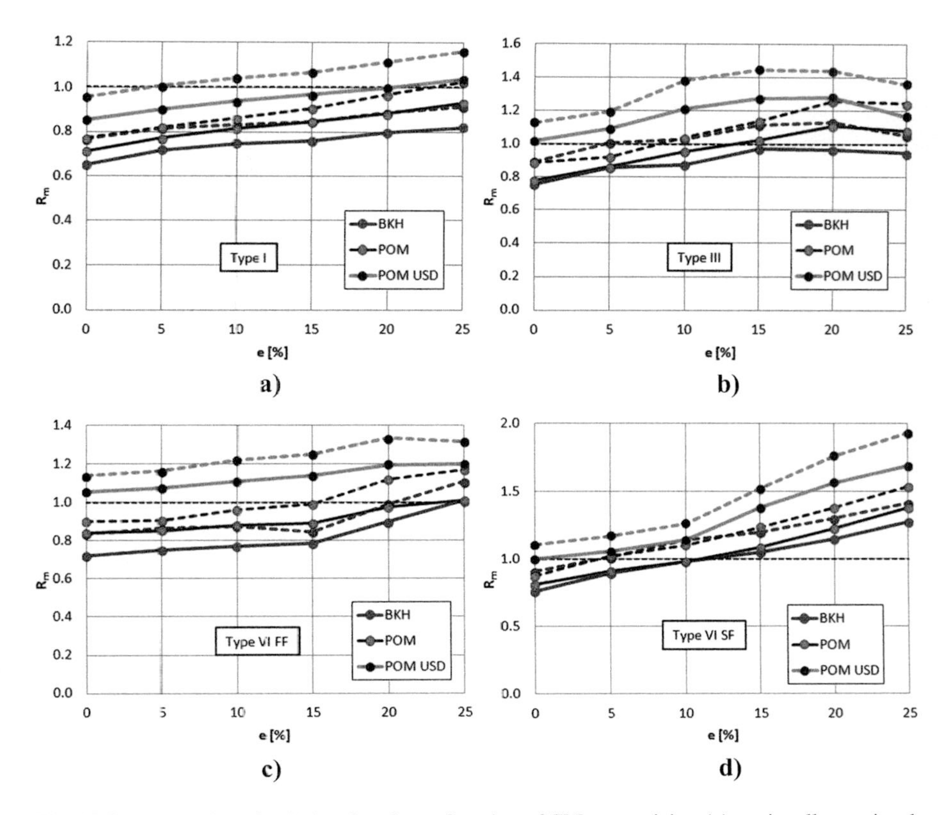

Fig. 15.5 Mean value of R (R_m) safety factor function of CM eccentricity: (**a**) torsionally restricted; central core type (**b**) without perimeter frames; (**c**) with flexible perimeter frames; (**d**) with stiff perimeter frames

Due to the fact that for eccentricities greater than 15% the torsionally restricted structure becomes sensitive to torsion (P100-1/2013) the study was repeated for a reduced behaviour factor value q = 3. In this case, the R safety factor has quite the same mean value as for q = 4.6 (Fig. 15.7a), only the coefficient of variation drops. Structural elements in longitudinal direction are more loaded, but still do not yield. For the torsionally restrained structure the behaviour factor reduction is not recommended, as the higher structural costs are not justified.

For the central core structure (type III) increasing strength produces decreasing displacement demand and safety factor R for all hysteretic models used (Fig. 15.7b). Therefore, it can be concluded that decreasing behaviour factor is recommended for central core structures.

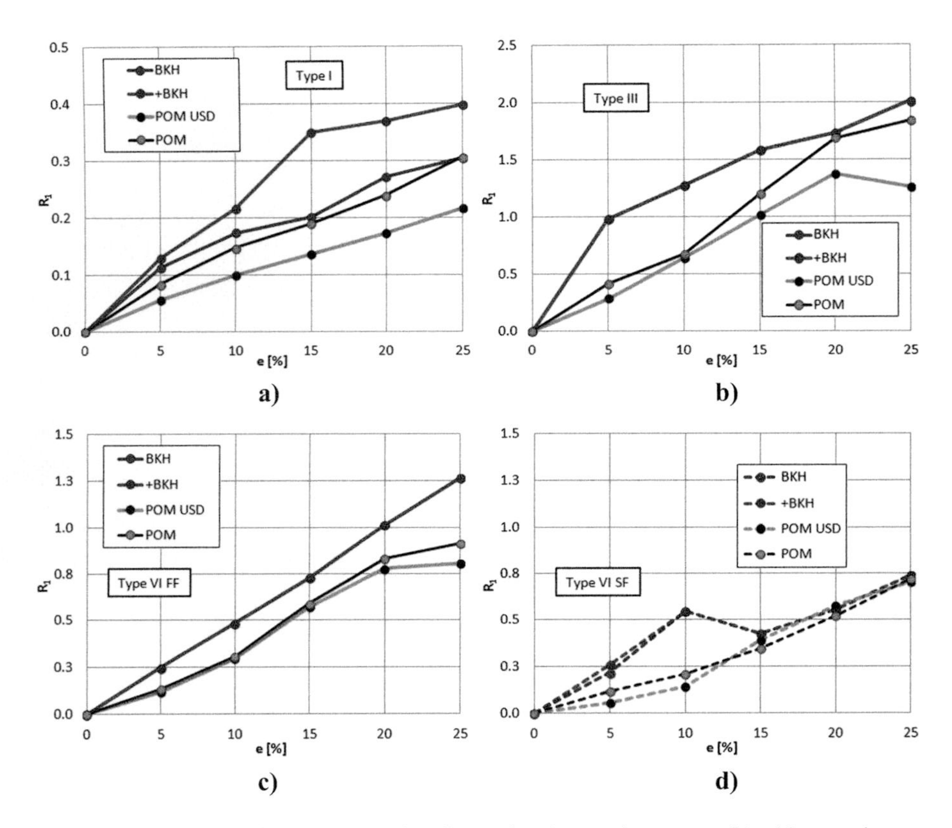

Fig. 15.6 R_1 drift amplification: (**a**) torsionally restricted; central core type (**b**) without perimeter frames; (**c**) with flexible perimeter frames; (**d**) with stiff perimeter frames

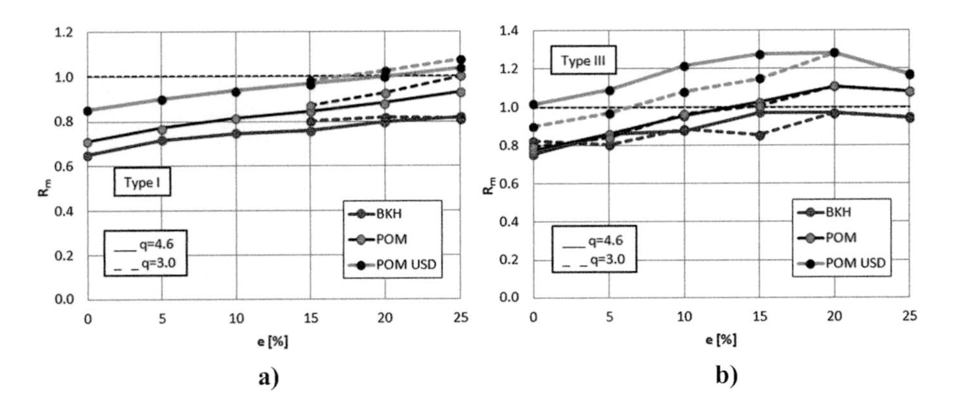

Fig. 15.7 Mean value of R safety factor for two values of behaviour factor (**a**) torsionally restricted structure; (**b**) central core structure without perimeter frames

15.4 Results from Dynamic Nonlinear Computation on 3D Models (Method b)

Dynamic analysis of the MDOF structures was performed in PERFORM 3D software (CSI 2006). Only torsionally restricted (type I) and central core structure with perimeter frames (type VI) were modelled.

The nonlinearity of beams and columns was modelled with plastic hinges lumped at the end of elements. The hysteretic behaviour of the hinges is Takeda type. The nonlinearity of walls was modelled with nonlinear panel elements. The panel has a nonlinear behaviour in combined bending and axial force, and a linear behaviour for the action of shear. The nonlinearity is introduced by the constitutive curves of materials. In general, a wall is modelled with a single panel in a story. Therefore, in the median height of the story there is a fiber section, and each fiber has a material associated with it. The plastic length of the hinge is equal to the story height.

For the steel, an elastic-perfect plastic curve was used. Tensile strength of concrete was neglected. Compressive curve of concrete was considered trilinear, with elastic stiffness 80% from the uncracked stiffness in order to take into account plasticity produced by other phenomena such as shrinkage.

In order to calibrate the model, a pushover analysis was performed. The analysis revealed that, although the fundamental period of the building was not the same as for the 3DOF model due to fiber hinge modelling of walls, the period secant to yielding point is equal to that of 3DOF model (Fig. 15.8). Furthermore, the analysis provided capable displacements of each strength plane, based on capable plastic rotations of walls and columns.

Figure 15.9 presents the comparison of safety factors provided by equivalent 3DOF model and by MDOF model. It can be observed that the results are quite similar for all hysteretic models used for 3DOF model. The most appropiate hysteretic model for simplified 3DOF model was proven to be Takeda, as it can be observed from the plots in Fig. 15.9 (red vs. black curves).

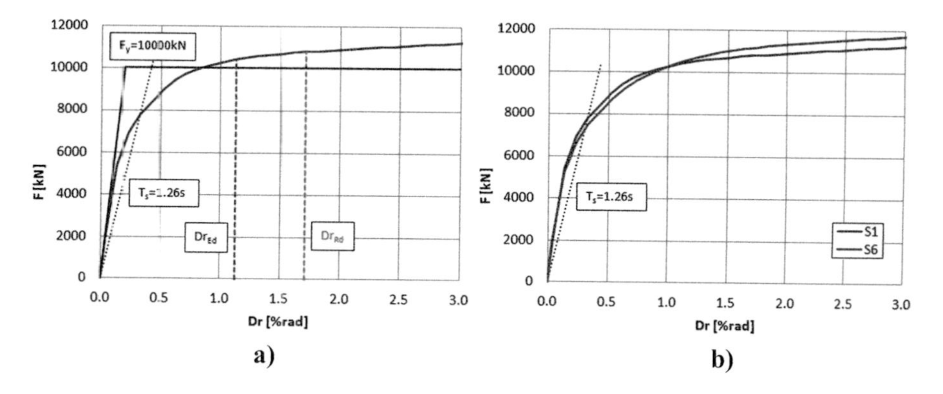

Fig. 15.8 (**a**) Pushover curve of the base building (type I); (**b**) Comparison of the pushover curves for S1 and S6 buildings

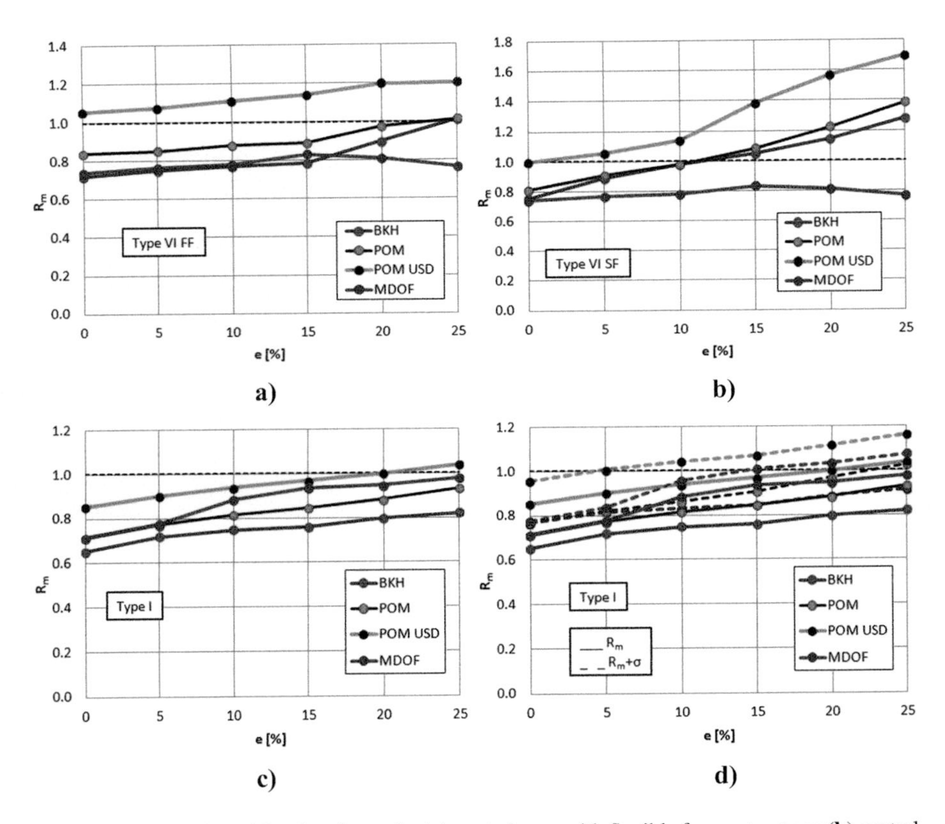

Fig. 15.9 Mean value of R safety factor for (**a**) central core with flexible frame structure; (**b**) central core with stiff frame structure; (**c**) torsionally restricted structure; (**d**) torsionally restricted structure (mean plus standard deviation)

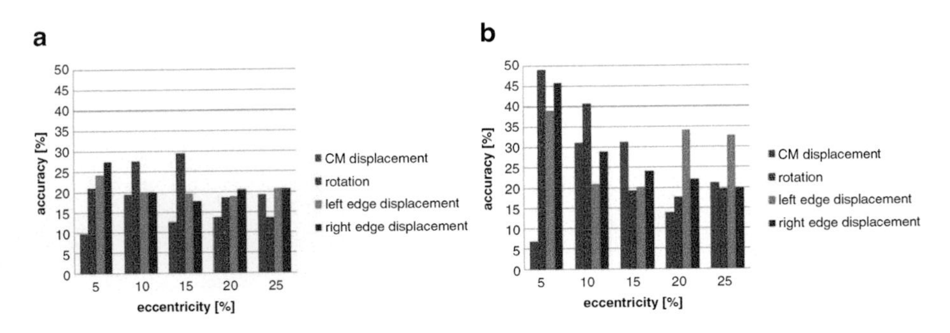

Fig. 15.10 Results from the SESA method considering a reduced stiffness for the perimeter frames: (**a**) stiff frames; (**b**) flexible frames

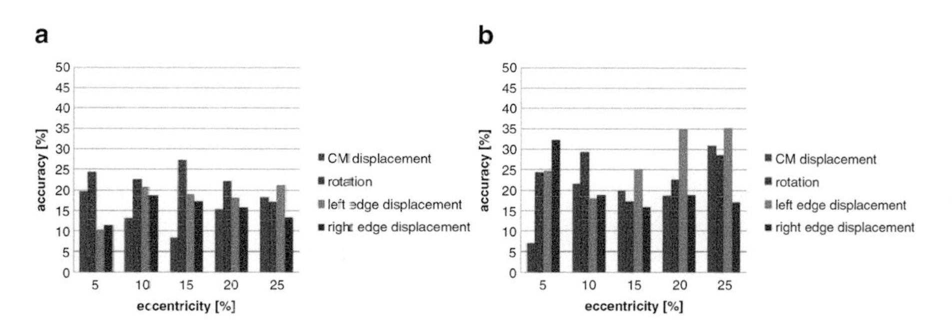

Fig. 15.11 Results from the SESA method considering the elastic stiffness for the perimeter frames: (**a**) stiff frames; (**b**) flexible frames

15.5 Results from SESA (Method c)

Results from the simplified SESA method are presented in Figs. 15.10 and 15.11.

SESA estimates the structural seismic response of a single story irregular system (Zamfirescu 2000a, b), using modal response spectrum analysis. The inelastic behaviour is considered with the help of the capacity spectrum method, where the nonlinear system is equated to an elastic system, both having the same translational behaviour. A linear equivalent system results which has a secant stiffness indicated to the maximum displacement. Viscous damping characteristics are set equivalent to the hysteretic damping properties of the initial system. The main advantage of the SESA method is its simplicity (it applies spectrum analysis), nevertheless it may indicate the displacement amplification given by torsional effects (Goel and Chopra 1990).

By iteration the equivalent damping ratio is chosen such that the displacement of both systems (the equivalent linear one and the inelastic system) are equal.

The SESA method was applied to the central core structure with perimeter frames (type VI) in order to evaluate in a praxis suitable way the perimeter frame participation to the total structural seismic response.

Frames were considered once with a reduced stiffness, accounting for the cracked concrete due to yielding of reinforcement (as provided in nowadays seismic design codes) and once with their elastic stiffness.

Results are mean values for the whole set of 9 accelerograms.

SESA shows the same response trend as the dynamic nonlinear analysis on 3DOF models. Results show a better accuracy when considering the elastic stiffness of the frames, proving that frames participate with their whole stiffness to the torsional response, even if for translation their stiffness is low compared to that of the walls.

SESA generally underestimates the left side floor displacements and overestimates the right side floor displacements by up to 25% (20%) for flexible (stiff) frames.

15.6 Conclusions

There are several conclusions to be pointed out, regarding the study of torsion in structures and the behaviour of buildings to seismic actions:

(i) 3DOF model is an excellent tool for studying torsion of buildings and provides results similar to MDOF model;
(ii) POM without unloading stiffness degradation is recommended for the analysis of 3DOF model;
(iii) flexible perimeter frames improve significant the behaviour of central core structures;
(iv) increasing structural strength improves the behaviour of central core structures and is not justified for torsionally restricted structures;
(v) SESA shows the same deformation trend compared to dynamic nonlinear analysis on 3DOF models.

References

CSI (2006) Perform components and elements for perform 3D and perform collapse, computers and structures, Berkeley, CA
EN 1998-1: 2004 Design of structures for earthquake resistance. General rules, seismic actions and rules for buildings, 45–69
Goel RK, Chopra AK (1990) Inelastic seismic response of one – story, asymmetric – plan systems. College of Engineering, University of California at Berkeley, Report No. UBC/EERC – 90/14
Gutunoi A (2014) Aspecte specifice ale răspunsului seismic de torsiune în domeniul neliniar. Dissertation, Technical University of Civil Engineering Bucharest (in Romanian)
Köber D, Zamfirescu D (2009) Simplified methods used for evaluation of the displacement gain due to general torsion. Sci J Math Modell Civil Eng 5(2):32–51
Köber D, Zamfirescu D (2010) Effects of general torsion on structural displacements. In: Proceedings 14 ECEE, Ohrid, Macedonia, ISBN:978-608-65185-1-6
P100-1/2013 Cod de proiectare seismică. Prevederi de proiectare pentru clădiri, 32–61 (in Romanian)
Postelnicu T, Damian I, Zamfirescu D, Morariu E, (2012) Proiectarea structurilor de beton armat în zone seismice, MarLink (in Romanian)
Zamfirescu D (2000a) TORSDIN – program de calcul dinamic neliniar al structurilor cu 3 GLD
Zamfirescu D (2000b) SINEL – program de calcul al spectrelor neliniare de răspuns

Chapter 16
Seismic Design Particularities of a Five Story Reinforced Concrete Structure, Irregular in Plan and Elevation

Dietlinde Köber

Abstract The design of structures subjected to torsional effects is highly complex, and becomes more difficult to assess when elevation irregularity is also present. This paper has as main objective to present the particularities of design for a plan as well as elevation irregular structure (11.0 × 15.0 m layout and 15.50 m total height), built in Bucharest (corner period of 1.6 s and high displacement demand). Architectural considerations, owners wishes and structural needs join together in a complex layout: uneven in plane distribution of vertical earthquake resistant structural elements (RC walls and frames); eccentric vertical stair case circulation (coupled RC walls); one structural wall present only at ground level; at the last level the layout shrinks to the staircase circulation perimeter. In order to meet the code requirements the design had to be based on modal analysis because of the interference of translational and torsional movements. As a praxis suitable computation tool static nonlinear analysis was performed (using the SAP software, CSI analysis reference manual, University Avenue. Computers and Structures Inc., Berkeley, 1995. RC walls are modeled as frame elements with point hinges. The structural performance is evaluated in terms of structural displacements (in the center of mass and on the floor edges). The order of plastic hinge formation and their damage degree is traced. The efficiency of the chosen structural layout with respect to the reduction of general torsion effects is pointed out and the behaviour factor option is outlined. Linear dynamic analysis is performed also, provididng an upper bound for the expected displacement amplification in the nonlinear range of behaviour.

Keywords Design · Nonlinear behaviour · Static nonlinear analysis · Torsional effects

---

D. Köber (✉)
Technical University of Bucharest, Bucharest, Romania

D. Köber et al. (eds.), *Seismic Behaviour and Design of Irregular and Complex Civil Structures III*, Geotechnical, Geological and Earthquake Engineering 48,
https://doi.org/10.1007/978-3-030-33532-8_16

16.1 Introduction

In engineering praxis often plan or elevation irregular structures are built. According to seismic code regulations in this case design should be performed by modal analysis and a reduction of 20% of the behaviour factor value is recommended if elevation irregularity is present. Therefor usually no supplementary check in the nonlinear range of behaviour is considered.

This paper presents a both plan and elevation irregular structure that is about to be built in Bucharest. Its complex geometry, the changes in stiffness along the elevation, the layout shrink at last level and the plan layout extension from the ground level up, are all structural particularities for which the question arises how they will influence the structural behaviour in the nonlinear range.

Although the limitations of the modal static nonlinear (push-over) analysis for plan irregular structures (MPA) are known, (Fajfar et al. 2005), it remains a praxis suitable method that offers an insight to the nonlinear structural behaviour. The main disadvantage of the MPA is the fact that the seismic input respects an elastic modal distribution (Bhatt et al. 2010; Belejo and Bento 2014). Nevertheless, the MPA method may indicate the "weak" structural elements and the overall ductility demand for the structure.

For the analysed structure, expectations from modal analysis are confronted with static nonlinear computation results and conclusions regarding the design are drawn.

The 3D structural model was investigated for different angles of incidence of the seismic input.

For seismic input in X direction (direction with the highest eccentricity) the torsional displacement amplification was computed also by linear dynamic analysis, as an upper bound for the expected torsional amplification in the nonlinear range of behaviour (Fajfar et al. 2005).

16.2 Description of the Analysed Structure

The analysed structure has plan dimensions of approximately 11 × 15 m for the first four levels and 3 × 4 m at the last level (where only the vertical circulation wall assembly remains).

It is plan as well as elevation irregular. The plan irregularity is due to the position of the vertical circulation wall assembly and to the structural wall situated only on one side in X direction at underground level.

The irregularity in elevation is due to the structural retreat of about 38% in X direction and 45% in Y direction at the last floor. Supplementary elevation irregularity is encouraged by a 10% structural extension in X direction from the ground floor to the top, which is not supported at underground level.

In order to check plan irregularity, the rule which compares the mean floor displacement for two opposite corners with the maximum displacement at one corner

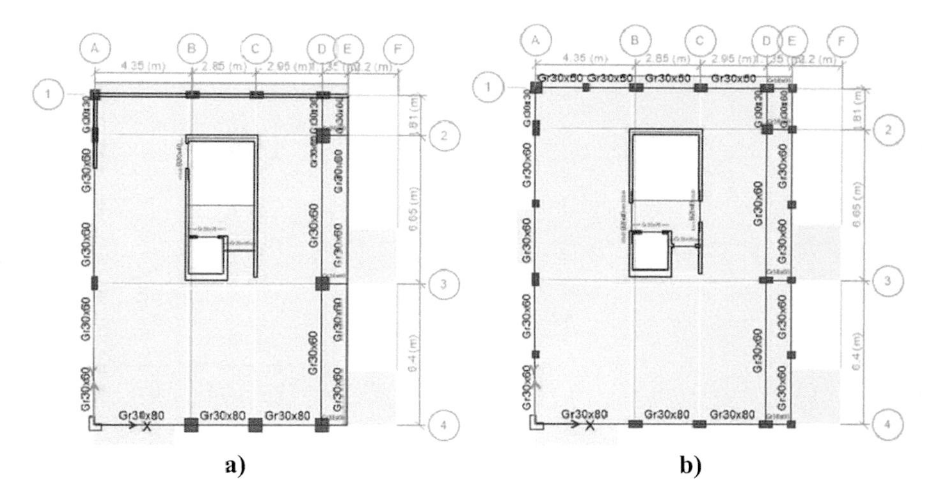

Fig. 16.1 Plan layout: (**a**) underground level; (**b**) current level

was used, stating that plan irregularity is given if the difference exceeds 35%, (P100-1/2013). The displacement values were considered from modal analysis. An upper bound of these results was given by linear dynamic analysis.

Following this rule, the first structural approach was characterized by a displacement difference of 45% in X direction and 27% in Y direction. The second eigen mode was predominantly torsion. Given the architectural layout limitations, following structural changes were performed in order to reduce plan irregularity:

- beams in elevation 4 (the flexible side) were enlarged (from 25×40 to 30×80 at underground level and 30×60 at all other levels).
- coupling beams in Y direction where reduced (from 25×95 to 20×40)
- walls of the vertical circulation were reduced in Y direction from 25 to 20 cm.

The final structure (see Fig. 16.1) is characterized by a displacement difference of 29% in X direction and 6% in Y direction (under consideration of 5% accidental eccentricity). The first two eigen modes are predominantly translation, with mass participating ratios of 60% in X direction and 67% in Y direction.

Linear dynamic analysis shows a maximum expected displacement difference of 35% (it is a mean value for the considered accelerogram set).

Floors are supported by beams only along the perimeter of the structure. Due to architectural reasons most columns change their shape from the underground to the ground level.

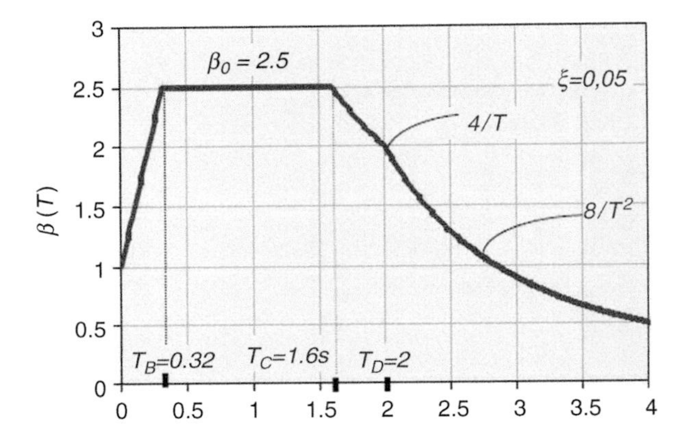

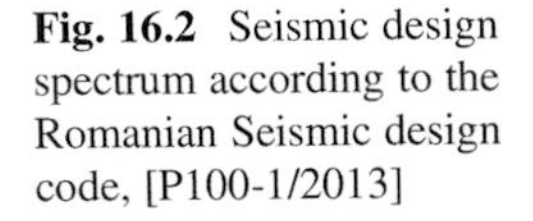
Fig. 16.2 Seismic design spectrum according to the Romanian Seismic design code, [P100-1/2013]

16.3 Seismic Particularities and Methods of Analysis

The structure presented in this paper is about to be built in Bucharest (maximum expected ground acceleration of 0.30 g), a region characterized by large corner period values (Tc = 1.6 s). The seismic design spectrum according to the Romanian Seismic Design Code is shown in Fig. 16.2, where *β(T)* is the ratio between the response acceleration and the peak ground acceleration.

Design was performed according to modal analysis results. In order to check the nonlinear behaviour of the structure, static nonlinear analysis was applied, as a praxis suitable method. A triangular distribution of forces was considered, at incidence angles of 45 and 90 degrees.

Plastic hinges were modelled as point hinges [EN 1998-3] and structural walls as frame elements. Due to its geometry, the wall situated at underground level in the northern part of the layout was modelled as shell element and no plastic deformation was considered for it.

The design of the structure was performed according to modal analysis results and considering a behaviour factor value of 4.0. A reduction of 20% of the behaviour factor was applied according to code regulations for elevation irregular structures.

According to modal analysis results, the walls are capable of taking over the entire base shear force. Columns are reinforced for minimum detailing requirements (1% longitudinal reinforcement).

In order to have an upper bound for the expected torsional displacement amplification, linear dynamic analysis has been performed considering a set of five spectrum compatible accelerograms and two natural inputs (Vrancea '77 NS, Vrancea '77EW, Incerc registration).

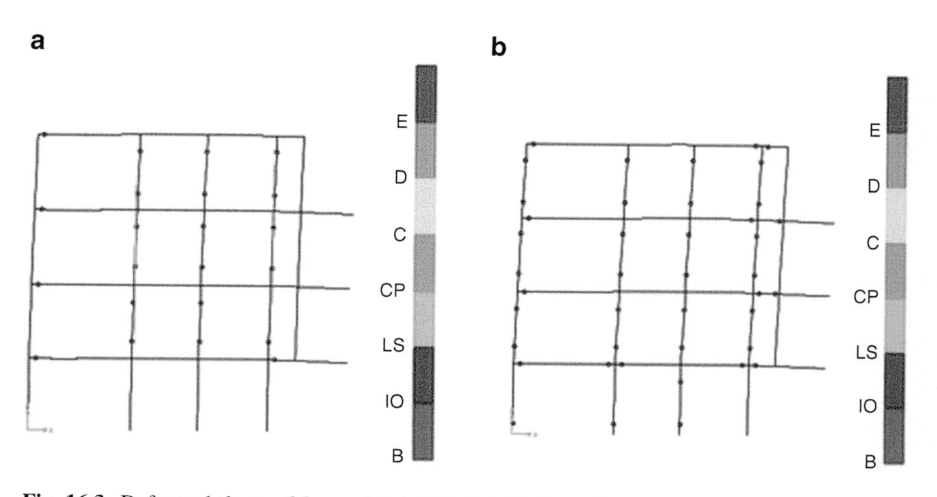

Fig. 16.3 Deformed shape of frame 4: (**a**) at first yield; (**b**) at first structural failure

16.4 Observations Regarding the Nonlinear Behaviour

16.4.1 General Remarks

The expected behaviour of the analysed structure in the hypothetical situation of in plane regularity, may be resumed as follows:

- plastic hinge formation at ground level for walls and columns, due to the greater stiffness of the underground level
- maximum displacement demand in the structural elements of the ground level
- important amount of plastic hinge formation in walls and less in columns and beams

Due to in plane irregularity, those expectations change and following observations regarding the nonlinear structural behaviour can be made:

- plastic hinge formation throughout the elevation, less at underground level
- maximum displacement demand on the flexible side of the structural layout and at the last level
- important amount of plastic hinge formation in columns and less in beams (although capacity design has been applied), see Fig. 16.3.

In Fig. 16.3 the pink dots indicate plastic hinge yielding lower than the Serviceability Limit State threshold and the blue dots yielding above this limit but lower than the Ultimate Limit State threshold.

Due to the layout shrink at the top of the structure, the walls at the bottom of the last level experience more plastic deformation than the ones at the levels below.

The highest torsional movement is registered for earthquake in X direction, due to the fact that the greatest eccentricity is present along Y direction.

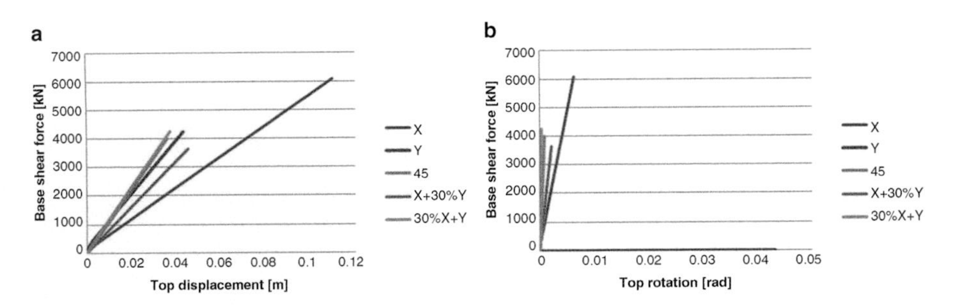

Fig. 16.4 Structural results: (**a**) base shear force-top displacement; (**b**) base shear force-top rotation

Table 16.1 Structural over strength (static nonlinear analysis results at first failure versus static nonlinear analysis results at first yield)

	Maximum base shear force/design base shear force	Maximum top displacement/top displacement at first yield
X-direction	3,6	12,0
Y-direction	2,5	9,0
45 degrees	2,3	4,7
X + 30%Y	2,1	6,0
30%X + Y	2,5	6,5

16.4.2 Push-over Curves

Figure 16.4 shows the relation between the base shear force and the deformation at the last complete level (displacement or rotation).

The slope of the force-displacement diagram indicates that the overall yielding of the structural elements is low. The resistance of the structure is high and limited ductility demand is requested from the plastic hinges.

The structure has lowest stiffness for seismic input in X direction. Therefor under this loading situation maximum displacement at top level is expected. Highest structural stiffness is given in Y direction.

In terms of structural rotations at top level the highest demand is expected for combined earthquake input (45 degrees or 30%X + Y). For seismic input in Y direction almost no torsional effects are present.

The base shear force value at which the first structural failure (in the staircase walls at the bottom of the last level) occurs, is shown in Table 16.1 for all considered incidence angles of the seismic input. The minimum over strength is registered for predominant earthquake in X direction and 30% earthquake in Y direction.

Table 16.2 Ductility demand

	Level	Element
X-direction	4	Staircase
	2,3	Columns axe 4
Y-direction	4,3	Staircase
	1	Wall axe C
45 degrees	4, 3	Staircase, coupling beam
X + 30%Y	4	Staircase
30%X + Y	All	Staircase
	1	Wall axe C

Nevertheless all values exceed 2.0 showing at least a double structural strength compared to the design base shear force.

Comparing the displacement at first yield and the one at the registration of the first structural failure, the choice for the behaviour factor may be verified. Values are shown in Table 16.1. The lowest behaviour factor can be observed for earthquake at 45 degrees.

16.4.3 Ductility Demand

Table 16.2 shows the structural elements which experience the highest ductility demand for each incidence angle of the seismic input. Levels are denoted with 0–4 from the underground level up.

The element with the highest ductility demand is the staircase at the bottom section of the last level. This section is the first which exceeds its capacity, irrespective of the incidence angle considered for the seismic input.

The coupling beams connected to the elevator and staircase walls are loaded only for Y direction (or 30%X + Y) and do not exceed their capacity prior to the staircase section at the last level.

In X direction, columns on the flexible side of the structure experience also high displacement demands, due to the torsional component. They also don't exceed their capacity prior to the staircase section at the last level.

Among the inclined seismic input cases, the highest ductility demand is registered for the 30%X + Y load case and affects the structural walls. In Y direction the eccentricity is small and the deformation of the structure is mostly translation. For translation the walls are most loaded because they are the stiffer vertical elements.

16.4.4 Level to Level Deformations

In order to understand the overall seismic response of the analysed structure, translation in X and Y direction at first yield and at first structural failure are related

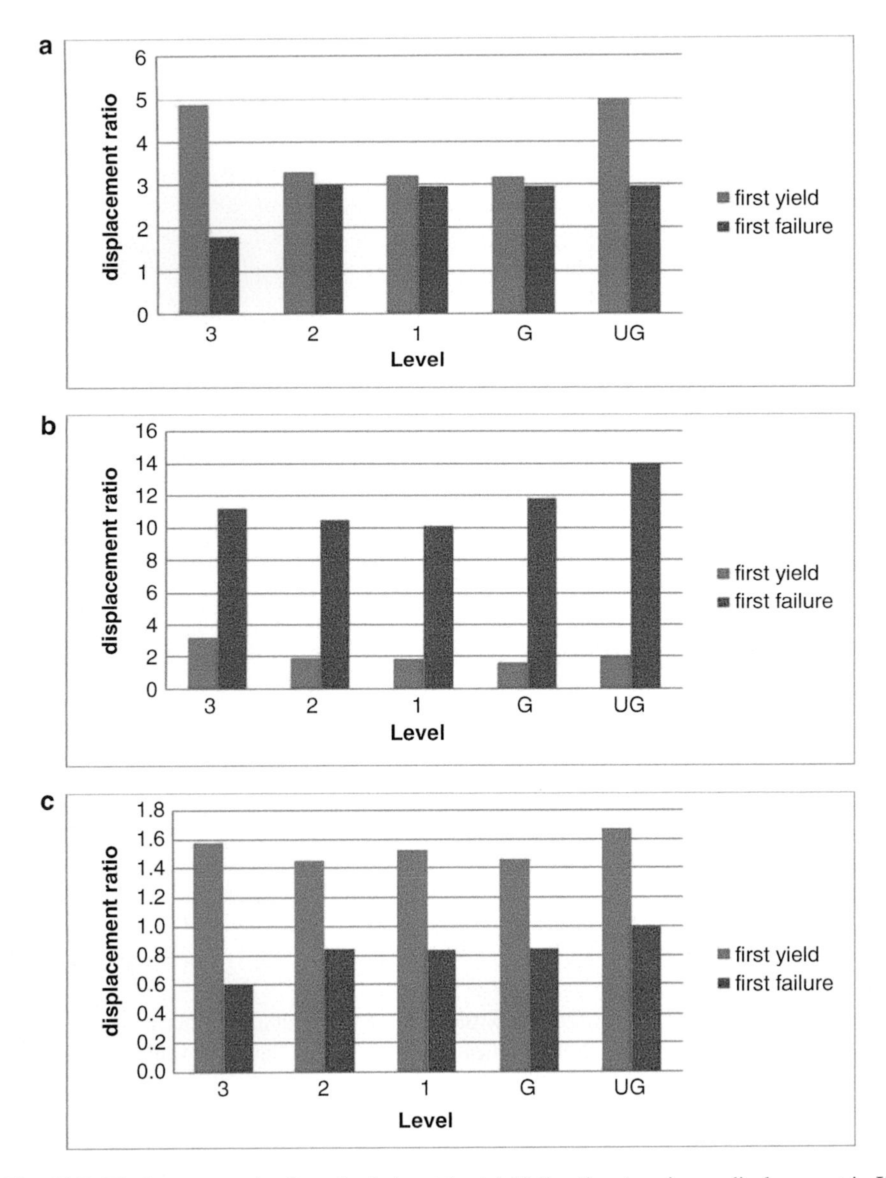

Fig. 16.5 Displacement ratios for seismic input in: (**a**) X direction (maximum displacement in X direction versus maximum displacement in Y direction); (**b**) Y direction (maximum displacement in Y direction versus maximum displacement in X direction); (**c**) 45 degrees (maximum displacement in X direction versus maximum displacement in Y direction)

and shown in Fig. 16.5 at each level. The displacements correspond to the point for which maximum displacement values are registered and is kept the same at all levels (except the last level). For X and Y direction greater values of the displacement ratio mean that the translational behaviour is predominant (for unidirectional seismic

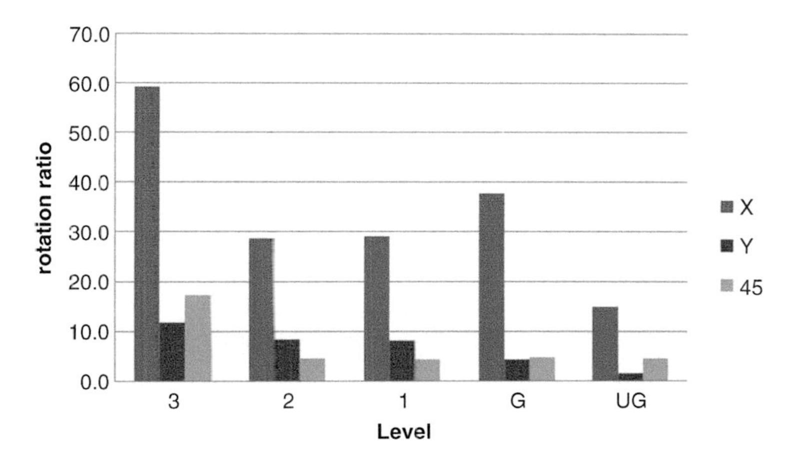

Fig. 16.6 Ratios between rotations at first yield and at first structural failure

input the displacement in the direction of the seismic input is higher than the displacement perpendicular to it).

Regarding the seismic input in X direction, at first yield the stiffness difference between the first two and the last two levels shows more translational behaviour than at the levels G, 1 and 2. Once the structure yields more, higher rotations are expected at the last level.

Regarding the seismic input in Y direction, the translational behaviour is more present after the structure yields more. The difference to the X direction may be explained by the fact that in Y direction the secondary displacement (perpendicular to the direction of the seismic input) does not rise with the seismic input.

Regarding the seismic input at 45 degrees, at first yield displacements in X direction are greater than displacements in Y direction. The predominance changes when structural yielding becomes more intense. This observation may be explained by the fact that in the elastic range of behaviour torsion is more violent than after yielding and causes greater displacements in X direction, which is more sensitive to torsional effects than the Y direction.

Figure 16.6 shows ratios between structural rotations at first failure and first yield, for all levels and seismic input in X direction, Y direction and at 45 degrees.

Due to the structural eccentricity, maximum rotations are registered for seismic input in X direction. No monotonic variation can be seen. For seismic input at 45 degrees structural rotations are quite constant for all levels having constant plan layout. For seismic input in Y direction a monotonic decrease of the structural rotation with the level number is present. The 45 degrees seismic input is more influenced by the stiffness in Y direction.

16.5 Conclusions

The design according to modal analysis results is conservative. The maximum expected base shear force at first structural failure is at least twice the design base shear force, depending on the direction of the seismic input. This force level is reached at displacement values that are lower than the displacement demand for the considered structure.

Nevertheless, referring the maximum expected displacement at first structural failure and the displacement at first yield, behaviour factor values of at least 4,7 are computed. The lowest behaviour factor value is determined for earthquake at 45 degrees.

Due to the fact that the structural strength at first structural failure is at least twice the design base shear force and that the choice for the behaviour factor is confirmed by the push-over curve, the displacement demand is not a concluding measure for the structural behaviour, in this case.

According to seismic design, the walls of the vertical circulation tube may take over the entire base shear force, frames being not active for horizontal forces. The static nonlinear analysis shows that frames are inactive when translation governs the structural behaviour (seismic input in Y direction), they become active due to torsion (seismic input in X direction). As expected, the most loaded frame is the one in axe 4 (the farest from the vertical circulation tube, where the centre of stiffness is situated).

Although capacity design was applied throughout the design process, columns tend to yield earlier than beams. This may be explained by the fact that having quite large spans, the beams have a limited contribution to the frame stiffness.

Due to plan irregularity high deformations are expected at the upper levels, even if the stiffness shift from the underground to the ground level would make us believe that at ground level high plastic deformations will be concentrated.

Due to elevation irregularity, the last level experiences the highest displacement demand and at its bottom the first structural failure is registered in the walls of the vertical circulation.

The static nonlinear analysis is performed by the SAP software only until the first failure occurs, not until structural instability appears. This may be explained by the fact that the structure is loaded in order to reach the control displacement indicated by the user, and as long as for higher plastic deformation frames become more active, shifting back the overall deformed shape of the structure, the control displacement will be never reached. Nevertheless the static nonlinear analysis works far enough in order to check the correct design of the considered structure.

Linear dynamic analysis results regarding the torsional displacement amplification for seismic input in X direction offer an upper bound of 35% for the expected displacement amplification in the nonlinear range of behaviour. From modal analysis the displacement amplification is of 27%.

Acknowledgements I wish to acknowledge the help of my colleagues from SC Consild SRL for their valuable contribution to the design of the structure presented in this paper.

References

Belejo A, Bento R (2014) Application of nonlinear static procedures for the seismic assessment of a 9-storey asymmetric plan building. Seismic behaviour and design of irregular and complex civil structures II. ISBN:978-3-319-14245-6, 123-134

Bhatt C, Bento R, Pinho R (2010) Verification of nonlinear static procedures for a 3D plan-irregular building in Turkey. In: Proceedings of the 14th European conference on earthquake engineering, Ohrid, Macedonia

CSI analysis reference manual (1995) University Avenue. Computers and Structures Inc., Berkeley

EN 1998-3: 2005 Design of structures for earthquake resistance. Assessment and retrofitting of buildings, 35–42

Fajfar P, Marusic D, Perus I (2005) Torsional effects in the pushover-based seismic analysis of buildings. J Earthq Eng 9(6):831–854

P100 -1/ (2013) Cod de proiectare seismică. Prevederi de proiectare pentru clădiri, 32–61

Chapter 17
Effect of the Mechanical Properties of Concrete on the Seismic Assessment of RC Irregular Buildings

Mario De Stefano, Marco Tanganelli, and Stefania Viti

Abstract In this work the role of the effective concrete strength on the seismic assessment of RC buildings has been investigated. The concrete strength has been described after a wide investigation on the structural elements of a case-study, i.e. an existing building located in Sansepolcro (Italy), currently used as a Hospital. Alternative representations of the concrete strength distribution have been compared, consisting respectively of: assuming a uniform strength distribution for the entire structure (as required by the International Codes), assuming a *mean* storey strength at each storey and assigning to each tested member the strength value found by the experimental survey.

The seismic performance of the building has been found with reference to three different limit states, by representing the structural response through a nonlinear static analysis and comparing the maximum chord rotation and shear stress to the limit values corresponding to the assumed concrete strength of each member. The effects of the actual distribution of the concrete strength have been finally checked by comparing the seismic performance obtained through the three different assumptions for concrete strength distribution.

Keywords Existing RC buildings · Concrete compressive strength · Experimental assessment of the concrete strength · Assessment of the seismic performance of RC buildings · Randomness in the mechanical properties of concrete

17.1 Introduction

The assessment of RC buildings is usually made through numerical analyses, whose results depend on the adequacy of the adopted structural model and the accuracy of the mechanical properties assumed for materials. The performance level of

M. De Stefano · M. Tanganelli · S. Viti (✉)
Department of Architecture (DiDA), University of Florence, Florence, Italy
e-mail: mario.destefano@unifi.it; marco.tanganelli@unifi.it; viti@unifi.it

D. Köber et al. (eds.), *Seismic Behaviour and Design of Irregular and Complex Civil Structures III*, Geotechnical, Geological and Earthquake Engineering 48,
https://doi.org/10.1007/978-3-030-33532-8_17

buildings, anyway, is achieved through conventional procedures, i.e. following the instructions provided by the Technical Codes. According to Eurocodes, the mechanical properties assumed for material must be quantified as a function of the knowledge levels achieved for the structure, and – consequently – by reducing the material capacity through proper safety factors. Whichever the knowledge level is, the stress capacity assumed for material is considered to be constant for all the structural members. This assumption, however, is far away to be realistic, since RC structures, especially when constructed without any specific attention to the quality procedures, can present a high variability in mechanical properties, particularly the concrete ones.

In recent years, many Authors have investigated the concrete strength characterization, focusing their attention both on the choice of the parameters to assume in the characterization (Masi and Vona 2009) and on the variability of the strength, with the consequent assumption of a suitable value for analysis (Franchin et al. 2007, 2009; Fardis 2009; Masi et al. 2008; Rajeev et al. 2010; Jalayer et al. 2008; Cosenza and Monti 2009; Marano et al. 2008; Monti et al. 2007; D'Ambrisi et al. 2013; De Stefano et al. 2013a, b, 2014, 2015a, b, 2016).

In this paper the seismic response of an existing RC 4-storey building has been found by accounting for the actual distribution of the concrete strength. A detailed investigation has been performed on the structural elements, including 8 cores and 44 SonRebs.

Alternative representations of the strength distribution have been compared. The first one follows the Code provisions, and it consists of assuming a uniform strength distribution, equal to the *mean* value, in all members. The second one, instead, consists in assuming, at each storey, the *mean* strength of such storey, found through the experimental tests. The third one, finally, consists of assigning to each tested member the strength value found by the experimental survey, leaving the *mean* strength to the other ones.

The capacity of the building has been evaluated through a nonlinear static analysis, considering the seismic input related to the Life Safety limit state. The seismic performance of the building has been assessed by comparing the achieved maximum response, expressed in terms of chord rotation and shear stress, to the limit values found according to the assumptions made for the concrete strength. The effects of the actual distribution of the concrete strength have been checked by comparing the seismic performance obtained by the three different assumptions for concrete strength among resisting elements.

17.2 Case-Study

The case study, shown in Fig. 17.1, is a RC framed building, made in the 70s, located in Sansepolcro (Italy), and belonging to a Hospital complex. The building, 4-storey height, is rectangular in plan, with sides of 30.15 m and 18.90 m respectively, and 3 and 9 bays in the two directions (see Fig. 17.2).

Fig. 17.1 Views of the case-study

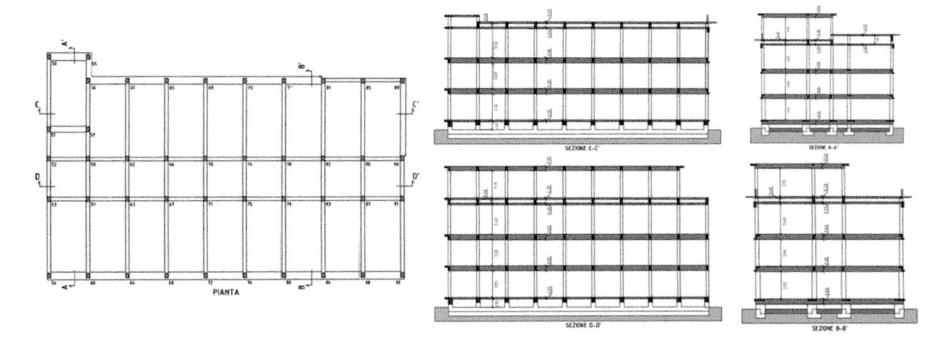

Fig. 17.2 Case-study: plan and sections

The RC structure is made by 10 transversal frames along the E-W direction, and by three RC frames along the N-S direction. It adjoins three other buildings; all buildings have their own structure, and they are separated to each other by joints, which do not entirely comply the seismic requirements. In this work the seismic behavior of the case-study has been considered as independent from the one of the adjacent buildings.

The columns have different sections at each storey, ranging between 30 × 50 cm at the first level, and 30 × 35 cm at the top one. The transversal beams (EW direction), which sustain the floors, have a cross section of 30 × 60 cm, with the exception of some beams on the external wall, which have a special S shape to sustain the outside infill panels. The longitudinal beams (NS direction) have a lower depth at the first level, whilst they have the same dimensions of the transversal ones (30 × 60 cm) at the upper storeys. Ceiling is made of two different floors, next each other, while the flat roofs includes 1 m long cantilever along the entire perimeter.

17.3 Concrete Characterization

17.3.1 The Concrete Characterization

The experimental investigation on the concrete strength has been performed by combining both destructive and not-destructive tests. The non-destructive test

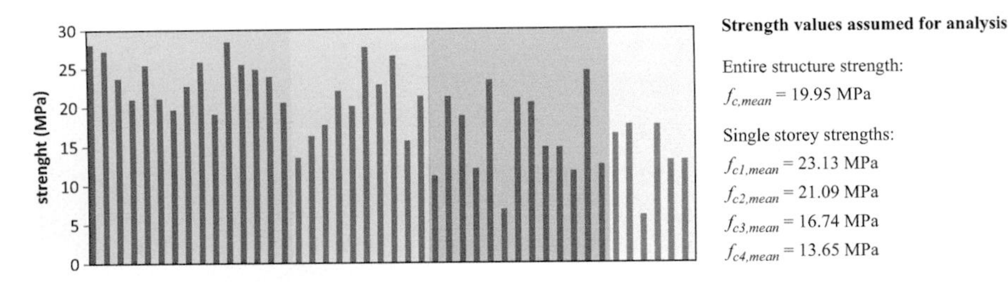

Fig. 17.3 Strength values obtained through the SonReb test and final strength values assumed for analysis

consists of 44 SonReb (SONic + REBound) lectures. The SonReb test provides an evaluation of the concrete strength ($f_{c,\ SR}$) through the cross-examination of the values of the speed of ultrasound waves (V_{us}) and the values of sclerometric index rebound (I_R).

Eight of the elements checked through the SonReb test have been subjected to the core crashing. The core sample testing, indeed, is considered to be the most effective procedure for the concrete mechanical characterization. The crushing test made on the (cylindrical) samples has been performed after a careful preparation of the sample (UNI 12390-2 (2002), 12390-2 (2009), 12504-1 (2009), 12390-1 (2012)), and by adopting a constant loading rate (UNI 2002); the obtained results ($f_{c,core}$) must be arranged to account for the possible uncertainties of the testing procedure and to the sample shape. Even in this case, different formulations (British Standard 1983; Concrete Society 1987; Cestelli Guidi and Morelli 1981) can be adopted to relate the experimental values to the one to assume for analysis. In this work the relationship proposed by Masi (2005) has been adopted, since it leads to refer to the cylinder strength of the samples, without requiring the preliminary transformation to cube values. The strength values provided by the core crushing and from the Masi relationship have been shown in Fig. 17.5.

The experimental data coming from the SonReb and the core crushing tests have been compared, in order to calibrate the SonReb results. Different formulations (Giacchetti and Lacquaniti 1980; Gašparik 1992; Di Leo and Pascale 1994; Cristofaro 2009) can be adopted to determine the concrete strength from the SonReb lectures; in most cases, the proposed correlation approaches present the same Eq. (17.1), differing only for the numerical parameters a, b and c.

$$f_{c,SR} = a \cdot (V_{us})^\wedge b \cdot (I_{\mathrm{R}})^\wedge c \tag{17.1}$$

In this work a proper (ad-hoc) relationship has been adopted, having the expression reported in Eq. 17.1, and the following numerical parameter: a $= 10^{-6962}$, b $= 1.518$, c $= 1.87$, found through the comparison between the experimental data coming from SonReb and core crushing. In Fig. 17.3 the values of the cylinder strength found by applying the ad-hoc relationship to the data provided by the SonReb analysis are shown. Each strength value is the mean of the couple of values found for the couple of lectures in the same element. The mean strength of the

Table 17.1 Considered concrete strength distributions

Considered cases	1st case	2nd case	3rd case
Concrete strength assumptions	$f_{c,mean}$ to all elements	$f_{c,mean,storey}$ to all elements of the same storey	$R_{c,lect}$, to the tested elements, $R_{c,mean}$ to all the other ones

sample data has been found by averaging the *mean* of the data found by the core crushing (arranged through Masi 2005) and the *mean* of the SonReb cylinder strength, excluding the elements checked by the core test. In Fig. 17.3 the strength values obtained for the entire structure ($f_{c,mean}$) and of each storey ($f_{c1,mean}$, $f_{c2,mean}$, $f_{c3,mean}$, $f_{c4,mean}$) are listed.

17.3.2 The Assumptions Made in the Analysis

Both European (Eurocode 8, EC8) and Italian (NTC 2008) Technical Codes suggest to describe the material strength through its *mean* value, eventually reduced by a proper Confidence Factor, *CF*, depending on the knowledge level achieved on the structure. In this work a *CF* equal to 1 has been assumed for analysis, in order to check the effects of the concrete strength distribution, by comparing a uniform strength (described through the *mean* value) to a more realistic distribution, described after the experimental data.

To this purpose three different cases have been examined regarding the concrete strength distribution. The first one consists of assuming the same strength value, i.e. the *mean* concrete strength, $R_{c,mean}$, for all the structural elements of the building. The second case consists of assuming a uniform strength value for each storey ($f_{c,mean,storey}$). Such strength, in turn, has been found as the *mean* of all the tested element belonging to the storey. Finally, the third case consists in describing all the tested elements through their effective strength, i.e. the strength values provided by each lecture ($f_{c,lect}$). All the other elements, instead, have been described through the *mean* value, $f_{c,mean}$, of the entire building. In Table 17.1 the three assumptions considered for the concrete strength distributions have been listed.

17.4 The Analysis

A nonlinear static analysis has been performed to describe the seismic performance of the structure. Two different horizontal patterns, respectively proportional to the storey masses and to the first vibrational mode have been considered. For each horizontal pattern, a $\pm$ 5% eccentricity has been introduced in the two directions, according to EC8 prescriptions. In Table 17.2 the considered cases of the analysis have been listed.

Table 17.2 Performed analysis

Mass-proportional horizontal pattern											
X-direction						Y-direction					
E = 0		E = +5%		E = −5%		E = 0		E = +5%		E = −5%	
+way	−way	+way	−way	+way	−way	+way	−way	+way	−way	+way	−way
MX + E0	MX−E0	MX + E+	MX−E+	MX + E−	MX−E−	MY+E0	MY−E0	MY+E+	MY−E+	MY+E−	MY−E−
1st mode horizontal pattern											
X-direction						Y-direction					
E = 0		E = +5%		E = 0		E = +5%		E = 0		E = +5%	
+way		−way		+way		−way		+way		way	
1X + E0	1X−E0	1X + E+	1X−E+	1X + E−	1X−E−	1Y + E0	1Y−E0	1Y + E+	1Y−E+	1Y + E−	1Y−E−

Fig. 17.4 3D views of the structural model

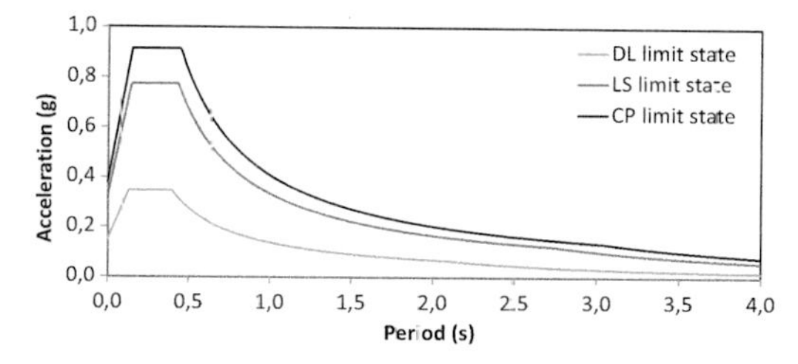

reference quantities	limit states		
	DL	LS	CP
a_g	0,109	0,227	0,287
F_0	2,327	2,371	2,396
T^*_C	0,273	0,295	0,310
S_T	1,000	1,000	1,000

a_g = ground acceleration, F_0 = amplification factor on the rock-site, T^*_C = beginning period of the velocity-constant branch, S_T = topographic amplification factor

Fig. 17.5 Elastic spectra of the case-study

The analysis has been performed through the software SeismoStruct (Seismosoft 2013), by representing the cross sections of each member through a fiber, and subdivided the members themselves into four segments. The Mander et al. model (1988) has been assumed for the core concrete, a three-linear model has been assumed for the unconfined concrete, and a bilinear model has been assumed for the reinforcement steel. The stiffness of floor slabs has been considered by introducing a rigid diaphragm. In Fig. 17.4 some 3d views of the structural model have been shown.

The seismic input has been represented by the elastic spectrum provided by the Code for the building site. Since the building is located in Sansepolcro (Italy), the Italian Code NTC 2008 has been considered for the spectrum definition. The PGA at the rock outcrop, according to the Italian soil classification, is equal to 0.227 g for a Return Period, *RP*, equal to 475 years, i.e. for a probability of occurrence equal to 10% in 50 years (Montaldo et al. 2007), expressing the seismic input for Life Safety (*LS*) limit state.

The site Complex has been object of a careful geological investigation (Tanganelli et al. 2016; Viti et al. 2017), which provided controversial information regarding the soil-type. In this study a soil-class B, according to the EC8 and NTC 2008 classification, has been assumed, since it is the most conservative hypothesis evidenced by the investigation. In Fig. 17.5 the elastic spectra assumed for analysis have been shown, together with the main parameters of the foundation soil.

The response of the structure after the assumed seismic input has been compared to the limit values provided by EC8; namely, as regards the Life Safety limit state, the considered limit condition is the attainment of the limit chord rotation θ_{LS},

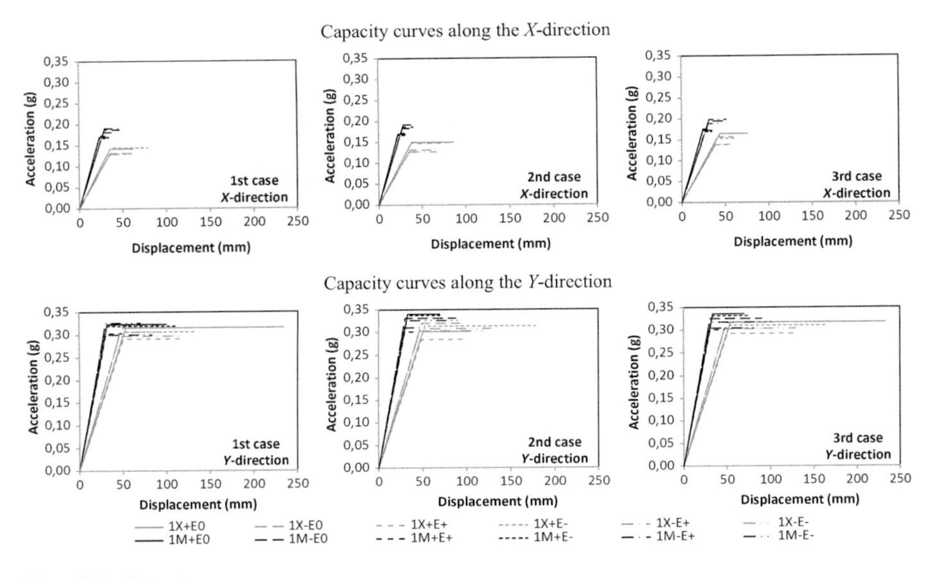

Fig. 17.6 Capacity curves

defined as the ¾ of the ultimate rotation θ_U, quantified according to the EC8 (and NTC 2008 as well) provisions. (EC8, Annex A Eq. A1). The required Code verification includes even the check of the maximum shear force attained in the members. In this study, however, the shear force has not been checked, since similar investigations (Tanganelli et al. 2013; La Brusco et al. 2015) made on buildings belonging to the same complex, evidence level of shear largely exceeding the limit values, regardless of the specific features and mechanical properties of the buildings.

17.5 The Results

17.5.1 Seismic Response (Global Behavior)

The capacity curves found for the case-study are shown Fig. 17.6. Each curve stops at the achievement of the first limit condition, i.e. the limit chord rotation in the structure or a 20% drop in the shear force. The bi-linearization of the curves has been made according to the EC8 provision, assuming the same maximum shear force of the MDOF system and imposing to the elastic branch to pass for the point with a 60% of the maximum shear.

As can be noted, the structure evidences a different capacity along the two directions. All columns of the building, indeed, are oriented along the *Y*-direction, providing a higher stiffness and ductility to the structure. The considered cases of analysis, i.e. the assumed strength distributions, do not seem to affect very much the results in terms of capacity. The two considered horizontal patterns differ mostly for the elastic stiffness of the structure. The maximum capacity, in terms of

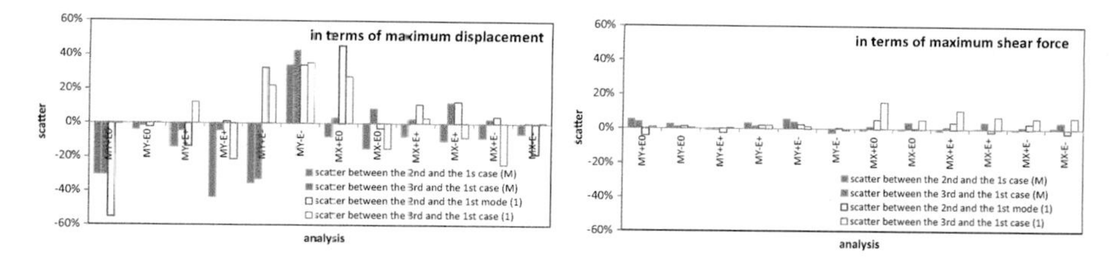

Fig. 17.7 Scatter in the capacity related to the considered cases of analysis

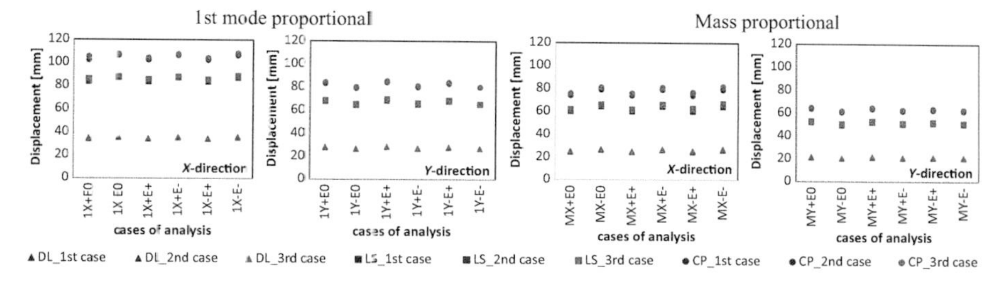

Fig. 17.8 Maximum displacement requested by each model

displacement and shear force, provided by each analysis has been compared and shown in Fig. 17.7, where the scatter among the three considered hypotheses assumed for the concrete strength distributions have been evidenced.

Figure 17.8 shows the maximum displacement requested by each model for three different seismic intensities, found by intersecting the capacity curves to the seismic spectra representing, respectively, the *DL*, *LS* and *CP* limit states.

As can be noted, the three considered strength distributions provide almost identical results. The models, however, evidence a different behavior in the two directions. Namely, the capacity of the structure along the *X*-direction is much lower than the one along the *Y*-direction, whilst the seismic demand is higher along the *X*-direction.

17.5.2 Seismic Performance (Local Behavior)

As shown in Sect. 17.5.1, the assumed strength distributions do not affect the global response of the structure. The global stiffness indeed, as well as the seismic demand, are almost identical for the three considered cases. The seismic performance, however, could be more sensitive to such assumptions, since it is related to the achievement of the limit condition in the weakest section of the building. In this study, the performance has been checked with reference to the attainment of the limit chord rotation in the columns only. The limit condition, therefore, has been defined as the yield chord rotation, ϕ_y, (EC8, Annex A, Eq. A10b) for the *DL* limit state, and as ϕ_{LS} and ϕ_{CP}, respectively, for the *LS* and *CP* limit states. The quantity ϕ_{CP} has been found as the ultimate chord rotation (EC8, Annex A, Eq. A1), whilst ϕ_{LS} is

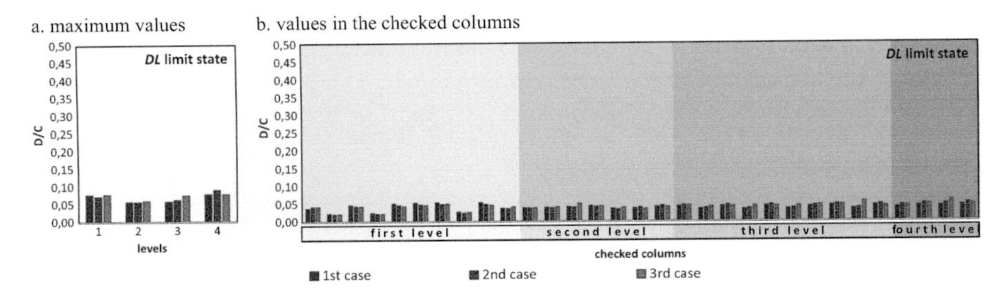

Fig. 17.9 Seismic performance of the structure: *DL* limit state

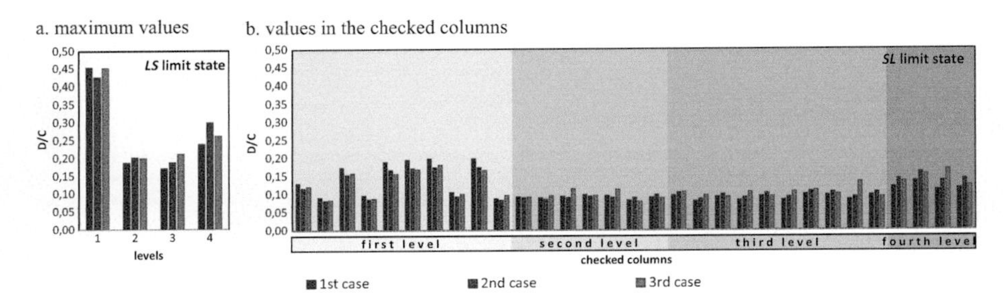

Fig. 17.10 Seismic performance of the structure: *LS* limit state

defined as 2/3 ϕ_{CP}. In all cases, the chord rotation has been found by assuming the actual amount of axial load in the columns, since this quantity proved to affect the seismic capacity of the members (Mariani et al. 2015).

The seismic performance of the building has been measured through the performance index (D/C), expressed as the ratio between the demanded chord rotation (D) and the limit one (C) in each member of the structure. The reliability of the structure is related to the most involved cross section: at the exceeding of the limit condition in one section, indeed, the seismic performance of the entire structure is considered unacceptable. In this paragraph, for sake of brevity, the performance of the building refers to the seismic response of the structure in the *Y*-direction, under horizontal forces proportional to the 1st vibrational mode.

Figure 17.9a, 17.10a and 17.11a show the maximum values of the performance index achieved in the columns at each storey for the three considered limit states. As can be noted, the three considered strength distributions induce a difference in the performance index, even if it is well below the unity in all cases.

In the same figures, the values of the performance index found for each checked columns have been shown. The differences induced by the three cases of analysis in the single checked columns, in some members, are higher than the ones found by checking the maximum response of each storey. Even in this case, however, the performance index of the columns is well below the unity. It should be noted that the checked columns are located in the central part of the building plan. The maximum response of the building, instead, is experienced by the side columns, since they are subjected to additional torsional effects due to the building shape.

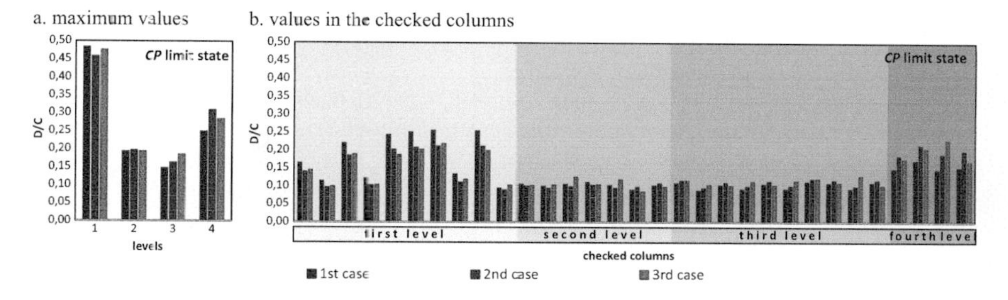

Fig. 17.11 Seismic performance of the structure: *CP* limit state

17.6 Conclusive Remarks

In this work the effect of the concrete strength distribution on the seismic assessment of RC buildings has been investigated with reference to a real building located in Sansepolcro (Italy), currently used as an Hospital. A wide investigation has been made on the mechanical properties of the concrete, including destructive (cores) and non-destructive (SonReb) tests. The results of the investigation has been arranged according to the current research proposals, obtaining reliable values of strength for all the tested members.

Three alternative representations of the concrete strength distribution have been proposed, consistent with the results of the investigation. The first assumption consist of assuming an uniform strength distribution for the entire structure (as required by the International Codes); the second one assumes a *mean* storey strength at each storey, whilst the third one assigns to each tested member the strength value found by the experimental survey.

Both the global response of the structure and its seismic performance have been checked.

The seismic response of the building has been found by intersecting the capacity curves provided by a nonlinear static analysis to the seismic spectra provided by the Code to represent three different limit states: a serviceability (*DL*) and two ultimate (*LS, CP*) ones. The seismic response of the building does not evidence a relevant sensitivity to the considered cases of strength distribution, whilst it largely varies as a function of the direction of analysis and to the assumed horizontal patterns.

The seismic performance of the building has been found with reference to the chord rotation of the columns, by introducing a performance index (D/C) defined as the ratio between the demanded rotation (D) to the available one (C). The case-study evidence a satisfactory performance for both the investigated limit states. The considered strength distribution induced a not negligible effect in the seismic performance of the structure, despite it keeps, in all cases, well below the required level.

References

British Standard 1881 (1983) Testing concrete. Part 120: Methods for determination of the compressive strength of concrete cores. British Standard Institute, London

Cestelli Guidi M, Morelli G (1981) Valutazione della resistenza dei calcestruzzi sulle strutture finite. L'Industria Italiana del Cemento n°3, Italia, pp 195–206 (in Italian)

Concrete Society (1987) Technical report n. 11. Concrete core testing for strength, 1987

Cosenza E, Monti G (2009) Assessment and reduction of the vulnerability of existing reinforced concrete buildings. Atti Convegno finale ReLuis, pp 51–110

Cristofaro MT (2009) Metodi di valutazione della resistenza a compressione del calcestruzzo di strutture in c.a. esistenti, Tesi di Dottorato, Universitá di Firenze (in Italian)

D'Ambrisi A, De Stefano M, Tanganelli M, Viti S (2013) The effect of common irregularities on the seismic performance of existing RC framed buildings. In: Lavan O, De Stefano M (eds) Seismic behaviour and design of irregular and complex civil structures, Geotechnical, geological and earthquake engineering 24. Springer, Dordrecht. https://doi.org/10.1007/978-94-007-5377-8_4

De Stefano M, Tanganelli M, Viti S (2013a) Effect of the variability in plan of concrete mechanical properties on the seismic response of existing RC framed structures. Bull Earthq Eng. https://doi.org/10.1007/s10518-012-9412-5

De Stefano M, Tanganelli M, Viti S (2013b) On the variability of concrete strength as a source of irregularity in elevation for existing RC buildings: a case study. Bull Earthq Eng 11(5):1711–1726

De Stefano M, Tanganelli M, Viti S (2014) Variability in concrete mechanical properties as a source of in-plan irregularity for existing RC framed structures. Eng Struct. https://doi.org/10.1016/j.engstruct.2013.10.027

De Stefano M, Tanganelli M, Viti S (2015a) Torsional effects due to concrete strength variability in existing buildings. Earthq Struct 8(2):379–399

De Stefano M, Tanganelli M, Viti S (2015b) Seismic performance sensitivity to concrete strength variability: a case-study. Earthq Struct 9(2):321–337, https://doi.org/10.12989/eas.2015.9.2.321

De Stefano M, Tanganelli M, Viti S (2016) Concrete strength variability as a source of irregularity for existing RC structures. Seismic Behaviour and Design of Irregular and Complex Civil Structures II, Geotechnical, Geological and Earthquake Engineering 40:149–159

Di Leo A, Pascale G (1994) Prove non distruttive sulle costruzioni in cemento armato. In: Proc "Convegno sistema Qualità e prove non Distruttive per l'affidabilità e la sicurezza delle strutture civili", Bologna SAIE, ottobre, Italia

Fardis MN (2009) Seismic design, assessment and retrofitting of concrete buildings: based on EN-Eurocode8. Springer

Franchin P, Pinto PE, Rajeev P (2007) Confidence factor? J Earthq Eng 14/7:989–1007

Franchin P, Pinto PE, Rajeev P (2009) Confidence in the confidence factor. In: Proceedings of "Eurocode 8 perspectives from the Italian standpoint workshop, Napoli, Italy, pp 25–38

Gasparik J (1992) Prove non distruttive nell'edilizia. Quaderno didattico AIPnD, Brescia, 1992 (in Italian)

Giacchetti R, Lacquaniti L (1980). Controlli non distruttivi su impalcati da ponte in calcestruzzo armato. – Nota tecnica 04, Università degli Studi di Ancona, Facoltà di Ingegneria, Istituto di Scienza e Tecnica delle Costruzioni, Italia

Jalayer F, Iervolino I, Manfredi G (2008) Structural modeling: uncertainties and their influence on seismic assessment of existing RC structures. Struct Saf

La Brusco A, Mariani V, Tanganelli M, Viti S, De Stefano M (2015) Seismic assessment of a real RC asymmetric hospital building according to NTC 2008 analysis methods. Bull Earth Eng 13(10):2973–2994

Mander JB, Priestley MJN, Park R (1988) Theoretical stress-strain model for confined concrete. J Struct Div ASCE 114:1804–1826

Marano GC, Quaranta G, Mezzina M (2008) Hybrid technique for partial safety factors calibration. In: Proceedings of Reluis2Rm08 "Valutazione e riduzione della vulnerabilità sismica di edifici esistenti in c.a.", Roma, 29–30 maggio, Cosenza E, Manfredi G, Monti G (eds) Polimetrica International Scientific Publisher. ISBN:978-88-7699-129-5, pp 109–120

Mariani V, Tanganelli M, Viti S, De Stefano M (2015) Combined effects of axial load and concrete strength variation on the seismic performance of existing RC buildings. Bull Earthquake Eng 14:805–819. https://doi.org/10.1007/s10518-015-9858-3

Masi A (2005) La stima della resistenza del calcestruzzo in situ mediante prove distruttive e non distruttive, Il Giornale delle Prove Non Distruttive, Monitoraggio, Diagnostica, nr 1, pp 23–32 (in Italian)

Masi A, Vona M (2009) Estimation of the in-situ concrete strength: provisions of the European and Italian seismic codes and possible improvements provisions of the European and Italian seismic codes and possible improvements. In: Proceedings of the "Eurocode 8 perspectives from the Italian standpoint workshop, Napoli, Italy, pp 67–77

Masi A, Vona M, Manfredi V (2008) A parametric study on RC existing buildings to compare different analysis methods considered in the European seismic code (EC8-3). In: Proceedings of 14th WCEE, Beijing, China

Montaldo V, Meletti C, Martinelli F, Stucchi M, Locati M (2007) On-line seismic hazard data for the new Italian building code. J Earthq Eng 11:119–132

Monti G, Alessandri S, Goretti A (2007) Livelli di conoscenza e fattori di confidenza. XII Convegno ANIDIS, L'ingegneria sismica in Italia. Pisa 10–14 Giugno 2007 (in Italian)

Rajeev P, Franchin P, Pinto PE (2010) Review of confidence factor in EC8-Part 3: A European code for seismic assessment of existing buildings, international conference on sustainable built environment ICSBE 2010, Kandy, Sri Lanka

Seismosoft (2013) Seismostruct version 5.2.2 – a computer program for static and dynamic nonlinear analysis of framed structures. Available online from www.seismosoft.com

Tanganelli M, Viti S, De Stefano M, Reinhorn A (2013) Influence of infill panels on the seismic response of existing RC buildings: a case study. In: Lavan O, De Stefano M (eds) Seismic behaviour and design of irregular and complex civil structures, Geotechnical, geological and earthquake engineering, vol 24. Springer, Dordrecht. https://doi.org/10.1007/978-94-007-5377-8_9

Tanganelli M, Stefania V, Valentina M, Maria P (2016) Seismic assessment of existing RC buildings under alternative ground motion ensembles compatible to EC8 and NTC 2008. Bull Earthquake Eng:1–22, ISSN:1570-761X. https://doi.org/10.1007/s10518-016-0028-z

UNI EN 12390-2. Prova sul calcestruzzo indurito – Resistenza alla compressione – Specifiche per macchine di prova, 2002 (in Italian)

UNI EN 12390-2, Prove sul calcestruzzo indurito. Confezione e stagionatura dei provini per prova di resistenza, Maggio 2009 (in Italian)

UNI EN 12504-1. Prove sul calcestruzzo nelle strutture – Carote. Prelievo, esame e prova di compressione, Maggio 2009 (in Italian)

UNI EN 12390-1, Prova sul calcestruzzo indurito. Forma, dimensioni ed altri requisiti per provini e casseforme, Ottobre 2012 (in Italian)

Viti S, Tanganelli M, D'Intinosante V, Baglione M (2017) Effects of the soil characterization on the seismic input. J Earthq Eng (Aug. 2017)

Chapter 18
An Assessment of American Criterion for Detecting Plan Irregularity

V. Alecci, M. De Stefano, S. Galassi, M. Lapi, and M. Orlando

Abstract The European seismic code 8 (Eurocode 8) classifies buildings as plan-wise regular according to four criteria which are mostly qualitative and a fifth one which is based on parameters such as stiffness, eccentricity and torsional radius that can be only approximately defined for multi-story buildings. Therefore, such plan-regularity criteria need to be improved. ASCE seismic code, according to a different criterion, considers plan irregularity when the maximum story drift, at one end of the building structure, exceeds more than 1.2 times the average of the story drifts at the two ends of the structure under building static analysis. Nevertheless, both the ASCE approach and the threshold value of 1.2 need to be supported by adequate background studies, based also on nonlinear seismic analysis. In this paper a numerical analysis is carried out, by studying the seismic response of an existing r.c. school building. Static analysis is developed by progressively shifting the centre of mass, until the ratio between the maximum lateral displacement of the floor at the level considered and the average of the horizontal displacements at extreme positions of the floor at the same level matches and even exceeds the value of 1.2. Then, nonlinear dynamic analyses are carried out to check the corresponding level of response irregularity in terms of uneven plan distribution of deformation and displacement demands and performance parameters. The above comparison leads to check the suitability of the ASCE approach and, in particular, of the threshold value of 1.2 for identifying buildings plan irregularity.

Keywords Plan irregularity · Eurocode 8 · ASCE seismic code · Threshold value assessment · r.c. frame structures

V. Alecci (✉) · M. De Stefano · S. Galassi
Department of Architecture (DiDA), University of Florence, Florence, Italy
e-mail: valerio.alecci@unifi.it; mario.destefano@unifi.it; stefano.galassi@unifi.it

M. Lapi · M. Orlando
DICEA – Department of Civil and Environmental Engineering, University of Florence, Florence, Italy
e-mail: massimo.lapi@unifi.it; maurizio.orlando@unifi.it

D. Köber et al. (eds.), *Seismic Behaviour and Design of Irregular and Complex Civil Structures III*, Geotechnical, Geological and Earthquake Engineering 48,
https://doi.org/10.1007/978-3-030-33532-8_18

18.1 Introduction

In-plan irregularity is very common in existing buildings and it is one of the most frequent sources of severe damage during earthquake (De Stefano and Pintucchi 2008). In 1985, during the earthquake of Mexico City, 42% of damaged or collapsed structures were corner buildings. Furthermore, many buildings failed in torsion due to asymmetric layout of masonry walls (Rosenlueth and Meli 1986). The need to define adequate provisions accounting for torsional effect due to structural asymmetry was reconfirmed by this event (Tso 1990).

During the 1970s–1980s the equivalent static analysis represented the most common method for computing seismic loads. Such a method, applied to asymmetric structures, underestimates the actions on flexible side elements, as it does not take into account the dynamic amplification of the torsional response. For this reason, researchers suggested the introduction of a design eccentricity in order to provide a torsional moment in correspondence to each story of the building. In the last decades, several studies on the design eccentricity were carried out (Tso 1990; Chandler and Hutchinson 1987; Chopra and Goel 1991; De Stefano et al. 1993; Wong and Tso 1995; Chandler 1995; De Stefano and Rutenberg 1997; Chandler 1997; Harasimowicz and Goel 1998; Anastassiadis et al. 1998). Later, spatial models and modal analysis became widely used by designers. Thus, the elastic response determination of in-plan irregular building turned easier.

In recent years, researchers focused on the inelastic response of in-plan irregular buildings. The response of asymmetric buildings was investigated at varying of several parameters like Centre of Mass (CM), eccentricity (Perus and Fajfar 2005), uneven distribution of concrete strength (De Stefano et al. 2013; Athanatopoulou et al. 2015), torsional stiffness and periods of vibration (Goel and Chopra 1990). Furthermore several studies were developed focusing on the application of non-linear static analysis to in-plan irregular buildings (D'Ambrisi et al. 2009; Magliulo et al. 2012; Bosco et al. 2013; Manoukas et al. 2012).

This paper is focused on the evaluation of the ASCE torsional provision (ASCE Standard 7-10) (ASCE, Minimum Design Loads for Buildings and Other Structures, ASCE/SEI 7 2010). The response of a case study building is investigated in order to evaluate the torsional irregularity provided by the code and the corresponding uneven distribution of inelastic demand which is detected using the non-linear dynamic analysis method.

18.2 Torsional Provisions

The previous generation of seismic codes was providing design eccentricities for equivalent static analysis (Eq. 18.1) as follows:

$$e_d = \alpha \cdot e_0 + \beta \cdot b \tag{18.1}$$

where e_d is the design eccentricity, e_0 is the actual distance between the Centre of Mass (CM) and the Centre of Rigidity (CR) measured orthogonally to the loading direction; b is the building dimension perpendicular to the direction of excitation, α is a coefficient accounting for dynamic amplification of the torsional response and β is a coefficient accounting for aleatoric position of the CM.

In Table 18.1 coefficients α and β are provided according to Eurocode 8 1993 (EC8-93) (European Committee for Standardization, Eurocode 8 1993), National Building Code of Canada 1995 (NBCC-95) (Associate Committee on the National Building Code, National Building Code of Canada 1995) and ASCE Standard 7-10 (ASCE 7-10) (ASCE, Minimum Design Loads for Buildings and Other Structures, ASCE/SEI 7 2010), where A_x is the amplification factor provided by ASCE 7-10 (defined in the following) and e_2 is the additional eccentricity provided by EC8-93, equal to the smaller of the following values:

$$e_2 = 0.1 \cdot (L + B) \cdot \sqrt{10 \cdot e_0 / L} \leq 0.1 \cdot (L + B)$$

$$e_2 = \frac{1}{2 \cdot e_0} \cdot \left[\ell_s^2 - e_0^2 - r^2 + \sqrt{\left(\ell_s^2 - e_0^2 - r^2\right)^2 + 4 \cdot e_0^2 \cdot r^2} \right]$$

in which:

- e_0 is the distance between CR and CM, measured along the x direction, which is orthogonal to the direction of analysis considered;
- ℓ_s is the radius of gyration of the floor mass in plan (square root of the ratio of the polar moment of inertia of the floor mass in plan (a) with respect to the centre of mass of the floor to (b) the floor mass (b));
- r is the square root of the ratio of the torsional stiffness to the lateral stiffness in the y direction ("torsional radius").

In current code provisions, such as Eurocode 8 2004 (EC8-04) (European Committee for Standardization, Eurocode 8 2004) and National Building Code of Canada 2010 (NBCC-10) (Associate Committee on the National Building Code, National Building Code of Canada 2010), in case of torsional irregularity only 3-D dynamic analysis is allowed. Even adopting spatial model and CQC modal combination the

Table 18.1 Design eccentricity coefficients

Analyzed elements	Flexible side elements		Stiff side elements	
Code	α	β	α	β
EC8-93	$1.0 + e_2/e_0$	0.05	1.0	−0.05
NBCC-95	1.5	0.10	0.5	−0.10
ASCE 7-10	1.0	$0.05 \cdot A_x$	1.0	$-0.05 \cdot A_x$

accidental eccentricity is considered accounting for the random position of CM, and the design eccentricity is computed by Eq. 18.2:

$$e_d = e_0 + \beta \cdot b \tag{18.2}$$

However, the criteria for assessing in-plan irregularity are still very important; indeed it affects the choice of the method of analysis and the definition of the behaviour factor.

Qualitatively, in-plan irregularity is due to asymmetric distributions of mass and stiffness, but quantitatively a definition universally shared does not exist. Eurocode 82,004 (EC8-04) provides a list of conditions to classify a building as regular in plan:

- With respect to the lateral stiffness and mass distribution, the building structure shall be approximately symmetrical in plan with respect to two orthogonal axes.
- The plan configuration shall be compact, i.e. each floor shall be delimited by a polygonal convex line. If in plan set-backs (i.e. re-entrant corners or edge recesses) exist, regularity in plan may still be considered as being satisfied provided that these set-backs do not affect the floor in-plan stiffness and that for each set-back, the area between the outline of the floor and a convex polygonal line enveloping the floor does not exceed 5% of the floor area.
- The in-plan stiffness of the floors shall be sufficiently large in comparison with the lateral stiffness of the vertical structural elements, so that the deformation of the floor shall have a small effect on the distribution of the forces among the vertical structural elements. In this respect, the L, C, H, I, and X plan shapes should be carefully examined, notably as concerns the stiffness of the lateral branches, which should be comparable to that of the central part, in order to satisfy the rigid diaphragm condition. The application of this paragraph should be considered for the global behaviour of the building.
- The slenderness λ = Lmax /Lmin of the building in plan shall not be higher than 4, where Lmax and Lmin are respectively the in plan larger and smaller dimension of the building, measured in two orthogonal directions.
- At each level and for each direction of analysis x and y, the structural eccentricity e_0 and the torsional radius r shall conform to the conditions (18.3) and (18.4):

$$e_o \leq 0.30 \cdot r \tag{18.3}$$

$$r \geq \ell_s \tag{18.4}$$

The first four criteria are almost qualitative, while the fifth one, if strictly applied, is valid only for single storey buildings. For multi-storey buildings only approximate definitions of the centre of stiffness and the torsional radius are possible (European Committee for Standardization, Eurocode 8 2004). Similarly, ASCE Standard 7-10 (ASCE, Minimum Design Loads for Buildings and Other Structures, ASCE/SEI 7 2010) provides a list of conditions to detect horizontal irregularity in buildings, as follows:

- Torsional Irregularity: it is defined to exist if the maximum story drift, computed including accidental torsion with $A_x = 1.0$, at one end of the structure transverse to an axis, is more than 1.2 times the average of the story drifts at the two ends of the structure. Torsional irregularity (Eq. 18.5) requirements apply only to structures in which the diaphragms are rigid or semirigid:

$$\delta_{max} \geq 1.2 \cdot \delta_{avg} \tag{18.5}$$

where:

$\delta_{avg} = \frac{\delta_A + \delta_B}{2}$,
is the average deflection determined by an elastic analysis (Fig. 18.1);

$1 < A_x = \left[\frac{\delta_{max}}{1.2 \cdot \delta_{avg}}\right]^2 \leq 3$, is the amplification factor of the accidental torsional moment.

- Extreme Torsional Irregularity: it is defined to exist if the maximum story drift, computed including accidental torsion with $A_x = 1.0$, at one end of the structure transverse to an axis, is more than 1.4 times the average of the story drifts at the two ends of the structure. Extreme torsional irregularity requirements (Eq. 18.6) apply only to structures in which the diaphragms are rigid or semirigid:

$$\delta_{max} \geq 1.4 \cdot \delta_{avg} \tag{18.6}$$

- Re-entrant Corner Irregularity: it is defined to exist where both plan projections of the structure beyond a re-entrant corner are greater than 15% of the plan dimension of the structure in the given direction.

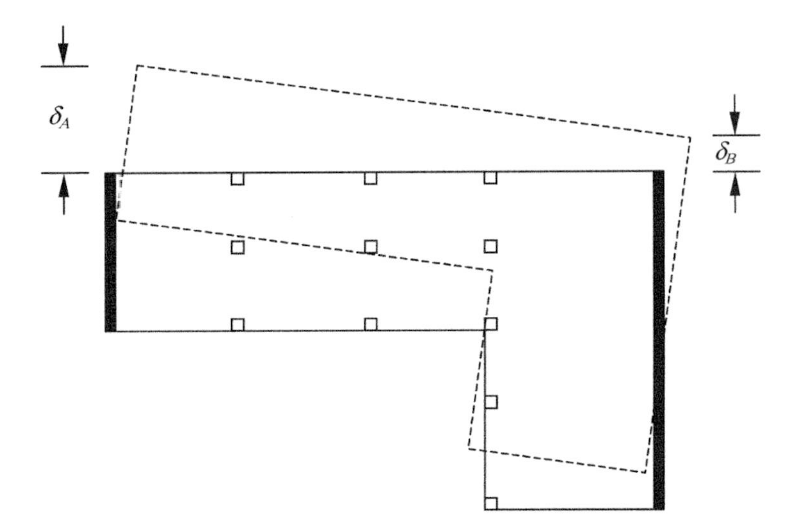

Fig. 18.1 Determination of the average deflection δ_{avg}, adapted from (Magliulo et al. 2012)

- Diaphragm Discontinuity Irregularity: it is defined to exist where there is a diaphragm with an abrupt discontinuity or variation in stiffness, including one having a cut-out or open area greater than 50% of the gross enclosed diaphragm area, or a change in effective diaphragm stiffness of more than 50% from one story to the next.
- Out-of-Plane Offset Irregularity: it is defined to exist where there is a discontinuity in a lateral force-resistance path, such as an out-of-plane offset of at least one of the vertical elements.
- Nonparallel System Irregularity: it is defined to exist where vertical lateral force-resisting elements are not parallel to the major orthogonal axes of the seismic force-resisting system.

The ASCE Code (ASCE 7-10) (ASCE, Minimum Design Loads for Buildings and Other Structures, ASCE/SEI 7 2010) provides a simpler approach for detecting torsional irregularity. Such a method involves directly the structural response of the building and it does not require the knowledge of specific characteristics (that are mandatory in Eurocode 8), such as the torsional radius, that can be only approximately defined for multi-storey buildings. The criterion provides the implementation of a linear static analysis taking into account the accidental eccentricity ($\beta = 5\%$). As shown previously, when the maximum story drift is more than 1.2 times the average drift, the torsional irregularity is detected. The use of this elastic index (i.e. the ratio between the maximum and the average elastic drift) appears to be very simple and effective, however some further researches are needed. Both the ASCE approach and the threshold value of 1.2 must be supported by adequate background studies.

With this aim, in this paper a numerical analysis is carried out, by studying the seismic response of an existing r.c. school building. As provided by ASCE 7-10, static elastic analyses are performed varying, step by step, the CM position and checking the ratio between the maximum and the average lateral displacements ($\delta_{max}/\delta_{avg}$). Lastly, nonlinear dynamic analyses are carried out in order to evaluate the corresponding level of response irregularity in terms of uneven distribution of inelastic demand.

18.3 Case Study

The school building under analysis is about 40 years of age and it is situated on a flat ground in the Municipality of Prato (PO), Italy. It consists of two independent blocks, the school and the gym.

The school block has been chosen as case study for the seismic analyses because it is made of a r.c. framed structure (conversely, the gym block has a precast structure).

Fig. 18.2 School building: (**a**) frontal elevation where there are the portico and the entrance hall; (**b**) and (**c**) longitudinal elevations

The school block (Fig. 18.2) has an elongated rectangular plan, approximately 49 by 13 m, and it is a three storey building above the ground level. The floor-to-floor height is 3.30 m. At the ground floor, along the shorter side of the plan, there is a portico that covers the first span of the frames and that provides access to the entrance hall. After the entrance hall, a central corridor leads to the classes and offices along the two long sides of the plan. This distribution pattern is repeated, almost unchanged, at the upper floors as well. The external cladding, in precast panels, as well as the interior walls which divide the classes, made of hollow bricks, follow the frame scheme and they also have ribbon windows outwards.

The building structure (Fig. 18.3) is entirely made of r.c. columns and beams, which form only three plane frames running along the longitudinal direction. In the transversal direction there are not brace frames, excepting for the frames at the ends of the structure. The floor slabs can be considered as rigid diaphragms.

The structural symmetry of the plan fails, in particular, at the ground floor where, in order to obtain a larger room devoted to refectory, the central longitudinal frame is devoid of a column. Therefore, the frame beam has a double-span and supports the column of the upper floor at mid-span. Probably due to this structural weakness, the rectangular beam cross section is 25 by 90 cm, unlike the other one are flat beams with a cross section of 80 by 22 cm. Cross section 25 by 90 cm was also used for the corresponding beams at the two upper floors. The storey floors are made of casted-in-place joists and hollow core slabs.

Instead, the rafters and the edge beams supporting the attic floor have a section of 12 by 40 cm and form a hip roof. The covering is made of insulated sandwich panels. The attic floor is not serviceable.

The columns have three different cross sections: 25 by 30 cm, 25 by 40 cm and 25 by 55 cm. In the two longitudinal edge frames, the longer section side is arranged orthogonally to the frames, while in the central frame the longer side follows the frame direction.

A finite element structural model has been performed and the analyses have been carried out by the software SAP2000. In addition to the weight of the structural elements depending on their geometry, the weights of the non-modeled structural

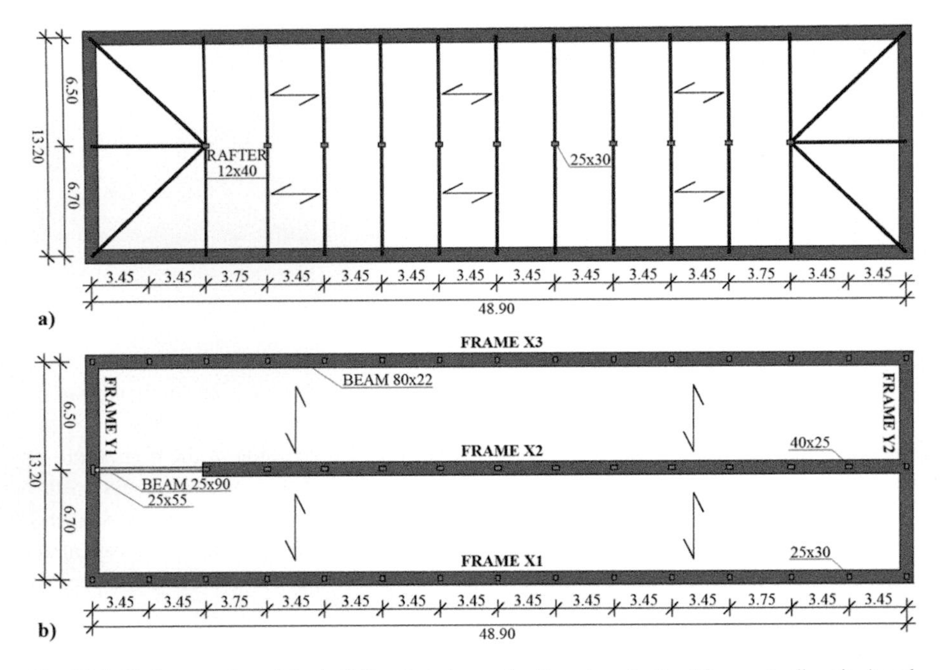

Fig. 18.3 Reference plan of the building that shows the three longitudinal frames (*x* direction) and the two transversal frames (*y* direction): (**a**) roof plan, (**b**) floor plan

elements, such as the storey and the roof floors (G_1), the permanent non-structural loads (G_2), such as the partition walls, the screed and the tiles, and the accidental loads (Q) have been inputted as external loads. In the following, the seismic combination (7) for the ultimate limit states is provided according to the Italian Code (NTC 2008) (DM.LL.PP. 2008):

$$E + G_1 + G_2 + 0.6 \cdot Q \tag{18.7}$$

where E is the seismic action along the considered direction. The seismic base shear F_h has been computed in accordance with Eq. 18.8, provided by the NTC 2008:

$$F_h = S_d(T_1) \cdot \frac{W}{g} \cdot \lambda \tag{18.8}$$

where:

$S_d(T_1)$
: is the design response spectrum as a function of the fundamental period T_1 of the building and obtained from the elastic response spectrum $S_e(T_1)$;

T_1 is the fundamental period of the building;
W is the effective seismic weight;
g is the gravity acceleration;
λ factor equal to 0.85 in the case of a multi-storey building (with storey number greater than or equal to 3).

The elastic response spectrum $S_e(T_1)$ used to calculate the seismic input for SLV verifications, assuming a building life of 50 years and a use class III ($C_u = 1.5$), has been obtained considering the effective municipality and building site (Fig. 18.4).

In the structural model, rigid diaphragms have been considered at each building floor, so as to easily define the center of masses (CM) position.

Firstly, the model has been seismically analyzed by the equivalent static method (Pugi and Galassi 2013), by applying the horizontal action at the CM of each storey floor. The seismic force, computed according to the formulae of the Italian NTC 2008 (DM.LL.PP. 2008), has been assumed acting in the direction of the shorter building side (y direction), where only the two edge frames exist, in order to catch the greatest displacements.

Then, non-linear dynamic analyses have been performed using a natural earthquake accelerogram (Fig. 18.5).

The adopted accelerogram is the record 147ya of Friuli (aftershock) earthquake happened in 1976, September 15th. The recorded magnitude (Mw) is 6 and the Peak Ground Acceleration (PGA) is equal to 2.32 m/s^2.

This accelerogram was chosen because it is spectrum compatible with the adopted elastic response spectrum. In the following, a comparison between the elastic response spectrum from the Italian NTC 2008 and the 147ya spectrum is proposed (Fig. 18.6).

The non-linear dynamic analyses were conducted using lumped plasticity. The plastic hinges were developed according to the ASCE 41-13 (ASCE, Seismic Evaluation and Retrofit of Existing Buildings (ASCE 41-13), ASCE/SEI 4 2014) and the

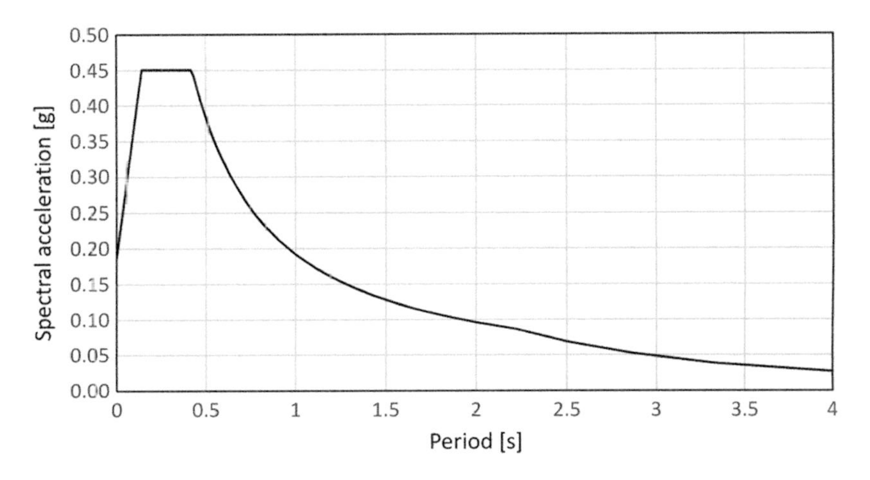

Fig. 18.4 Elastic response spectrum

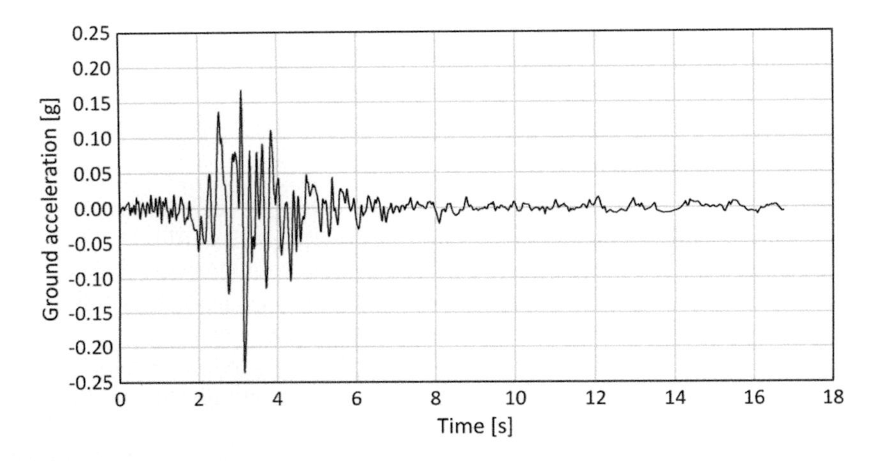

Fig. 18.5 Record 147ya – Friuli (aftershock), 1976, September 15th

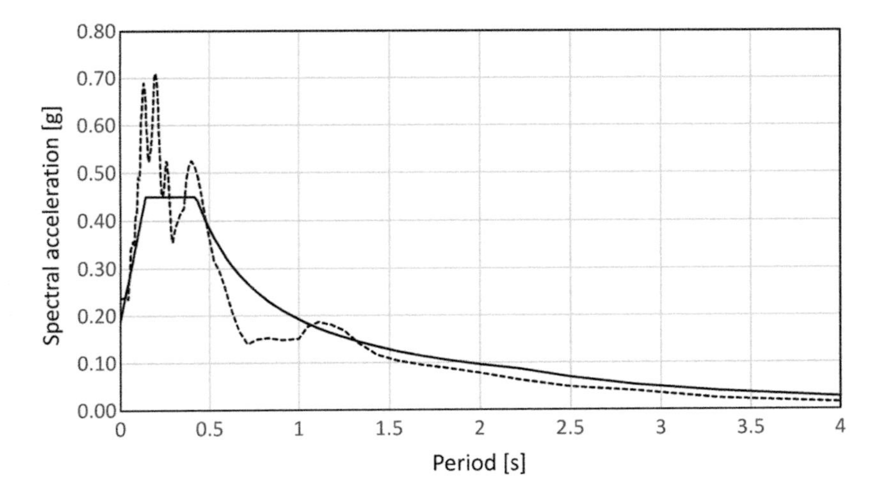

Fig. 18.6 Comparison between the elastic response spectrum (NTC 2008) and the spectrum obtained by the record 147ya of Friuli

inelastic demand (D) is expressed in terms of the plastic rotation at the chord (θ_{pl}). The capacity (C) is given by the Limit Life Safety (LS) as defined by ASCE 41-13.

18.4 Results and Discussion

In this section the results of the elastic analyses are shown. The modal response is investigated and the results are presented in Table 18.2. In Table 18.2, the periods of vibration, the percentage of participating mass u_x along the longitudinal direction of

Table 18.2 Results of the modal analysis

Mode	Period	u_x	u_y	r_z	$\Sigma(u_x)$	$\Sigma(u_y)$	$\Sigma(r_z)$
1	1.497	0.00	0.80	0.01	0.00	0.80	0.01
2	1.042	0.00	0.00	0.81	0.00	0.80	0.82
3	0.689	0.87	0.00	0.00	0.87	0.80	0.82
4	0.347	0.00	0.15	0.00	0.87	0.95	0.82
5	0.296	0.00	0.00	0.13	0.87	0.95	0.95
6	0.242	0.10	0.00	0.00	0.97	0.95	0.95
7	0.168	0.00	0.04	0.00	0.97	1.00	0.96
8	0.158	0.02	0.00	0.02	0.99	1.00	0.98
9	0.151	0.01	0.00	0.02	1.00	1.00	1.00

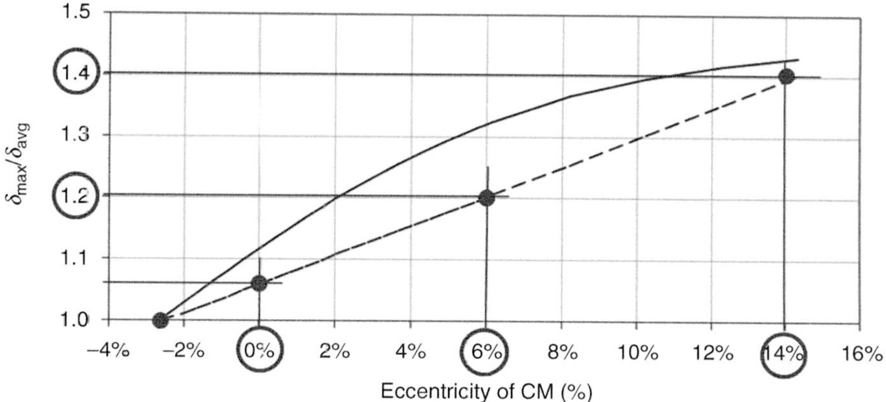

Fig. 18.7 Ratios between maximum δ_{max} and average displacement δ_{avg}, at varying of CM position. Comparison between dynamic analysis (continuous line) and static analysis (dotted line). The position of CM is given in % of the building longitudinal dimension (L = 48.85 m); for *e* = 0% CM is in the actual position

the building (x direction), the percentage of the participating mass u_y along the transversal direction of the building (y direction) and the percentage of the mass participating to the torsional mode r_z are listed for each mode of vibration considered.

Referring to the spectral ordinate $S_d(T_1)$, associated to the fundamental period of vibration along the y direction, the static forces at each floor level are calculated. Then equivalent static force analyses are performed in y direction, at varying of the CM position along the x direction. The same procedure is repeated for the response spectrum analysis and the ratios between the maximum and the average lateral displacements ($\delta_{max}/\delta_{avg}$), calculated with both methods at varying of the CM position, are presented in Fig. 18.7. The effect of variation of the periods of vibration, induced by the displacement of CM, is considered negligible.

Two considerations arise from the previous results. When the CM is placed in the actual position, the floor displacement includes rotational components (indeed $\delta_{max}/\delta_{avg} \neq 1$): this means that CM and CR are not aligned (CR is detected when

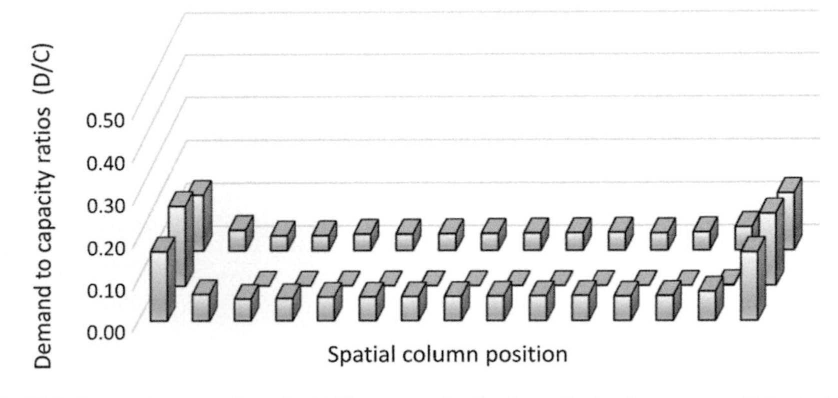

Fig. 18.8 Demand to capacity ratios D/C: uneven distribution of inelastic response, CM coincident with CR

$\delta_{max}/\delta_{avg} = 1$). The two methods, the equivalent static force and the response spectrum analysis, give almost the same results in term of displacement when CM is close to CR. However, increasing the distance between CM and CR the agreement between the results gets worse but this difference is in accordance with several researches and it is widely discussed in the section "Torsional Provisions".

Then non-linear dynamic analyses are performed in order to assess the uneven distribution of inelastic response at varying of CM position. In particular four analyses are performed: the first considers CM coincident with CR, the second considers CM in its actual position, the third and fourth consider CM in the positions that give $\delta_{max}/\delta_{avg}$ equal to 1.2 and 1.4 respectively in the equivalent static force analysis (see Fig. 18.7).

In Fig. 18.8 the ratios between inelastic demand D and capacity C (D/C), of all columns at ground floor, are shown in order to evaluate the damage levels. As previously said, in *y* direction the only frames are placed on the left and right side of the building. All the other columns, placed between these two frames, have to be considered as cantilever members since such columns are not connected by beams. This fact explains why the columns placed on the left and right side of the building are much more affected by the inelastic demand than the others.

Furthermore, the columns placed in the centerline have the longer side of the cross section oriented along the *x* direction, instead the columns placed on the first and on the third line have the longer side of the cross section oriented along the *y* direction. For this reason, the inelastic demand of the center line columns is lower than that of the columns placed on the first and third line, due to the greater shear force brought by the latter. Although the uneven distribution of stiffness, the demand to capacity ratios are symmetrically distributed and, excepted for the abrupt variation shown at the extremities, such ratios are almost constant.

When the CM is placed in its actual position the scenario is different because the symmetrical distribution of the ratios D/C is lost. The inelastic demands on the

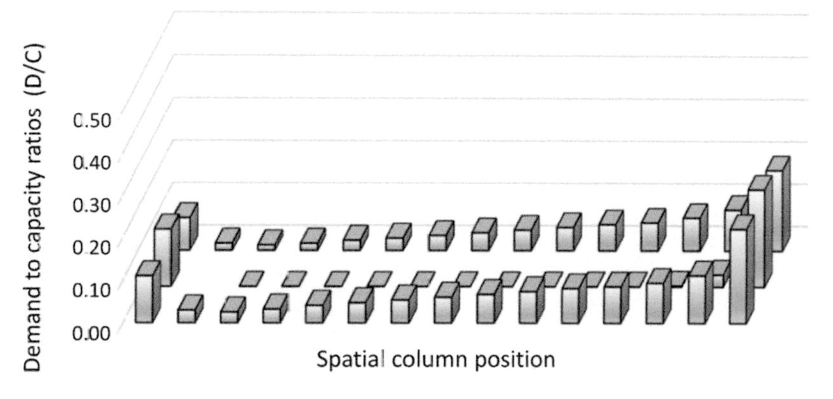

Fig. 18.9 Demand to capacity ratios D/C: uneven distribution of inelastic response, CM placed in the actual position

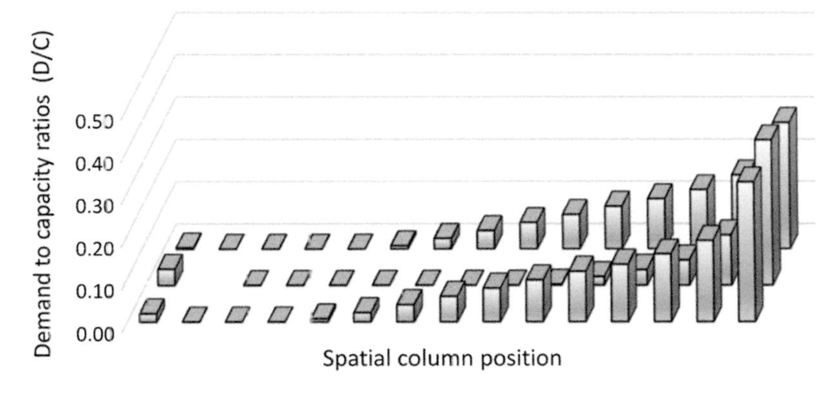

Fig. 18.10 Demand to capacity ratios D/C: uneven distribution of inelastic response, CM placed with 6% of eccentricity in order to obtain $\delta_{max}/\delta_{avg} = 1.2$ in the equivalent static analysis

flexible side are greater than those shown on the stiff side. Moving from the stiff to the flexible side the inelastic demand increases progressively, as shown in Fig. 18.9.

As previously shown (Fig. 18.7), assuming an eccentricity of 6% for the CM, the ratio $\delta_{max}/\delta_{avg}$, obtained by performing an equivalent static analysis, becomes equal to 1.2. In this configuration the inelastic demand appears clearly irregular. In approximately 40% of the building (see Fig. 18.10) the ratios D/C are almost zero, while in the other part the inelastic demand increases rapidly moving to the flexible side.

When CM is placed with 14% of eccentricity the ratio $\delta_{max}/\delta_{avg}$, calculated by an equivalent static analysis, is equal to 1.4 (see Fig. 18.7). Adopting this configuration, the inelastic response is strongly irregular. As shown in Fig. 18.11, approximately 50% of the building provides no inelastic demand. Conversely, moving to the flexible side the plastic demand increases rapidly providing high values of the ratio D/C.

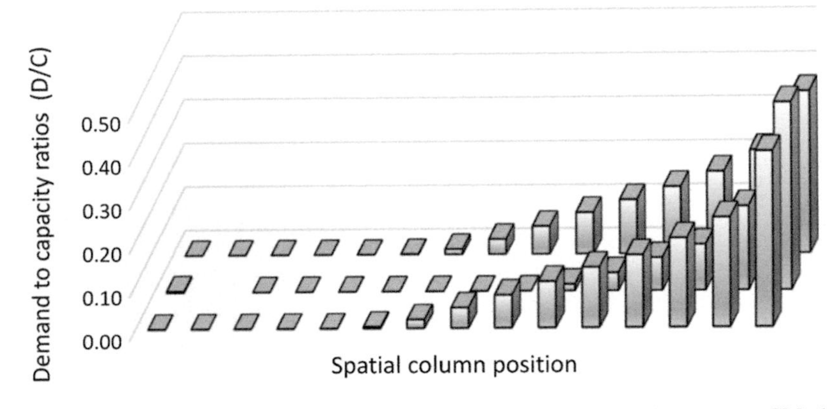

Fig. 18.11 Demand to capacity ratios D/C: uneven distribution of inelastic response, CM placed with 14% of eccentricity in order to obtain $\delta_{max}/\delta_{avg} = 1.4$ in the equivalent static analysis

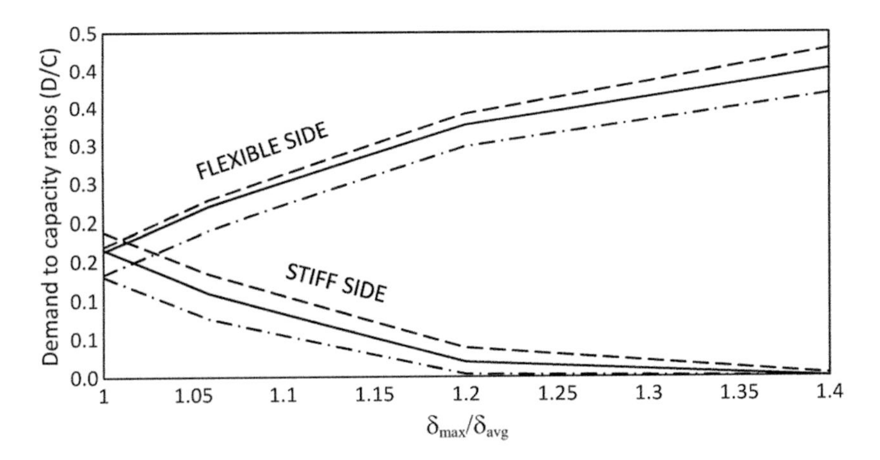

Fig. 18.12 Demand to capacity ratios D/C at varying of $\delta_{max}/\delta_{avg}$ provided by equivalent static analysis. Comparison between the inelastic response of the column placed on the stiff side and that provided by the column on the flexible side

In Fig. 18.12 the demand to capacity ratios of the three columns on stiff and flexible side (respectively left columns of frame Y1 and right columns of frame Y2 in Fig. 18.3) are plotted at varying of $\delta_{max}/\delta_{avg}$ calculated by the equivalent static analysis. In this case, it is worth noting that the inelastic demand is highly dependent on $\delta_{max}/\delta_{avg}$ and that the demand decreases on the stiff side and increases on the flexible side.

In Fig. 18.13 the demand to capacity ratios are normalized by D/C calculated in configuration CM $\equiv$ CR where $\delta_{max}/\delta_{avg}$ is equal to 1. In correspondence of $\delta_{max}/\delta_{avg} = 1.2$ the normalized D/C on the stiff side is about equal to zero and on the flexible side it is close to 2.

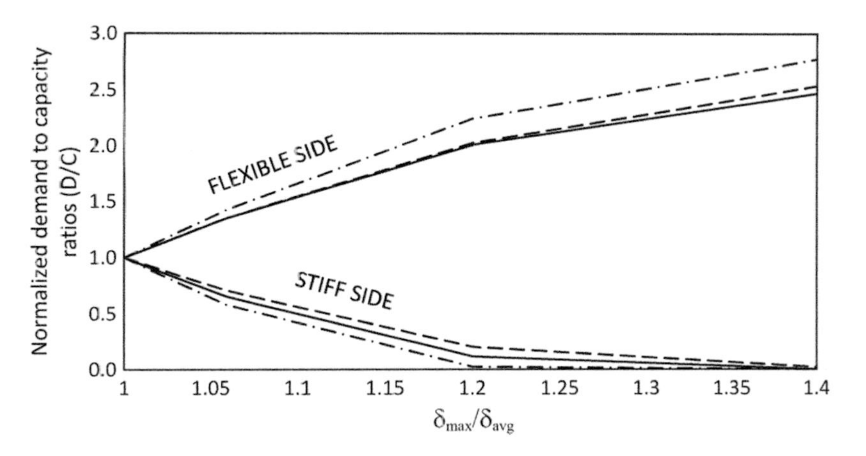

Fig. 18.13 Normalized demand to capacity ratios D/C at varying of $\delta_{max}/\delta_{avg}$ provided by equivalent static analysis. Comparison between the inelastic response of the column placed on the stiff side and that provided by the column on the flexible side

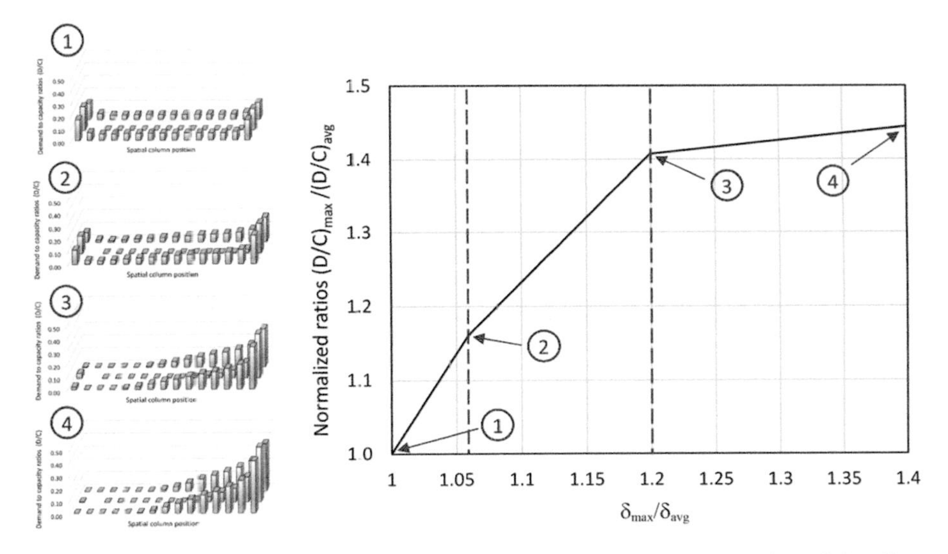

Fig. 18.14 Normalized demand to capacity ratios $(D/C)_{max}/(D/C)_{avg}$ at varying of $\delta_{max}/\delta_{avg}$ provided by the equivalent static analysis

Subsequently, in order to investigate the scattering of the damage levels over the building plan, the maximum $(D/C)_{max}$ and average $(D/C)_{avg}$ demand to capacity ratios of all columns at ground floor was computed in the four cases previously studied (Figs. 18.8, 18.9, 18.10 and 18.11).

In Fig. 18.14 the inelastic indices $(D/C)_{max}$ / $(D/C)_{avg}$ were normalized by the corresponding $(D/C)_{max}$ / $(D/C)_{avg}$ computed in the configuration in which CM ≡ CR (i.e. $\delta_{max}/\delta_{avg} = 1$).

Figure 18.14 clearly shows that the value $\delta_{max}/\delta_{avg} = 1.2$ may be considered as a threshold value beyond which plan scattering of damage level D/C does not increase significantly. Of course, this result may be affected by the particular condition of the analyzed building and, therefore, it needs to be validated by means of further investigations on different building types.

18.5 Conclusions

This paper is focused on the evaluation of the ASCE code approach for detecting building plan irregularity. The response of an existing r.c. school building, taken as reference case study, is investigated in order to evaluate how well torsional irregularity detected by the code corresponds to uneven distribution of inelastic demand found by non-linear dynamic analyses. In particular, for the studied case, the threshold value of 1.2 for identifying buildings plan irregularity seems to be appropriate when plan scattering of damage levels is considered. In this specific case the value of 1.2 seems already to characterize an extreme torsional irregularity condition and, therefore, the value 1.4 provided by the ASCE code may be too large. However, it is worth noting that this trend could also arise from the specificity of the building under investigation and, therefore, a larger number of existing buildings with different structural arrangements and plan configurations are currently under analysis to provide support to this thesis.

Acknowledgements The financial support of ReLUIS through the Progetto Esecutivo Convenzione DPC/ReLUIS 2017 – PR2 Strutture in cemento armato – is gratefully acknowledged.

References

Anastassiadis K, Athanatopoulou A, Makarios T (1998) Equivalent static eccentricities in the simplified methods of seismic analysis of buildings. Earthquake Spectra 14(1):1–34

ASCE, Minimum Design Loads for Buildings and Other Structures, ASCE/SEI 7 (2010) Reston, Virginia 20191: American Society of Civil Engineers

ASCE, Seismic Evaluation and Retrofit of Existing Buildings (ASCE 41-13), ASCE/SEI 4 (2014) American Society of Civil Engineers

Associate Committee on the National Building Code, National Building Code of Canada (1995)

Associate Committee on the National Building Code, National Building Code of Canada (2010)

Athanatopoulou A, Manoukas G, Throumoulopoulos A (2015) Parametric study of inelastic seismic response of reinforced concrete frame buildings. In: Zembaty Z, De Stefano M (eds): Geotechnical, geological and earthquake engineering, vol. 40(15), Seismic behaviour and design of irregular and complex civil structures II, Springer, pp 171–180

Bosco M, Ghersi A, Marino EM, Rossi PP (2013) Comparison of nonlinear static methods for the assessment of asymmetric buildings. Bull Earthq Eng 11(6):2287–2308

Chandler AM (1995) Influence of accidental eccentricity on inelastic seismic torsional effects in buildings. Eng Struct 17(3):167–178

Chandler AM (1997) Performance of asymmetric code-designed buildings for serviceability and ultimate limit states. Earthq Eng Struct Dyn 26(7):717–735
Chandler AM, Hutchinson GL (1987) Evaluation of code torsional provisions by a time history approach. Earthq Eng Struct Dyn 15(4):491–516
Chopra AK, Goel RK (1991) Evaluation of torsional provisions in seismic codes. J Struct Eng 117 (12):3762–3782
D'Ambrisi A, De Stefano M, Tanganelli M (2009) Use of pushover analysis for predicting seismic response of irregular buildings: a case study. J Earthq Eng 13(8):1089–1100
De Stefano M, Pintucchi B (2008) A review of research on seismic behaviour of irregular building structures since 2002. Bull Earthq Eng 6(2):285–308
De Stefano M, Rutenberg A (1997) A comparison of the present SEAOC/UBC torsional provisions with the old ones. Eng Struct 19(8):655–664
De Stefano M, Faella G, Ramasco R (1993) Inelastic response and design criteria of plan-wise asymmetric systems. Earthq Eng Struct Dyn 22(3):245–259
De Stefano M, Tanganelli M, Viti S (2013) Effect of the variability in plan of concrete mechanical properties on the seismic response of existing RC framed structures. Bull Earthq Eng 11 (4):1049–1060
DM.LL.PP (2008) 14 gennaio 2008, Nuove norma tecniche per le costruzioni (NTC), G.U. n. 29 del 4 febbraio 2008
European Committee for Standardization, Eurocode 8 (1993) Design provisions for earthquake resistance of structures part 1–2: general rules – General rules for building. Brussels
European Committee for Standardization, Eurocode 8 (2004) Design of structures for earthquake resistance – part 1: general rules, seismic actions and rules for buildings Eurocode
Goel RK, Chopra AK (1990) Inelastic seismic response of one-story, asymmetric-plan systems. Wiley, Richmond
Harasimowicz AP, Goel RK (1998) Seismic code analyisis of multi-storey asymmetric buildings. Earthq Eng Struct Dyn 27(2):173–185
Magliulo G, Maddaloni G, Cosenza E (2012) Extension of N2 method to plan irregular buildings considering accidental eccentricity. Soil Dyn Earthq Eng 43:69–84
Manoukas G, Athanatopoulou A, Avramidis I (2012) Multimode pushover analysis for asymmetric buildings under biaxial seismic excitation based on a new concept of the equivalent single degree of freedom system. Soil Dyn Earthq Eng 38:88–96
Perus I, Fajfar P (2005) On the inelastic torsional response of single-storey structures under bi-axial excitation. Earthq Eng Struct Dyn 34(8):931–941
Pugi F, Galassi S (2013) Seismic analysis of masonry voussoir arches according to the Italian building code. Int J Earthq Eng 30(3):33–55
Rosenlueth E, Meli R (1986) The 1985 Earthquake: causes and effects in Mexico City. Concr Int 8 (5):23–34
Tso WK (1990) Static eccentricity concept for torsional moment estimations. J Struct Eng 116 (5):1199–1212
Wong CM, Tso WK (1995) Evaluation of seismic torsional provisions in uniform building code. J Struct Eng 121(10):1436–1442

Chapter 19
Effects of Modelling Assumptions on the Plan Irregularity Criteria for Single Storey Buildings

N. Postolov, R. Volcev, K. Todorov, and Lj. Lazarov

Abstract For single-storey buildings, EN 1998-1 allows determination of the centre of lateral stiffness and the torsional radius by considering the translational stiffness represented by the moments of inertia of the cross-section of the vertical elements. Additionally, stiffness of the beam or shear deflections can affect the frame lateral stiffness changing the position of centre of stiffness or torsional radius. In order to investigate the influence of these two parameters, six different single storey buildings with different degree of plan irregularity were examined. All of these buildings were analysed for four different modelling assumptions, regarding the stiffness characteristic of the structural elements. Additionally, hand calculation of these structural features, according the recommendation of EN 1998-1 was carried out. Obtained results show that certain parameters have significant influence for structures with lower degree of plan irregularity, while some other parameters are more influential for structures with higher degree of irregularity. Hand calculated values for eccentricity and torsional radius are most conservative compared to the corresponding ones obtained from numerical analysis. Applied criteria for consideration of structural regularity in plan, prescribed in EN 1998-1, are compared with the criteria given by ASCE 7-10 and NZS 1170.5-2004. In order to compare the seismic response of considered single storey structures with different degree of irregularity in plan, detailed nonlinear time history analyses for seven different acceleration histories, applied in direction perpendicular to the axes of irregularity, scaled to three different levels of seismic hazard, were performed.

Keywords Criteria for regularity in plan · Single storey buildings · Modelling assumptions · Nonlinear seismic assessment

---

N. Postolov (✉) · R. Volcev · K. Todorov · L. Lazarov
Faculty of Civil Engineering, University "Ss. Cyril and Methodius", Skopje, Republic of Macedonia
e-mail: postolov@gf.ukim.edu.mk; volcev@gf.ukim.edu.mk; todorov@gf.ukim.edu.mk; lazarov@gf.ukim.edu.mk

D. Köber et al. (eds.), *Seismic Behaviour and Design of Irregular and Complex Civil Structures III*, Geotechnical, Geological and Earthquake Engineering 48,
https://doi.org/10.1007/978-3-030-33532-8_19

19.1 Introduction

Irregularity in plan can significantly affect the desired ductile response of structures exposed to earthquake loading. Buildings with eccentricity between the centre of the mass and centre of stiffness or with lack of minimal torsional rigidity can undergo coupled lateral and torsional motions during earthquakes, which can significantly increase the seismic demand, especially at perimeter frames. For these reasons most of the seismic design codes contain provisions for control of structural irregularities. If the prescript criteria for regularity are not satisfied, certain restrictions related to the selected method or numerical model for seismic analysis have to be done. Moreover, due to potentially uncontrollable torsion oscillation and reduced ductile response, design seismic forces have to be obtained for a lower reduction factor.

High seismic vulnerability of plan irregular structures was a motivation for many researches in the past. An extensive review of different structural irregularities and systematization of conducted researches can be found in Rutenberg (2002), De Stefano and Pintucchi (2008), Varadharajan et al. (2013). Many of these researches, Cosenza et al. (2000), Humar and Kumar (2000), Zheng et al. (2004), Rasulo et al. (2004), Özhendekci and Polat (2008), are related to the analysis of codes provisions for plan irregularity.

19.2 Codes Provisions for Plan Irregularity

According to EN 1998-1 (CEN 2004), a building can be characterized as regular in plan, if six different conditions are satisfied, at all storey levels. Some of these conditions are qualitative, and can be checked in the preliminary design stage, but some of them that are based on the eccentricity between the centre of mass and the centre of stiffness or torsional radius, Eq. 19.1, are quantities that have to be calculated additionally. In-depth discussion of the conditions for plan regularity according to EN1998-1 (CEN 2004), can be found in Penelis and Penelis (2014), Fardis et al. (2015).

$$\begin{gathered} e_x \leq 0.3 r_x; e_y \leq 0.3 r_y \\ r_x \geq l_s; r_y \geq l_s \end{gathered} \tag{19.1}$$

For single storey buildings these characteristics are uniquely defined and EN1998-1 allows to be calculated through the moments of inertia of the cross section of vertical elements. In general, some additional parameters, like beams stiffness or shear deflections can affect the position of centre of stiffness or torsional radius. In multi storey buildings, Eurocode 8 allows simplified definition for classification of structural regularity in plan and for the approximate analysis of torsional effects only for buildings in which all lateral load resisting systems are running from the foundation to the top and have similar deformation patterns under lateral loads.

Moreover, Eurocode 8 accepts that in frames and in systems of slender walls with prevailing flexural deformations, the position of the centre of stiffness and the torsional radius of all stories may be calculated as those of the moments of inertia of the cross-section of the vertical elements.

ASCE 7-10 distinguishes structures as torsional or extreme torsional irregular, depending on differences between the maximum story drift, computed including accidental torsion with 5% accidental eccentricity, and the average story drift. Torsional irregularity is defined to exist where the maximum story drift, at one end of the structure transverse to an axis is more than 1.2 times the average story drift at the two ends of the structure. If this ratio exceeds 1.4 times the average story drifts, then extreme torsional irregularity exists.

Similar to ASCE 7-10, NZS 1170.5-2004 defines that torsional sensitivity shall be considered to exist when the largest ratio between maximum storey displacement at the extreme points of the structure, at each level, in the direction of earthquake induced by equivalent static actions acting with accidental eccentricity of 10% and average of the displacement at the extreme points at same level in both orthogonal direction exceeds 1.4. This requirement, with the exception of assumed torsional eccentricity, is identical with the condition for extreme torsional irregularity given in ASCE 7-10, so the structure that will be torsional sensitive in New Zealand will have an extreme torsional irregularity according to ASCE 7-10.

19.3 Structural Characteristics for Control of Plan Irregularity

19.3.1 Centre of Stiffness

At single storey buildings the centre of stiffness is uniquely defined and depend on the position of elements which contribute to the lateral stiffness of structure. At orthogonal buildings, it can be easily obtained as a centroid of lateral stiffness of individual frames perpendicular to the considered direction.

$$x_{CS} = \frac{\sum K_{Yi} \cdot x}{\sum K_{Yi}} \qquad y_{CS} = \frac{\sum K_{Xi} \cdot y}{\sum K_{Xi}} \tag{19.2}$$

Lateral stiffness of the frames can readily be determined only for the case of absolutely rigid or absolutely flexible beams, Fig. 19.1. If it is assumed that beams are absolutely rigid, and all columns are made from same material and have a same height h, then the Eq. 19.2 will get a well-known form:

$$x_{CS} = \frac{\sum \frac{12EI_x}{h^3} \cdot x}{\sum \frac{12EI_x}{h^3}} = \frac{\sum I_x \cdot x}{\sum I_x} \qquad y_{CS} = \frac{\sum I_y \cdot y}{\sum I_y} \tag{19.3}$$

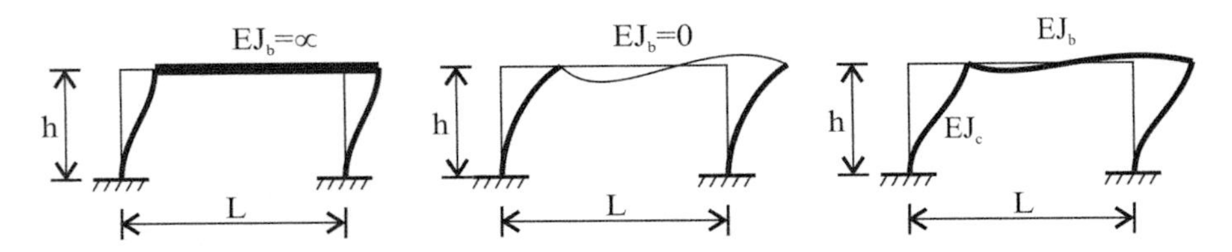

Fig. 19.1 Influence of beam rigidity on the horizontal stiffness of portal frame

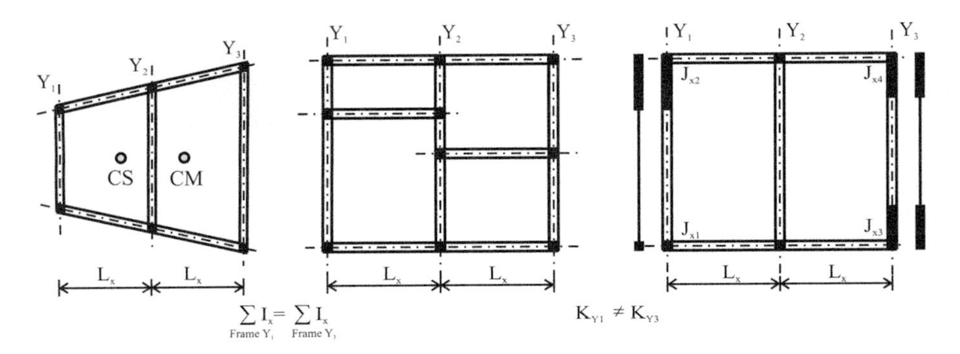

Fig. 19.2 Influence of beams stiffness on the plan irregularity

However, in certain characteristic cases, as presented at Fig. 19.2, the sum of the moments of inertia of the cross-sections of the vertical elements within the frame Y1 and Y3 are the same, $\sum$Ix(frameY1) = $\sum$Ix(frameY3), but due to the influence of the beams rigidity, the frames Y1 and Y3 have different lateral stiffness, $K_{Y1} \neq K_{Y3}$.

Neglecting the effect of beams is reasonable in the case of orthogonal plan disposition, where the beams have same length and same geometrical characteristics. Also, this assumption can be acceptable in the situation of plastic hinges formation. Namely, beam plastic mechanism lead to pin – pin beam to column connection, which is almost identical to beams with no rotational stiffness, while formation of plastic hinges at the both column ends will lead to situation of significantly larger beam stiffness, compared with the stiffness of columns.

For elastic analysis of multi bay frames with different span length, or in the case of non-orthogonal structures, Fig. 19.2, the lateral stiffness of the frames can be computed with the standard procedures of static condensation with respect to horizontal degree of freedom. Lateral stiffness of the one bay frame, including the contribution of beam can be calculated according to Eq. 19.4.

$$k_Y = \frac{24EI_c}{h^3}\frac{12\rho + 1}{12\rho + 4} \tag{19.4}$$

where ρ = (EIb/L)÷ (EIc/h) is beam to column stiffness ratio

It must be noted that in this equation, contribution of shear deformations to the lateral stiffness of columns are also neglected. For structural walls, where slenderness ratio is lower than 4–5, shear deformation can be significant compared with the flexural one, so the contribution of shear should be taken into account in the calculation of lateral stiffness. In case of absolutely flexible beams, equivalent lateral stiffness of frames, taking into account the contribution of shear deformation can be calculated according to Eq. 19.5.

$$k_Y = \sum \frac{1}{\frac{h^3}{3EI} + \frac{h}{GAs}} \tag{19.5}$$

Compared with the exposed analytical approach for determination of centre of stiffness, numerical approach based on the results from FEM analysis, offer a universal solution, taking into account the features of different particular cases.

For the spatial structural model with rigid plate in their plain, position of centre of stiffness can be easily obtained through few steps of numerical procedure. For a applied moment of rotation around the vertical axis z, the rigid plate will rotate around the centre of rigidity (pole of rotation) for the angle $\partial\varphi$, Fig. 19.3. All points on the rigid plate will receive displacement proportional to the rotation and distance to the pole of rotation ρ.

$$\partial u = \rho \cdot \partial \phi \tag{19.6}$$

If a centre of mass with coordinates CM (x_{CM}, y_{CM}) is selected as a reference point, the displacement of this point with respect to two orthogonal axes can be expressed as a product of the rotation angle and the normal distance from the reference point to the pole of rotation, which represent the centre of rigidity.

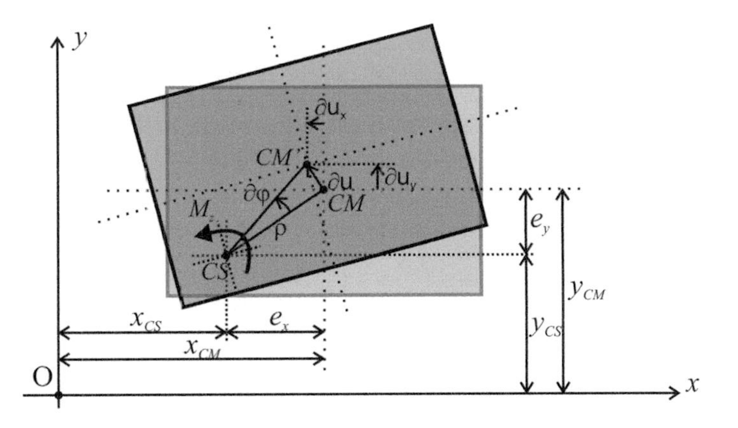

Fig. 19.3 Plate rotation around the centre of stiffness

$$\begin{aligned} \partial u_x &= -e_y \cdot \partial \phi \\ \partial u_y &= e_x \cdot \partial \phi \end{aligned} \tag{19.7}$$

In this equations e_x and e_y are eccentricities between the centre of stiffness and centre of mass along the two orthogonal axes, and they can be express as the difference between the coordinates of mass and stiffness centre.

$$\begin{aligned} e_x &= x_{CM} - x_{CS} \\ e_y &= y_{CM} - y_{CS} \end{aligned} \tag{19.8}$$

Hence, the coordinates of the centre of stiffness can be expressed through the known coordinates of the centre of mass and certain eccentricities with respect to the two orthogonal axes.

$$\begin{aligned} x_{CS} &= x_{CM} - e_x = x_{CM} - \partial u_y / \partial \phi \\ y_{CS} &= y_{CM} - e_y = y_{CM} + \partial u_x / \partial \phi \end{aligned} \tag{19.9}$$

19.4 Torsional Radius

Torsional radius as a structural characteristic represents the potential of torsional vibration of structures exposed to earthquake ground motion. For single story structures torsional radius is defined as the square root of the ratio of the torsional stiffness K_{Rz} with respect to the centre of lateral stiffness, to the storey lateral stiffness K_X or K_Y.

$$r_x = \sqrt{\frac{K_{Rz}}{K_Y}} \quad r_y = \sqrt{\frac{K_{Rz}}{K_X}} \tag{19.10}$$

If the effects of beams and shear deformation of vertical element are neglected, as was explained in Sect. 19.3.1, the torsional radius can be obtained on the basis of the moments of inertia of the cross-section of the vertical elements.

$$r_x = \sqrt{\frac{\sum J_x \cdot x_{iCS}^2 + J_y \cdot y_{iCS}^2}{\sum J_x}} \quad r_y = \sqrt{\frac{\sum J_x \cdot x_{iCS}^2 + J_y \cdot y_{iCS}^2}{\sum J_y}} \tag{19.11}$$

In the above equation x_{iCS} and y_{iCS} represent the coordinate of individual elements with respect to centre of stiffness, while the Jx and Jy denote the moments of inertia of the cross section of vertical elements for bending about an axis parallel to global direction x or y, respectively.

For the control of the criteria for torsional flexibility, according to EN1998-1, torsional radius is defined with respect to the centre of lateral stiffness, which is reasonable in the case of nearly double symmetric structures. In the case non-symmetrical systems, control of the criteria for torsional flexibility need to be done with the torsional radius defined with respect to the centre of mass.

$$r_{mx} = \sqrt{{r_x}^2 + {e_x}^2} \quad r_{my} = \sqrt{{r_y}^2 + {e_y}^2} \tag{9.12}$$

For the non-symmetrical systems with extreme eccentricity, ratio between torsional radius with respect to the centre of mass and centre of stiffness is higher, which leads to fulfilling the requirement for torsional inflexibility given in EN1998-1.

19.5 Description of the Analysed Structures

In order to investigate the influence of modelling assumption on the eccentricity between centre of stiffness and centre of mass and torsional radius, six different single storey buildings with different degree of plan irregularity were analysed. All analysed structures are rectangular in plan and consist of three frames in x, and five frames in the global y direction. The distance between the frames in both directions is 5 m, while the story height is 3 m. At the first model (M0), all columns have a 40 cm square cross section. At other five models (M1 to M5), the middle column in the one exterior frame in the global y direction, was replaced with the structural wall with a thickness of 25 cm and a length of 75, 100, 150, 200 and 300 cm, Fig. 19.4. All beams are with rectangular cross section 30/50 cm. In the mathematical model interior beams are modelled as a T section with effective flange width of 180 cm, while exterior as Γ section with effective flange width of a 110 cm and thickness of 15 cm.

All of these buildings were analysed for four different modelling assumptions, regarding the stiffness characteristic of structural elements. At the first modelling assumption, equivalent with approximate hand calculation, beams were absolutely rigid and shear deformation was neglected (NS_BN). At the second model shear stiffness was included, while the beams were absolutely rigid (SI_BN). The third model represents an opposite situation where the beams are modelled with the realistic elastic rigidity, while the deformation induced by shear deformation was neglected (NS_BE). And finally, fourth most realistic model is when both beam stiffness and shear deformation were included in the analysis (SI_BE). Reduction of lateral and torsional stiffness, determined in respect to the centre of stiffness, for the considered models with different modelling assumptions is presented in Fig. 19.5. The influence of the beams rigidity on the lateral and torsional stiffness of the analyzed structures is greater compared with the influence of shear deformations. An exception to this is noted at the structure M5 (structure with largest wall length), where shear deformation has a greater impact on the reduction of the lateral stiffness compared with the rigidity of the beams.

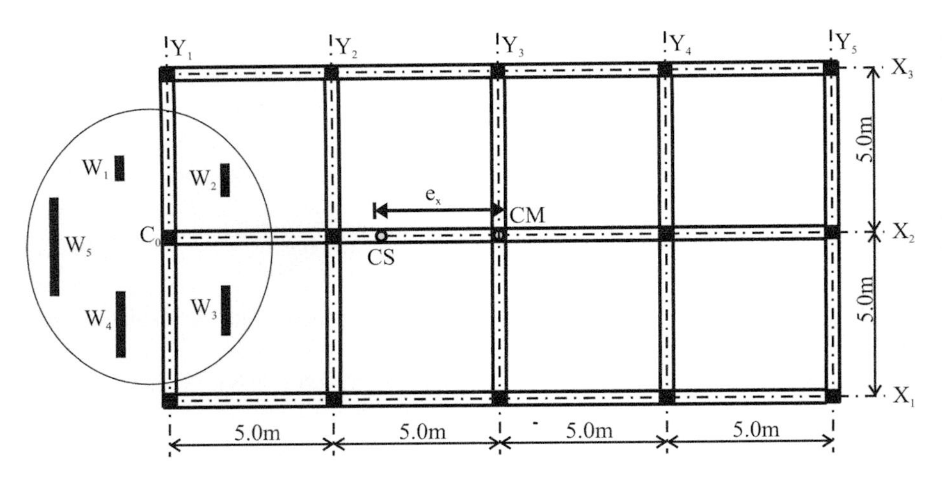

Fig. 19.4 Plan view of analysed structures

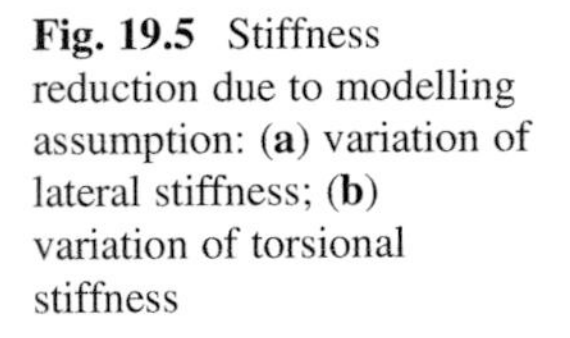

Fig. 19.5 Stiffness reduction due to modelling assumption: (**a**) variation of lateral stiffness; (**b**) variation of torsional stiffness

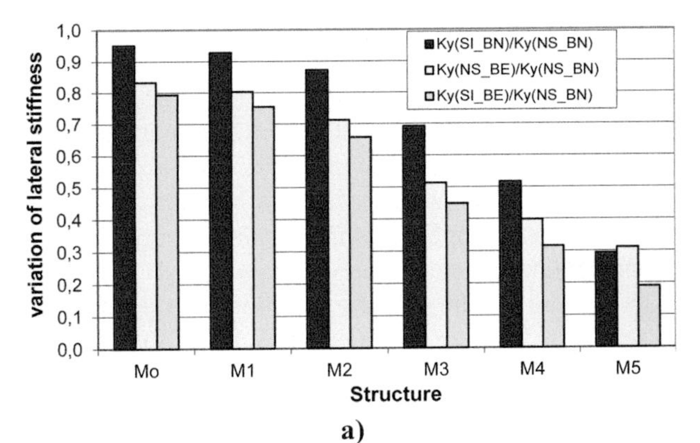

a)

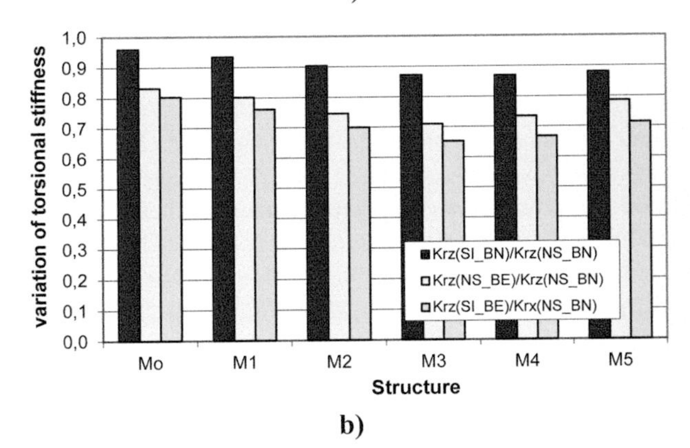

b)

Reduction of lateral stiffness ranges from 5% to 20% in the structure with columns of up to 70% to 80% for the model with an eccentric wall with a length of 300 cm. The lower degree of reduction in torsional stiffness at structures with eccentric walls is due to the fact that in these structures the distance between the eccentric wall and the centre of stiffness decreases, leading to a lower participation of these elements in the total torsional stiffness.

19.6 Comparison of Criteria for Regularity in Plan

The various degree of stiffness reduction of the individual elements leads to a variation in the eccentricity between the centre of stiffness and the centre of the mass, as well as the changes in the torsional radius, and therefore also affects the criteria for plan irregularity given in the Eurocode 8, Fig. 19.6

The results obtained by the analysis of the elastic beams and included shear deformation, which is closest to the realistic behaviour of the structure, lead to

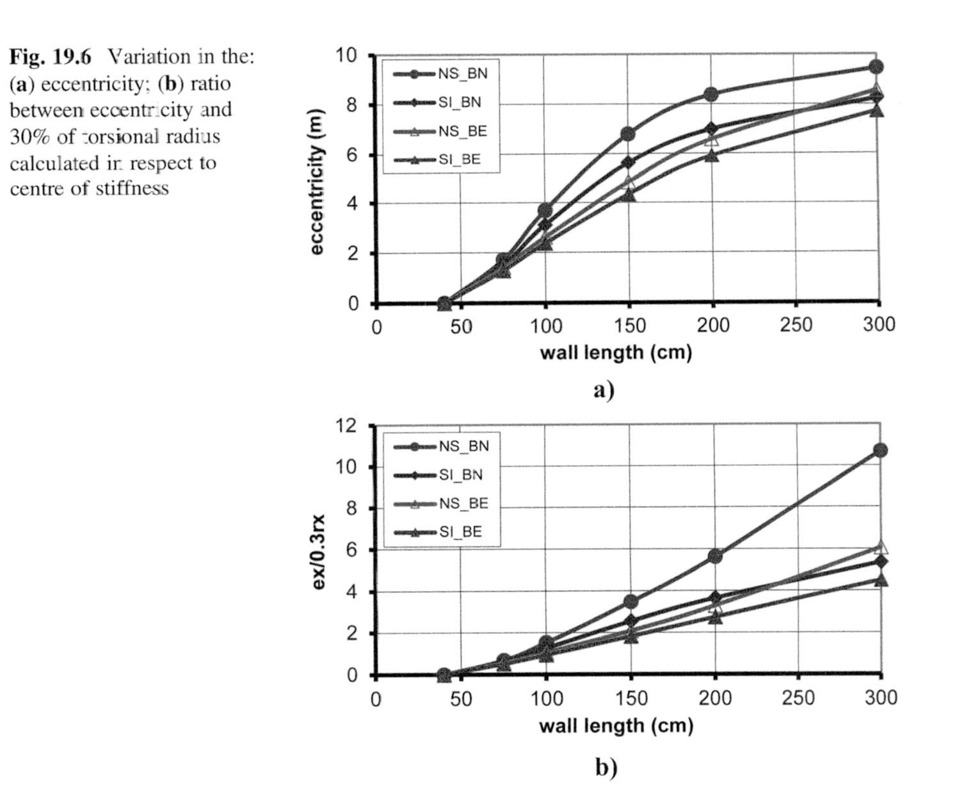

Fig. 19.6 Variation in the: (**a**) eccentricity; (**b**) ratio between eccentricity and 30% of torsional radius calculated in respect to centre of stiffness

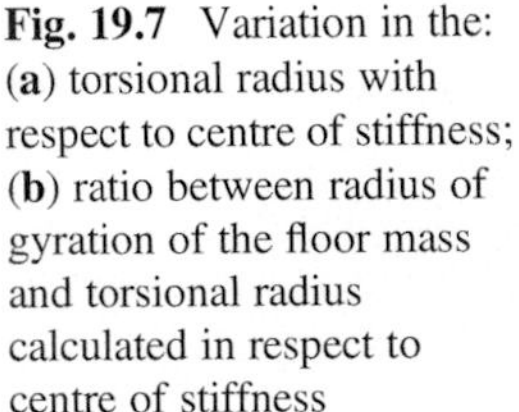

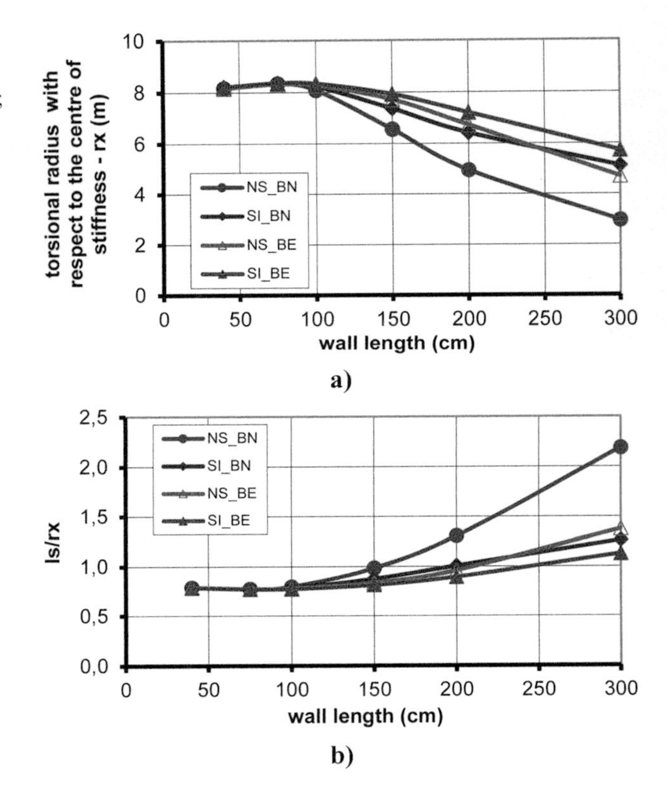

Fig. 19.7 Variation in the: (**a**) torsional radius with respect to centre of stiffness; (**b**) ratio between radius of gyration of the floor mass and torsional radius calculated in respect to centre of stiffness

smallest value of eccentricity. The most conservative results are obtained for the case of absolutely rigid beams and neglected shear deformation. For the structure with the highest degree of irregularity this assumption leads to twice the ratio between the eccentricity and 30% of torsional radius. From the analysis with neglected shear deformation and elastic beams, it can be noted that the deformation characteristics of the beams have more influence in structures with a lower degree of irregularity. Oppositely shear stiffness, due to presence of walls with lower aspect ratio, have more influence in structures with a higher degree of irregularity.

The influence of the beams stiffness and shear induced deformations on the torsional radius determined with respect to the centre of stiffness and to the centre of the mass as well as the ratio of the radius of gyration of the floor mass in plan and the torsional radius, which is one of the conditions for plan regularity according to EC8 is presented in Figs. 19.7 and 19.8.

In the case where the torsional radius is determined with respect to the centre of stiffness, Fig. 19.7, the assumption of rigid beams and neglecting the shear deformation leads to the smallest values of the torsion radius, and hence to more rigorous requirements regarding the satisfaction of the criteria of plan regularity. When the torsional radius is determined with respect to the centre of mass, Fig. 19.8, due to the opposite influence of the stiffness assumptions in terms of eccentricity, no significant influence of the modelling assumptions was observed. The criteria for torsional

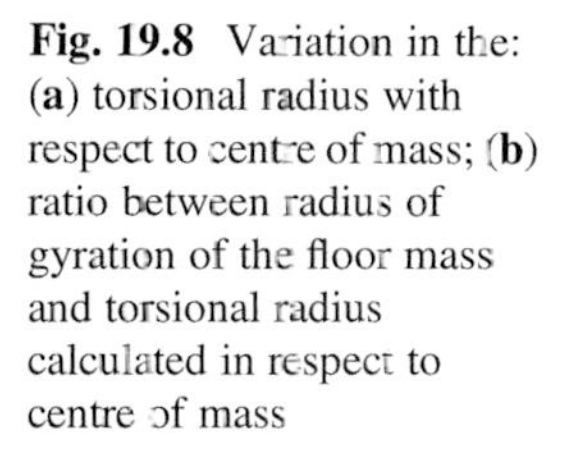

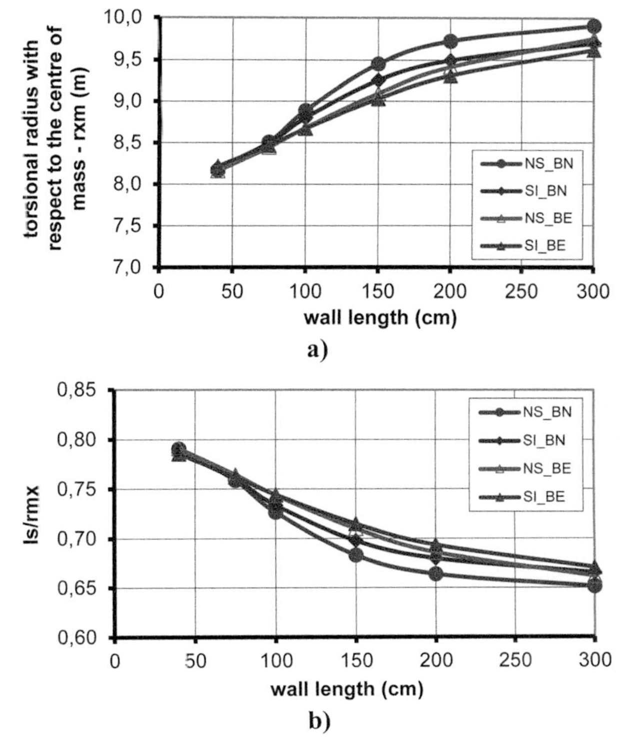

Fig. 19.8 Variation in the: (**a**) torsional radius with respect to centre of mass; (**b**) ratio between radius of gyration of the floor mass and torsional radius calculated in respect to centre of mass

flexibility is satisfied for all analysed structures. The greatest ratio between the radius of the gyration of the floor mass and the torsional radius is observed for the models where elastic beams and shear deformation are taken into account.

Comparison of the criteria for the plan regularity, i.e. the torsional sensitivity, for the analysed structures with different degree of irregularity, determined according to EN1998-1 (CEN 2004), ASCE 7–10 (ASCE 2010) and NZS 1170.5:2004 (SNZ 2004) are presented in Fig. 19.9. The values of eccentricity, torsional radius and ratio between maximal and average displacement, determined for the model with the "real" stiffness characteristics, are normalized with the criteria for irregularity defined in the corresponding seismic codes. So, the normalized criteria value higher than 1 indicate that the regularity requirements has not been met. In structures with a lower degree of irregularity (M1 and M2), eccentricity is less than 30% of the torsional radius. Thus, the criteria for plan regularity at these structures, according to the conditions of EC8 is satisfied.

In structures with higher degree of irregularity (M3, M4 and M5) this ratio significantly increase, which leads to exceeding the criteria for plan regularity by 4.5 times at the structure with a wall length of 3 m. Ratios between radius of gyration of floor mass and torsional radius determined with respect to the centre of mass is less than 1 at all analysed structures. Criteria for torsional flexibility according to ASCE7-10 and NZS 1170.5:2004 is exceeded at all analysed structures, with the exception of the structure M1 which meet the criteria for extreme torsion irregularity.

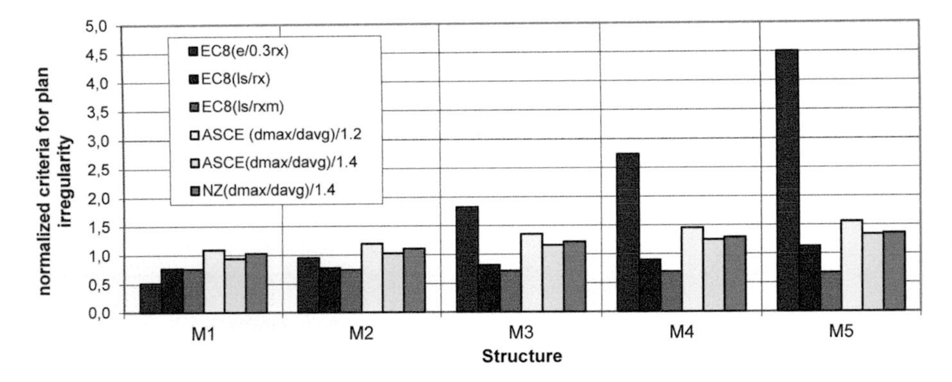

Fig. 19.9 Normalized criteria for plan irregularity for analysed structures according different codes requirement

Unlike the criteria according to the EC8, the criteria according to ASCE7-10 and NZ1170.5 have a significantly slower growth with an increasing of the degree of irregularity.

19.7 Nonlinear Seismic Assessments of Analysed Structures

In order to compare the seismic response of single storey structures with different degree of irregularity in plan, detailed nonlinear time history analyses for seven different input acceleration histories (Imperial Valley 1940,Victoria Mexico – Chihuahua 1980, Chi Chi Taiwan 06 – 1999, Duzce – Turkey 1999, Petrovac – Monte Negro 1979, Loma Prieta 1989, San Fernando 1971) applied in direction perpendicular to the axes of irregularity, scaled to three different levels of seismic hazard (PGA 0,18 g, 0,36 g and 0,54 g), were performed. Figure 19.10 shows the normalised elastic spectra of the acceleration histories and the mean spectra.

All analysed structures are same as in previous analyses. All beams are reinforced with longitudinal reinforcement 3∅14 mm at the bottom, 5∅14 mm at the top and ∅8/15 mm in the effective flange width. The columns are reinforced with 8∅16 mm. The structural walls are reinforced in accordance with EN 1998-1 part 5.3. The analyses are carried out with software Seismostruct. The structural elements are modelled with inelastic force – based plastic hinges frame elements. The length of the plastic hinge is 10% of the length of the element.

Figure 19.11 shows the mean and median total displacement at the most distant frame from the centre of rigidity. The maximal mean displacement of the irregular structures is 10% higher that the displacement of the regular structure (M0).

The obtained maximal displacements for all levels of seismic hazard are greater than the yield displacements for all analyzed structures. The first plastic hinge most commonly occurs in the columns of the frame Y5, at lateral displacement in range of

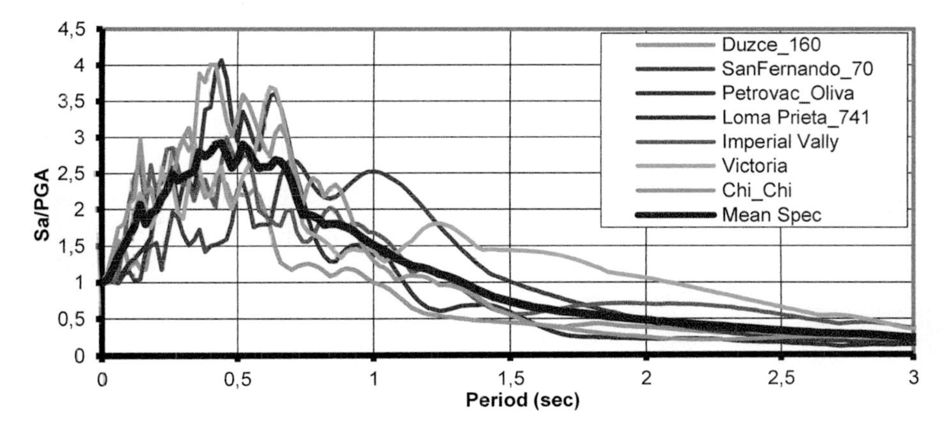

Fig. 19.10 Normalised elastic spectra and mean spectra

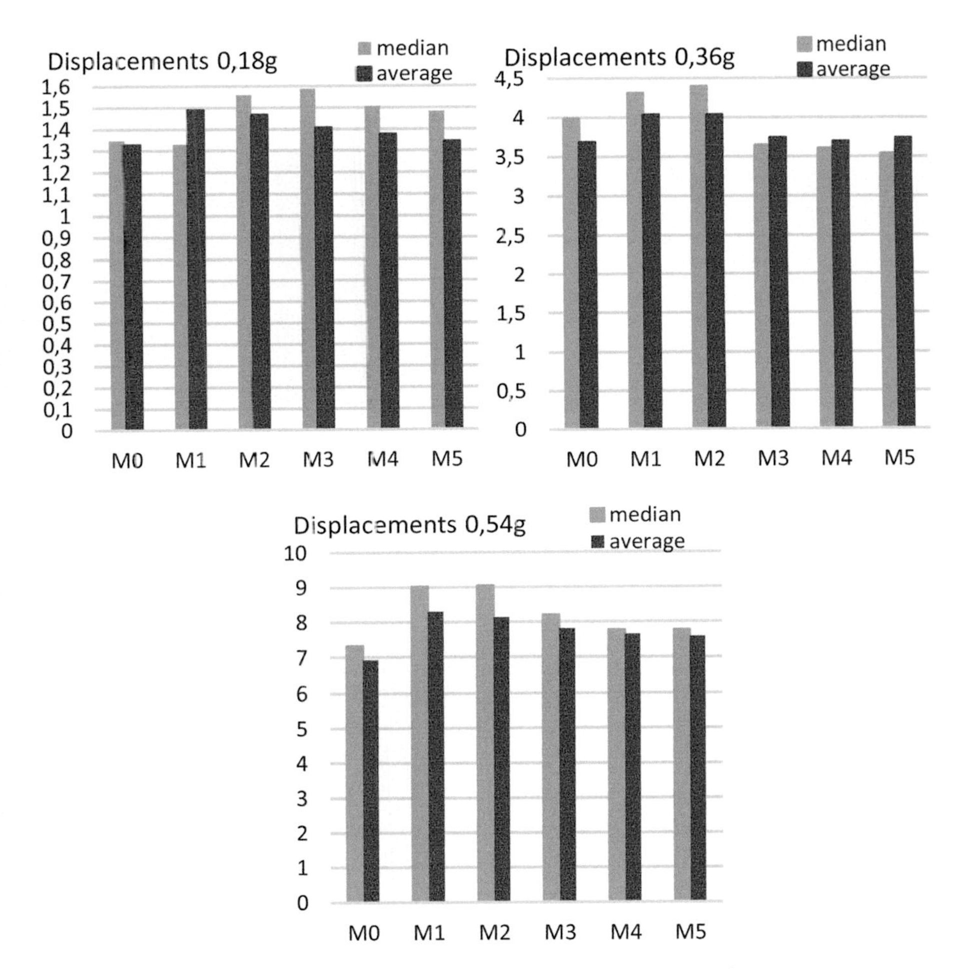

Fig. 19.11 Mean and median displacement for PGA 0,18 g, 0,36 g and 0,54 g

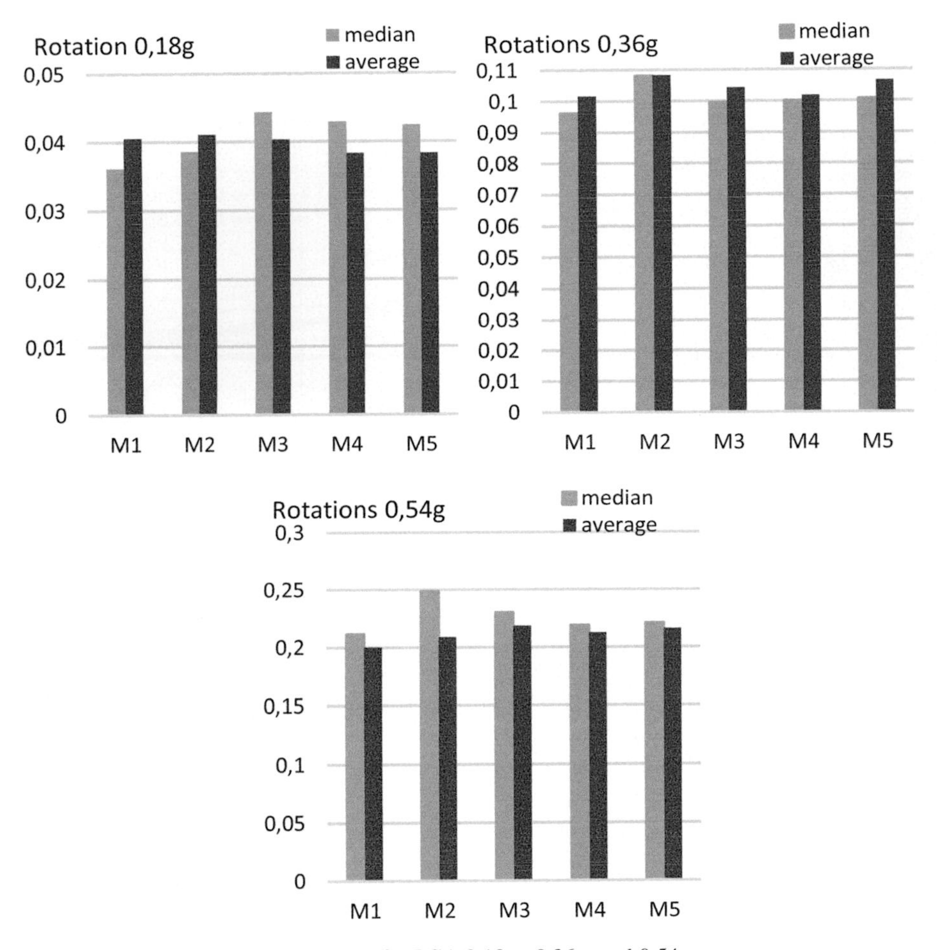

Fig. 19.12 Mean and median rotation for PGA 0,18 g, 0,36 g and 0,54 g

0.83–0.9 cm. In some cases, (about 20% of the analyzes carried out), the first plastic hinge occurs in the elements of another frame, most often in the wall of the most rigid frame Y1. However, in these cases, the displacement amplification on the flexible side of the structure results with the appearance of plastic hinges in the elements of the frame Y5 in some of the following time steps of the analysis. Ductility demand for analysed structures ranges from 1.5 to 1.7 for PGA 0.18 g, 4.1 to 4.7 for PGA 0.36 g, or 8 to 9 for PGA 0.54 g.

Figure 19.12 shows the mean and median maximal rotation of the slab. It can be noticed that the value of the rotation does not vary for different degree of irregularity.

19.8 Final Remarks

The position of centre of stiffness and torsional radius can vary depending on the assumptions in the mathematical model. The approximate determination of these parameters for single storey buildings, by considering the translational stiffness represented by the moments of inertia of the cross-section of the vertical elements, neglecting the effect of beams, is most conservative and can be used in the preliminary design stage. It is advisable to control these parameters in the later design stage.

For structures with larger eccentricities between the centre of stiffness and the centre of mass EC8 prescribes more rigorous criteria for plan regularity compared with the ASCE7-10. However, degree of irregularity does not implicate proportional increasing of seismic demand, unlike ASCE7-10, where the degree of irregularity leads to amplification of accidental torsional moment.

From the nonlinear analyses it can be noticed that structures with different degree of irregularity, according to EC8, have similar behaviour when are subjected to earthquake motion. This is due to the increased load capacity of the frame where the structural walls are located.

References

ASCE (2010) Minimum design loads for buildings and other structures. ASCE/SEI Standard 7-10

CEN – European Committee for Standardization (2004) Eurocode 8: design of structures for earthquake resistance – part 1. European standard EN 1998 – 1

Cosenza E, Manfredi G, Realfonzo R (2000) Torsional effects and regularity conditions in R/C buildings. In: Proceedings of the 12th world conference on earthquake engineering, Auckland, New Zealand, 2000, paper no. 2551

De Stefano M, Pintucchi B (2008) A review of research on seismic behavior of irregular building structures since 2002. Bull Earthq Eng 6(2):282–308

Fardis MN, Carvalho EC, Fajfar P, Pecker A (2015) Seismic design of concrete buildings to Eurocode 8. CRC Press, Taylor & Francis Group, 6000 Broken Sound Parkway NW

Humar J, Kumar P (2000) A new look at the torsion design provisions in seismic building codes. In: Proceedings of the 12th world conference on earthquake engineering, Auckland, New Zealand, 2000, paper no. 1707

Özhendekci N, Polat Z (2008) Torsional irregularity of buildings. The 14th world conference on earthquake engineering, October 12–17, 2008, Beijing, China

Penelis GG, Penelis GG (2014) Concrete buildings in seismic regions. CRC Press Taylor & Francis Group, 6000 Broken Sound Parkway NW

Rasulo A, Paolacci F, De Stefano M (2004) Criteria for plan regularity based on seismic reliability analysis. 13th world conference on earthquake engineering, Vancouver, BC, Canada August 1–6, 2004, paper no. 2293

Rutenberg A (2002) EAEE Task Group (TG) 8: behavior of irregular and complex structures—progress since 1998. In: Proceedings of the 12th European conference on earthquake engineering, CD ROM. London, September

SNZ (2004) New Zealand Standard NZS 1170.5:2004 structure design, part 5: earthquake actions – New Zealand. Standards New Zealand, Wellington
Varadharajan S, Sehgal VK, Saini B (2013) Review of different structural irregularities in buildings. J Struct Eng 39(5):393–418, December 2012–January 2013 No. 39–51
Zheng N, Yang Z, Shi C, Chang Z (2004) Analysis of criterion for torsional irregularity of seismic structures. 13th world conference on earthquake engineering, Vancouver, BC, Canada, August 1–6, 2004, paper no. 1465

Chapter 20
Numerical Study on Seismic Response of a High-Rise RC Irregular Residential Building Considering Soil-Structure Interaction

Tomasz Falborski

Abstract The objective of the present study is to investigate the importance of soil-structure interaction effects on the seismic response of a high-rise irregular reinforced-concrete residential building. In order to conduct this research, a detailed three-dimensional structure model was subjected to various earthquake excitations, also including a strong mining tremor. Soil-foundation flexibility was represented using the spring-based solutions, incorporating foundation springs and dashpots. For each soil type analyzed in this study, the foundation stiffness was calculated using the static stiffness, embedment correction factors, and dynamic stiffness modifiers. The influence of diverse soil conditions (represented by their average effective profile velocities and shear moduli) on the dynamic characteristics (e.g. fundamental vibration period) and seismic response (e.g. peak lateral accelerations) of the structure model was investigated and discussed. The numerical analysis results clearly demonstrate that the seismic performance of the building to the strong earthquake shaking can be significantly affected by the soil-structure interaction effects.

Keywords Soil-structure interaction · Soil flexibility · Seismic response · High-rise RC building · Irregular structures

20.1 Introduction

Dynamic response of a building structure subjected to strong seismic motions may be affected by many different factors including base isolation (see, for example, Falborski and Jankowski 2013, 2016, 2017a, b), structural pounding (see, for

T. Falborski (✉)
Faculty of Civil and Environmental Engineering, Gdansk University of Technology, Gdansk, Poland
e-mail: tomfalbo@pg.gda.pl

D. Köber et al. (eds.), *Seismic Behaviour and Design of Irregular and Complex Civil Structures III*, Geotechnical, Geological and Earthquake Engineering 48,
https://doi.org/10.1007/978-3-030-33532-8_20

example, Jankowski and Mahmoud 2015, 2016; Naderpour et al. 2016; Sołtysik et al. 2016, 2017), or damage level (see, for example, Ebrahimian et al. 2017). Among these effects the interaction between the structure foundation and the underlying soil is considered one of the most important contributors (see, for example, Gazetas 1991; Wolf 1985; Mylonakis and Gazetas 2000; Stewart et al. 1999a, b; Veletsos and Prasad 1989). Even though the soil-foundation flexibility may significantly modify the dynamic characteristics, and thus alter seismic response of a structure (see, for example, Stewart and Fenves 1998; Wong et al. 1988), many studies either do not utilize soil-structure interaction, which may result in misrepresentation of the actual building behaviour, or are conducted for strongly simplified stick models, according to which the buildings can be idealized as multi-degree-of-freedom systems (see, for example, Falborski and Jankowski 2017a, b), which may not be fully appropriate for complex and irregular structures. Therefore, the present study is designed to analyze the seismic performance of a three-dimensional numerical model of a high-rise irregular reinforced-concrete residential building considering the interaction between the structure foundation and the underlying soil. Diverse soil conditions were represented by their average effective profile velocities and shear moduli. Soil-foundation flexibility was utilized using spring-based solutions, incorporating foundation springs and dashpots, as it is the most commonly adopted approach for idealizing the soil-foundation interface in current engineering practice.

20.2 Building Model

The structure model considered in the present study is a high-rise reinforced-concrete irregular shear wall building, supported on a mat foundation. The 13-storey structure with an interior shear wall core is 37.50 m tall from the ground surface to the roof. The height of the embedded basement is 3.0 m. The height of the ground storey is 3.0 m, and all other above-grade stories are 2.50 m. The height of the rooftop structure (i.e. emergency exit) is 2.0 m. The overall plan dimensions of the structure are 26 m wide by 28 m long, although the typical floor is irregular in plan. The gross dimensions of the foundation mat are 28 m wide by 30 m long. Detailed numerical model of the high-rise building considered in the present study was developed using 4-node shell elements (nearly 21,000 in total) available in an educational version of Autodesk Robot Structural Analysis Professional 2016.

20.3 Site Conditions

Characterization of site conditions is an integral component and an essential first step in many seismic analyses incorporating soil-structure interaction. Accurate estimation of soil properties requires knowledge of both structural and soil dynamics principles. Therefore, geotechnical data (e.g. shear wave velocity and soil profiles,

Table 20.1 Geotechnical characteristics of the soil types considered

Soil type	Shear wave velocity v_s (m/s)	Mass density ρ (kN/m^3)	Poisson's ratio ν (–)
I. Dense soil/Soft rock	400	20	0.40
II. Stiff soil	250	18	0.35
III. Soft soil	180	16	0.30

soil mass density etc.) should be evaluated in close collaboration with geotechnical engineers, most preferably, through site-specific measurements, which would provide the most accurate results. If the geophysical measurements are conducted in free-field conditions, shear wave velocity should be increased to account for the presence of overburden pressures caused by the added weight of the structure. Moreover, for soil profiles that vary with depth, values of overburden-corrected shear wave velocities should be averaged to obtain so-called *average effective profile velocity*, which is one of the most valuable indicators of dynamic properties of the soil in many seismic analyses incorporating soil flexibility. More detailed procedures and guidelines for estimation of shear wave velocities may be found, for example, in guidelines prepared by Pacific Earthquake Engineering Research Center (2012).

In order to investigate the importance of soil-structure interaction effects on the seismic response of a high-rise irregular reinforced-concrete residential building considered in the present study, three different site conditions were utilized. Soil properties are represented by shear wave velocities, mass densities, and Poisson's ratios. Geotechnical characteristics are briefly summarized in Table 20.1.

20.4 Soil-Structure Flexibility

Soil-structure interaction may be implemented in response history analyses through many different methodologies, either incorporating foundation springs and dashpots or modelling the soil beneath the structure as a solid continuum with finite elements (see, for example, Stewart et al. 2003; Wong and Luco 1986; Mylonakis et al. 2006). In the present study soil-foundation flexibility was utilized using spring-based solutions proposed by Pais and Kausel (1988), which are identified among the most commonly adopted equations in current engineering practice. This approach includes calculations of soil springs to consider translational and rotational degrees of freedom, and dashpots to address soil damping effects. For each site condition analyzed in this study, characteristics of springs and dashpots were developed by first calculating translational (k_x, k_y) and rotational (k_{xx}, k_{yy}) stiffnesses for rectangular rigid foundation as well as dashpot coefficients (c_x, c_y, c_{xx}, c_{yy}). The foundation stiffness was calculated using the static stiffness, embedment correction factors η, and dynamic stiffness modifiers α. The base spring stiffness was subtracted from the overall horizontal stiffness to determine the portion of the horizontal stiffness attributed to passive pressure resistance against basement walls (total translational stiffness is larger due to embedment). According to the recommendations specified in documents prepared by National Earthquake Hazards Reduction Program (2012) and Applied Technology Council (2005), the value of shear modulus was reduced to account for large strain effects associated with nonlinear behaviour. The reduction factor of 0.50 was used for the 1994 Northridge earthquake, whereas a modulus reduction factor of 1.0 was assumed for the 2002 Polkowice mining tremor.

Vertical springs and dashpots were distributed over the footprint of the foundation, allowing the foundation to deform in a natural manner. Edge intensities were adjusted to match the overall rocking stiffness values, as the vertical soil reaction is not uniform, and tends to increase near the edges of the foundation. Correction factors were determined using the equations presented by Harden and Hutchinson (2009). Corner intensities ware calculated as the average of edge intensities in both directions. Table 20.2 summarizes the calculation results conducted for the 1994 Northridge earthquake with a shear modulus reduction factor of 0.50. For the 2002 Polkowice mining tremor values would be doubled, as the reduction factor is 1.0.

It should be noted that both horizontal and vertical springs are elastic with no compression capacity limit and zero tension capacity.

20.5 Modal Analysis

The next step was to conduct modal analysis to investigate the influence of soil-structure interaction on dynamic characteristics of the analyzed high-rise building. Fundamental modes of vibration for the fixed-base building are presented in Fig. 20.1. Results obtained for all site conditions considered are briefly summarized in Table 20.3.

20.6 Seismic Analysis

The final step of the current investigation was to conduct response history analyses for all soil types considered. In addition to the gravity load, the high-rise building model was subjected to the 1994 Northridge earthquake (Santa Monica station, CSMIP[1] No. 24538, record scaled by factor 0.5), and the 2002 Polkowice mining tremor, as an example of so-called induced seismicity, which has recently become an issue of major concern of both academic and professional communities in Poland (see, for example, Zembaty 2004). Seismic records in NS direction were applied along the y global axis, whereas records in the EW direction were used along the X

[1]California Strong Motion Instrumentation Program

Table 20.2 Stiffness and damping parameters used for the 1994 Northridge earthquake

Spring / Dashpot	Soil type	Static stiffness (kN/m) or (kNm/rad)	η (–)	α (–)	β_{sur} (–)	β_{emb} (–)	Dynamic stiffness (kN/m) or (kNm/rad)	Dashpot coefficient (kg/s) or (kg·m²/s)
Horizontal total x-direction $k_{x,total}$, $c_{x,total}$	I	$1.383 \cdot 10^7$	1.28	1	0.139	0.185	$1.77 \cdot 10^7$	$6.021 \cdot 10^8$
	II	$4.714 \cdot 10^6$			0.23	0.292	$6.035 \cdot 10^6$	$3.24 \cdot 10^8$
	III	$2.108 \cdot 10^6$			0.329	0.407	$2.699 \cdot 10^6$	$2.02 \cdot 10^8$
Horizontal total y-direction $k_{y,total}$, $c_{y,total}$	I	$1.387 \cdot 10^7$	1.28	1	0.139	0.185	$1.776 \cdot 10^7$	$6.057 \cdot 10^8$
	II	$4.728 \cdot 10^6$			0.229	0.292	$6.053 \cdot 10^6$	$3.255 \cdot 10^8$
	III	$2.114 \cdot 10^6$			0.328	0.407	$2.707 \cdot 10^6$	$2.027 \cdot 10^8$
Horizontal base x-direction $k_{x,base}$, $c_{x,base}$	I	$1.383 \cdot 10^7$	–	1	0.139	–	$1.383 \cdot 10^7$	$3.549 \cdot 10^8$
	II	$4.714 \cdot 10^6$			0.23		$4.714 \cdot 10^6$	$1.996 \cdot 10^8$
	III	$2.108 \cdot 10^6$			0.329		$2.108 \cdot 10^6$	$1.278 \cdot 10^8$
Horizontal base y-direction $k_{y,base}$, $c_{y,base}$	I	$1.387 \cdot 10^7$	–	1	0.139	–	$1.387 \cdot 10^7$	$3.549 \cdot 10^8$
	II	$4.728 \cdot 10^6$			0.229		$4.728 \cdot 10^6$	$1.996 \cdot 10^8$
	III	$2.114 \cdot 10^6$			0.328		$2.114 \cdot 10^6$	$1.278 \cdot 10^8$
Vertical z-direction k_z, c_z	I	$1.885 \cdot 10^7$	1.139	0.99	–	0.259	$2.126 \cdot 10^7$	$1.014 \cdot 10^9$
	II	$6.116 \cdot 10^6$		0.975		0.397	$6.795 \cdot 10^6$	$4.967 \cdot 10^8$
	III	$2.617 \cdot 10^6$		0.954		0.556	$2.844 \cdot 10^6$	$2.91 \cdot 10^8$
Rocking about x-axis k_{xx}, c_{xx}	I	$3.407 \cdot 10^9$	1.256	0.961	–	0.01	$4.114 \cdot 10^9$	$7.631 \cdot 10^9$
	II	$1.106 \cdot 10^9$		0.91		0.035	$1.264 \cdot 10^9$	$8.152 \cdot 10^9$
	III	$4.731 \cdot 10^8$		0.849		0.084	$5.048 \cdot 10^8$	$7.793 \cdot 10^9$
Rocking about y-axis k_{yy}, c_{yy}	I	$3.578 \cdot 10^9$	1.253	0.958	–	0.01	$4.293 \cdot 10^9$	$8.067 \cdot 10^9$
	II	$1.161 \cdot 10^9$		0.903		0.036	$1.314 \cdot 10^9$	$8.649 \cdot 10^9$
	III	$4.968 \cdot 10^8$		0.84		0.086	$5.227 \cdot 10^8$	$8.275 \cdot 10^9$

where:
k_x, k_y – translational stiffness for rectangular rigid foundations
k_{xx}, k_{yy} – rotational stiffness for rectangular rigid foundations
c_x, c_y – translational dashpot coefficient
c_{xx}, c_{yy} – rotational dashpot coefficient
η – embedment correction factor for static stiffness of rigid foundations
α – dynamic stiffness modifier for rigid foundations
β_{sur} – radiation damping ratio for rigid foundations
β_{emb} – radiation damping ratio for embedment foundations

global axis. The damping ratio of 5% was taken for the analyzed high-rise building, as it is a typical value assumed for concrete structures.

Peak lateral accelerations in NS and EW directions for different site conditions are briefly reported in Table 20.4. Acceleration-time history plots for two radically different site conditions (i.e. building fixed at base and erected on soft soil) are

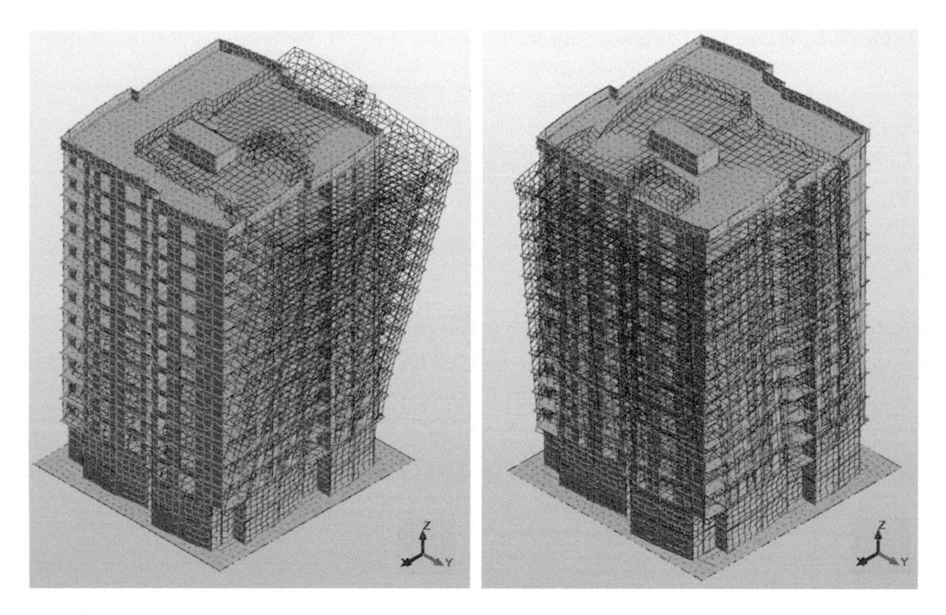

Fig. 20.1 Fundamental modes of vibration for the analyzed high-rise building: 1st longitudinal (left) and 1st transverse (right)

Table 20.3 Modal analysis results

Site condition	Mode 1			
	Longitudinal		Transverse	
	Frequency (Hz)	Period (sec)	Frequency (Hz)	Period (sec)
Fixed at base	1.73	0.58	2.17	0.46
I. Dense soil/Soft rock	1.59	0.63	1.93	0.52
II. Stiff soil	1.53	0.65	1.82	0.55
III. Soft soil	1.42	0.70	1.63	0.61

Table 20.4 Results of modal analysis

Seismic event	PGA (× g)		Peak lateral acceleration at roof level in NS direction (× g)				Peak lateral acceleration at roof level in EW direction (× g)			
			Site condition				Site condition			
	NS	EW	Fixed base	I	II	III	Fixed base	I	II	III
Northridge	0.45	0.19	0.82	0.77	0.74	0.60	0.45	0.40	0.38	0.25
Polkowice	0.17	0.10	0.33	0.30	0.29	0.27	0.19	0.15	0.14	0.12

where PGA denotes *Peak Ground Acceleration*

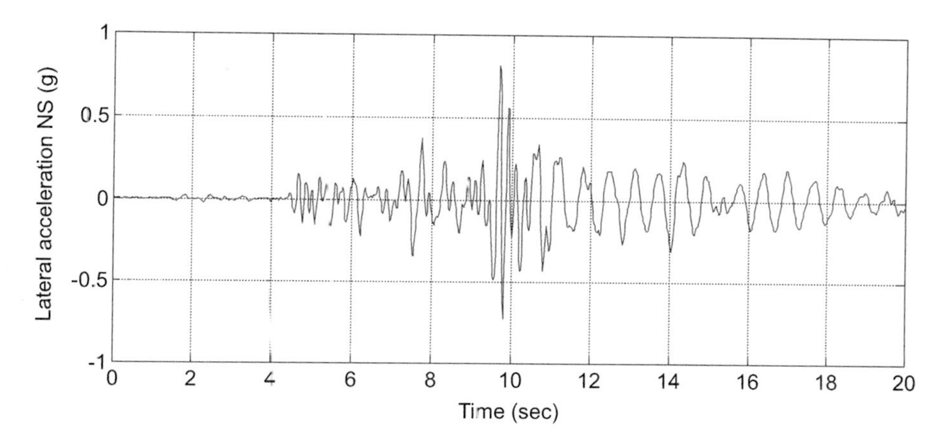

Fig. 20.2 Acceleration-time history record in NS direction for the Northridge earthquake (fixed base building)

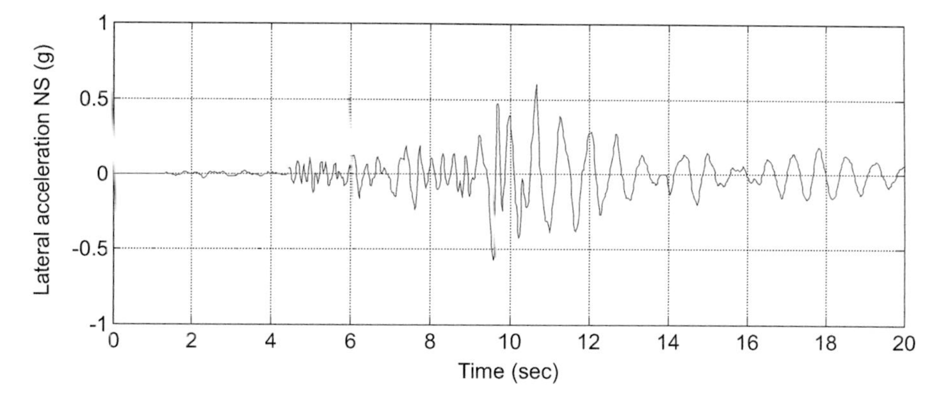

Fig. 20.3 Acceleration-time history record in NS direction for the Northridge earthquake (soft soil)

presented in Figs. 20.2, 20.3, 20.4, 20.5, 20.6, 20.7, 20.8 and 20.9. Additionally, the Fast Fourier Transform (FFT) functions calculated for the NS direction are shown in Figs. 20.10 and 20.11.

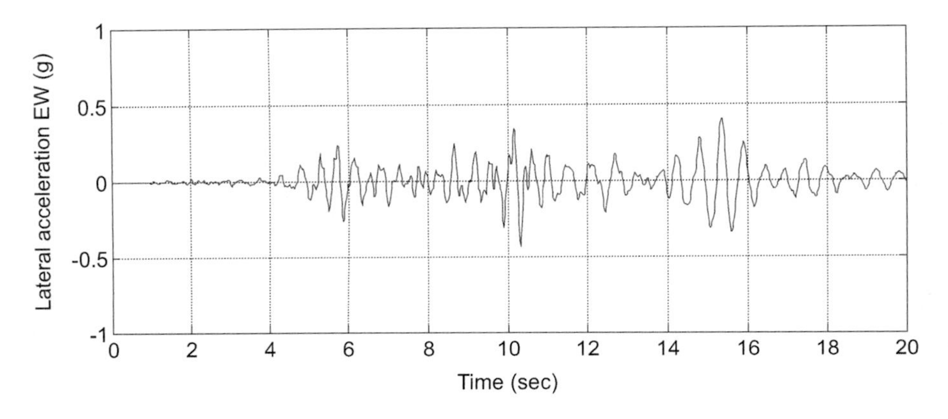

Fig. 20.4 Acceleration-time history record in EW direction for the Northridge earthquake (fixed base building)

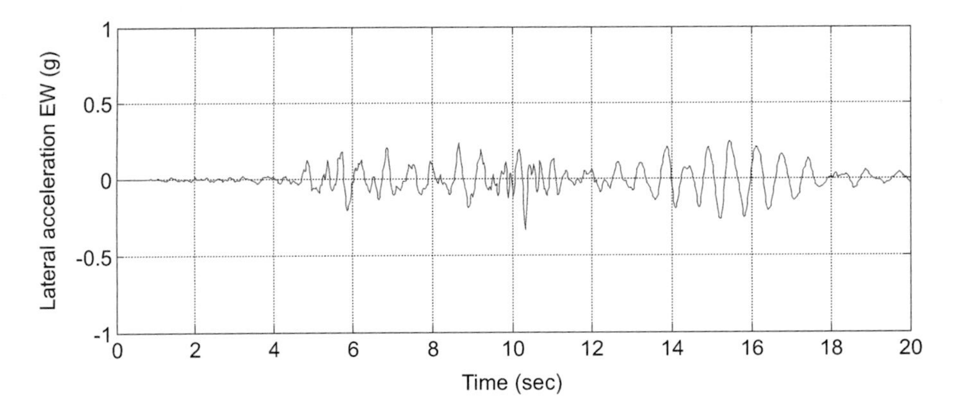

Fig. 20.5 Acceleration-time history record in EW direction for the Northridge earthquake (soft soil)

20.7 Conclusions

The present study was designed to examine the soil-structure interaction effects on the seismic response of the three-dimensional high-rise irregular reinforced-concrete residential building. Three different site conditions were utilized to investigate the importance of the soil flexibility in seismic analyses. The numerical investigation resulted in the following conclusions:

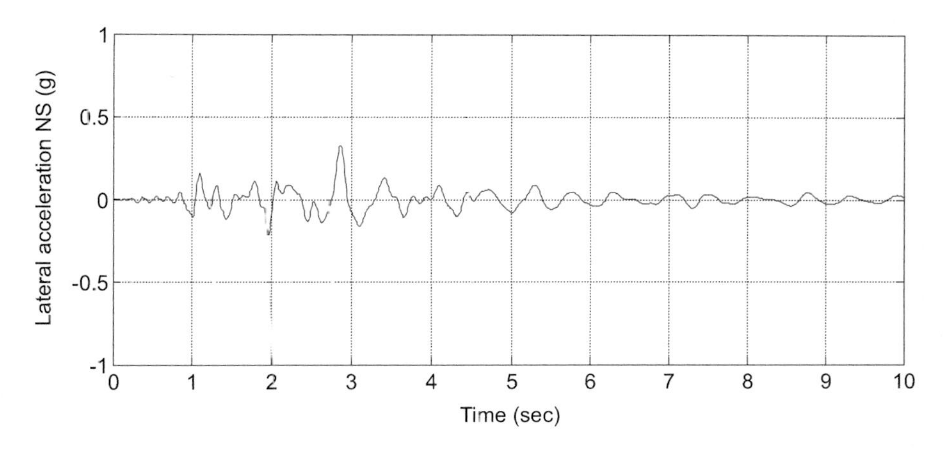

Fig. 20.6 Acceleration-time history record in NS direction for the Polkowice mining tremor (fixed base building)

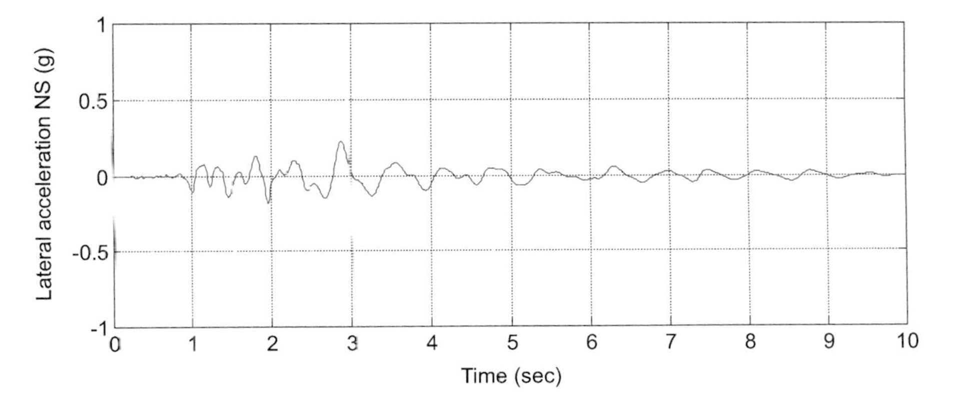

Fig. 20.7 Acceleration-time history record in NS direction for the Polkowice mining tremor (soft soil)

(a) Close inspection of Table 20.3 clearly indicates that soil flexibility can significantly modify the dynamic characteristics of the building structure by lengthening its fundamental period (the lower the fundamental frequency, the longer the fundamental period of vibration). For site condition III (i.e. soft soil) the reduction in the fundamental frequencies in NS and EW directions, when compared to fixed-base model, are 18% (1.73 Hz decreased to 1.42 Hz) and 25% (2.17 Hz was reduced to 1.63 Hz), respectively. This effect (referred to as the period shift effect) can also be observed in Figs. 20.10 and 20.11.

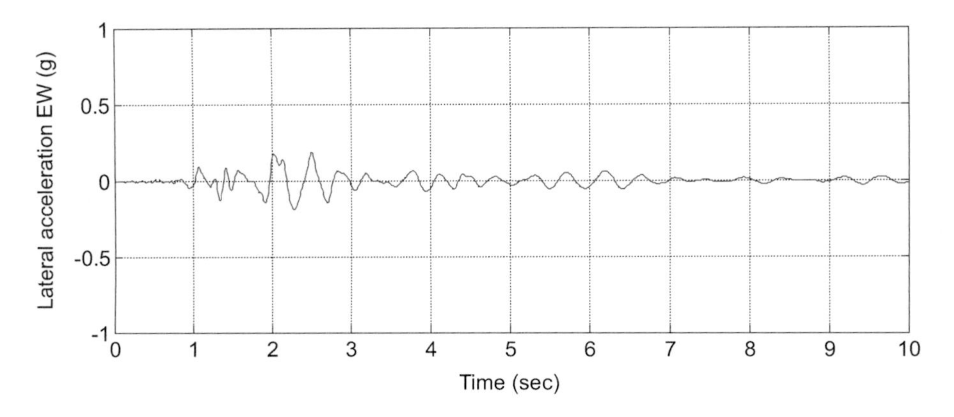

Fig. 20.8 Acceleration-time history record in EW direction for the Polkowice mining tremor (fixed base building)

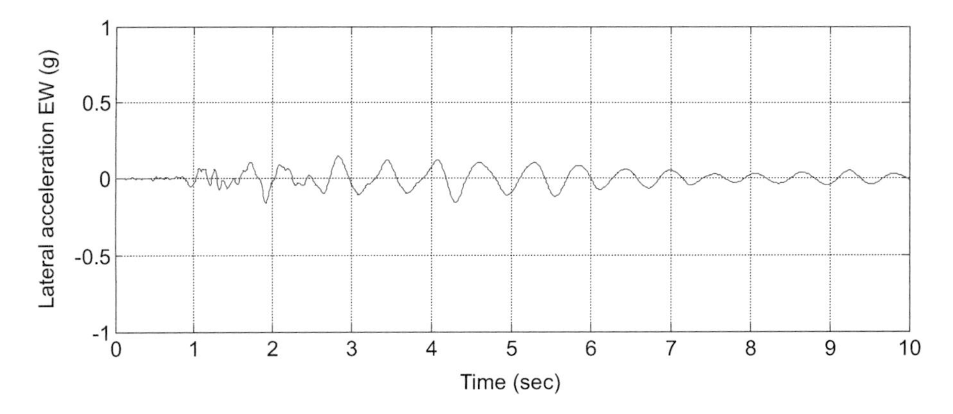

Fig. 20.9 Acceleration-time history record in EW direction for the Polkowice mining tremor (soft soil)

(b) Table 20.4 explicitly demonstrates the decrease in lateral accelerations due to soil flexibility. For site condition III (soft soil) the reduction in computed accelerations in the NS and EW directions for the 1994 Northridge earthquake, when compared to the fixed base structure, is 27% (0.82 g was reduced to 0.60 g) and 44% (0.45 g decreased to 0.25 g), respectively, whereas for the 2002 Polkowice mining tremor the reduction in NS and EW directions, if compared to fixed-base structure, is 18% (0.33 g was reduced to 0.27 g) and 37% (0.19 g decreased to 0.12 g), respectively.

Conducted numerical investigation, even though utilizing simple engineering approach of simulating the soil-foundation flexibility with springs and dashpots,

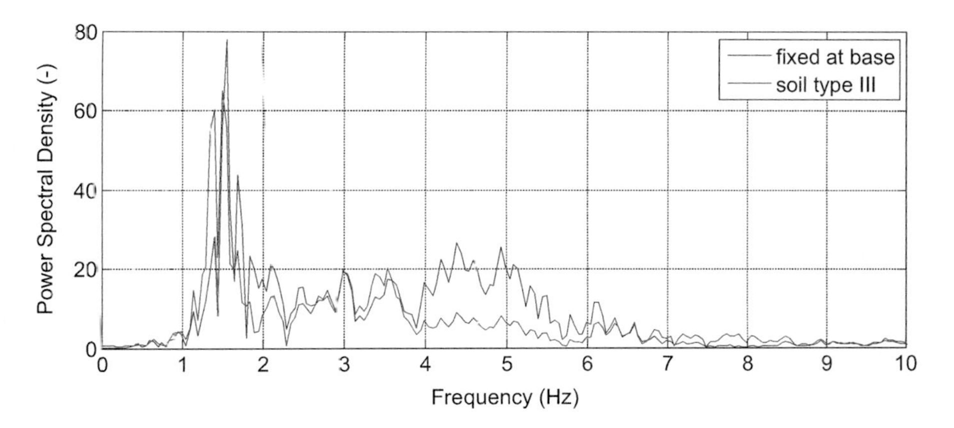

Fig. 20.10 FFT functions for the acceleration-time history records in NS direction (Northridge earthquake)

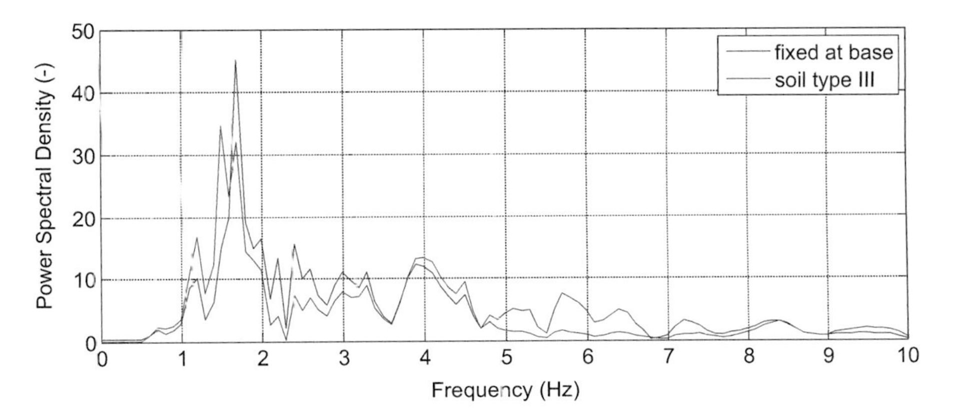

Fig. 20.11 FFT functions for the acceleration-time history records in NS direction (Polkowice mining tremor)

clearly demonstrates the significance of soil-structure interaction in seismic analyses. Ignoring the effects of the surrounding soil may provide satisfactory results only for soft and hard rocks. For site condition I (dense soil/soft rocks), fundamental period and peak lateral accelerations, when compared to fixed base structure, were only modestly affected. As expected, the effects of soil-foundation flexibility become more significant as the stiffness of the underlying soil decreases. Results of the conducted investigation explicitly show that site conditions, particularly soft soils, should be taken into account at the design stage building structures, especially complex and irregular ones.

References

Applied Technology Council ATC (2005) Improvement of nonlinear static seismic analysis procedures, FEMA440 report. Redwood City, CA

Ebrahimian M, Todorovska MI, Falborski T (2017) Wave method for structural health monitoring: testing using full-scale shake table experiment data. J Struct Eng 143(4). https://doi.org/10.1061/(ASCE)ST.1943-541X.0001712

Falborski T, Jankowski R (2013) Polymeric bearings – a new base isolation system to reduce structural damage during earthquakes. Key Eng Mater 569–570:143–150

Falborski T, Jankowski R (2016) Behaviour of asymmetric structure with base isolation made of polymeric bearings. In: Zembaty Z, De Stefano M (eds) Geotechnical, geological and earthquake engineering 40: seismic behaviour and design of irregular and complex civil structures II. Springer, Cham, pp 333–341

Falborski T, Jankowski R (2017a) Experimental study on effectiveness of a prototype seismic isolation system made of polymeric bearings. Appl Sci 7(8):808. https://doi.org/10.3390/app8030400

Falborski T, Jankowski R (2017b) Numerical evaluation of dynamic response of a steel structure model under various seismic excitations. Proc Eng 172:277–283

Gazetas G (1991) Foundation vibrations. In: Fang H-Y (ed) Foundation engineering handbook. Springer, New York

Harden CW, Hutchinson TC (2009) Beam-on-nonlinear-Winkler-foundation Modeling of shallow, rocking-dominated footings. Earthquake Spectra 25:277–300

Jankowski R, Mahmoud S (2015) Earthquake-induced structural pounding. Springer, Cham

Jankowski R, Mahmoud S (2016) Linking of adjacent three-storey buildings for mitigation of structural pounding during earthquakes. Bull Earthq Eng 14:3075–3097

Mylonakis G, Gazetas G (2000) Seismic soil-structure interaction: beneficial or detrimental? J Earthq Eng 4:277–301

Mylonakis G, Nikolaou S, Gazetas G (2006) Footings under seismic loading: analysis and design issues with emphasis on bridge foundations. Soil Dyn Earthq Eng 26(9):824–853

Naderpour H, Barros RC, Khatami SM, Jankowski R (2016) Numerical study on pounding between two adjacent buildings under earthquake excitation. Shock and Vibration 2016, article ID 1504783. https://doi.org/10.1155/2016/1504783

National Earthquake Hazards Reduction Program NEHRP (2012) Soil-structure interaction for building structures

Pacific Earthquake Engineering Research Center (2012) Guidelines for estimation of shear wave velocity profiles

Pais A, Kausel E (1988) Approximate formulas for dynamic stiffness of rigid foundations. Soil Dyn Earthq Eng 7(4):213–227

Sołtysik B, Falborski T, Jankowski R (2016) Investigation on damage-involved structural response of colliding steel structures during ground motions. Key Eng Mater 713:26–29

Sołtysik B, Falborski T, Jankowski R (2017) Preventing of earthquake-induced pounding between steel structures by using polymer elements – experimental study. Proc Eng 199:278–283

Stewart JP, Fenves GL (1998) System identification for evaluating soil–structure interaction effects in buildings from strong motion recordings. Earthq Eng Struct Dyn 27(8):869–885

Stewart JP, Fenves GL, Seed RB (1999a) Seismic soil-structure interaction in buildings I: analytical methods. J Geotech Geoenviron 125(1):26–37

Stewart JP, Fenves GL, Seed RB (1999b) Seismic soil-structure interaction in buildings II: empirical findings. J Geotech Geoenviron 125(1):38–48

Stewart JP, Kim S, Bielak J, Dobry R, Power MS (2003) Revisions to soil-structure interaction procedures in NEHRP design provisions. Earthquake Spectra 19(3):677–696

Veletsos AS, Prasad AM (1989) Seismic interaction of structures and soils: stochastic approach. J Struct Eng 115(4):935–956

Wolf JP (1985) Dynamic soil-structure interaction. Prentice-Hall, Upper Saddle River

Wong HL, Luco JE (1986) Dynamic interaction between rigid foundations in a layered half-space. Soil Dyn Earthq Eng 5(3):149–158

Wong HL, Trifunac MD, Luco JE (1988) A comparison of soil-structure interaction calculations with results of full-scale forced vibration tests. Soil Dyn Earthq Eng 7(1):22–31

Zembaty Z (2004) Rockburst induced ground motion – a comparative study. Soil Dyn Earthq Eng 24(1):11–23

Chapter 21
Optimum Torsion Axis of Multi-storey Buildings Based on Their Dynamic Properties

Grigorios Manoukas and Asimina Athanatopoulou

Abstract The objective of the present paper is the determination of the optimum torsion axis of multi-storey asymmetric in plan buildings on the basis of their dynamic properties. For this purpose, a three-storey reinforced concrete diaphragm system is analyzed by means of linear time-history analysis for both uniaxial and biaxial horizontal seismic excitations. The mass centers of the diaphragms are successively transposed and the resulting floor rotation angles for each case are computed. The position of the mass center which leads to the minimization of the sum of the squares of the floor rotation angles designates the location of the optimum torsion axis. The results are verified by means of modal analysis and compared with those resulting from the relevant methodology prescribed by the Greek seismic code. All three methods produce results that do not differ significantly, so the approximate procedure suggested by the Greek seismic code can be rigorously applied in order to determine the optimum torsion axis of asymmetric buildings.

Keywords Optimum torsion axis · Multi-storey buildings · Time-history analysis · Floor rotation

21.1 Introduction

All the modern seismic codes (e.g. CEN 2004; EPPO 2003) adopt the linear static procedure as an alternative method of analysis appropriate for certain categories of buildings. A critical point of this procedure is the representation of the torsional behaviour of the buildings which is achieved by introducing proper static eccentricities from a suitable reference point.

Anastassiadis et al. (1998) studied this issue using single-storey models and developed analytical formulae which allow the calculation of additional

G. Manoukas · A. Athanatopoulou (✉)
Department of Civil Engineering, Aristotle University of Thessaloniki, Thessaloniki, Greece
e-mail: minak@civil.auth.gr

D. Köber et al. (eds.), *Seismic Behaviour and Design of Irregular and Complex Civil Structures III*, Geotechnical, Geological and Earthquake Engineering 48,
https://doi.org/10.1007/978-3-030-33532-8_21

eccentricities, so that the maximum static displacements on both sides of the floor as well as the static rotation of the floor are equal to the respective ones obtained by means of response spectrum analysis. The additional eccentricities are measured with reference to the elastic centre, which always exists in single-storey systems and possesses at the same time all the properties of the centre of rigidity, shear centre and twist centre.

The aforementioned concept can be rigorously extended to special categories of multi-storey diaphragm systems in which the rigidity, shear and twist centres coincide in each floor (elastic centre of the floor), while the elastic centres of all the floors lie in the same vertical axis (elastic axis of the building). These categories include doubly symmetric in plan buildings and isotropic buildings (i.e. buildings having vertical resisting elements with proportional stiffness matrices (Makarios and Anastassiadis 1998a, b)).

However, the vast majority of real buildings do not belong to these categories. In order to overcome this problem, Makarios and Anastassiadis (1998a, b) proposed the concept of 'optimum torsion axis' as a reference line for the calculation of structural eccentricity in the general case of multi-storey asymmetric buildings. For a given static load pattern, the optimum torsion axis is defined as the vertical line connecting the points of the floor diaphragms, where the horizontal forces must be applied, so that the sum of the squares of the floor rotations θ_i is minimized ($i = 1, 2, \ldots N$, where N the number of the building diaphragms):

$$\Sigma\theta_i^2 = min \tag{21.1}$$

Furthermore, in order to simplify the determination of the optimum torsion axis location, they developed an approximate methodology which has been adopted by the Greek seismic code (EPPO 2003). In particular, based on extensive parametric studies, they suggested that, when the minimum $\Sigma\theta_i^2$ is attained, the diaphragm rotation becomes zero at the level $z = 0.8H$ from the base, where H is the height of the building.

Marino and Rossi (2004) examined the same problem from an analytical point of view and proposed mathematical expressions to define the exact location of the optimum torsion axis for buildings having the principal axes of the resisting elements parallel to a given orthogonal coordinate system. Generalizing this approach, Doudoumis and Athanatopoulou (2008) proved that the aforementioned analytical methodology can be applied to all asymmetric buildings without any particular restriction.

All the aforementioned studies aim to determine the optimum torsion axis under horizontal static loads. The objective of the present paper is the determination of the optimum torsion axis on the basis of the dynamic properties of buildings. For this purpose, a three-storey reinforced concrete diaphragm system is analyzed by means of linear time-history analysis for both uniaxial and biaxial horizontal seismic excitations. The mass centers of the diaphragms are successively transposed and the resulting floor rotation angles for each case are computed for the critical orientation of the seismic excitation using a relevant procedure developed by Athanatopoulou (2005). The position of the mass centre which leads to the minimization of the sum of the squares of the floor rotation angles designates the location of the optimum torsion axis. The results are verified by means of modal analysis and compared with those resulting from the relevant methodology prescribed by the Greek seismic code. Hence, interesting conclusions are derived.

21.2 Structural Models

The structural models examined in the framework of the present study are based on an archetype three-storey asymmetric in plan reinforced concrete building. The floor plan of the building as well as the position of the optimum torsion axis (OTA) according to the Greek seismic code is shown in Fig. 21.1. All storey heights are 3 m. The slab thickness is equal to 15 cm. All beams have a height of 60 cm and a thickness of 25 cm. The columns are square shaped with dimension of 40 cm. The length of the walls is equal to 1.5 m (W1), 2 m (W2, W3) or 3 m (W4) and their thickness is equal to 25 cm. All the vertical resisting elements are fixed at base. The mass of each floor is taken equal to 1 t/m^2. The mass centre of each floor of the building is transposed to 25 different locations (Fig. 21.2). Hence, 25 building models are produced and analyzed by means of linear time-history analysis according to the following section.

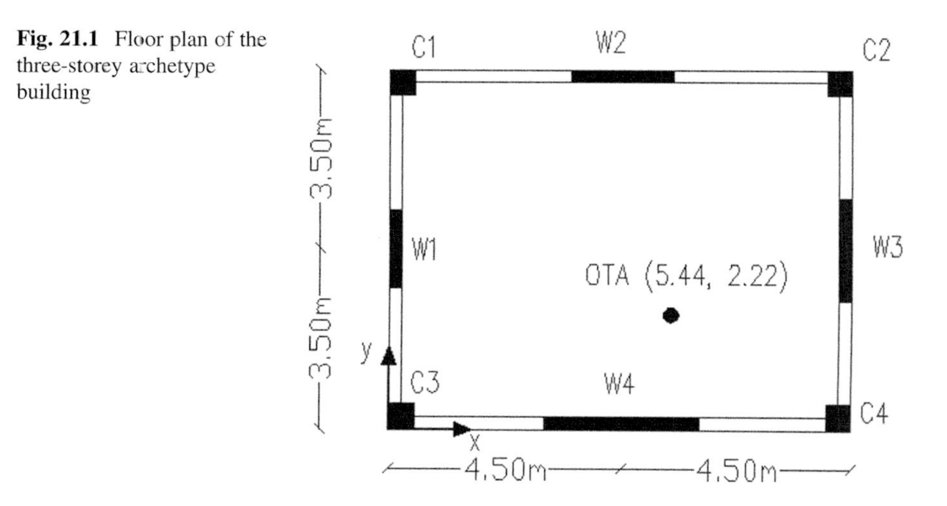

Fig. 21.1 Floor plan of the three-storey archetype building

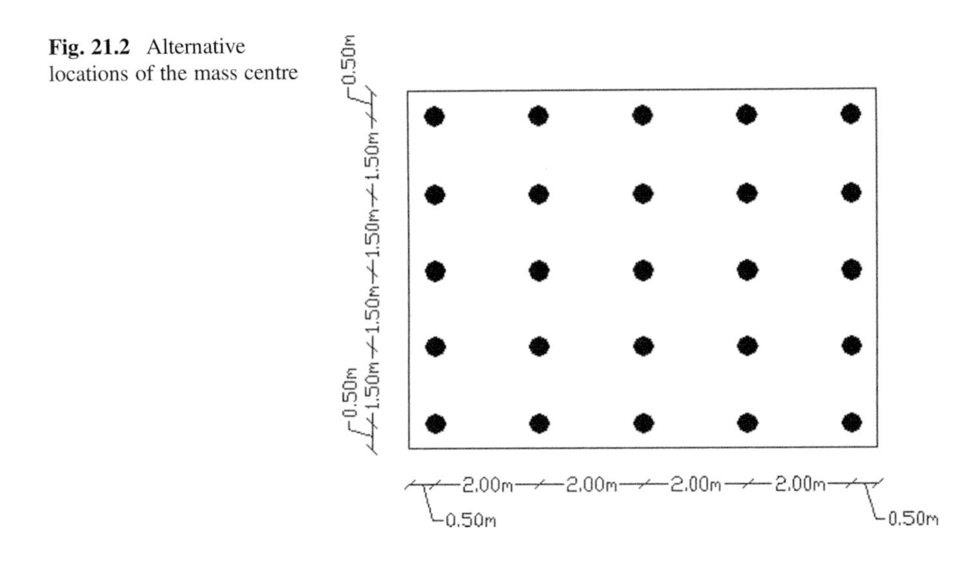

Fig. 21.2 Alternative locations of the mass centre

21.3 Analysis Process

21.3.1 Ground Motions

The whole investigation conducted here comprises a number of 7 unscaled accelerograms, which is considered adequate to obtain concrete conclusions. These accelerograms (Table 21.1) are obtained from the PEER strong motion database (2003) and tabulated in table C-3 of Appendix C of FEMA 440 project (ATC 2005).

21.3.2 Uniaxial Seismic Excitation

Firstly, the building models are analyzed by means of linear time-history analysis for uniaxial seismic excitation. In particular, each of the recorded accelerograms $\ddot{u}_g(t)$is considered to act separately along x (load case '0') and y axis (load case '90') as it is shown in Fig. 21.3. If the values of a scalar response quantity for load cases '0' and '90' are denoted as $R_{'0'}(t)$ and $R_{'90'}(t)$ respectively, as it has been demonstrated by Athanatopoulou (2005), the maximum value R_{max} of the response quantity for the critical orientation of the seismic excitation is:

$$R_{max} = max\sqrt{R^2_{'0'}(t) + R^2_{'90'}(t)} \tag{21.2}$$

The time instant t_{cr} that R_{max} occurs as well as the angle θ_{cr} which defines the critical orientation of the seismic excitation could also be determined applying

Table 21.1 Ground motions

Number	Date	Earthquake name	Magnitude (Ms)	Station name	Component (deg)	PGA (cm/s^2)
1	28/06/92	Landers	7.5	Yermo, Fire Station	270	240.0
2	28/06/92	Landers	7.5	Palm Springs, airport	90	87.2
3	28/06/92	Landers	7.5	Pomona, 4th and locust, free field	0	65.5
4	17/01/94	Northridge	6.8	Los Angeles, Hollywood storage Bldg.	360	381.4
5	17/01/94	Northridge	6.8	Santa Monica City Hall	90	866.2
6	17/01/94	Northridge	6.8	Los Angeles, N. Westmoreland	0	393.3
7	17/10/89	Loma Prieta	7.1	Gilroy 2, Hwy 101 Bolsa Road Motel	0	394.2

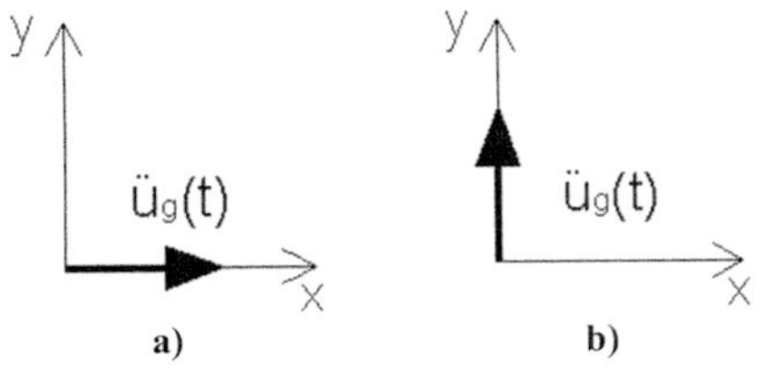

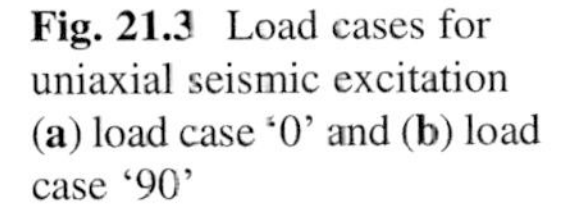
Fig. 21.3 Load cases for uniaxial seismic excitation (**a**) load case '0' and (**b**) load case '90'

analytical formulae (Athanatopoulou 2005). However, this is beyond the objective of the present paper.

The sum of the squares of the diaphragm rotations $\Sigma\theta_i^2(t)$ is computed for each load case with the aid of the program SAP2000. Then, applying Eq. 21.2 the maximum value $max\Sigma\theta_i^2$ is calculated. The resulting values for the 25 analyzed building models are shown in Figs. 21.4, 21.5, 21.6, 21.7, 21.8, 21.9 and 21.10 where each model is characterised by its centre of mass (CM) coordinates in the coordinate system given in Fig. 21.1.

The curves given in Figs. 21.4, 21.5, 21.6, 21.7, 21.8, 21.9 and 21.10 are idealized to parabolic curves governed by equations just like Eq. 21.3:

$$max\,\Sigma\theta_i^2 = ax^2 + bx + cy^2 + dy + e \tag{21.3}$$

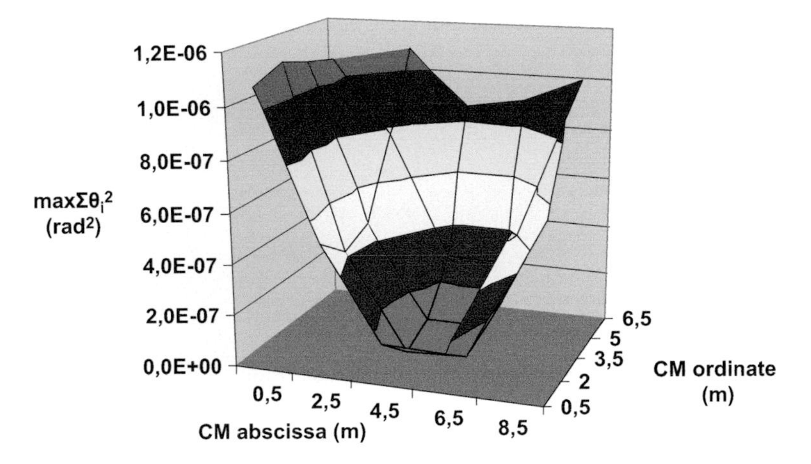

Fig. 21.4 Maximum sum of the squares of diaphragm rotations $max\Sigma\theta_i^2$ – Ground motion 1

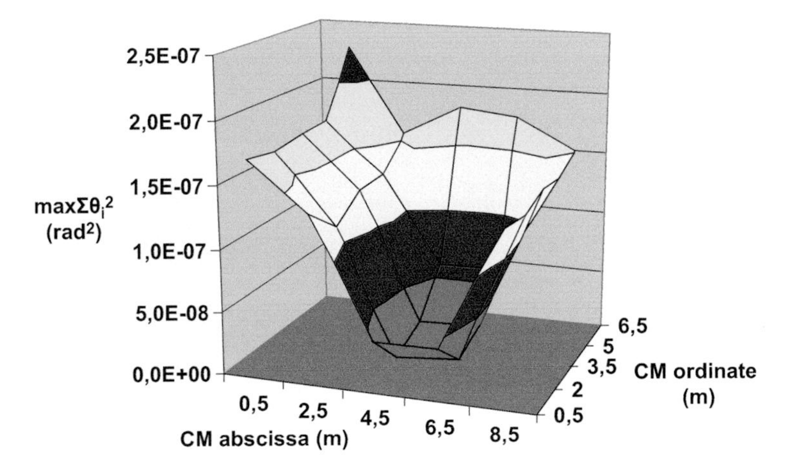

Fig. 21.5 Maximum sum of the squares of diaphragm rotations $max\Sigma\theta_i^2$ – Ground motion 2

where a, b, c, d, e are constant factors (different for each ground motion). Applying well known principles of mathematics, the coordinates of the point that leads to the minimization of $max\Sigma\theta_i^2$ are estimated for each ground motion (Table 21.2). The mean values of the coordinates indicate the location of the optimum torsion axis (OTA). It is apparent that the resulting coordinates are much closed to those determined according to the Greek seismic code (see Fig. 21.1).

In order to verify the results, the centre of mass of the archetype building is transposed to the location of the optimum torsion axis and a modal analysis is conducted. The resulting mode shape vectors (Table 21.3) resemble those of a

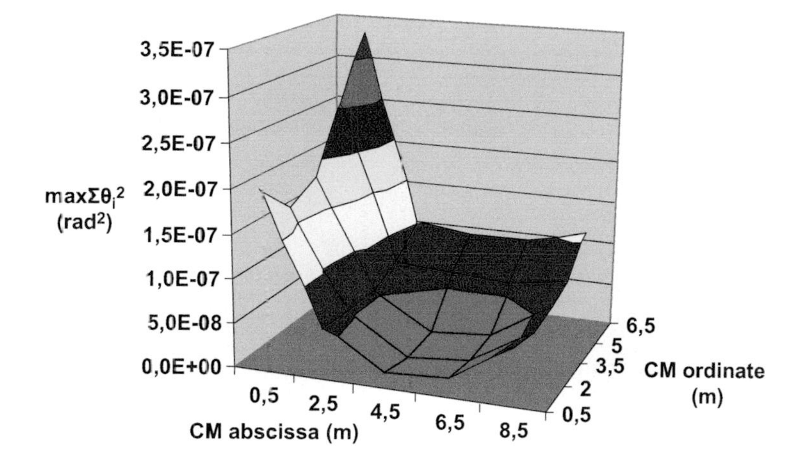

Fig. 21.6 Maximum sum of the squares of diaphragm rotations $max\Sigma\theta_i^2$ – Ground motion 3

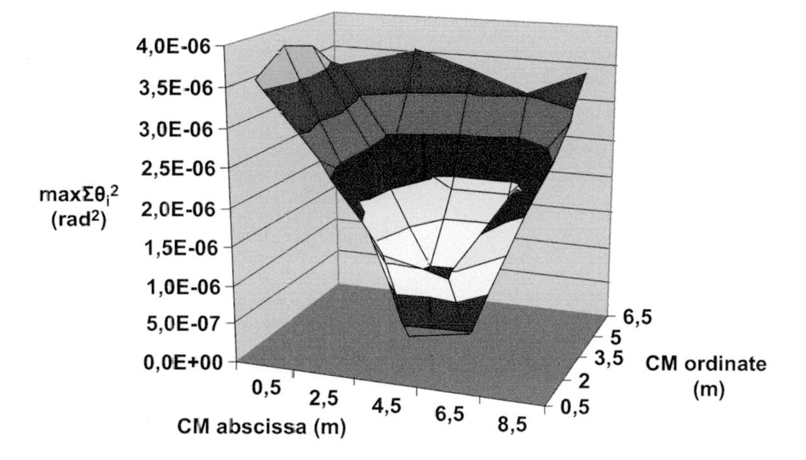

Fig. 21.7 Maximum sum of the squares of diaphragm rotations $max\Sigma\theta_i^2$ – Ground motion 4

doubly symmetric building without coupling between lateral and torsional response. In particular, in modal shape vectors 1, 4 and 7 the terms corresponding to rotations and translations along x axis are negligible and obviously modes 1, 4 and 7 dominate the response under seismic excitation along y axis. On the other hand, in modal shape vectors 2, 5 and 8 the terms corresponding to rotations and translations along y axis are negligible and obviously modes 2, 5 and 8 dominate the response under seismic excitation along x axis. These conclusions are also confirmed by the modal participating mass ratios (Table 21.4). Finally, modes 3, 6 and 9 are mainly rotational.

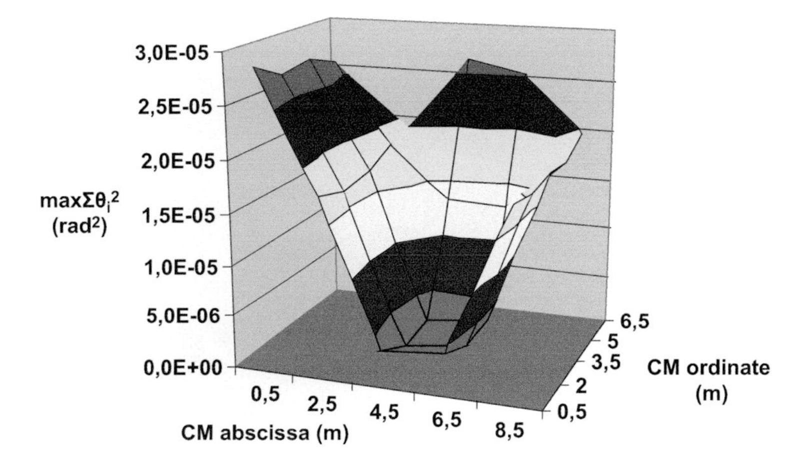

Fig. 21.8 Maximum sum of the squares of diaphragm rotations $max\Sigma\theta_i^2$ – Ground motion 5

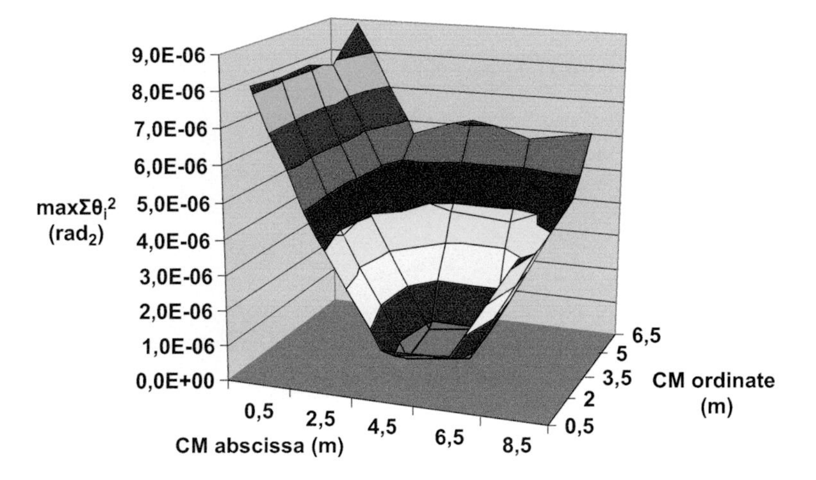

Fig. 21.9 Maximum sum of the squares of diaphragm rotations $max\Sigma\theta_i^2$ – Ground motion 6

21.3.3 Biaxial Seismic Excitation

The whole process is identical to that followed for uniaxial seismic excitation. The only difference is the definition of load cases '0' and '90' which comprise concurrent action of each accelerogram along x and y axes (Fig. 21.11).

The maximum sum of the squares of the diaphragm rotations $max\Sigma\theta_i^2$ for the 25 analyzed building models is shown in Figs. 21.12, 21.13, 21.14, 21.15, 21.16, 21.17 and 21.18.

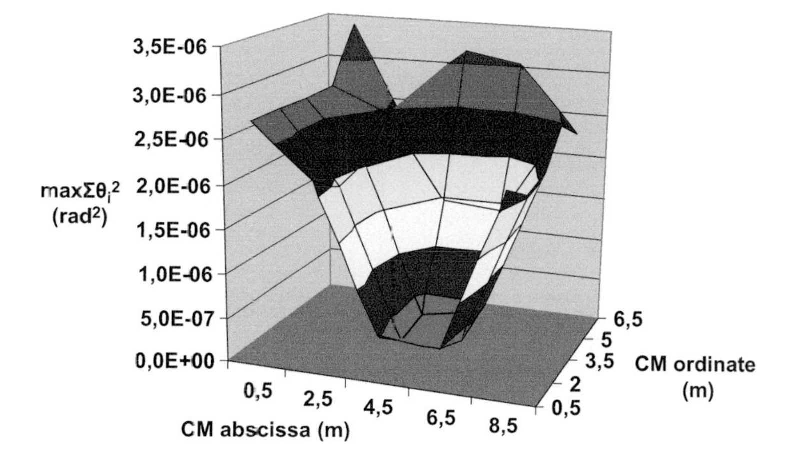

Fig. 21.10 Maximum sum of the squares of diaphragm rotations $max\Sigma\theta_i^2$ – Ground motion 7

Table 21.2 Coordinates of the optimum torsion axis – uniaxial excitation

Ground motion number	OTA abscissa (m)	OTA ordinate (m)
1	5.43	2.32
2	5.41	2.28
3	5.30	2.35
4	5.38	2.23
5	5.40	2.39
6	5.37	2.44
7	5.39	2.46
Mean	5.38	2.35

The coordinates of the point that leads to the minimization of $max\Sigma\theta_i^2$ for each ground motion and the corresponding mean values are shown in Table 21.5. It is apparent that the resulting values are much closed to those calculated for uniaxial excitation as well as to those determined according to the Greek seismic code.

In order to verify the results, the centre of mass of the archetype building is transposed to the location of the optimum torsion axis and a modal analysis is conducted. The resulting mode shape vectors (Table 21.6) resemble those of a doubly symmetric building without coupling between lateral and torsional response. In particular, in modal shape vectors 1, 4 and 7 the terms corresponding to rotations and translations along x axis are negligible and obviously modes 1, 4 and 7 dominate the response under seismic excitation along y axis. On the other hand, in modal shape vectors 2, 5 and 8 the terms corresponding to rotations and translations along y axis are negligible and obviously modes 2, 5 and 8 dominate the response under seismic excitation along x axis. These conclusions are also confirmed by the modal participating mass ratios (Table 21.7). Finally, modes 3, 6 and 9 are mainly rotational.

Table 21.3 Mode shape vectors (CM ≡ OTA for uniaxial excitation)

Displacements	Floor	Mode 1	Mode 2	Mode 3	Mode 4	Mode 5	Mode 6	Mode 7	Mode 8	Mode 9
Translation x	1	0.0003	0.2069	−0.0011	0.0039	1.0000	0.3260	0.0062	1.0000	0.1302
	2	0.0004	0.5951	0.0253	0.0038	0.9409	0.2160	−0.0004	−0.8307	−0.1351
	3	−0.0002	1.0000	0.1148	−0.0041	−0.7668	−0.1728	−0.0007	0.2874	0.0488
Translation y	1	0.2384	0.0005	−0.0383	1.0000	−0.0078	−0.0047	1.0000	−0.0052	0.1399
	2	0.6397	0.0006	−0.0361	0.9156	−0.0067	0.2070	−0.8278	0.0039	−0.1121
	3	1.0000	−0.0008	0.0518	−0.8241	0.0047	−0.1051	0.2913	−0.0013	0.0365
Rotation	1	−0.0012	−0.0025	0.2313	−0.0088	−0.0216	1.0000	−0.0052	−0.0170	1.0000
	2	−0.0018	−0.0062	0.6298	−0.0044	−0.0213	0.9188	0.0172	0.0073	−0.8301
	3	−0.0004	−0.0075	1.0000	0.0154	0.0244	−0.8075	−0.0103	−0.0016	0.2912

Table 21.4 Modal participating mass ratios (CM ≡ OTA for uniaxial excitation)

Excitation	Mode 1	Mode 2	Mode 3	Mode 4	Mode 5	Mode 6	Mode 7	Mode 8	Mode 9
x	0.00	0.77	0.00	0.00	0.18	0.00	0.00	0.04	0.00
y	0.80	0.00	0.00	0.16	0.00	0.00	0.04	0.00	0.00

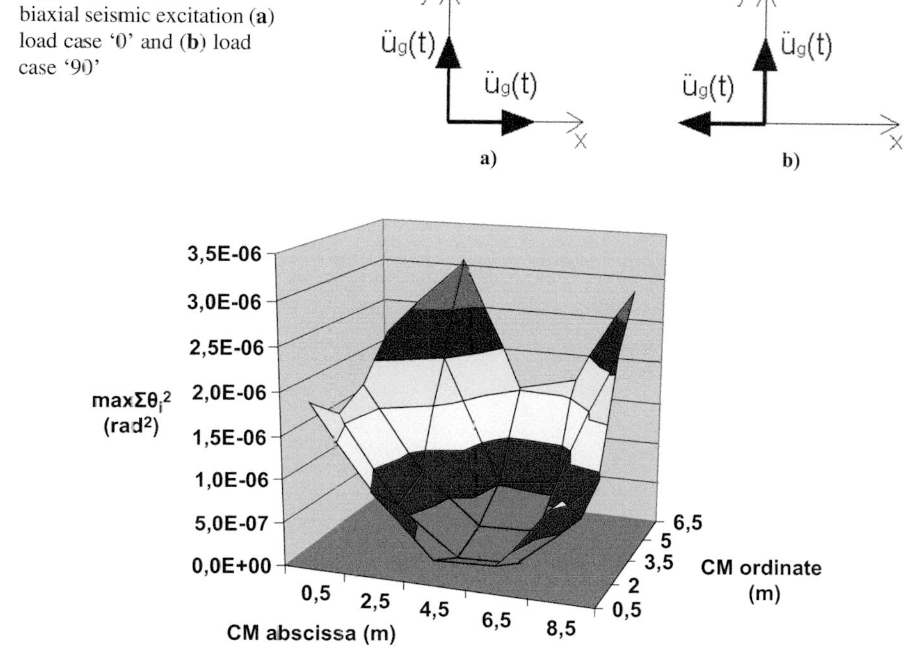

Fig. 21.11 Load cases for biaxial seismic excitation (**a**) load case '0' and (**b**) load case '90'

Fig. 21.12 Maximum sum of the squares of diaphragm rotations $max\Sigma\theta_i^2$ – Ground motion 1

21.4 Conclusions

The objective of the present paper is the determination of the optimum torsion axis on the basis of the dynamic properties of buildings. For this purpose, a three-storey reinforced concrete diaphragm system is analyzed by means of linear time-history analysis. The location of the optimum torsion axis for both uniaxial and biaxial horizontal seismic excitation is very close to the location resulting from the simplified procedure prescribed by the Greek seismic code. The transposition of the mass centre to the determined optimum torsion axis leads to dynamic properties which resemble those of doubly symmetric systems. The method presented herein proves that in asymmetric multistorey buildings there is an optimum torsional axis. This axis can be determined with the aid of real accelerograms or static lateral forces using exact method or approximate method. All three methods produce results that do not differ significantly. So the approximate procedure suggested by the Greek seismic

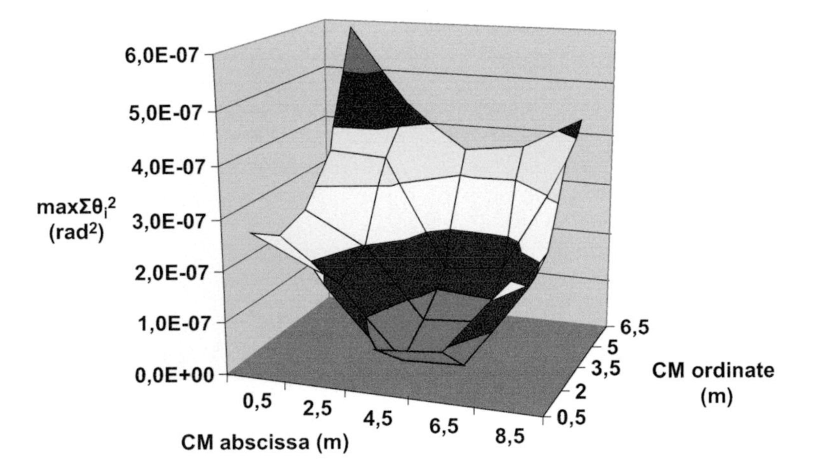

Fig. 21.13 Maximum sum of the squares of diaphragm rotations $max\Sigma\theta_i^2$ – Ground motion 2

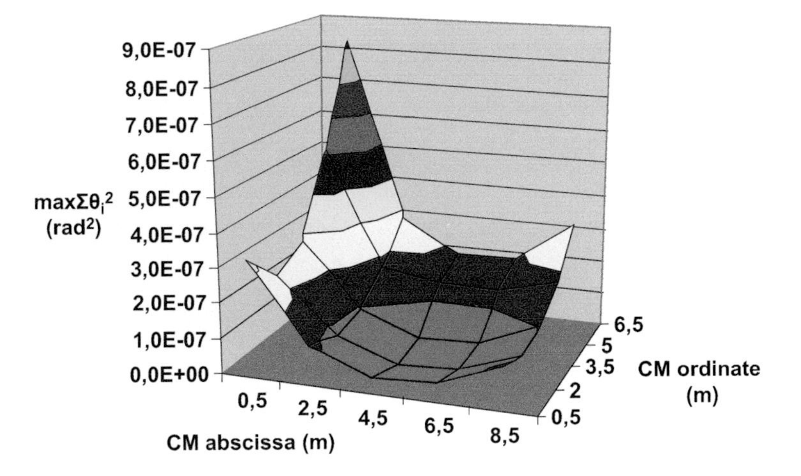

Fig. 21.14 Maximum sum of the squares of diaphragm rotations $max\Sigma\theta_i^2$ – Ground motion 3

code can be rigorously applied in order to determine the optimum torsion axis of asymmetric buildings. However, the generalization of this conclusion requires further investigations, comprising applications to a large variety of structural systems with more complex configuration.

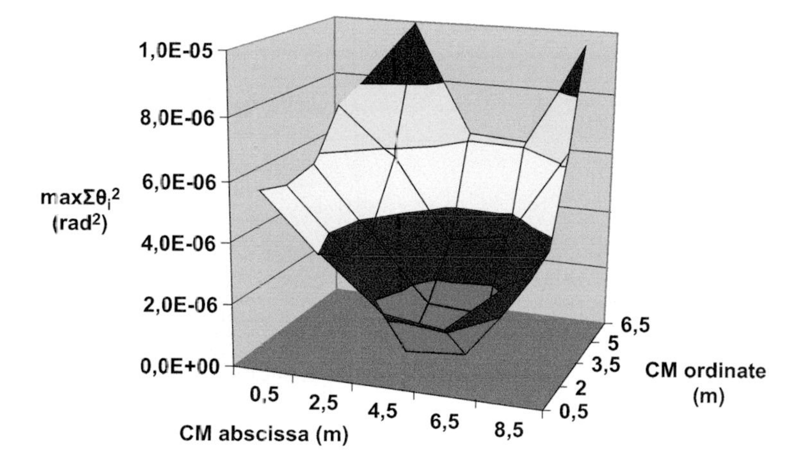

Fig. 21.15 Maximum sum of the squares of diaphragm rotations $max\Sigma\theta_i^2$ – Ground motion 4

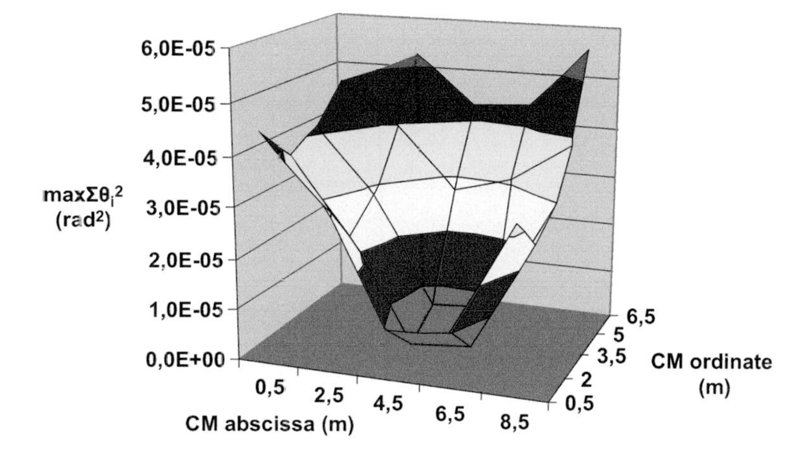

Fig. 21.16 Maximum sum of the squares of diaphragm rotations $max\Sigma\theta_i^2$ – Ground motion 5

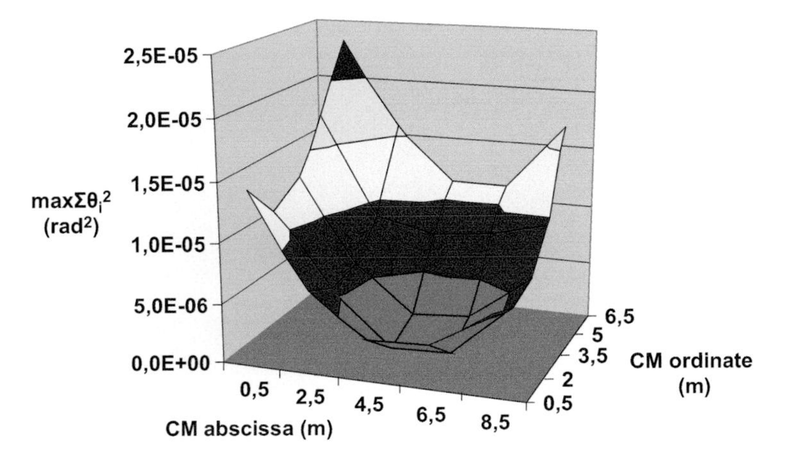

Fig. 21.17 Maximum sum of the squares of diaphragm rotations $max\Sigma\theta_i^2$ – Ground motion 6

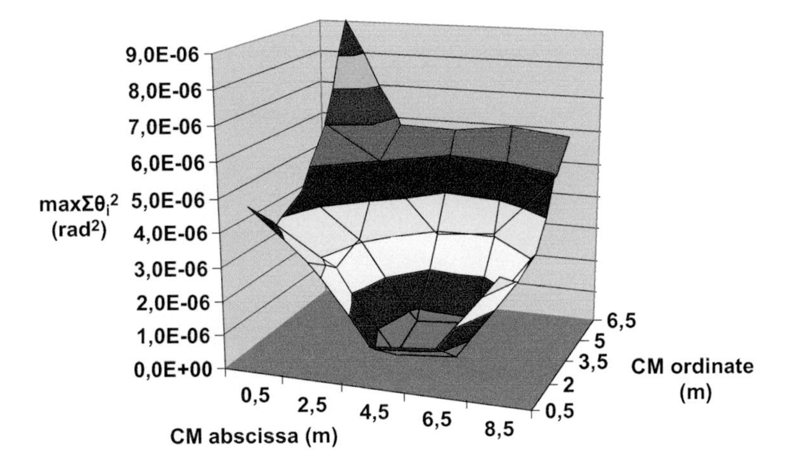

Fig. 21.18 Maximum sum of the squares of diaphragm rotations $max\Sigma\theta_i^2$ – Ground motion 7

Table 21.5 Coordinates of the optimum torsion axis – biaxial excitation

Ground motion number	OTA abscissa (m)	OTA ordinate (m)
1	5.43	2.21
2	5.40	2.17
3	5.30	2.35
4	5.37	2.24
5	5.40	2.21
6	5.37	2.23
7	5.39	2.27
Mean	5.38	2.24

Table 21.6 Mode shape vectors (CM ≡ OTA for biaxial excitation)

Displacements	Floor	Mode 1	Mode 2	Mode 3	Mode 4	Mode 5	Mode 6	Mode 7	Mode 8	Mode 9
Translation x	1	0.0001	0.2066	−0.0170	0.0015	1.0000	0.2015	0.0026	1.0000	−0.0022
	2	0.0000	0.5946	−0.0251	0.0016	0.9398	0.0986	0.0013	−0.8311	−0.0252
	3	−0.0005	1.0000	0.0226	−0.0024	−0.7657	−0.0770	−0.0009	0.2875	0.0108
Translation y	1	0.2384	0.0004	−0.0384	1.0000	−0.0044	0.0002	1.0000	−0.0006	0.1402
	2	0.6397	0.0006	−0.0362	0.9156	−0.0029	0.2041	−0.8278	0.0002	−0.1123
	3	1.0000	−0.0001	0.0519	−0.8241	0.0021	−0.1043	0.2914	0.0000	0.0365
Rotation	1	−0.0012	−0.0007	0.2315	−0.0088	−0.0104	1.0000	−0.0055	−0.0049	1.0000
	2	−0.0018	−0.0012	0.6301	−0.0044	−0.0111	0.9192	0.0170	−0.0028	−0.8291
	3	−0.0004	0.0005	1.0000	0.0154	0.0147	−0.8087	−0.0101	0.0019	0.2907

Table 21.7 Modal participating mass ratios (CM ≡ OTA for biaxial excitation)

Excitation	Mode 1	Mode 2	Mode 3	Mode 4	Mode 5	Mode 6	Mode 7	Mode 8	Mode 9
x	0.00	0.77	0.00	0.00	0.19	0.00	0.00	0.04	0.00
y	0.80	0.00	0.00	0.16	0.00	0.00	0.04	0.00	0.00

References

Anastassiadis K, Athanatopoulou A, Makarios T (1998) Equivalent static eccentricities in the simplified methods of seismic analysis of buildings. Earthquake Spectra 13:1–34

ATC (2005) Improvement of nonlinear static seismic analysis procedures, FEMA440 report. Redwood City, CA

Athanatopoulou AM (2005) Critical orientation of three correlated seismic components. Eng Struct 27:301–312

CEN, Comité Européen de Normalisation (2004) Eurocode 8: design of structures for earthquake resistance. Part 1: general rules, seismic actions and rules for buildings. EN 1998–1:2004. Brussels, Belgium

Doudoumis IN, Athanatopoulou AM (2008) Invariant torsion properties of multi-storey asymmetric buildings. Struct Des Tall Spec Build 17:79–97

EPPO, Earthquake Planning and Protection Organization (2003) Greek seismic code: earthquake resistant design of structures. Athens, Greece

Makarios T, Anastassiadis K (1998a) Real and fictitious elastic axis of multi-storey buildings: theory. Struct Des Tall Build 1:33–45

Makarios T, Anastassiadis K (1998b) Real and fictitious elastic axis of multi-storey buildings: applications. Struct Des Tall Build (1):57–71

Marino E, Rossi P (2004) Exact evaluation of the location of the optimum torsion axis. Struct Des Tall Build 13:277–290

Strong Motion Database (2003) Pacific Earthquake Engineering Research Centre (PEER). http://peer.berkeley.edu/smcat/. Accessed 1 July 2017

Chapter 22
Seismic Behaviour of 3D R/C Irregular Buildings Considering Complex Site Conditions

Ioanna-Kleoniki Fontara, Konstantinos Kostinakis, and Asimina Athanatopoulou

Abstract Local site conditions generate large amplifications as well as spatial variations in the seismic motions that must be accounted for in the earthquake resistant design of structures. The present paper aims to evaluate the influence of complex site effects on the non-linear response of irregular buildings. To achieve this purpose, site dependent ground motions are produced via Boundary Element Method (BEM). An ensemble of nine earthquakes recorded at the outcropping rock are considered as an input at the seismic bed of complex geological profiles, and acceleration time histories at the ground surface are computed. Several complex geological configurations are considered, taking into account the following key parameters: (i) canyon topography, (ii) layering and (iii) material gradient effect. Two 5-storey buildings are considered: a symmetric and an asymmetric in plan building. A series of Nonlinear Time History Analyses are conducted. The results of this study demonstrate that the presence of local site conditions influence the inelastic dynamic response of buildings.

Keywords Irregular buildings · Complex site effects · Seismic damage · Nonlinear time history analysis

22.1 Introduction

During an earthquake, seismic waves radiate from a fault and travel through the earth's crust. As seismic waves travel through the bedrock and the soil deposits, various complex geological profiles produce local distortions in the incoming wave

I.-K. Fontara
Department of Civil Engineering, Technische Universität, Berlin, Germany

K. Kostinakis (✉) · A. Athanatopoulou
Department of Civil Engineering, Aristotle University of Thessaloniki, Thessaloniki, Greece
e-mail: kkostina@civil.auth.gr

D. Köber et al. (eds.), *Seismic Behaviour and Design of Irregular and Complex Civil Structures III*, Geotechnical, Geological and Earthquake Engineering 48,
https://doi.org/10.1007/978-3-030-33532-8_22

field which may lead to large amplifications as well as strong spatial variations in frequency, amplitude and phase in the ground motions recorded along the surface.

For the dynamic analysis of structures, ground motion records are required as an input to the structural model under investigation. A simple, yet widely accepted approach is to assume that local soil conditions resemble those at the site where ground motions were first recorded. However, there is still a lack of real ground motion records able to describe all required seismological and geological conditions. A large number of studies has indicated the influence of 2D site effects on ground motions (e.g. Aki 1993; Bard 1994; Chavez-Garcia and Faccioli 2000). It has been demonstrated that several physical phenomena associated with 2D analysis like propagation of locally generated surface waves and possible 2D resonance, may greatly influence the computed seismic field.

Irregular in plan Reinforced Concrete (R/C) buildings are among the most common structural systems in many countries with high seismicity. The observation of post-earthquake damages on R/C structures has led to the conclusion that the presence of structural irregularities may significantly alter the seismic performance of buildings due to earthquake induced torsion, leading in many cases to negative effects. Therefore, a proper simulation and analysis of non-symmetric buildings is required taking into account the torsional response of non-symmetric buildings under earthquake excitation.

The aim of the present work is to evaluate the influence of complex site effects on the non-linear seismic response of irregular in plan multistory buildings. First, site dependent ground motions are produced considering 2D analysis of the soil profile via Boundary Element Method (BEM) numerical technique. An ensemble of nine earthquake excitations recorded at the outcropping rock are considered as an input at the seismic bed of complex geological profiles, and acceleration time histories at the ground surface are computed via BEM. Several complex geological configurations are considered, taking into account the following key parameters: (i) canyon topography, (ii) layering and (iii) material gradient effect. Next, two 5-storey buildings with structural systems that consist of members in two perpendicular directions (x and y) are considered. The first one is doubly-symmetric and the second is irregular in plan according to EC8. A series of Nonlinear Time History Analyses (NTHA) are conducted using the aforementioned family of site dependent ground motions. For each earthquake record the expected structural damage state of each building is determined in terms of the Maximum Interstorey Drift Ratio (MIDR). The results of this study demonstrate the importance of local site conditions on the structural damage of irregular buildings.

22.2 Site-Specific Ground Motions

In the present study, BEM numerical schemes based on a library of fundamental solutions and Green's function for inhomogeneous media (Fontara 2015) are used to model the seismic wave propagation through complex geological profiles so as to

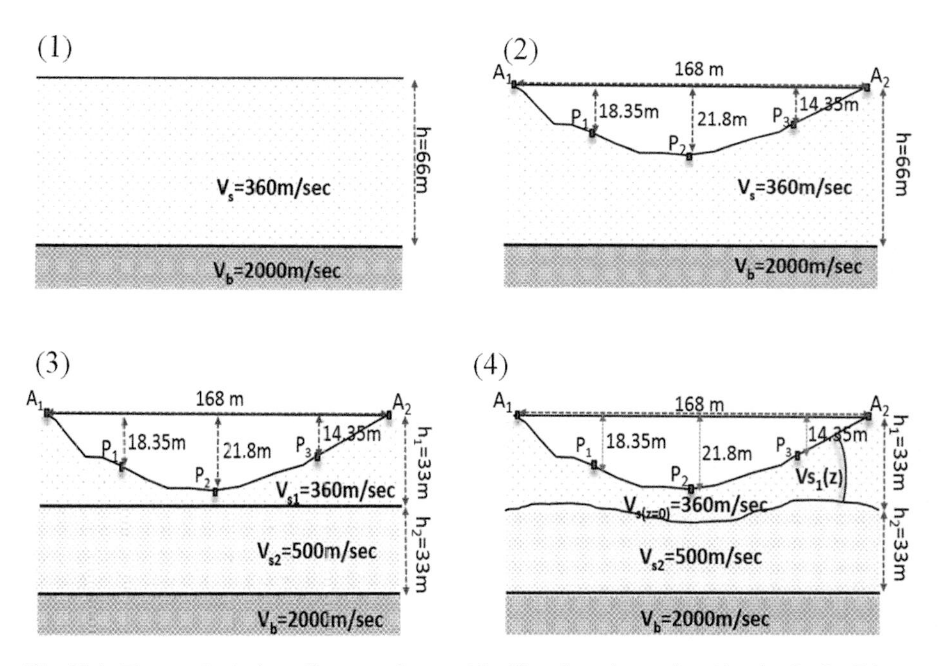

Fig. 22.1 Four geological profiles, namely type (1)–(4) and receiver points (A_1, A_2, P_1, P_2, P_3) on which the structures are assumed to be located

recover ground motion records that account for local site conditions. In particular, consider 2D wave propagation in viscoelastic, isotropic half-plane consisting of parallel or non-parallel inhomogeneous layers with a free surface and sub-surface relief of arbitrary shape. The dynamic disturbance is provided by either an incident SH wave or by waves radiating from an embedded seismic source.

The above mentioned methodology is applied to four different hypothetical geological profiles on which the R/C buildings in question are considered to be located, see Fig. 22.1. The examined geological key parameters are: (a) canyon topography; (b) layering; (c) material gradient. In particular, the site is represented by the following configurations: (1) a homogeneous layer with flat free surface producing a uniform excitation at the free surface; (2) a homogeneous layer with a valley in which the structures are considered to be located at different points along the free surface A_1, A_2, P_1, P_2 and P_3; (3) a double homogeneous layer deposit as a damped soil column with a valley at the surface; (4) a two-layer damped soil column with a valley at the surface, in which the top layer is continuously inhomogeneous with expressing an arbitrary variation in the wave speed depth profile (Fontara et al. 2015).

The bottom layer is homogeneous and the interface between the first and the second layer is irregular. All geological profiles are overlying elastic bedrock. The first geological configuration that produces a uniform excitation pattern ridge is considered as a reference case. A suite of nine earthquake excitations given in Table 22.1 that are recorded at the outcropping rock on site class A (according to

Table 22.1 Data of earthquake records

No	Date	Earthquake name	Station name	Closest distance (Km)	Component (deg)	PGA (g) (uncorrelated)
1	9/2/1971	San Fernando	Cedar Springs, Allen Ranch	89.4	95	0.020
2	9/2/1971	San Fernando	Pasadena – Old Seismo Lab	21.5	180	0.205
3	18/10/1989	Loma Prieta	Piedmont Jr High School Grounds	73.0	315	0.099
4	18/10/1989	Loma Prieta	Point Bonita	83.5	207	0.076
5	18/10/1989	Loma Prieta	SF – Pacific Heights	76.1	270	0.070
6	18/10/1989	Loma Prieta	SF – Rincon Hill	74.1	0	0.102
7	18/10/1989	Loma Prieta	So. San Francisco, Sierra Pt.	63.2	115	0.110
8	17/1/1994	Northridge	Wonderland Ave	20.3	95	0.160
9	17/1/1994	Northridge	Vasquez Rocks Park	23.6	0	0.152

FEMA classifications) are considered as an input at the seismic bed level for all geological profiles. These records were drawn from the PEER (2003) strong motion database.

In what follows, the influence of site effects on ground motions recorded at different points along the free surface is investigated. Figure 22.2 plots the acceleration response spectra of San Fernando No1 ground motion recorded at different points A_1, A_2, P_1, P_2, P_3 along the free surface of geological profile Type (2) that accounts for canyon topography (Fig. 22.2a); Type (3) that accounts for canyon topography and layering effect (Fig. 22.2b); and Type (4) that accounts for canyon topography, layering and material gradient effect (Fig. 22.2c). The above mentioned geological profiles are compared with the reference case of type (1) that produces uniform excitation.

Observe that spectral acceleration values can differ significantly when they are recorded at different points along the surface topography. The seismic signal depends strongly on the presence of canyon topography. From Fig. 22.2a we can see that spectral accelerations are more pronounced for low values of period when

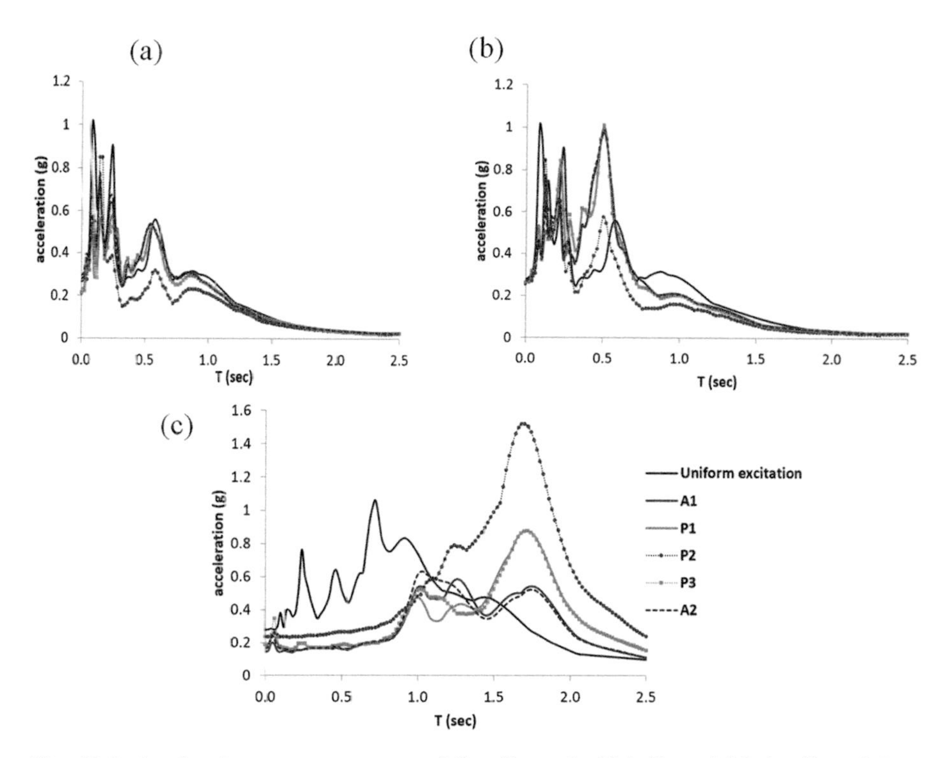

Fig. 22.2 Acceleration response spectra of San Fernando No1 Ground Motion Recorded at different receiver points (A_1, A_2, P_1, P_2, P_3) at the free surface of geological profile (**a**) type (2); (**b**) type (3) and (**c**) type (4) and compared with uniform excitation type (1)

they are recorded at the bottom of the canyon, while high period values lead to significant spectral acceleration at the edge of the canyon. The influence of the combined effects of soil layering and canyon topography structure on ground motions is shown in Fig. 22.2b. The shape of the response spectra is now modified, while an expected increase of the peaks and a shifting to the right (higher periods) due to the layering effect is clearly depicted. The combined influence of canyon topography, layering and material gradient effect on the ground motion is illustrated in Fig. 22.2c. As previously mentioned, in this case the top layer has a continuous variation of the wave speed with depth, avoiding this way the great wave speed contrast between the first and the second layer of the previously examined case. In addition, we also introduce here a spatial irregularity in the interface between the two soil layers. The presence of material gradient increases the material stiffness gradually, the soil becomes stiffer and the dynamic characteristics of the geological profile are modified. As a result, the spectral acceleration values are de-amplified across the entire range of periods between 0 and 1.1 s.

22.3 R/C Buildings Modeling

Next we focus on the nonlinear response of two 3D R/C buildings, one with symmetric and one with asymmetric wall system. Geometrical and material properties for the analysis and design of the examined buildings are given in Figs. 22.3 and 22.4. The buildings have five stories and structural system that consists of elements in two perpendicular directions (x and y axes). More specifically, the following buildings are investigated:

- Symmetric Wall System SWS (Fig. 22.3): Double-symmetric building with walls that take approximately 65%–70% of the base shear along both x and y axes (wall system according to the structural types of the EC8 2004). The building is regular in plan and in elevation.
- Asymmetric Wall System AWS (Fig. 22.4): Asymmetric building with walls that take approximately 65%–70% of the base shear along both x and y axes (wall system according to the structural types of the EC8. The ratio e_{0X}/r_X along x-axis is equal to 0.4 (>0.3), and the respective ratio along y-axis is equal to 0.58 (>0.3). So the building is irregular in plan, but is regular in elevation according to EC8 (2004).

In Table 22.2 all the common design data of the examined buildings are presented. Based on the above data, the process of calculating the upper limit value of the behavior factor q of EC8 (2004) led to the values which are shown in Figs. 22.3 and 22.4 for the two buildings. These values were used for the design of the examined buildings.

The two structures were analyzed using the modal response spectrum analysis, as described in EC8 (2004). The R/C structural elements were designed following the clauses of EC2 (2004) and EC8 (2004). It should be noted that the choice of the

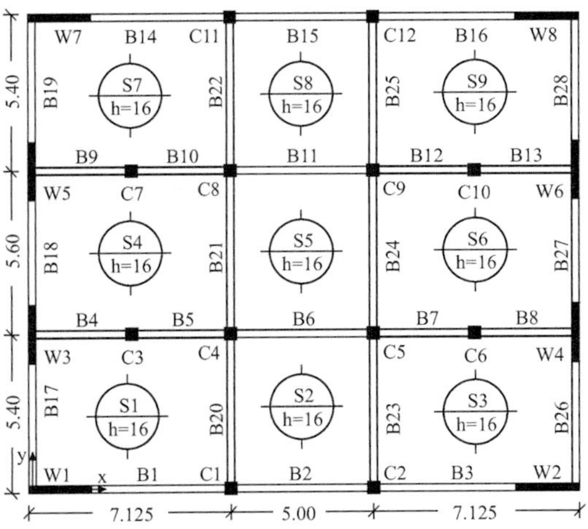

Dimensions of the cross-sections of R/C members:

Storey	Beams
1st	25/55
2nd-3rd	25/55 (B18,B27: 25/45)
4th	25/50 (B18,B27: 25/45)
5th	20/45

Storey	Columns	Walls
1st	C1,C2,C11,C12: 35/35 C3,C6,C7,C10: 30/30 C4,C5,C8,C9: 40/40	W1,W2, W7,W8: 220/25
2nd - 3rd	35/35 (C3,C6,C7,C10: 30/30)	W3,W4, W5,W6: 200/25
4th - 5th	30/30	

Behavior factor: q=3.0

Fig. 22.3 5-Story Symmetric Building (SWS). Plan view, geometrical and design parameters

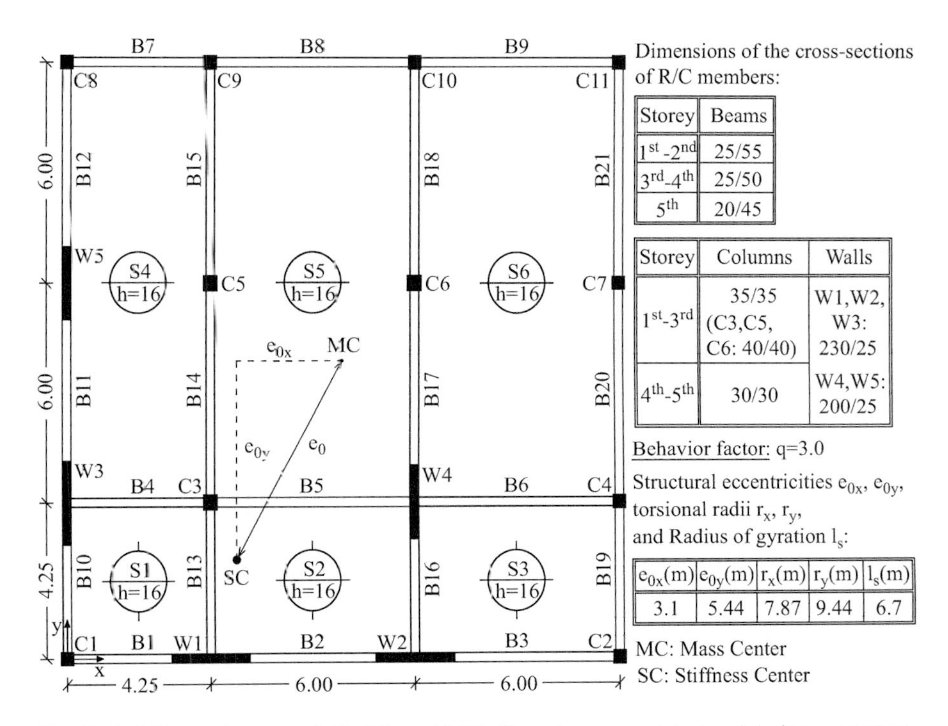

Fig. 22.4 5-Storey Asymmetric Building (AWS). Plan view, geometrical and design parameters

Table 22.2 Common design data for the two buildings

Storeys' heights H_i	Ductility class	Concrete	Steel
3.2 m	Medium (DCM)	*C20/25*	*S500B*
		$E_c = 3 \cdot 10^7$ kN/m^2 $\nu = 0.2$ $w = 25$ kN/m^3	$E_s = 2 \cdot 10^8$ kN/m^2 $\nu = 0.3$ $w = 78.5$ kN/m^3
Slab loads	**Masonry loads**	**Design spectrum (EC8)**	
Dead: G = 1.0 kN/m^2 *Live*: Q = 2.0 kN/m^2	*Perimetric beams*: 3.6 kN/m^2 *Internal beams*: 2.1 kN/m^2	*Reference PGA*: $a_{gR} = 0.24$ g *Importance class*: II $\rightarrow \gamma_I = 1$ *Ground type*: C	

dimensions of the structural members' cross-sections as well as of their reinforcement was made bearing in mind the optimum exploitation of the structural materials (steel and concrete). The professional computer program RA.F (T.O.L. -Engineering Software House 2014) was used for the design of the buildings.

For the modeling of the buildings' nonlinear behavior, plastic hinges located at the column and beam ends as well as at the base of the walls were used. The material inelasticity of the structural members was modeled by means of a Modified Takeda hysteresis rule (Otani 1974). It is important to notice that the effects of axial load-biaxial bending moment ($P\text{-}M_1\text{-}M_2$) interaction at column and wall hinges are taken into consideration by means of the $P\text{-}M_1\text{-}M_2$ interaction diagram which is implemented in the software used to conduct the analyses (Carr 2006). The yield moments as well as the parameters needed to determine the $P\text{-}M_1\text{-}M_2$ interaction diagram of the vertical elements' cross sections are determined using appropriate software (Imbsen Software Systems, XTRACT: Version 3.0.5 2006).

22.4 Nonlinear Dynamic Response of the Buildings

A series of Nonlinear Time History Analyses (NTHA) were conducted under the suite of the nine ground motions presented in Table 22.1 for the following cases: (i) recorded at the surface of geological profile (1) (ii) recorded at points A_1, A_2, P_1, P_2 and P_3 along the surface of geological profile (2) (iii) recorded at points A_1, A_2, P_1, P_2 and P_3 along the surface of geological profile (3) and, (iv) recorded at points A_1, A_2, P_1, P_2 and P_3 along the surface of geological profile (4). For each ground motion, the damage state of the buildings was determined.

In the present research, the seismic damage of the buildings is expressed in the form of the Maximum Interstorey Drift Ratio (MIDR), which is a global, structural and deterministic damage index. The MIDR, which is generally considered an effective indicator of global structural and nonstructural damage state of R/C buildings (e.g. Naeim 2001), has been used by many researchers for the assessment of the inelastic response of structures (e.g. Elenas and Meskouris 2001). It corresponds to the maximum drift among the perimeter frames.

22.5 Analyses Results

The influence of the site effects on the structural response of the examined symmetric and asymmetric buildings is demonstrated in Figs. 22.5 and 22.6. More specifically, the MIDR of the symmetric and asymmetric building located at the examined geological profiles (Fig. 22.1) and subjected to the nine earthquake records given in Table 22.1 are given. Figures 22.5 and 22.6 present indicatively the results for the two buildings located at the geological profile (3). Note that for comparison reasons, in order to examine the influence of the site effects on the dynamic response of the buildings, in the above mentioned figures the results produced for the reference geological profile (1) are also presented. Moreover, for every building and geological profile, the average values of the MIDR over all nine earthquake records are presented, in order to generalize trends.

For all cases examined here, the asymmetric building produces higher values of MIDR than the corresponding symmetric one. The difference between the MIDR produced by the symmetric and the asymmetric building can significantly depend on

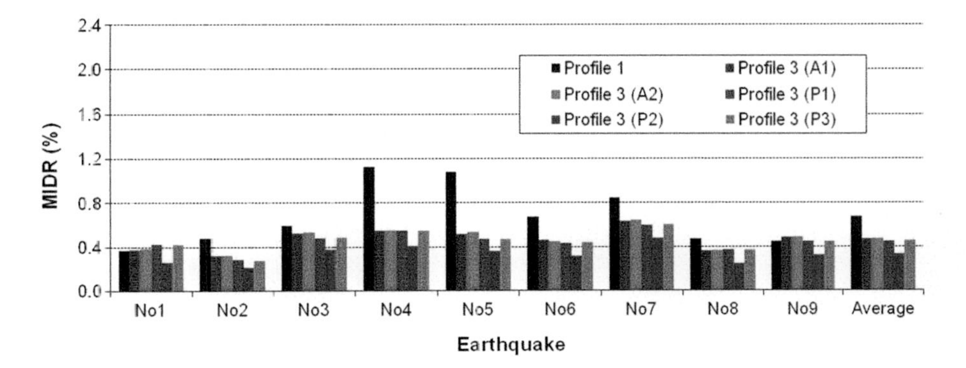

Fig. 22.5 Maximum Interstorey Drift Ratio (MIDR) of the 5-Storey Symmetric Building (SWS) due to ground motions recorded at different points (A_1, A_2, P_1, P_2, P_3) along the surface of profile type (3)

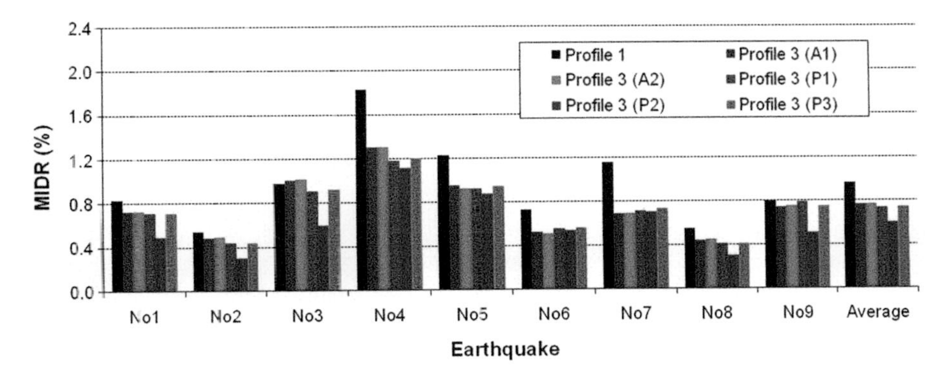

Fig. 22.6 Maximum Interstorey Drift Ratio (MIDR) of the 5-Storey Asymmetric Building (AWS) due to ground motions recorded at different points (A_1, A_2, P_1, P_2, P_3) along the surface of profile type (3)

the earthquake record and the point where the buildings are located (A_1, A_2, P_1, P_2 or P_3) regarding their position in the canyon topography.

Investigating the influence of site effects on the symmetric building, we can see that for almost all ground motions the consideration of site effects in the numerical analysis is beneficial leading to lower values of MIDR. Moreover, the analyses revealed that for the most earthquake records point P_2 leads to the smallest values of MIDR. Therefore, can be said, that a building located at the edge of a canyon is more prone to damages rather than located at the bottom of the canyon. Similar conclusions can be extracted for the case of the asymmetric building. We see that for the majority of the ground motions the consideration of site effects in the numerical analysis of the building is beneficial leading to lower design values. Moreover, the analyses showed that for the most earthquake records, point P_2 leads to the smallest values of MIDR.

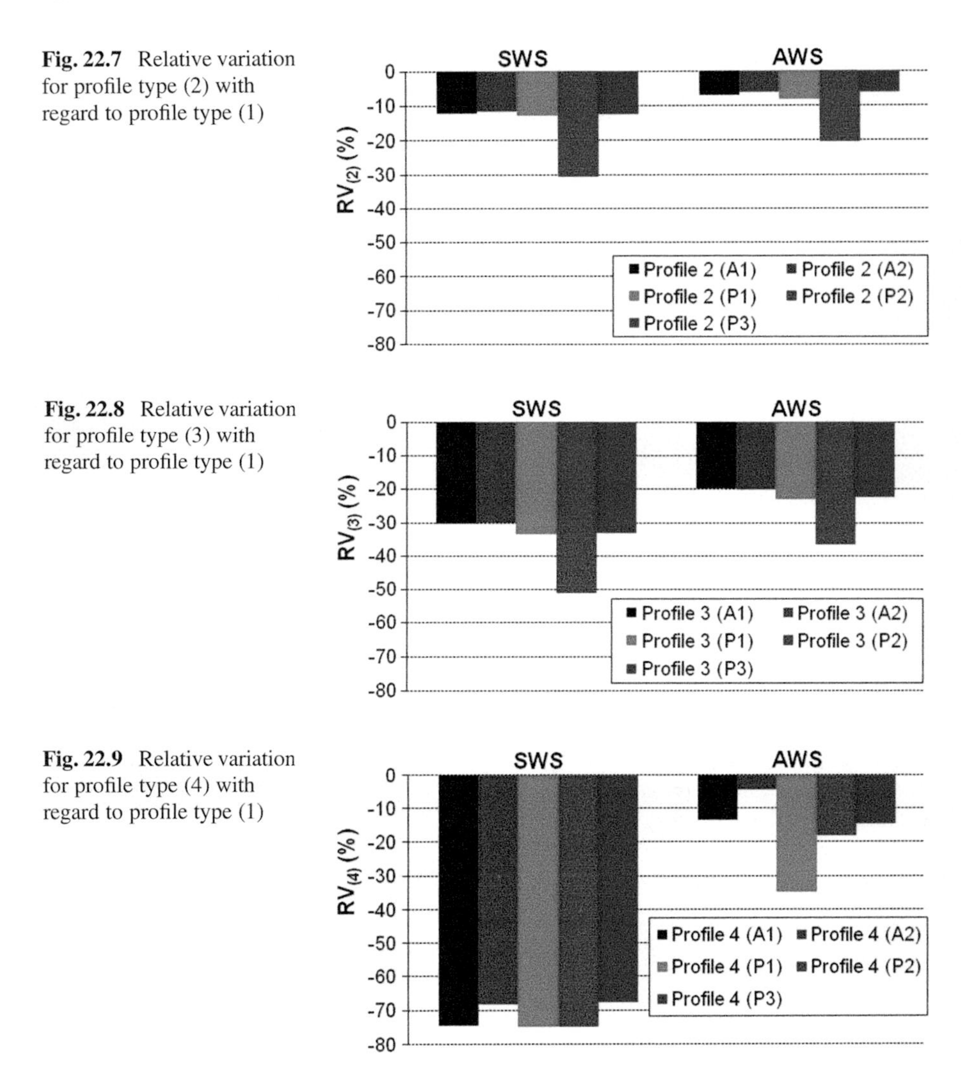

Fig. 22.7 Relative variation for profile type (2) with regard to profile type (1)

Fig. 22.8 Relative variation for profile type (3) with regard to profile type (1)

Fig. 22.9 Relative variation for profile type (4) with regard to profile type (1)

Furthermore, to facilitate comparisons, the Relative Variation ($RV_{(i)}$) for profile types i (where i: (2), (3) or (4)) with regard to profile type (1) is introduced:

$$RV_{(i)} = \frac{averageMIDR_{Profile\ (i)} - averageMIDR_{Profile\ (1)}}{averageMIDR_{Profile\ (1)}} \cdot 100(\%) \qquad (22.1)$$

where: $averageMIDR_{Profile(i)}$ is the average value of the MIDR over all nine earthquake records corresponding to geological profile (i). In Figs. 22.7, 22.8 and 22.9 the values of $RV_{(i)}$ for the three geological profiles accounting for site effects are illustrated.

The study of Figs. 22.7, 22.8 and 22.9 reveals that the consideration of site effects in the numerical analysis of the buildings is beneficial leading to negative values of RV in every case. This conclusion is valid for both the symmetric and the asymmetric building. It is noticeable that the RV can reach the value of −75% (profile (4), points P_1 and P_2). A comparison between the two buildings shows that site effects can be more beneficial in case of the symmetric building. Comparing different geological profiles examined here, we see that in the case of the symmetric building, the most beneficial effects are produced by the profile (4), whereas profile (2) is the least beneficial one. In the case of the asymmetric building, the most beneficial effects are produced by the profile (3). Furthermore, the analyses revealed that in case of the profiles (2) and (3), site effects are more beneficial when the buildings are located at the point P_2, whereas, in case of the profile (4), the most beneficial effects are produced for points A_1, P_1 and P_2 (symmetric building) and for point P_1 (asymmetric building).

22.6 Conclusions

In the present study, the influence of local site conditions on the seismic behavior of 3D reinforced concrete multistory buildings is investigated. Site specific ground motions are generated from complex geological profiles that account for canyon topography, layering and material gradient effect using BEM numerical schemes. Following that, a series of nonlinear dynamic analyses of two buildings, one symmetric and one asymmetric, are conducted under the above mentioned site specific ground motions. The comparative assessment of the results has led to the following conclusions:

- Higher values of MIDR are produced in the asymmetric building than in the corresponding symmetric one. The difference between the symmetric and the asymmetric building can be significantly depending on the earthquake record, the geological profile and the point where the buildings are located.
- For almost all ground motions and geological profiles the consideration of site effects in the numerical analysis of symmetric and asymmetric buildings is beneficial leading to lower design values of MIDR. However, in case of the asymmetric building, there are very few earthquake records that led to detrimental results when the influence of site effects was taken into account in the dynamic analysis.

References

Aki K (1993) Local site effects on weak and strong ground motion. Tectonophysics 218:93–111

Bard P-Y (1994) Effects of surface geology on ground motion: recent results and remaining issues. Paper presented at the 10th European Conf. on Earthq. Engng, Vienna

Carr AJ (2006) Ruaumoko – a program for inelastic time-history analysis: program manual. Department of civil engineering. University of Canterbury, New Zealand

Chavez-Garcia FJ, Faccioli E (2000) Complex site effects and building codes: making the leap. J Seismol 4(1):23–40

EC2 (Eurocode 2) (2004) Design of Concrete Structures, Part 1–1: general rules and rules for buildings, European Committee for Standardization

EC8 (Eurocode 8) (2004) Design of structures for earthquake resistance – Part 1: general rules, seismic actions and rules for buildings. European Committee for Standardization

Elenas A, Meskouris K (2001) Correlation study between seismic acceleration parameters and damage indices of structure. Eng Struct 23:698–704

Fontara IK (2015) Simulation of seismic wave fields in inhomogeneous half-plane by non-conventional BEM. Christian-Albrechts-University in Kiel, Germany, 1, ISSN 2365-7162

Fontara I-K, Dineva P, Manolis G, Parvanova S, Wuttke F (2015) Seismic wave fields in continuously inhomogeneous media with variable wave velocity profiles. Arch Appl Mech 86 (1):65–88

Imbsen Software Systems, XTRACT: Version 3.0.5. (2006) Cross-sectional structural analysis of components, Sacramento

Naeim F (ed) (2001) The seismic design handbook. Kluwer Academic, Boston

Otani A (1974) Inelastic analysis of RC frame structures. J Struct Div ASCE 100(7):1433–1449

PEER (Pacific Earthquake Engineering Research Centre) (2003) Strong Motion Database. http://peer.berkeley.edu/smcat/

T.O.L. -Engineering Software House (2014) RAF Version 4.4: structural analysis and design software. Iraklion, Crete, Greece

Chapter 23
Application of Artificial Neural Networks for the Assessment of the Seismic Damage of Buildings with Irregular Infills' Distribution

Konstantinos Kostinakis and Konstantinos Morfidis

Abstract One of the most common structural systems in earthquake prone areas are Reinforced Concrete (R/C) buildings with masonry infills. The observation of post-earthquake damages has led to conclusion that the masonry infills can greatly modify the seismic performance of these buildings. In the context of the direct assessment of the buildings' seismic vulnerability, many researches have been conducted aiming to use the capacities of artificial intelligence, such as the Artificial Neural Networks (ANNs). The present study examines the influence of the infills' irregular distribution on the seismic damage level of R/C buildings using Multilayer Feedforward Perceptron (MFP) ANNs. More specifically, a 5-storey R/C building with a large number of different masonry infills' distributions possessing several degrees of irregularities is analyzed by means of Nonlinear Time History Analysis for 65 actual ground motions. The optimum configured and trained networks are applied for the rapid estimation of the damage of the examined building. The results of these applications show that the best configured and trained networks are capable to adequately estimate the damage of the R/C buildings with asymmetry caused by the irregular location of masonry infills.

Keywords Masonry infills · Irregular buildings · Artificial Neural Networks · Seismic damage · Nonlinear time history analysis

---

K. Kostinakis (✉)
Department of Civil Engineering, Aristotle University of Thessaloniki, Thessaloniki, Greece
e-mail: kkostina@civil.auth.gr

K. Morfidis
E.P.P.O.-I.T.S.A.K., Thessaloniki, Greece

D. Köber et al. (eds.), *Seismic Behaviour and Design of Irregular and Complex Civil Structures III*, Geotechnical, Geological and Earthquake Engineering 48,
https://doi.org/10.1007/978-3-030-33532-8_23

23.1 Introduction

One of the most common structural systems in countries with regions of high seismicity is Reinforced Concrete (R/C) buildings with masonry infills. The infill walls are not accounted for in analytical models because they are usually considered as non-structural elements. However, the post-earthquake damages' observation has led to the conclusion that the presence of infills can greatly modify the seismic behavior of R/C buildings (e.g. EERI 2000; Ricci et al. 2010; Palermo et al. 2014). In particular, experimental as well as numerical research studies have concluded that a uniform masonry infills' distribution may lead to the increase of lateral stiffness and robustness, thus altering the dynamic characteristics of structures and leading to a lower vibration period (e.g. Bertero and Brokken 1983). On the contrary, if the infills are non-uniformly placed, negative effects, such as the soft story mechanism or non-controlled torsional effects, can be induced (e.g. Negro and Colombo 1997; Yuen and Kuang 2015). The impact of the infills' asymmetric/irregular placement on the seismic damage level of R/C buildings can be especially crucial for the Performance-Based Earthquake Engineering, which aims at the assessment of the damage risk of buildings due to future seismic events.

Performance-Based Earthquake Engineering has received significant attention during the past years. Numerous studies have dealt with methods of predicting the damage level of R/C buildings subjected to strong earthquakes. However, for these methods there is a significant shortcoming, namely they can use only a very limited number of structural or seismic parameters. The above shortcoming can be dealt by utilizing methods based on Artificial Intelligence, such as the Artificial Neural Networks (ANNs). Although the ANNs have been developed on an idea that dates back to the 1940s, the first thorough investigation dealing with the use of them for the direct assessment of the level of seismic damage was published in the middle of 1990s (Molas and Yamazaki 1995). Since then, this approach has been the subject of a large number of researches, which led to highly important and interesting results. These results designate the ability of ANNs to predict the potential seismic damage of buildings in an approximate but generally reliable way (e.g. Lautour and Omenzetter 2009; Rofooei et al. 2011; Vafaei et al. 2013).

The aim of the present study is to evaluate the impact of the infills' irregular distribution on the seismic damage level of R/C buildings using Multilayer Feedforward Perceptron (MFP) ANNs. More specifically, for the needs of the present investigation, a 5-storey building was chosen. Then, a large number of different infills' distributions with several degrees of irregularities for the chosen building was examined. The infilled buidings were then analyzed by means of Nonlinear Time History Analysis for 65 actual ground motions. For the estimation of the expected seismic damage state the Maximum Interstorey Drift Ratio (MIDR) (Naeim 2001) was computed. Using the results of these analyses a data set for the training of the ANNs was created. As input parameters of ANNs, structural parameters (the structural eccentricity due to the irregular location of infills and the percentage of the infills along the two orthogonal structural axes of the building), as well as seismic parameters (14 widely used strong motion measures, e.g. Kramer 1996) were considered. For the purposes of the current study MFPs with one hidden layer were used. The number of neurons and their activation functions in the hidden layer which lead to the optimum performance were also investigated. The optimum configured networks were then applied for the rapid evaluation of the damage (MIDR) for the examined 5-storey infilled building subjected to earthquakes different from the earthquakes which were used for the training data set generation (study of the optimum configured networks' generalization ability). The results of the applications show that the best configured and trained networks are capable to adequately estimate the seismic damage of the R/C buildings with asymmetry caused by the irregular location of infills.

23.2 Artificial Neural Networks – General

The Artificial Neural Networks (ANNs) are complex computational tools that are capable to handle problems utilizing the general rules of the human brain functions (see e.g. Haykin 2009). Thus, using ANNs it is possible to approximate the solution of problems such as the pattern recognition and the function approximation problem.

The ANNs' function is based on the combined action of interconnected processing units that are called artificial neurons (Fig. 23.1a). For more details about the function of artificial neurons see e.g. (Haykin 2009). Figure 23.1b presents the typical configuration of a MFP type ANN with four layers of neurons (input layer, two hidden layers and output layer).

The solution of problems through the utilization of ANNs is accomplished if they have been trained using the training algorithms. These algorithms are procedures which require a set of n input vectors $\mathbf{x}$ and the corresponding to them n output

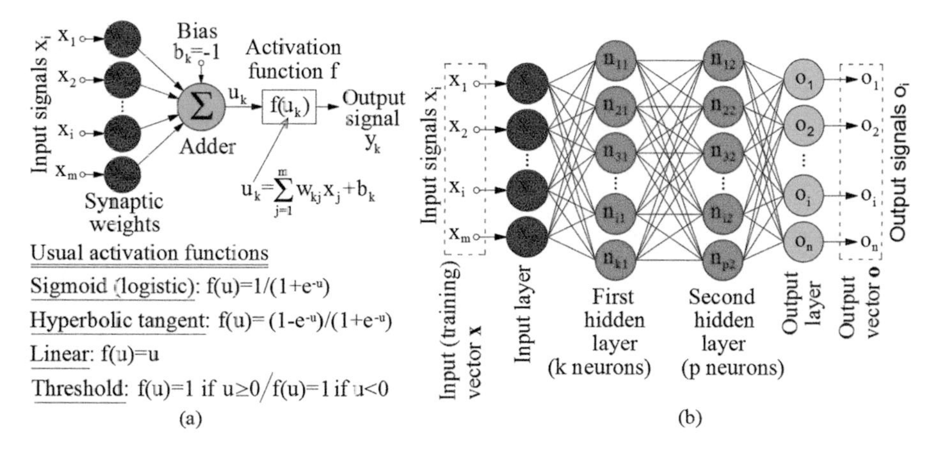

Fig. 23.1 The Typical Artificial Neuron (**a**) and Typical Configuration of a Multilayer Feedforward Perceptron (MFP) Network (**b**)

vectors **d** that called target vectors. The *n* pairs of vectors **x** and vectors **d** constitute the training data set. During the training procedure the values of the synaptic weights w_{ki} (see Fig. 23.1a) are successively altered until the error vector that is produced by the ANN is minimized.

23.3 Input and Output Parameters

The ANNs possess the inherent ability to handle and approach the solution of multiparametric problems. The parameters that describe the problem of the prediction of the seismic damage of R/C buildings can be categorized in two categories: the structural parameters and the seismic parameters.

23.3.1 Input Parameters – Structural Parameters

The structural parameters are utilized for the description of the buildings' response due to earthquake excitations. In the present study, the configured ANNs concern buildings with specific structural characteristics (i.e. structural system, total height, structural eccentricity, number and dimensions of R/C walls, configuration in plan, constructional materials, external loads etc.). For this reason, the only structural parameters which are altered and must be examined as regards their influence on the seismic damage state of buildings are four: the ratios of the masonry infills along the two perpendicular structural axes x and y (ratio $n_{mas,x}$ and ratio $n_{mas,y}$), and the eccentricities e_{0x} and e_{0y} of floors due to the configuration of the masonry infills along the axes x and y (Fig. 23.2).

More specifically, a 5-storey double-symmetric in plan R/C building, with plan view and geometrical and design parameters supplied in Fig. 23.2 was studied. The building's structural system consists of structural members in two perpendicular directions and is regular in elevation and in plan according to the criteria set by EC8 (EC8 (Eurocode 8) 2005). The building was designed for static vertical loads and for seismic loads utilizing the modal response spectrum analysis, as described in EC8. The R/C structural members were designed according to the provisions of EC2 (EC2 (Eurocode 2) 2005) and EC8.

The effects caused by the arbitrary placement of masonry infills were taken into account adopting a large number of different distributions of the infills. It must be noted that the same configuration of the infills was used for all the five stories of the building. Every infilled building can be characterized by certain structural parameters. For the needs of the present study, as mentioned above, the following structural parameters are adopted:

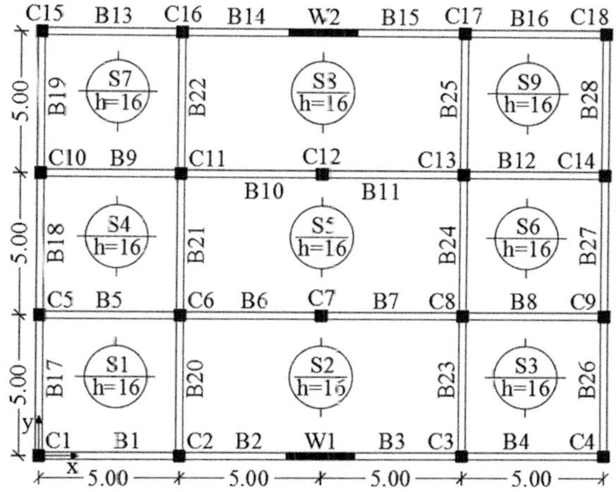

Dimensions of the cross-sections of R/C members:

Storey	Beams	Columns	Walls
1st-2nd	25/55	40/40	W1
3rd	25/50	(C7,C12:35/35)	240/25
4th	25/50	35/35	W2
5th	20/45	30/30	240/25

Behavior factor: q=3.0
Storeys' heights: H_i=3.2m
Ductility class: Medium (DCM)
Materials: Concr. C20/25, Steel S500B
Slab loads:
G=1kN/m² (Dead), Q=2kN/m² (Live)
Masonry loads (Perimetric / Internal):
3.6kN/m² / 2.1kN/m²
Design spectrum: EC8-1
a_{gR}=0.24g, γ_I=1, Ground type: C

Fig. 23.2 Plan view and geometrical parameters of the studied R/C building

- Structural eccentricities e_x and e_y (= the distance between the mass centre of the infilled building and the mass centre of corresponding bare building) along the axes x and y respectively.
- Ratios of the masonry infills along the structural axes x and y (Eq. 23.1):

$$n_{mas,x} = \frac{W_{mas,x}}{maxW_{mas,x}} \qquad n_{mas,y} = \frac{W_{mas,y}}{maxW_{mas,y}} \tag{23.1}$$

where: $W_{mas,x}$ is the weight of the masonry infills along the axis x, $maxW_{mas,x}$ is the maximum weight of the infills along the axis x (i.e. the masonry infills' weight in the case in which they are present at all spans parallel to the axis x), $W_{mas,y}$ is the weight of the masonry infills along the axis y; $maxW_{mas,y}$ is the maximum weight of the masonry infills along the axis y (i.e. the masonry infills' weight in the case in which they are present at all spans parallel to the axis y). The explanation of the above described terms is better illustrated with the aim of example of Fig. 23.3a.

It must be noticed that the different distributions of the infills were chosen arbitrarily, in order to cover a wide range of all possible values of the above structural parameters. Thus, 141 different infills' distributions were considered. Concerning the modeling of each masonry infill, in the present study, the single equivalent diagonal strut model is adopted. In particular, each infill panel was modeled as single equivalent diagonal strut with stress-strain diagram based on the model proposed by Crisafulli (1997) (Fig. 23.3b).

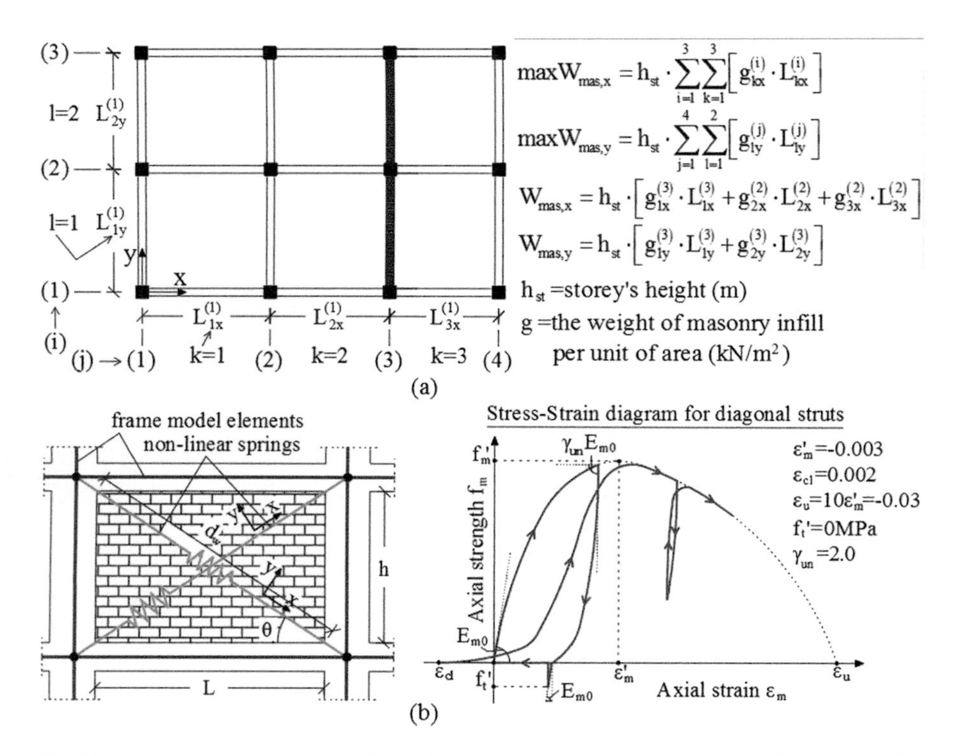

Fig. 23.3 Procedure of calculation of the selected input structural parameters (**a**), Simulation of the Masonry infill response using the model proposed by Crisafulli (**b**)

23.3.2 Input Parameters – Seismic Parameters

As regards the seismic parameters which are used to describe the seismic excitations and their impact to structures, there are many definitions which are resulted from the analysis of accelerograms records (see e.g. Kramer 1996). For the investigation conducted in the present study, the 14 seismic parameters which are illustrated in Table 23.1 have been chosen.

A suite of 65 pairs of horizontal bidirectional seismic motions obtained from the PEER (2003) and the European (European Strong – Motion Database 2003) strong motion database was used for the analyses which were performed in order to generate the networks' training data set. The seismic excitations have been selected from worldwide well-known sites with strong seismic activity, are recorded on Soil Type C according to EC8 and have magnitudes (M_s) between 5.5 and 7.8.

Table 23.1 Seismic parameters and the range of their values in training data-set

	Parameter	Units	Range
1	Peak Ground Acceleration: **PGA**	g	0.004–0.822
2	Peak Ground Velocity: **PGV**	cm/s	0.86–99.35
3	Peak Ground Displacement: **PGD**	cm	0.36–60.19
4	Effective Peak Acceleration: **EPA**	g	0.003–0.63
5	Specific Energy Density: **SED**	cm^2/s	1.24–16762
6	Acceleration Spectrum Intensity: **ASI**	g·s	0.003–0.633
7	Cumulative Absolute Velocity: **CAV**	cm/s	14.67–2684
8	Housner Intensity: **HI**	cm	3.94–317
9	Arias Intensity: $\mathbf{I_a}$	m/s	≈0.0–5.59
10	V_{max}/A_{max} **(PGV/ PGA)**	s	0.03–0.33
11	Predominant Period: **PP**	s	0.07–1.26
12	Uniform Duration: **UD**	s	≈0.0–17.7
13	Bracketed Duration: **BD**	s	≈0.0–61.9
14	Significant Duration: **SD**	s	1.74–51.0

Table 23.2 Relation between MIDR and damage state

MIDR (%)	<0.25	0.25–0.50	0.50–1.00	1.00–1.50	>1.50
Damage level	Null	Slight	Moderate	Heavy	Destruction
	No damages	Repairable slight damages	Repairable significant damages	Non-repairable damages	

23.3.3 Output Parameters

Due to the fact that the problem which is investigated in the present study is the prediction of the seismic damage of R/C buildings using ANNs, one of the basic steps of the problem's formulation procedure is the selection of an appropriate damage index i.e. the selection of the networks' output. In the present study, the seismic damage level of R/C structures was expressed using the Maximum Interstorey Drift Ratio (MIDR). The MIDR is considered an efficient measure of global structural as well as nonstructural damage of R/C structures e.g. (Naeim 2001) and is calculated as the maximum drift among the perimeter frames. The relation between the MIDR values and the description of the seismic damage state of R/C buildings which was used in the present study is illustrated in the Table 23.2 (Masi et al. 2011).

23.4 Generation of the Training Dataset and ANNs' Configuration

23.4.1 Generation of the Training Data Set

In order to generate the data set needed for the ANNs' training, the 141 infilled structures presented above were analyzed by Nonlinear Time History Analysis (NTHA) for each one of the 65 earthquake records accounting for the design vertical loads of the buildings. Moreover, as the angle of incidence with regard to structural axes is unknown, the two horizontal accelerograms of each strong motion were applied along horizontal orthogonal axes forming with the structural axes angles $\theta = 0°, 90°, 180°$ and $270°$. Consequently, a total of 36,660 NTHA were conducted.

For the structures' nonlinear behavior modeling lumped plasticity models at the column and beam ends as well as at the base of the shear walls, were used. The length of the plastic hinges was computed using the relations proposed by Pauley and Priestley (1992). The inelastic behavior of the material was modeled using the Modified Takeda hysteresis rule (Otani 1974). The hysteretic behavior was considered to be perfectly plastic after yielding.

23.4.2 ANNs' Configuration and Training Algorithms

In general, the parameters which are needed for the MFP networks' configuration are the following: (a) the hidden layers' configuration; (b) the number of neurons in each hidden layer; (c) the neurons' activation functions; (d) the performance evaluation parameters; and (e) the method used to partition the data set in training, validation and testing subsets. In the present research, specific choices for some of the abovementioned parameters were made, while for some others more than one choice was made, in order to detect the network which extracts the optimum predictions (optimum configured network). These choices are given in the Fig. 23.4. Finally, regarding the training algorithms, two algorithms were used: the Levenberg-Marquardt algorithm (LM algorithm, Marquardt 1963) and the Scaled Conjugate Gradient algorithm (SCG algorithm, Moller 1993). The configuration and the training of ANNs was made using the neural network tool box of Matlab (2013).

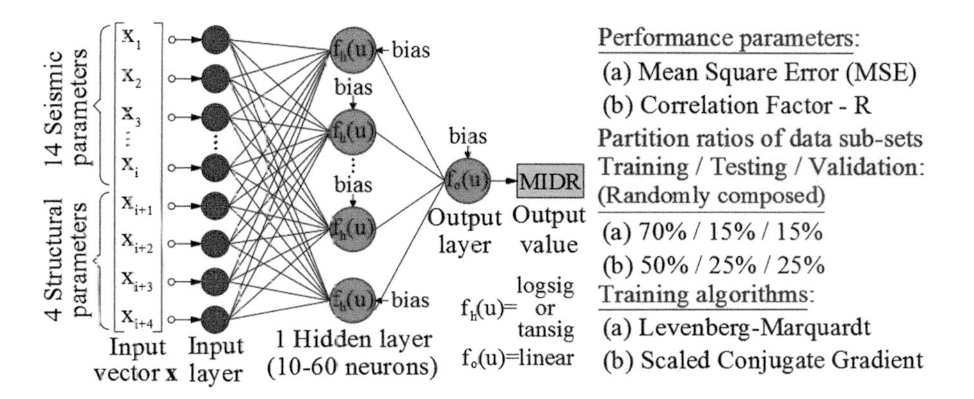

Fig. 23.4 Configuration of the MFP Networks and the other parameters which were examined in the present study

23.5 Analyses Results

23.5.1 Parametric Investigation for the Optimum Configuration of Networks

The parametric investigation was divided in two parts because of the utilization of two training algorithms (LM/SCG). In each one of these parts, two types of MFP networks were configured. The ANNs of the first type have activation functions tansig (tan) for the hidden layer's neurons. Respectively, the ANNs of the second type have logsig (log) functions. More specifically, 51 different versions of networks regarding the number of neurons ($10 \div 60$) in the hidden layer were configured for each one of the two network types. Therefore, $102(= 2 \times 51)$ different networks were configured in each one of the two parts of investigation. Each one of these networks was trained 75 times (i.e. $102 \times 75 = 7650$ trainings were performed in each part). This is done because differences in ANNs' performance are caused by synaptic weights' and biases' initial values (see e.g. Lautour and Omenzetter 2009), and also by the random composition of the three sub-sets of the total data set (Matlab 2013). From the 75 trainings of each one of the 102 configured ANNs the optimum ones were detected. More specifically, the trainings which led to the optimum values of the performance parameters MSE and R-factor on the basis of the testing sub-set and the total data set were detected. Thus, from the 7650 trainings of each part 408 ($=4 \times 102$) optimum trained ANNs were emerged considering one of the following criteria:

- min(MSE) for: (a) the testing sub-set and (b) the total data set,
- max(R) for: (a) the testing sub-set and (b) the total data set.

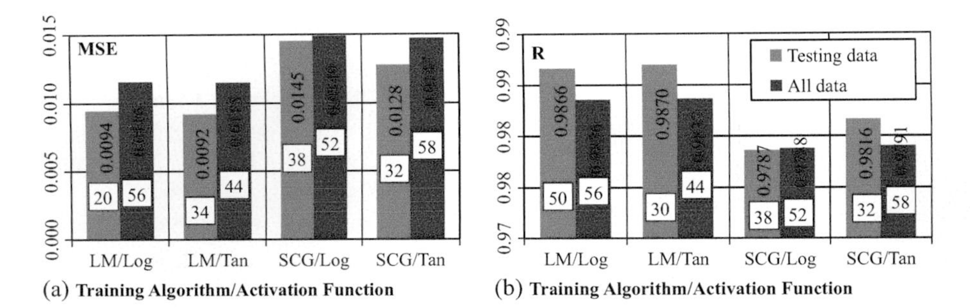

Fig. 23.5 Optimum values of MSE, R – Partition ratios of data sub-sets: 70%/15%/15%

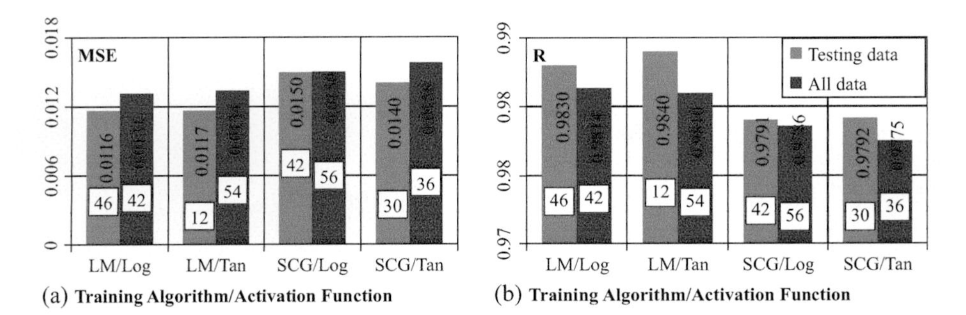

Fig. 23.6 Optimum values of MSE, R – Partition ratios of data sub-sets: 50%/25%/25%

The whole procedure which was described above was performed two times because the partitioning of the total data set in the three sub-sets was done using two different ratios, namely 70%/15%/15% and 50%/25%/25% (as mentioned in the Fig. 23.4).

Figures 23.5 and 23.6 show the results of the investigation for the optimum configured networks in the case in which the total data set is partitioned to the training, testing and validation sub-sets using the ratios 70%/15%/15% and 50%/25%/25% respectively. The numbers in the boxes in every bar of these figures' diagrams represent the corresponding number of the neurons in the hidden layer (optimum number of neurons in the hidden layer).From the study of Figs. 23.5 and 23.6 the next conclusions can be drawn:

- In general, the efficiency of the training procedures is significantly high because the maximum value of the MSE parameter does not exceed the value of 0.0158. Taking into consideration that the usual MIDR values in the case of significant earthquakes is about 0.5–1.5% (Moderate – Heavy damages, see Table 23.2), errors of this order of magnitude are practically negligible. Regarding the values of the R-factor they are in any case higher than 0.9775. This means that the MIDR values which are predicted by the optimum configured (and trained) networks have high degree of correlation with the corresponding values which are calculated performing NTHA.

- In any case the training utilizing the LM algorithm leads to more efficient networks. This conclusion regards the evaluation using the MSE parameter as well as the utilization of the R factor.
- The choice of activation function for the hidden layer's neurons is not so important parameter for the efficiency of the trained networks as is the choice of the training algorithm. The resultant percentages of differences in the values of the MSE parameter and the R-factor are generally lower than 2.7% in the case of utilization of the LM algorithm. The corresponding differences when the SCG algorithm is utilized are greater as regards the values of MSE parameters. However, these differences do not exceed the 6.7% with the exception of the MSE values which are extracted from the testing sub-set when the data set is partitioned to the three sub-sets using the ratio 70%/15%/15% (in this case the difference is 14.5%), (Fig. 23.5a).
- The partitioning ratio of the total data set to the three sub-sets (70%/15%/15% or 50%/25%/25%) does not change generally the above described conclusions. The main difference between these ratios is the reduction of differences between the MSE values which are extracted by the two used algorithms (see Figs. 23.5a and 23.6a).

In the Table 23.3 the best configured networks which are extracted from the above described parametric investigation procedure are summarized. These networks were used for the investigation of the generalization ability of ANNs (i.e. the quality of ANNs' predictions in the case of input data which are not included in the training data set).

In Fig. 23.7 the MIDR values, that were computed from NTHA, are plotted against the MIDR values predicted by the optimum configured ANNs of Table 23.3 (according to the minMSE criterion) for the samples of the total data set. From the study of diagrams of Fig. 23.7 we see that the vast majority of the data points sits very close to the straight diagonal reference line which is the line where the data points that fulfill the mathematic condition $MIDR_{NTHA} = MIDR_{ANN}$ are located. This means that the predicted by the networks MIDR values are in general not very different from the corresponding MIDR values that are computed using the NTHA.

This conclusion is also obvious from the study of the corresponding values of the R-factor. More specifically the R-factor's values are greater than 0.98 which means that the results which are extracted from the optimum configured networks are strongly related with the corresponding results of the NTHA.

23.5.2 Generalization Ability of the Optimum Configured Networks

In the following sub-section, the results of investigation of the optimum configured networks' (Table 23.3) generalization ability are presented and discussed. In order to check this ability 5 testing earthquakes different from the 65 ground motions which

Table 23.3 Configuration parameters of the best configured networks

	Criterion for the sub-set:		Partitioning ratio of the total data set	
			70%/15%/15%	50%/25%/25%
Number of neurons in the hidden layer	**Testing**	**minMSE**	34	46
		maxR	30	12
	Total	**minMSE**	44	42
		maxR	44	42
Training algorithm	**Testing**	**minMSE**	LM	LM
		maxR	LM	LM
	Total	**minMSE**	LM	LM
		maxR	LM	LM
Activation functions	**Testing**	**minMSE**	tansig/linear	logsig/linear
		maxR	tansig/linear	tansig/linear
	Total	**minMSE**	tansig/linear	logsig/linear
		maxR	tansig/linear	logsig/linear
ANN's name[a]	**Testing**	**minMSE**	LM/tan-34-MSE-pr70	LM/log-46-MSE-pr50
		maxR	LM/tan-30-R-pr70	LM/tan-12-R-pr50
	Total	**minMSE**	LM/tan-44-MSE-pr70	LM/log-42-MSE-pr50
		maxR	LM/tan-44-R-pr70	LM/log-42-R-pr50

[a]The best configured networks' names in the Table 23.3 are composed by the following parts: The used training algorithm ("LM" or "SCG") /the type of activation functions of the hidden layer's neurons ("tan" or "log") – the number of the hidden layer's neurons – the name of the used performance evaluation parameter ("MSE" or "R") – the used partitioning ratio of the total data set to the three sub-sets ("pr70" for partitioning ratio 70%/15%/15% or "pr50" for partitioning ratio 50%/25%/25%)

were utilized for the generation of the training data set, were selected. Moreover, 2 configurations of the masonry infills were taken into account. As in the case of the selected testing earthquakes, these selected configurations were different from the configurations that were utilized for the generation of the training data set. The studied building (Fig. 23.2) with the new configurations of masonry infills was analysed for the 5 testing earthquakes using NTHA. Thus, a testing data set consisting of 10 samples was generated.

Figure 23.8 presents the diagrams which illustrate the correlation degree between the predictions of MIDR values made by the best configured ANNs (Table 23.3), and the corresponding MIDR values which are calculated performing NTHA.

From the study of the Fig. 23.8 the following conclusions can be drawn:

(a) The best configured networks on the basis of ANNs' performance for the testing sub-set (Fig. 23.8a, c) export results which are in general well correlated with the corresponding results of NTHA. This conclusion holds independently of the used partitioning ratio of the total data set (using ratio 70%/15%/15% we

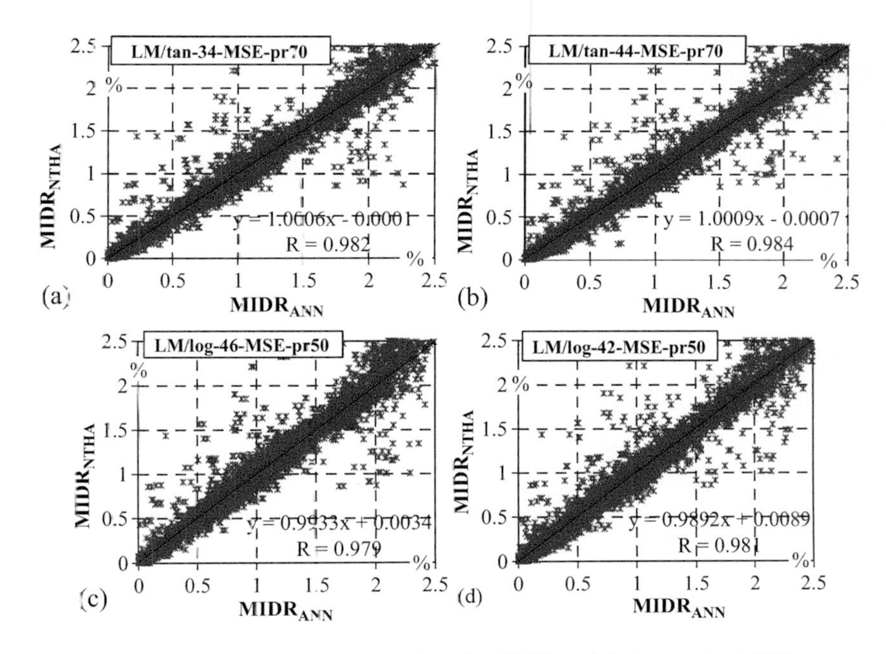

Fig. 23.7 Comparison of MIDR values predicted by NTHA and the best trained ANNs

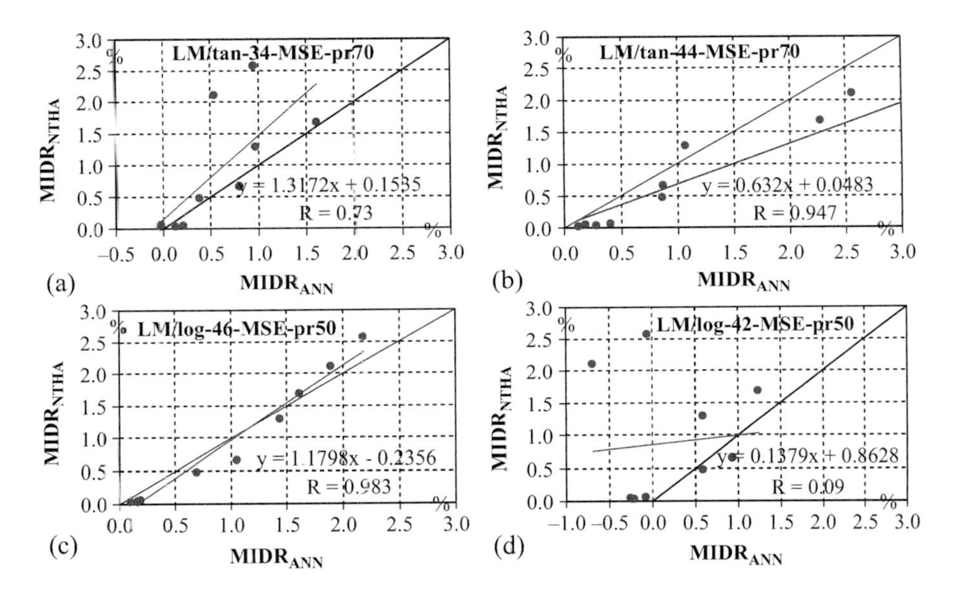

Fig. 23.8 Correlation degree between the MIDR values predicted by ANNs and by NTHA

exported results with R = 0.73 whereas using ratio 50%/25%/25% the corresponding value is 0.98). To the contrary, when the best configured networks are yielded using performance criterion on the basis of the total set the correlation degree is not always acceptable. Thus, in the case of partitioning of the total set using the ratio 50%/25%/25% ("LM/log-42-MSE-pr50" network – Fig. 23.8d) the correlation degree between ANN and NTHA results is absolutely unacceptable (R = 0.09).

(b) The "LM/log-46-MSE-pr50" network exports the best correlated results with the results of NTHA. This conclusion is due to the following facts:

  (i). The utilization of a criterion which is based on the testing sub-set leads to the best configured networks using samples of the total data set which are not used for the optimization of their synaptic weights' values. This means that the samples which are used for the assessment of the networks' performance are unknown to networks, and for this reason the utilization of the testing sub-set for the performance assessment of networks is the best choice for their generalization ability testing.

  (ii). Using the ratio 50%/25%/25% for the division of the total data set the number of samples of the testing sub-set is greater than the corresponding number when the ratio 70%/15%/15% is used. This means that the calculation of the performance parameter is based on a greater and wider data set. Thus, the reliability of the performance assessment as well as the validity of this assessment is greater. In other words, when the ratio 50%/25%/25% is used the testing of the generalization ability of networks is performed for a greater set of unknown samples, and for this reason a network which exports acceptable results for the corresponding testing data sub-set is expected to have better generalization ability than a network which exports high value for performance parameter on the basis on a testing sub-set which represents the 15% of the total data set.

23.6 Conclusions

In the present research, the influence of the infills' irregular distribution on the seismic damage level of R/C buildings using MFP ANNs with one hidden layer was investigated. The main conclusions that turned out are the following:

(a) The MFP networks with one hidden layer are capable to successfully approach the influence of eccentricity which is caused by the irregular distribution of masonry infills on the seismic response of R/C buildings. To this end, the networks must have as input parameters the ratios of the infills along two perpendicular axes as well as the eccentricities of floors due to the configuration of them along these axes.

(b) The LM training algorithm is more efficient than the SCG algorithm.

(c) The activation function for the hidden layer's neurons is not an important parameter as the choice of the training algorithm in this problem.

(d) The generalization ability of the networks depends on the data sub-set for which the evaluation parameter of ANNs is calculated, as well as on the ratio by which the total data set is partitioned to training, validation and testing sub-sets. More specifically, the best generalization is achieved by the network which is the best configured on the basis of the optimum value of the MSE that is computed for the samples of the testing sub-set. Moreover, the optimization of the generalization ability is achieved when the total data set is partitioned using the ratio 50%/25%/25%.

Finally, we must emphasize the usefulness of ANNs' utilization for the solution of the problem which was investigated in the present study, due to the capability of them to export instantly a reliable estimation of the influence of the selected masonry infills' configuration on the seismic response of new R/C buildings, in the framework of the design procedure. Thus, unfavorable configuration of masonry infills can be early avoided.

References

Bertero VV, Brokken S (1983) Infills in seismic resistant buildings. J Struct Eng ASCE 109 (6):1337–1361
Crisafulli FJ (1997) Seismic behaviour of reinforced concrete structures with masonry infills. Ph.D. Thesis, University of Canterbury, Christchurch
EC2 (Eurocode 2) (2005) Design of Concrete Structures, Part 1-1: General rules and rules for buildings. EN1992-1-1. Brussels, Belgium
EC8 (Eurocode 8) (2005) Design of structures for earthquake resistance – Part 1: General rules, seismic actions and rules for buildings. EN1998-1. Brussels, Belgium
EERI (2000) Kocaeli, Turkey earthquake reconnaissance report. Earthquake Spectra 16 (S1):237–379
European Strong – Motion Database (2003.): http://www.isesd.hi.is/ESD Local/frameset.htm
Haykin S (2009) Neural networks and learning machines, 3rd edn. Prentice Hall
Kramer SL (1996) Geotechnical earthquake engineering. Prentice-Hall
Lautour OR, Omenzetter P (2009) Prediction of seismic-induced structural damage using artificial neural networks. Eng Struct 31:600–606
Marquardt DW (1963) An algorithm for least squares estimation of non-linear parameters. J Soc Ind Appl Math 11(2):431–441
Masi A, Vona M, Mucciarelli M (2011) Selection of natural and synthetic accelerograms for seismic vulnerability studies on reinforced concrete frames. J Struct Eng 137:367–378
Matlab (2013) Neural networks toolbox user guide
Molas G, Yamazaki F (1995) Neural networks for quick earthquake damage estimation. Earthq Eng Struct Dyn 24:505–516
Moller MF (1993) A scaled conjugate gradient algorithm for fast supervised learning. Neural Netw 6:525–533
Naeim F (ed) (2001) The seismic design handbook. Kluwer Academic, Boston
Negro P, Colombo A (1997) Irregularities induced by nonstructural masonry panels in framed buildings. Eng Struct 19(7):576–585
Otani A (1974) Inelastic analysis of RC frame structures. J Struct Div ASCE 100(7):1433–1449
Palermo M, Hernandez RR, Mazzoni S, Trombetti T (2014) On the seismic behavior of a reinforced concrete building with masonry infills collapsed during the 2009 L'Aquila earthquake. Earthq Struct 6(1):45–69
Paulay T, Priestley MJN (1992) Seismic design of reinforced concrete and masonry buildings. Wiley, New York

PEER (Pacific Earthquake Engineering Research Centre) (2003): Strong Motion Database. http://peer.berkeley.edu/smcat/
Ricci P, De Luca F, Verderame GM (2010) 6th April 2009 L' Aquila earthquake, Italy: reinforced concrete building performance. Bull Earthq Eng 9(1):285–305
Rofooei FR, Kaveh A, Farahani FM (2011) Estimating the vulnerability of the concrete moment resisting frame structures using artificial neural networks. Int J Optim Civ Eng 3:433–448
Vafaei M, Adnan AB, Rahman ABA (2013) Real-time seismic damage detection of concrete shear walls using artificial neural networks. J Earthq Eng 17:137–154
Yuen YP, Kuang JS (2015) Nonlinear seismic response and lateral force transfer mechanisms of RC frames with different infill configurations. Eng Struct 91:125–140

Chapter 24
Dynamic Eccentricities in Pushover Analysis of Asymmetric Single-Storey Buildings

Athanasios P. Bakalis and Triantafyllos K. Makarios

Abstract A new version about a method of documented application of pushover analysis on reinforced concrete, asymmetric single-storey buildings has recently been presented. To rationally consider the coupling between torsional (about vertical axis) and translational vibrations, the equivalent lateral static floor force of the proposed pushover method is applied using suitable dynamic eccentricities which are added with the accidental ones in such a way that the final design eccentricities move the application point of the external, lateral, static, floor force away from the diaphragm Mass Centre. The appropriate dynamic eccentricities for the stiff and the flexible side of the floor plan are related to the so-called "Capable Near Collapse Principal Axes", reference to the Near Collapse limit state, derive from extensive parametric analysis, and are calculated by graphs and suitable equations. In the present work, a numerical example of a (double) asymmetric single-storey building is presented to clarify the step by step application of the proposed pushover analysis method. It is a torsional sensitive r/c building, designed in accordance with Eurocodes EN 1992-1 and 1998-1 for ductility class high (DCH). The proposed method is evaluated relative to the results of non-linear response history analysis (RHA). The final results show that the proposed method of pushover analysis predicts with safety the displacement of the stiff side of the building as well as that of the flexible side.

Keywords Capable near collapse Centre of stiffness of the building · Stiffness eccentricity · Strength eccentricity · Dynamic eccentricity · Torsionally-flexible building · Nonlinear static analysis · Pushover analysis · Response history analysis · Capable near collapse principle horizontal axes of the building

---

A. P. Bakalis · T. K. Makarios (✉)
Institute of Structural Analysis and Dynamics of Structure, School of Civil Engineering, Aristotle University of Thessaloniki, Thessaloniki, Greece
e-mail: abakalis@civil.auth.gr; makariostr@civil.auth.gr

D. Köber et al. (eds.), *Seismic Behaviour and Design of Irregular and Complex Civil Structures III*, Geotechnical, Geological and Earthquake Engineering 48,
https://doi.org/10.1007/978-3-030-33532-8_24

24.1 Introduction

To seismically assess the seismic bearing capacity of a building, the basic method of analysis proposed by all contemporary seismic codes is the static non-linear method of analysis (pushover). For the documented application of pushover analysis on asymmetric single-storey buildings, the lateral static floor load should be applied with eccentricity relative to the Mass Centre of the diaphragm and should also be properly oriented along each horizontal principal axis of the building (Bakalis and Makarios 2017, 2018). This is the only way to consider the phenomenon of the development of diaphragm torsional vibrations (about vertical axis) coupled to translational ones, due to the developing inertial torsional (about Z-axis) moment of the building floor, a phenomenon occurring both in linear and in non-linear areas (Bakalis and Makarios 2017, 2018; Bosco et al. 2012, 2015, 2017). Eurocode EN 1998-1 suggests that the position of the Mass Centre (where the external lateral static floor force is applying) must be moved by the floor accidental eccentricity, but in this way, there is still the inability to obtain the inertial torsional moment of the diaphragm. Both in recent work (Bakalis and Makarios 2017, 2018) and the current one, where a numerical example is presented, the use of suitable dynamic eccentricities is proposed, in order to determine the point of application of the lateral static floor force in the plan. The calculation of, first, the starting point for the measurement of the dynamic eccentricities, second, the appropriate orientation of the lateral static floor force and third, the magnitude of the dynamic eccentricities is the threefold objective of the present paper, where the theoretical analysis has been given in (Bakalis and Makarios 2018).

24.2 Methodology

The proposed pushover method of analysis aims at the Near Collapse (NC) limit state using the Displacement-Based concept. Thus, for a given elasto-plastic capacity curve of an r/c building, we are moving directly at the end of the plastic branch (NC), and there, we calculate the building Stiffness Centre, using the lateral strength (secant stiffness at yield) of the structural elements. Indeed, according to EN 1998-3, all building elements have to be supplied with their secant stiffness (EI_{sec}) at yield. That is, it is assumed that all element end-sections have simultaneously yielded, as this happens in an extreme possible (fictitious) Near Collapse state. In this fictitious state, the Centre of Stiffness CR_{sec} (at the last step where all element end sections have been yielded) is calculated and we call it "Capable Near Collapse Centre of Stiffness". It is worth noting that the abovementioned concept is an alternative procedure to bypass the use of the displacement ductility factor (or reduction factor R for large periods). The orientation of the "Capable Near Collapse Principal Axes" I_{sec} and II_{sec} resulting from the above model and the "Capable Near Collapse Torsional Radii" $r_{I,\ sec}$ and $r_{II,\ sec}$ in respect to the I_{sec} and II_{sec} axes are also determined. All the above-mentioned properties are calculated as a classic case on an asymmetric single-storey building in the linear area (Makarios and Anastassiadis 1998a, b; Makarios 2008).

From the extensive parametric investigation (Bakalis and Makarios 2018) on asymmetric single-storey r/c buildings as well as by the recently international literature review (Bosco et al. 2012, 2015, 2017), the main conclusions and the proposed methodology are summarized below:

1. The most appropriate point (independent of the loading) that must be the starting point for measuring the new dynamic eccentricities is the "Capable Near Collapse Centre of Stiffness" CR_{sec} resulting from the model whose all structural elements have been provided with their secant stiffness (EI_{sec}) in the yield state,
2. The most appropriate orientation of the horizontal static floor force of the pushover method is along the directions of the horizontal "Capable Near Collapse Principal Axes" I_{sec} and II_{sec} resulting for the above model of conclusion (1),
3. The control of the building torsional sensitivity must be performed in the above model of conclusion (1). The asymmetric single-storey buildings were divided into two categories: (a) buildings with torsional sensitivity when $r_{I,sec}$ or $r_{II,sec} \leq 1.10\ r_m$ applies and (b) buildings without torsional sensitivity when $r_{I,sec}$ and $r_{II,sec} > 1.10\ r_m$ applies, where $r_{I,sec}$ and $r_{II,sec}$ are the "Capable Near Collapse Torsional Radii" in respect to the I_{sec} and II_{sec} axes respectively, and r_m is the radius of gyration of the diaphragm, from the relation $r_m = \sqrt{J_m/m}$,
4. The magnitude of the appropriate dynamic eccentricities e_{stiff} and e_{flex}, for the building stiff and flexible side respectively, along each horizontal "Capable Near Collapse Principal Axis" I_{sec} or II_{sec}, has been determined from a statistical processing on the results of extended parametric analysis and is given through graphs (Figs. 24.1 and 24.2) and equations (Eqs. 24.1, 24.2, 24.3 and 24.4, prediction lines with a suitable standard deviation):

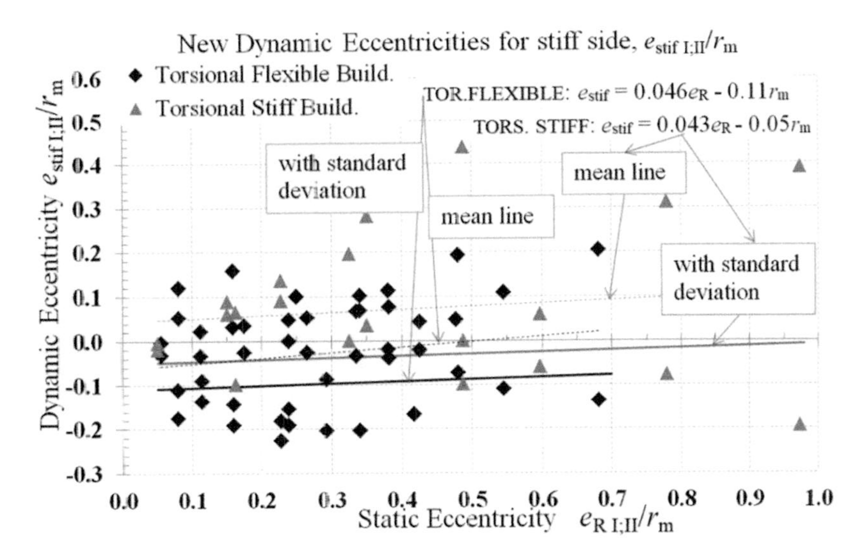

Fig. 24.1 Normalized dynamic eccentricities $e_{stif\ I;II}/r_m$ for the stiff side of plan

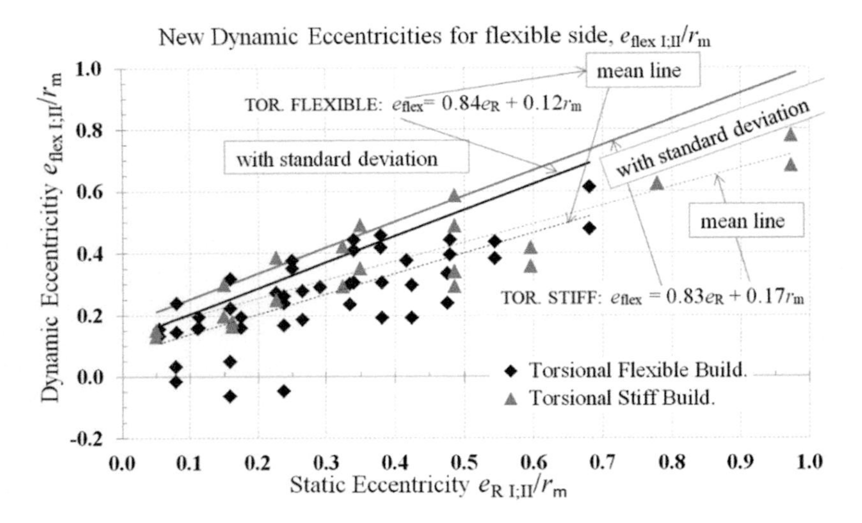

Fig. 24.2 Normalized dynamic eccentricities $e_{flex\ I;II}/r_m$ for the flexible side of plan

For buildings with Torsional Sensitivity, $r_{I,sec}$ or $r_{II,sec} \leq 1.10\ r_m$:

$$e_{stiff} = 0.046 \cdot e_R - 0.11 \cdot r_m \tag{24.1}$$

$$e_{flex} = 0.84 \cdot e_R + 0.12 \cdot r_m \tag{24.2}$$

For buildings without Torsional Sensitivity, $r_{I,sec}$ and $r_{II,sec} > 1.10\ r_m$:

$$e_{stiff} = 0.043 \cdot e_R - 0.05 \cdot r_m \tag{24.3}$$

$$e_{flex} = 0.83 \cdot e_R + 0.17 \cdot r_m \tag{24.4}$$

In Eqs. 24.1, 24.2, 24.3 and 24.4, e_R is the static (or stiffness) eccentricity, regarding the CR_{sec} position in the plan, relative to the examined horizontal axis I_{sec} or II_{sec},

5. The proposed method for the documented application of pushover analysis on asymmetric single-storey r/c buildings uses the design eccentricities. Noting that the design eccentricities e_1, e_2 (Eqs. 24.5 and 24.6) and e_3, e_4 (Eqs. 24.7 and 24.8) are used for loading along the "Capable Near Collapse Principal Axes" II_{sec} and I_{sec} respectively, all measured from the "Capable Near Collapse Centre of Stiffness" CR_{sec} and with positive direction towards the Mass Centre CM (see Fig. 24.4). The calculation of the design eccentricities is achieved by the simultaneous reception of the new dynamic eccentricities resulting from Eqs. 24.1, 24.2, 24.3 and 24.4 and of the floor accidental eccentricities e_a obtained in such a

way that the final position of the horizontal static floor force to be more eccentric relative to the Mass Centre CM:

$$e_1 = e_{\mathrm{flex},I} + e_{\mathrm{a},I} \tag{24.5}$$

$$e_2 = e_{\mathrm{stiff},I} - e_{\mathrm{a},I} \tag{24.6}$$

$$e_3 = e_{\mathrm{flex},II} + e_{\mathrm{a},II} \tag{24.7}$$

$$e_4 = e_{\mathrm{stiff},II} - e_{\mathrm{a},II} \tag{24.8}$$

In Eqs. 24.5, 24.6, 24.7 and 24.8, e_{stiff} and e_{flex} are the dynamic eccentricities from Eqs. 24.1, 24.2, 24.3 and 24.4, while the accidental eccentricities are determined from the equations "$e_{\mathrm{a,\,I\ or\ II}} = \pm\,(0.05 \sim 0.10) \cdot L_{\mathrm{I\ or\ II}}$" according to EN1998-1, where L_I and L_{II} are the maximum floor plan dimension normal to the loading direction.

6. Considering the two signs (±) of application of the lateral static floor forces, eight separate pushover analyses of the building are performed. The final results from the spatial action of the earthquake are computed as the general concept of Eurocode EN 1998-1, i.e. by the SRSS combinations on the results of the eight separate pushover analyses (sixteen loading combinations). These combinations are carried out in that step of the separate pushover analyses where the "seismic target-displacement" at the lateral load application point (and not at the Mass Centre CM of diaphragm, because at the CM the Principle of the Virtual Work is not true) is achieved, and from the sixteen combinations the envelope is taken.

24.3 Numerical Example

In this section, an example of a double-asymmetric single-storey r/c building is presented to illustrate clearly and in detail the application of the proposed pushover analysis method using design eccentricities and also for evaluation purposes.

24.3.1 Building Characteristics

Consider the asymmetric single-storey r/c building of Fig. 24.3 with construction materials C25/30 for the concrete and B500c for the steel of average strengths $f_{\mathrm{cm}} = 33$ MPa and $f_{\mathrm{ym}} = 550$ MPa respectively. The elastic and inertial characteristics of its non-linear model (in which all structural elements are supplied with their secant stiffness EI_{sec} at their yield state) are also presented. The building is characterized as torsional sensitive ($r_{\mathrm{I,sec}}/r_{\mathrm{m}} = 0.94 < 1.10$). The height of the buildings is 3 m.

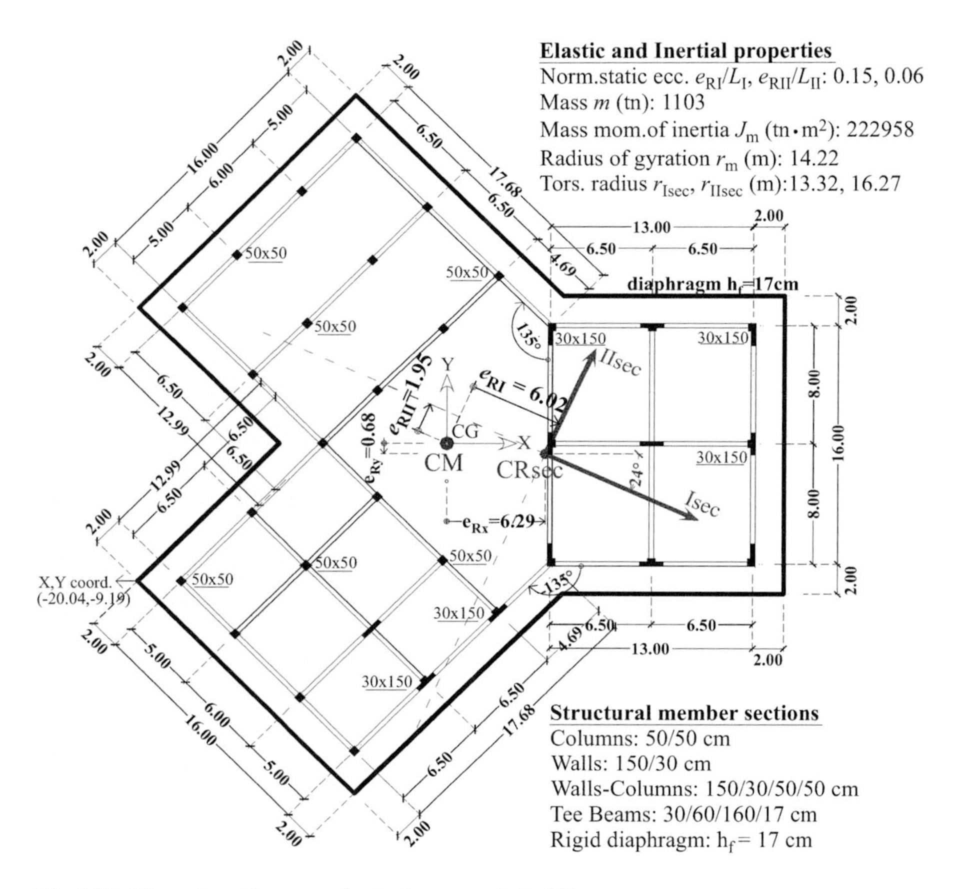

Fig. 24.3 Floor plan of asymmetric single-storey r/c building

24.3.2 Building Design

The design is performed according to the provisions of Eurocodes EN1992-1 and EN1998-1. The building system is characterized as wall-equivalent dual according to EN1998-1 (§5.1.2). The design model of the building is also torsional sensitive ($r_{I,des}/r_m = 0.96$). The building has an importance factor $\gamma_1 = 1$ and is designed for effective peak ground acceleration $\alpha_g = 0.24$ g, soil D, ductility class high (DCH) and total behavior factor $q = 3$. The details of the longitudinal and confinement steel reinforcements can be found in authors.

24.3.3 Non-linear Model

For the application of non-linear analyses, the elements of the building model are provided with the secant moments of inertia I_{sec} (at their yield). The secant stiffness

EI_{sec} is taken as a constant value over the entire length of each structural element, is equal to the numerical average of the EI_{sec} values of its two end cross-sections for positive and negative bending and is calculated by the following equation of EN 1998-3 (informative Annex A):

$$EI_{sec} = \frac{M_y}{\theta_y} \cdot \frac{L_V}{3} \tag{24.9}$$

For the determination of chord rotation at yield θ_y, the equations (A.10b) and (A.11b) of EN1998-3 are used for columns-beams and walls respectively. The curvature at yield φ_y and the yield moment M_y, at the end-sections of structural elements, are determined by the module Section Designer of the analysis program SAP2000 (Computers and Structures 2013). The unconfined and confined model for the concrete follows the constitutive relationship of the uniaxial model proposed by *Mander* et al. (1988), while the steel reinforcement is represented by the simple (parabolic at strain-hardening region) model of SAP2000. The axial force of the vertical resisting elements, that is used for the calculation of φ_y, is determined from the (seismic) combination $G + 0.3Q$, where G is the permanent and Q is the live vertical load, respectively. The shear span L_V was assumed equal to the half clear length L_{cl} of the structural elements along the frame bending planes except the strong direction of walls and the direction of columns with cantilever bending where it was considered equal to L_{cl}. The secant stiffness EI_{sec} at yield is calculated by Eq. 24.9 as percentage of the geometric stiffness EI_g. It is equal to an average of 11% for columns along frame bending planes, 17% for columns along cantilever bending planes, 11% for the strong wall direction, 12% for the weak wall direction and 10% for the beams, where the average modulus E_{cm} of concrete C25/30 was considered equal to 31 GPa. For the determination of the plastic capacity θ_{pl} in terms of chord rotations, the analysis software uses the relation $\theta_{pl} = (\varphi_u - \varphi_y) \cdot L_{pl}$, where L_{pl} is calculated from equation (A.9) of EN 1998-3. In the non-linear analysis model, point plastic hinges were inserted at each end-section of all structural elements. P-M_2-M_3 hinges and M_3 hinges are used for vertical elements and beams respectively.

The asymmetric singe-storey r/c building has static eccentricities $e_{RI} = 6.02$ and $e_{RII} = 1.95$ m (Makarios and Anastassiadis 1998a, b) along the horizontal axes I_{sec} and II_{sec} (position of the CR$_{sec}$ relative to CM) which are rotated by -24° relative to the Cartesian x, y axes (Makarios and Anastassiadis 1998a, b) (Fig. 24.3) and the building is characterized as torsional sensitive since $r_{I,sec}/r_m = 0.94 < 1.10$ applies (Makarios 2008), where the torsional radius $r_{I,sec}$ refers to CR$_{sec}$. The periods of the three coupled modes are $T_1 = 0.363$ sec, $T_2 = 0.326$ sec and $T_3 = 0.226$ sec. Also, the three uncoupled modes have periods $T_1 = 0.334$ sec, $T_2 = 0.292$ sec and $T_3 = 0.274$ sec with the second one being torsional.

In the non-linear model, the accidental eccentricities along the I_{sec} and II_{sec} axes are taken with a value equal to 5% of the maximum plan dimension (Fig. 24.4, $L_{I,sec} = 40.33$ m and $L_{II,sec} = 31.72$ m) normal to the loading direction:

$$e_{a,I} = \pm 0.05 \cdot L_I = \pm 0.05 \cdot 40.33 = \pm 2.02 \text{ m} \quad (24.10)$$

$$e_{a,II} = \pm 0.05 \cdot L_{II} = \pm 0.05 \cdot 31.72 = \pm 1.59 \text{ m} \quad (24.11)$$

24.4 Calculation of Dynamic and Design Eccentricities

The calculation of the dynamic eccentricities e_{stif} and $e_{flex,}$ along the horizontal "Capable Near Collapse Principal Axes" I_{sec} and II_{sec} as well as of the design eccentricities e_1, e_2 and e_3, e_4 along the I_{sec} and II_{sec} axes respectively, which are used for the application of the proposed pushover analysis method on the asymmetric single-storey building (see Fig. 24.4), is performed step by step as follows:

- Stiffness eccentricity (CR_{sec}): $e_{R_{Isec}} = 6.02$ and $e_{R_{IIsec}} = 1.95$ m
- Storey Mass: $m = 1103$ tn
- Mass moment of inertia: $J_m = 222958 \text{ tn} \cdot \text{m}^2$
- Radius of gyration: $r_m = \sqrt{J_m/m} = \sqrt{222958/1103} = 14.22$ m
- Min torsional radius: $r_{I,sec} = 13.32$ m
- Torsional Sensitivity: $r_{I,sec}/r_m = 0.94 < 1.10 \rightarrow$ Torsional sensitive
- Accidental Eccentricity (Eqs. 24.10 and 24.11): $e_{a,Isec} = 2.02$ m and $e_{a,IIsec} = 1.59$ m
- Dynamic Eccentricities (Eqs. 24.1 and 24.2):

$$e_{stif,Isec} = 0.046 \cdot e_{RI,sec} - 0.11 \cdot r_m = 0.046 \cdot 6.02 - 0.11 \cdot 14.22 = -1.29 \text{ m}$$

$$e_{stif,IIsec} = 0.046 \cdot e_{RII,sec} - 0.11 \cdot r_m = 0.046 \cdot 1.95 - 0.11 \cdot 14.22 = -1.47 \text{ m}$$

$$e_{flex,Isec} = 0.84 \cdot e_{RI,sec} + 0.12 \cdot r_m = 0.84 \cdot 6.02 + 0.12 \cdot 14.22 = 6.76 \text{ m}$$

$$e_{flex,IIsec} = 0.84 \cdot e_{RII,sec} + 0.12 \cdot r_m = 0.84 \cdot 1.95 + 0.12 \cdot 14.22 = 3.35 \text{ m}$$

- Design Eccentricities (Eqs. 24.5, 24.6, 24.7 and 24.8) measured from CR_{sec}. Signs (+) and (−) means towards the flexible side and stiff one respectively:

$$e_1 = e_{flex,Isec} + e_{a,Isec} = 6.76 + 2.02 = 8.78 \text{ m along } I_{sec} \text{ axis}$$

$$e_2 = e_{stiff,Isec} - e_{a,Isec} = -1.29 - 2.02 = -3.31\text{m along } I_{sec} \text{ axis}$$

$$e_3 = e_{flex,IIsec} + e_{a,IIsec} = 3.35 + 1.59 = 4.93\text{m along } II_{sec} \text{ axis}$$

$$e_4 = e_{stiff,IIsec} - e_{a,IIsec} = -1.47 - 1.59 = -3.06 \text{ m along } II_{sec} \text{ axis}$$

24.5 Seismic Assessment

In this section, the proposed method of documented application of pushover analysis on the asymmetric single-storey r/c building is described in detail. Other pushover methods are also referred for evaluation purposes. All results by pushover analysis compare with the seismic demand ones (target displacement) by nonlinear response history analysis (RHA).

24.5.1 Proposed Method of Pushover Analysis

According to the proposed method of pushover analysis, the procedure to be performed is shown in Fig. 24.4. In this figure the proposed methodology is appropriately formulated and performed as described below:

1) The appropriate design eccentricities along each horizontal "Capable Near Collapse Principal Axis" I_{sec} or II_{sec} are calculated (see Sect. 24.4). The design

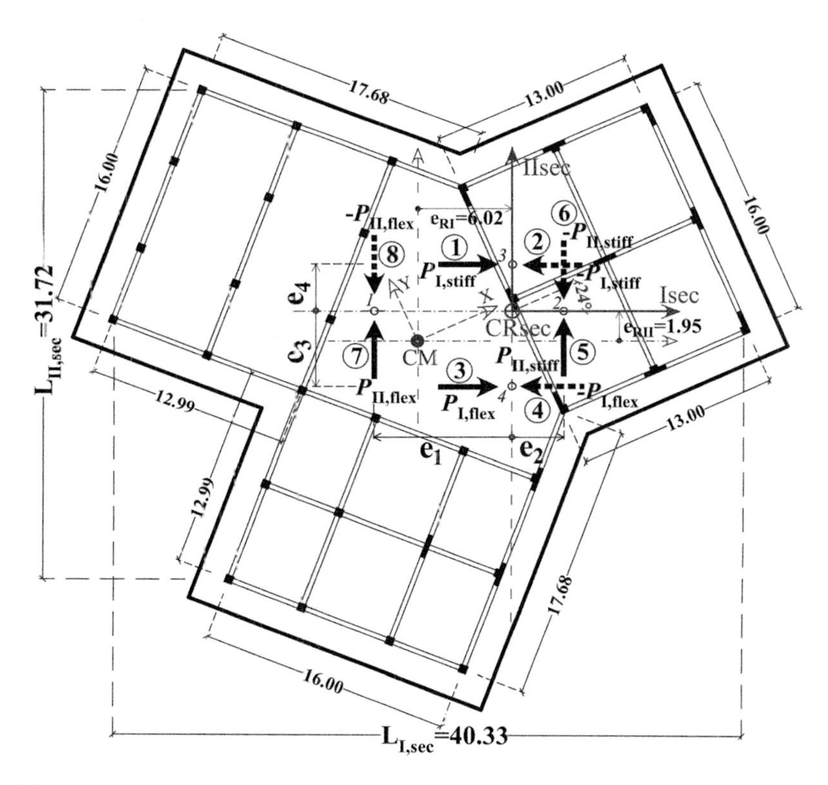

Fig. 24.4 Proposed method of pushover analysis on the asymmetric single-storey r/c building

eccentricities e_1, e_2 and e_3, e_4 are used for the positioning of the lateral static floor force along the II_{sec} and I_{sec} axes respectively,

2) In total, for both horizontal directions I_{sec} and II_{sec}, eight pushover analyses are performed considering the two signs (+, −) of application of the lateral static floor loads,
3) The displacement results along the horizontal "Capable Near Collapse Principal Axes" I_{sec} and II_{sec} of the eight separate pushover analyses are combined with the SRSS rule, in that step of the analyses where the seismic target-displacement at the lateral load application point is achieved, and from the sixteen combinations the envelope is taken.

For comparison purposes, the pushover analysis according to EN 1998-1, i.e. by applying the lateral static floor force on the position of the Mass Centre moved by the floor accidental eccentricity, is also executed. It is worthy note that the locations in the plan of the proposed pushover method lateral static forces are in fully disagreement with Eurocode EN 1998-1. Also, by the recently international literature using a fully different methodology, the pushover analysis according to the "corrective eccentricity method" by *Bosco* et al. (2017) is executed, where the corrective eccentricities e_{Isec} and e_{IIsec} for the building stiff sides (for loading along the horizontal axes II_{sec} and I_{sec} respectively) have been calculated, plus the accidental eccentricities. According to the method, for the flexible sides, only the accidental eccentricities are used. Thus, our investigative results are very compatible with *Bosco*'s ones.

In all the above-mentioned pushover methods (load numbering is the same for all methods as in Fig. 24.4) the sixteen SRSS combinations of the eight separate pushover analyses, from which the envelope of the displacement results along the horizontal axes I_{sec} and II_{sec} is calculated, are as follows: (1) $\oplus$ (5), (1) $\oplus$ (6), (1) $\oplus$ (7), (1) $\oplus$ (8) and (2) $\oplus$ (5), (2) $\oplus$ (6), (2) $\oplus$ (7), (2) $\oplus$ (8) and (3) $\oplus$ (5), (3) $\oplus$ (6), (3) $\oplus$ (7), (3) $\oplus$ (8) and (4) $\oplus$ (5), (4) $\oplus$ (6), (4) $\oplus$ (7), (4) $\oplus$ (8).

Finally, the extended N2 method of pushover analysis (Fajfar et al. 2005) is also considered. According to this method, the envelope of the results (deformations, displacements, stress) of the separate EN 1998-1 pushover analyses along the (±) examined direction is corrected by (amplification) factors determined with the application of a response spectrum analysis on the 3D building model ignoring any de-amplification due to torsion.

24.5.2 Non-linear Response History Analysis

In the context of the current work, the "seismic target-displacement" is calculated by performing non-linear response history analysis (RHA). According to EN 1998-1, the RHA is performed in a (non-linear) model resulting from the simultaneous movement of the Mass Centre CM by each accidental eccentricity (Eqs. 24.10 and 24.11) along the horizontal "Capable Near Collapse Principal Axes" I_{sec} and II_{sec}

which constitute the appropriate principal directions. Of the four sign combinations of the two accidental eccentricities $e_{a,I}$ and $e_{a,II}$, four displaced CM positions (models) are defined. Three pairs of horizontal accelerograms (Bakalis and Makarios 2018) consisting of five artificial accelerograms (created by SeismoArtif (Seismosoft 2016)) are used that have similar characteristics with the Hellenic tectonic faults (Makarios 2015) and are scaled to the Near Collapse state of the building. Each pair is rotated about the vertical axis successively per 22.5°, to find the worst seismic load state (Athanatopoulou and Doudoumis 2008) of the 16 RHA obtained for each pair. Finally, the envelope of the displacements along the I_{sec} and II_{sec} axes from all these RHA (192) was the "seismic target-displacement" for each control position in plan.

24.6 Results of Non-linear Analysis Methods

The seismic inelastic displacements of the building along the "Capable Near Collapse Principal Axes" I_{sec} and II_{sec} resulting from the non-linear analysis methods of Sects. 24.5.1 and 24.5.2 are presented here for comparison purposes. The results are illustrated in Fig. 24.5 in terms of plan inelastic displacement profile.

We observe that, relative to the displacement results from RHA (seismic target-displacement), the displacement of the stiff side $u_{II,sec}$ along the horizontal II_{sec} axis resulted from all pushover method of analysis is predicted with safety. However, the displacement of the stiff side $u_{I,sec}$ along the horizontal I_{sec} axis resulted from the pushover analysis according to EN 1998-1 is lower by 12%. Similarly, both the displacements $u_{I,sec}$ and $u_{II,sec}$ of the flexible sides along the horizontal axes I_{sec} and II_{sec} are a little lower by 4% and 2% respectively. We also notice that the extended N2 method gives conservative results for the displacements $u_{I,sec}$ of the stiff side

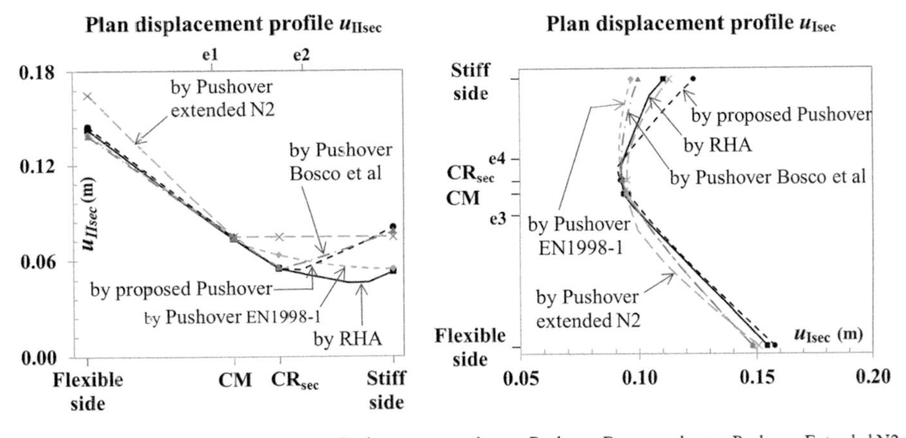

Fig. 24.5 Plan inelastic displacement profile along the "Capable Near Collapse Principal Axes" II_{sec} (left) and I_{sec}(right) resulted from the non-linear methods of analysis

along I_{sec} axis and $u_{II,sec}$ throughout the stiff side along II_{sec} axis. However, the displacement $u_{I,sec}$ throughout the flexible side along I_{sec} axis is lower by about 3%. Further, we observe that the proposed method of pushover analysis provides the displacement of the stiff side $u_{I,sec}$ along the horizontal I_{sec} axis with a safety margin of 11% and the displacements $u_{I,sec}$ and $u_{II,sec}$ of the flexible sides along the horizontal axes I_{sec} and II_{sec} also with a safety margin 2% and 1% respectively.

It is worthy noted that, relative to the displacement results from RHA, the "corrective eccentricity method" of pushover analysis gives non-conservative results by 10% for the displacement $u_{I,sec}$ of the stiff side along the I_{sec} axis. Similar results, such as the EN1998-1 pushover method, apply to the displacements $u_{I,sec}$ and $u_{II,sec}$ of the flexible sides along the I_{sec} and II_{sec} axes, i.e. lower by 4% and 2%, respectively. This is since only the accidental eccentricity is used for loading in order to predict the flexible side displacements.

24.7 Conclusions

In the current work, a proposed method of documented application of pushover analysis on asymmetric single-storey r/c buildings has been presented in detail. To clarify and evaluate the method, a single-storey r/c building has been assessed. The building is double-asymmetric and also is torsional sensitive. For the application of the method, the non-linear model of the building has been formed, in which the structural elements have been provided with their secant stiffness EI_{sec} (in their yield state). Then, the following have been calculated: (a) the plan floor position of the "Capable Near Collapse Centre of Stiffness" CR_{sec}, (b) the orientation of the horizontal "Capable Near Collapse Principal Axes" I_{sec} and II_{sec}, (c) the "Capable Near Collapse Torsional Radii" $r_{I,sec}$ and $r_{II,sec}$ relative to the horizontal axes I_{sec} and II_{sec}, and (d) the torsional sensitivity of the model according to the relationship $r_{I,sec}$ or $r_{II,sec} \leq 1.10 r_m$. Finally, using the previous data, the dynamic eccentricities plus the accidental ones, namely the design eccentricities, have been calculated from Eqs. 24.1, 24.2, 24.3, 24.4 and 24.5, 24.6, 24.7, 24.8 respectively. The process of applying the proposed method of pushover analysis, using the design eccentricities e_1, e_2 and e_3, e_4 along I_{sec} and II_{sec} axes respectively, has been illustrated in detail in Fig. 24.4. In the context of this work, the seismic target-displacement of each control point of the diaphragm has been calculated by RHA. The floor plan inelastic displacement profile along each horizontal axis I_{sec} and II_{sec} resulting from RHA has been compared with the corresponding ones from the proposed method of pushover analysis, from pushover analysis according to EN 1998-1, from the extended N2 pushover method (Fajfar et al. 2005) and from the pushover that use the "corrective eccentricities" (Bosco et al. 2017). The main conclusions are the following:

1) The pushover analysis method according to EN 1998-1 provides non-conservative results (by 12%) for the displacement of the building stiff

side along the horizontal I_{sec} axis. Also, the displacements of the flexible sides along both the I_{sec} and II_{sec} axes are a little lower. This is due to wrong location in the plan of the lateral static floor forces during the pushover procedure. On the contrary, the extended N2 pushover method gives in general very conservative results except for the flexible part of the plan along I_{sec} axis.

2) The application of the proposed method of pushover analysis (Fig. 24.4) predicts with safety the displacement of the building stiff side along the horizontal I_{sec} axis (by 11%) as well as the displacements of the building flexible sides along the horizontal axes I_{sec} and II_{sec} (by 1% and 2% respectively).
3) The "corrective eccentricity method" that have been proposed by *Bosco* et al. (2017), in which the floor lateral loading is applying with less eccentricity than the design eccentricity of the proposed pushover method, gives non-conservative results (by 10%) for the displacement of the stiff side along the I_{sec} axis. Also, the displacements of the flexible sides along both the horizontal axes I_{sec} and II_{sec} remain a little lower, as for EN1998-1.

Therefore, the proposed method of documented application of pushover analysis on asymmetric single-storey r/c buildings, using suitable design eccentricities, is a rational way to predict with safety the (real) coupling between the torsional vibrations (about vertical axis) with the translational ones under pure translational seismic excitation of building base, especially as regards the displacements of the stiff sides of the building. The proposed pushover method of the present paper is more accurate than the pushover method that uses the "corrective eccentricities"; however, the later method that is based on the corrective eccentricities certainly drives in compatible results with our parametric analysis. On the contrary, the locations in the plan of the lateral static floor forces into the frame of EN 1998-1 pushover analysis are fully inadequate but the extended N2 method gives in general conservative results.

References

Athanatopoulou A, Doudoumis I (2008) Principal directions under lateral loading in multistory asymmetric buildings. Struct Design Tall Spec Build 17(4):773–794

Bakalis A, Makarios T (2017) Dynamic eccentricities in pushover analysis of asymmetric single-storey buildings. In: Proccedings of eighth European workshop on the seismic behaviour of irregular and compex structures, Bucharest, Romania, 19–20 October 2017

Bakalis A, Makarios T (2018) Dynamic eccentricities and the "capable near collapse centre of stiffness" of r/c single-storey buildings in pushover analysis. Eng Struct 166:62–78. https://doi.org/10.1016/j.engstruct.2018.03.056

Bosco M, Ghersi A, Marino EM (2012) Corrective eccentricities for assessment by the nonlinear static method of 3D structures subjected to bidirectional ground motions. Earthq Eng Struct Dyn 41:1751–1773. https://doi.org/10.1002/eqe.2155

Bosco M, Ferrara GAF, Ghersi A, Marino EM, Rossi PP (2015) Predicting displacement demand of multi-storey asymmetric buildings by nonlinear static analysis and corrective eccentricities. Eng Struct 99:373–387. https://doi.org/10.1016/j.engstruct.2015.05.006

Bosco M, Ghersi A, Marino EM, Rossi PP (2017) Generalized corrective eccentricities for nonlinear static analysis of buildings with framed or braced structure. Bull Earthq Eng. https://doi.org/10.1007/s10518-017-0159-x

Computers and Structures (2013) SAP2000 v.16.0 – A structural analysis program

Fajfar P, Marusic D, Perus I (2005) Torsional effects in the pushover-based seismic analysis of buildings. J Earthq Eng 9(6):831–854

Makarios T (2008) Practical calculation of the torsional stiffness radius of multistorey tall buildings. J Struct Des Tall Spec Build 17(1):39–65

Makarios T (2015) Design characteristic value of the arias intensity magnitude for artificial accelerograms compatible with Hellenic seismic hazard zones. Int J Innov Res Adv Eng 2 (1):87–98

Makarios T, Anastassiadis K (1998a) Real and fictitious elastic Axis of multi-storey buildings: theory. Struct Des Tall Build 7(1):33–55

Makarios T, Anastassiadis K (1998b) Real and fictitious elastic Axis of multi-storey buildings: applications. Struct Des Tall Build 7(1):57–71

Mander JB, Priestley MJN, Park R (1988) Theoretical stress-strain model for confined concrete. J Struct Eng ASCE 114(8):1827–1849

Seismosoft (2016) SeismoArtif – A computer program for generating artificial earthquake accelerograms matched to a specific target response spectrum. www.seismosoft.com

Chapter 25
Suggestions for Optimal Seismic Design of Wall-Frame Concrete Structures

George K. Georgoussis

Abstract A seismic design methodology is described aiming to provide an insight into the parameters which mitigate the torsional response. The proposed methodology applies to wall-frame dual concrete systems and incorporates the concept that stiffness and strength are interrelated in concrete structures. In practice two conditions should be satisfied in order to obtain a virtually translational behavior: (i) the initially elastic response should be of minimum torsion and, (ii) the strength allocations in the lateral force resisting elements (LFRE) should be based on a static analysis of the symmetric counterpart structure. The first condition is satisfied when the stiffness centre (m-CR) of an equivalent single story system lies close to the mass axis of the real building. The second condition requires a static analysis under a lateral loading simulating the first mode of vibration of the uncoupled structure. As a planar static analysis is finally required, this enables the designer to assess the element flexural rigidities in relation to the assigned strengths to LFRE's. Only an estimate of the design shear is required at this stage, which may be assessed by experience or by means of Yield Point Spectra (YPS). The proposed methodology is presented in ten story dual concrete buildings, under the seismic excitation of Erzincan-NS 1992, where the base shear assigned to the frame sub-system varies from 20% to 45% of the total design shear.

Keywords Earthquake engineering · Inelastic concrete structures · Strength dependent stiffness · Asymmetric buildings · Modal center of rigidity · Yield point spectra · Torsional response

---

G. K. Georgoussis (✉)
Department of Civil Engineering Educators, School of Pedagogical and Technological Education (ASPETE), Attica, Greece
e-mail: ggeorgo@tee.gr

D. Köber et al. (eds.), *Seismic Behaviour and Design of Irregular and Complex Civil Structures III*, Geotechnical, Geological and Earthquake Engineering 48,
https://doi.org/10.1007/978-3-030-33532-8_25

25.1 Introduction

The current philosophy of seismic design of low to medium height building structures is based on the fundamental period of the structure and the acceleration spectrum of the country code. This is a force-based method for seismic design, based on the assumption that the period of vibration can be calculated from the dimensions of the building, independently of the strength of the concrete elements. This procedure is implemented by constructing inelastic spectra, which are obtained by dividing the elastic acceleration spectrum by a somewhat arbitrary reduction factor. In most seismic codes this factor depends on the type of the structure (e.g. moment resisting frames, shear walls, coupled walls) and the specified material damping ratio. The assumption of strength independent flexural stiffness of R.C. elements however, has been questioned in the last two decades after the work of Priestley and Kowalsky (1998), Priestley (2000), Paulay (2002, 2003), Priestley et al. (2007), etc., who demonstrated that the yield curvature of R.C. members depends on the depth of the member cross-sections and the yield strain of the reinforcing steel, but not on the amount of the reinforcement. The proposed equations for assessing yield curvatures of R.C. members are as follows

$$\Phi_{cy} = k_c \varepsilon_y / d_c \quad \text{for rectangular columns} \tag{25.1a}$$

$$\Phi_{wy} = k_w \varepsilon_y / d_w \quad \text{for walls} \tag{25.1b}$$

$$\Phi_{by} = k_b \varepsilon_y / d_b \quad \text{for beams} \tag{25.1c}$$

where ε_y is the yield strain of the longitudinal steel and, d_c, d_w and d_b are the depths of the column, wall and beam sections respectively. The coefficients k_c, k_w and K_b may be reasonably approximated (with an error of $\pm 10\%$), with the values $k_c = 2.12$, $k_w = 1.8$–2.0 (depending on the reinforcement details) and $K_b = 1.7$. An immediate consequence is that the period of vibration cannot be assessed early, in the preliminary stage of the design process, as the stiffness of the structural members are dependent on their strength. This is particularly important in the final stage of a practical design, where strength and stiffness must be adjusted to satisfy predetermined displacement limits. Another consequence of the above definition of yield curvatures of R.C. members, is that the yield drift of common types of frame structures is a relatively stable parameter. Priestley (1998) demonstrated that in R.C. frame structures the yield drift, θ_{fy} is approximately equal to 1%, when the steel yield strain is equal to $\varepsilon_y = 0.002$. A more detailed formula is provided by Georgoussis (2016b), as

$$\theta_{fy} = \frac{\varepsilon_y}{6} \left[\frac{k_b}{(d_b/l_b)} + \frac{k_c}{1.5(d_c/h)} \right] \tag{25.2}$$

where, in order to account for joint shear deformations and also for beam bar slip, the coefficient k_b of Eq. (25.1c) should be increased by 35% (Priestley 1998). Relatively constant values of yield drifts result in almost constant values of yield displacements (Gupta and Kunnath 2000) of frame structures, while the yield drifts of wall structures appear to be proportional to their aspect ratios (Paulay 2002; Tjhin et al. 2004).

The main objective of this paper is to propose a simple analytical solution which could be used by practicing engineers to assess with reasonable accuracy the yield displacement of the capacity curve of typical wall-frame buildings. It should be noted that such structural types are considerably effective in withstanding earthquake excitations and they are recommended by some country codes (e.g. the Greek Code for Earthquake Resistant Design (EAK 2000)). In these buildings, the soft story mechanism is prevented and they combine the merits of the two component sub-systems: wall and frame (e.g. Paulay and Priestley 1992; Garcia et al. 2010). Therefore, an accurate prediction of the aforementioned yield displacements of such structures, early in the design process, facilitates the overall design and enhances the utility of the yield point spectra. As demonstrated by Black and Aschheim (2000), these spectra may be used in a reversed process, that is, to evaluate the required lateral strength in relation to a specified overall ductility ratio.

The second objective of this study was to demonstrate that a configuration of minimum rotational response is feasible when the lateral strength has been specified. It was shown (Georgoussis 2009, 2010, 2015, 2016a) that the torsional response of common buildings, of a rather medium height, can be reasonably predicted on the grounds of a simplified analysis of an equivalent single story system. When the stiffness center (or, modal center of rigidity: m-CR) of the latter system lies close to the vertical axis passing through the floor masses of the real building, the response in the elastic domain is virtually translational. Besides, when the strength of the various structural members (beams, colulmns, etc) is determined from a planar static analysis, then the aforementioned response is preserved in the post elastic state of deformation. This is demonstrated in a parametric analysis on dual wall-frame systems with simple asymmetry, in which the stiffnesses of the various elements have been calculated in relation to their strength. In this analysis the shear sustained by the frame sub-system registered values ranging from 24% to 44% of the total strength.

In the numerical examples presented below, estimates of yield displacements, calculated on the grounds of the approximate method of the continuous medium (Heidebrecht 1975; Heidebrecht and Stafford Smith 1973), are also shown. This method provides an insight into the response of building systems and its simple equations may be easily used to calculate the basic dynamic quantities (frequencies, effective modal masses). In these examples, the same quantities are also computed by means of the accurate stiffness matrix method, following the formulation presented in an earlier paper (Georgoussis 2016b), and comparisons are made with the results of the aforementioned approximate methodology.

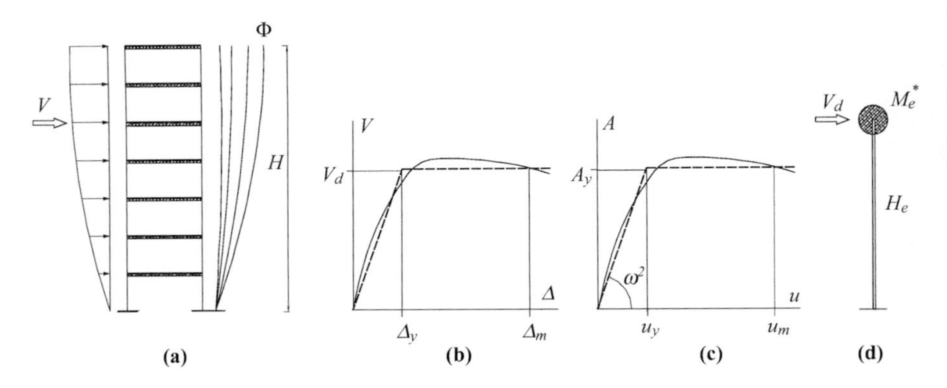

Fig. 25.1 Pushover procedure: (**a**) the building under increasing horizontal loading and displacement profiles; (**b**) pushover curve; (**c**) capacity curve; (**d**) the equivalent SDOF system

25.2 Constructing a Capacity Curve and Constant Ductility Yield Point Spectra

The procedure to construct the capacity curve of a multistory building is implemented by means of a pushover procedure which is schematically demonstrated in Fig. 25.1.

Assuming that the building is under an increasing horizontal loading vector proportional to $\mathbf{M\Phi}$ (i.e.: $\mathbf{V} = \alpha\mathbf{M\Phi}$, where $\mathbf{M}$ is the mass matrix and $\mathbf{\Phi}$ a specified displacement vector), the relationship between the base shear and roof displacement, which is usually defined as pushover curve (V-Δ in Fig. 25.1b) may be converted to the capacity curve (A-u, in Fig. 25.1c) using the following equations:

$$A = V/M_e^* \quad \text{and} \quad u = \Delta/\Gamma\Phi_r \tag{25.3a}$$

where M_e^*, u are the effective (modal) mass and displacement of the corresponding equivalent SDOF system respectively (Fig. 25.1d). The effective mass is equal to (Chopra 2008):

$$M_e^* = \left(\mathbf{\Phi}^{\mathrm{T}}\mathbf{M1}\right)^2/\mathbf{\Phi}^{\mathrm{T}}\mathbf{M\Phi} \tag{25.3b}$$

and, Φ_r (in the second of Eq. (25.3a)), is the value of the specified displacement vector $\mathbf{\Phi}$ at the top (roof) of the structure. Accordingly, Γ is the corresponding participation factor, i.e.:

$$\Gamma = \mathbf{\Phi}^{\mathrm{T}}\mathbf{M1}/\mathbf{\Phi}^{\mathrm{T}}\mathbf{M\Phi} \tag{25.3c}$$

where **1** is the unit vector. In Fig. 25.1b the actual pushover curve is approximated by a bilinear curve, which specifies the shear capacity, V_d, of the assumed structure. It follows that for a newly designed building V_d should be taken equal to the lateral

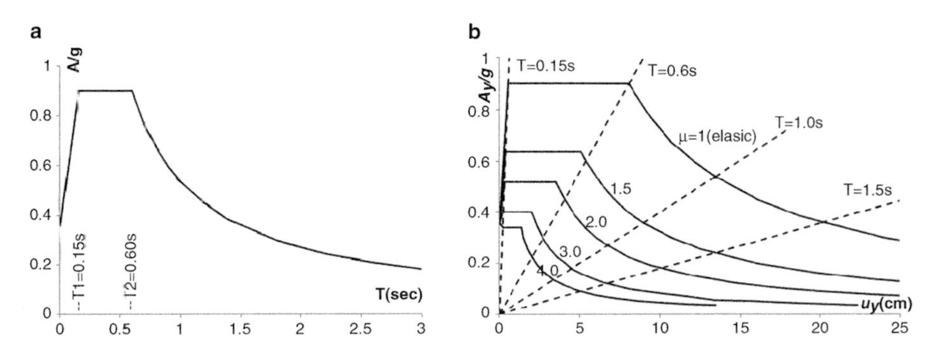

Fig. 25.2 (**a**) Typical elastic acceleration spectrum; (**b**) Yield point spectra for various values of ductility, μ

design shear. Similarly, the bilinear curve in Fig. 25.1c specifies the spectral values of yield acceleration, A_y, and yield displacement, u_y, of the equivalent, elastic-purely plastic, single-degree-of-freedom system (equivalent-SDOF), which is shown in Fig. 25.1d.

Using the equations above, yield point spectra of constant ductility may now be easily determined for a given elastic acceleration spectrum. Assume for example the elastic spectrum of Fig. 25.2a, which is recommended by many country codes and Eurocode 8-2004. It is normalized to 0.36 g PGA and the acceleration amplification factor is equal to 2.5. It is flat between the periods T_1 and T_2, linear in the ascending portion and hyperbolic in its descending branch. The corresponding yield point spectra are shown in Fig. 25.2b, determined according to the Newmark-Hall rules (in Chopra 2008). It is evident that an accurate assessment, early in the design process, of the aforementioned yield displacement, u_y, may be used as the starting point of a structural design to determine the required spectral acceleration, A_y, in relation to a specified ductility ratio μ. This is in fact the usefulness of the yield point spectra: they can be used in a reverse process to calculate the required strength for a given ductility ratio.

25.3 Yield Displacements of Dual System's Capacity Curves

In a resent paper, it was shown by means of the approximate method of the continuous medium that in wall-frame building structures, with structural members having strength dependent rigidities, that the fundamental frequency may be assessed from the equation (Georgoussis 2016b):

$$\omega^2 = 1.875^4 \left[\frac{1}{(\alpha H)^2} + \frac{1}{1.875^2}\right] \frac{\lambda\beta}{\theta_{fy} H} g \tag{25.4}$$

where, β is the seismic coefficient, which expresses the ratio of the lateral design strength, V_d, to the gravity load, W ($V_d = \beta W$), λ is the fraction that defines the portions of V_d which are sustained by the frame subsystem ($V_{df} = \lambda V_d$) and the wall subsystem ($V_{dw} = (1\text{–}\lambda)V_d$), and

$$\alpha H = \sqrt{\frac{GA}{EI}} H = \sqrt{\frac{3}{2} \frac{\lambda}{(1-\lambda)} \frac{k_w \varepsilon_y}{\theta_{fy}} \frac{H}{d_w}} \tag{25.5}$$

In this equation GA, EI are the shear and flexural rigidities of the frame and wall component subsystems respectively. In many country codes, as well as in Eurocode 8-2004, it is required that the fraction λ should be less than 0.5 and in practice it varies from 0.25 to 0.45. For these values of λ and wall aspect ratios, $A_w = H/d_w$, varying between 5 and 10, the parameter αH takes values from 0.95 to 2.10, when $k_w = 1.8$, $\theta_{fy} = 1\%$ and $\varepsilon_y = 0.002$. Accordingly, as it was shown in Georgoussis (2014), for values of αH in the aforementioned range the variation of the fundamental mode effective mass, M_e^*, is very limited: from 0.623 to 0.645 of the total mass, M_{tot}. Therefore, M_e^* may be taken equal to $0.635M_{tot}$ and consequently the yield displacement of the equivalent SDOF system, shown in Fig. 25.1(d), may be approximated by the following equation

$$u_y = \frac{V_d}{M_e^* \omega^2} = \frac{\beta}{0.635} \frac{g}{\omega^2} = 0.448 \frac{H}{\lambda/\theta_{fy} + 2.344(1-\lambda)/\left(k_w \varepsilon_y A_w\right)} \tag{25.6}$$

As it can be seen u_y is independent of the coefficient β, which expresses the magnitude of the design lateral loading, but dependends on the aspect ratio, A_w, and mainly on the fraction, λ. That is, the portion of the lateral force, which is assumed to be resisted by the frame sub-system. The advantage of Eq. (25.6) is that it can be used directly by practicing engineers to obtain reasonable estimates of yield displacements of individual frames and walls. That is, setting successively $\lambda = 1$ and $\lambda = 0$, the aforementioned displacements are found equal to

$$u_{fy} = 0.225\varepsilon_y \frac{l_b}{d_b} H, \qquad u_{wy} = 0.344\varepsilon_y \frac{H}{d_w} H \tag{25.7}$$

The expressions above are very close to those proposed by Priestley (1998), although different methodologies have been used: Eq. (25.7) is based on the approximate method of the continuous medium, while Priestley's equations are based on considerations of discrete member systems. The accuracy of the values given by Eq. (25.6) is investigated further below, in ten story discrete building models (Sect. 25.5) in comparison with the following data:

(i) the results given from Eq. (25.8), where the first mode quantities ω and M_e^*, have been evaluated by numerical analyses using the SAP2000-V16 software. In these analyses, the flexural rigidities of all structural members are strength dependent, determined according to the allocated seismic coefficient β, as described in Georgoussis (2016b).

$$u_y = \frac{V_d}{M_e^* \omega^2} = \frac{\beta}{M_e^*} \frac{W}{\omega^2} \tag{25.8}$$

(ii) the results obtained by pushover procedures. The assumed 10 story models are first detailed as inelastic systems, forming plastic hinges at the critical cross-sections when the external loading is equal to the assigned design shear V_d. The moment capacities of these hinges are also evaluated as described in Georgoussis (2016b). Under increasing lateral loads the approximate bilinear curve of Fig. 25.1c is easily drawn and therefore the yield displacement, u_y, can be easily assessed. In the proposed methodology, the external loading was assumed to be linearly distributed over the height of the structure in order to simulate a plastic beam-sway mechanism in the post elastic region.

25.4 Methodology to Minimize the Torsional Response of Dual Building Systems

The non-linear static (pushover) analysis, recommended by some design codes, is practically applicable to structures responding mainly in a translational mode of vibration (Fajfar and Gaspersic 1996). Therefore, the problem which arises is how to construct a building configuration responding in a somewhat translational manner. A simple method of designing such structural systems, with strength dependent rigidities, is presented in Georgoussis (2016b), and is briefly shown below using the approximate continuous medium approach.

A simple, but reasonably accurate method for predicting the first mode frequency of any structure can be achieved using Southwell's formula (Newmark and Rosenblueth 1971). Therefore, the frequency given by Eq. (25.4), may also be computed by the sum of the square values of the frequencies of the component elements. Each of these frequencies represents the frequency of the corresponding lateral load resisting element when it is assumed to carry, as a plane frame or wall, the floor masses of the real building. Therefore,

$$\omega^2 \approx \Sigma\omega_f^2 + \Sigma\omega_w^2 \tag{25.9a}$$

where the element frequencies, ω_f, (of the f-Frame which is a shear type bent), and ω_w (of the w-Wall which is a flexural type bent), for uniform over the height building systems, may be computed as follows (Chopra 2008; Clough and Penzien 1993):

$$\omega_f^2 = \frac{\pi^2}{4}\frac{GA_f}{\overline{m}H^2}. \qquad \omega_w^2 = 1.875^4\frac{EI_w}{\overline{m}H^4} \tag{25.9b}$$

In these equations, $\overline{m}$ is the mass per unit height and therefore the total mass of the building, M_{tot}, is equal to $\overline{m}H$.

A more accurate evaluation of the frequency, ω, (Eq. 25.9a), may be obtained when the sum of the squares of the element frequencies in the second part of this equation is replaced by the sum of the squares of the effective element frequencies. As shown in Georgoussis (2014) this replacement increases the accuracy of Southwell's formula, when the building system comprises very dissimilar bents (e.g. walls and frames). The effective element frequencies are defined as

$$\overline{\omega}_f^2 = \omega_f^2\frac{M_{ef}^*}{M_e^*} = \frac{\pi^2}{4}\frac{GA_f}{\overline{m}H^2}\frac{M_{ef}^*}{M_e^*}. \qquad \overline{\omega}_w^2 = \omega_w^2\frac{M_{ew}^*}{M_e^*} = 1.875^4\frac{EI_w}{\overline{m}H^4}\frac{M_{ew}^*}{M_e^*} \tag{25.9c}$$

where M_{ef}^* and M_{ew}^* are the effective masses of the fundamental mode of vibration of the *f*-Frame and *w*-Wall respectively, and are equal to 81% and 61.3% of the total mass, M_{tot} (Chopra 2008; Clough and Penzien 1993).

It was demonstrated (e.g. Georgoussis 2009, 2010) that when the stiffness center (modal center of rigidity: m-CR) of an equivalent single story system lies close to the vertical axis passing through the floor masses, the response of the building is essentially translational This distance in a coordinate system where the reference axes origin coincides with the centre of mass (CM), as shown in Fig. 25.3a, equals to

$$x_{m-CR} = \frac{\Sigma\left(x_w\overline{\omega}_w^2 + x_f\overline{\omega}_f^2\right)}{\Sigma\left(\overline{\omega}_w^2 + \overline{\omega}_f^2\right)} \tag{25.10}$$

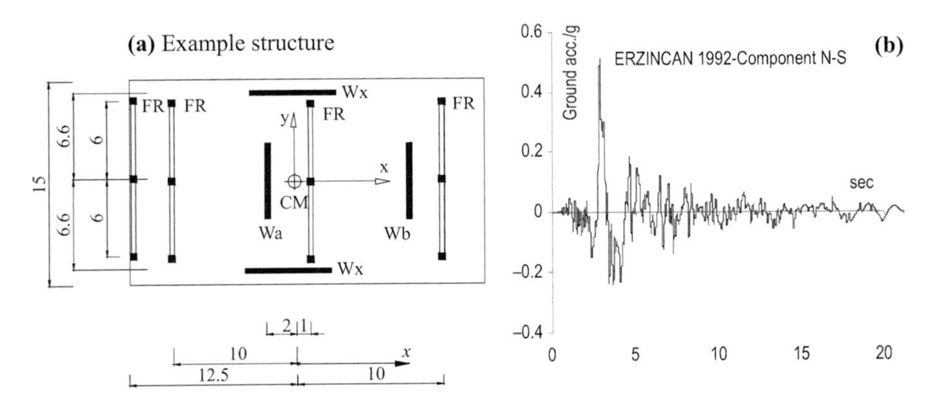

Fig. 25.3 (**a**) The example structure (all dimensions in meters); (**b**) the ground motion assumed

where x_w is the distance of the *w*-Wall from the center of mass and x_f is the distance of the *f*-Frame from the same centre. A quick estimate of the effective element frequencies $\overline{\omega}_f^2$ and $\overline{\omega}_w^2$ and therefore $x_{m\text{-}CR}$, in the case of dual systems with identical frames (and walls), that is, when $\overline{\omega}_f^2$ (or $\overline{\omega}_w^2$) is the same for all frames (or walls), may be easily obtained by firstly calculating the ratio $\Sigma\overline{\omega}_f^2/\Sigma\overline{\omega}_w^2$. A combination of Eq. (25.9c) with Eq. (25.5) gives the following expression

$$\frac{\Sigma\overline{\omega}_f^2}{\Sigma\overline{\omega}_w^2} = \frac{\pi^2}{4\times 1.875^4}(aH)^2\frac{0.81}{0.613} \tag{25.11}$$

which depends merely on the parameter αH. When the equation above, is further combined with Eq. (25.4) which, according to Southwell's formula provides the sum of $\overline{\omega}_f^2$ or $\overline{\omega}_w^2$, the calculation of the effective element frequencies becomes a routine procedure. Consequently, the location of m-CR, is easily determined and when the designer, at an early stage of the structural process, has the choice to decide on the locations of a few lateral load resisting bents, it is rather easy to 'move' m-CR close to CM (so that $x_{m\text{-}CR}$ equals to zero).

25.5 Design Examples

The response of a ten story concrete building, uniform over the height with simple asymmetry (Fig. 25.3a) is investigated under a purely translational earthquake excitation along the y-axis: the component NS of the Erzincan 1992 earthquake shown in Fig. 25.3b.

The example building is a common wall-frame structure in the y-direction and a slender wall structure along the x-direction. It is designed to resist a base shear equal to 20% of its weight (e.g. $V_d = 0.2\ W$) and three different models are investigated. The shear force allocated to the frame subsystem in each of these models is equal to 24% (Model 1), 36% (Model 2) and 44% (Model 3) of V_d. The floor mass was taken equal to $m = 305.8$ kNs^2/m and the radius of gyration about the center of the floor mass (CM) is $r = 8.416$ m. Equal story heights of 3.5 m are assumed and the elastic modulus of concrete was taken equal to 30×10^6 kN/m^2. The lateral resistance in the y-direction is comprises six resisting elements: two of them are slender shear walls (Wa, Wb: of a cross section 40×500 cm) and four of them are moment resisting frames (FR) consisting of three columns (of a section 60×60 cm), 6 m apart, connected by lintel beams 30×60 cm. The lateral resistance in the x-axis (axis of symmetry) is relied on a pair of slender shear walls (Wx) 40×600 cm, which are located symmetrically to the x-axis at distances $\pm$ 6.6 m (Fig. 25.3a).

The strength allocation in the various lateral load resisting elements, in each of the aforementioned models, is based on considerations of the symmetrical counterpart structure (i.e. the structure in which all decks are restrained against rotations). The moment resisting frames in each model are detailed to resist equal magnitudes of the

Table 25.1 Properties of analyzed models

	Model 1	Model 2	Model 3
Continuous medium method			
αH	1.07	1.43	1.69
ω	4.303/s ($T = 1.460$ s)	4.307/s ($T = 1.459$ s)	4.313/s (T = 1.457 s)
u_y	0.167 m	0.166 m	0.166 m
A_y	0.315 g	0.315 g	0.315 g
Stiffness matrix method			
M_{by}	315 kNm	472.5 kNm	577.5 kNm
M_{cy} (interior/exterior)	472.5/236.2 kNm	708.7/354.4 kNm	866.2/433.1 kNm
M_{wy}	53,200 kNm	44,800 kNm	39,200 kNm
I_{be}/I_{bg}	0.254	0.381	0.466
I_{ce}/I_{cg} (interior/exterior)	0.206/0.103	0.309/0.155	0.378/0.189
I_{we}/ I_{wg}	0.591	0.498	0.435
ω	4.040/s ($T = 1.555$ s)	4.098/s ($T = 1.533$ s)	4.121/s ($T = 1.525$ s)
M_e^*/M_{tot}	0.660	0.667	0.673
$u_y = V_d/M_e^*\omega^2$	0.182 m	0.175 m	0.172 m
$A_y = u_y\omega^2$	0.303 g	0.300 g	0.297 g

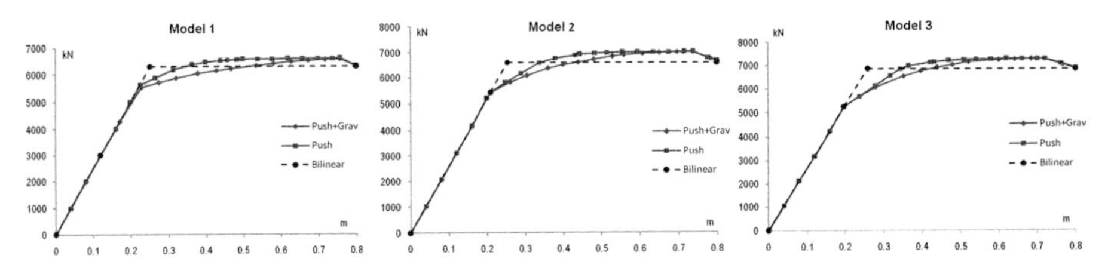

Fig. 25.4 Pushover curves of the assumed example models

assigned shear to the frame subsystem (1/4 of the allocated shear) and similar assumptions were made for the strength details in each of the two walls Wa and Wb. They share equally the shear force allocated to the wall subsystem. The yield drift calculated according to Eq. (25.2) is found equal to $\theta_y = 1.04\%$ when $k_b = 1.7 \times 1.35$, $k_c = 2.12$ and $\varepsilon_y = 0.002$. Comparing the results of the approximate continuous approach and those of the accurate stiffness matrix method in Table 25.1 (details in Georgoussis (2016b)), it may be concluded that the approximate approach underestimates the fundamental frequency, T, by less than 6%, the yield displacement, u_y, by less than 8% and overestimates the yield acceleration, A_y, by less than 6%.

The same data are also calculated by pushover analyses on the symmetrical counterpart structures. The load vector required for the pushover procedure was assumed to be of linear form (vector $\mathbf{\Phi}$ in Fig. 25.1a) to simulate displacements of a plastic beam-sway mechanism, and the results of these analyses are shown in Fig. 25.4. The pushover procedure was terminated at a story drift equal to the code limit of 2.5% (at the same displacement level the base shear was dropped approximately by 5% of its maximum value). In the aforementioned figure, the red curves, termed 'push', have been obtained by neglecting the gravity loads on the beams of the frames, while the blue curves, termed 'push+grav', have been drawn assuming a uniform load on the frame beams equal to 42 kN/m. Both curves initially indicate a rather extended elastic branch and only in Model 1 the initial linear branch of the 'push+grav' curve is slightly reduced. In any case, it may be seen that there is no significant difference between the 'push' and 'push+grav' curves and, because of their extended initial elastic branch, a 'bilinear' curve has been drawn with the same initial slope, representing both pushover curves (red and blue) in the inelastic range. Accordingly, for each model, the yield acceleration and frequency (period) of the equivalent-SDOF system (Fig. 25.1d), are as follows (Georgoussis 2016b):

$$\begin{aligned} &\text{Model1}: A_y = 0.267\ \text{g}, \quad u_y = 0.176\ \text{m}, \quad \omega = 3.858/\text{s}\ (T = 1.628\ \text{s}) \\ &\text{Model2}: A_y = 0.280\ \text{g}, \quad u_y = 0.178\ \text{m}, \quad \omega = 3.928/\text{s}\ (T = 1.599\ \text{s}) \\ &\text{Model3}: A_y = 0.291\ \text{g}, \quad u_y = 0.181\ \text{m}, \quad \omega = 3.971/\text{s}\ (T = 1.582\ \text{s}) \end{aligned} \tag{25.12}$$

Comparing the data shown above with those of the stiffness matrix method in Table 25.1, it may be seen that the deviation between the two procedures is less than 10%.

The rotational behavior of the example building models and particularly the accuracy of the proposed procedure for predicting a configuration of minimum torsional response, was determined as follows: all models were analyzed for any possible location of Wb along the x-axis, while all the other bents are assumed to be located at fixed positions, as shown in Fig. 25.3a. Using the approximate method of the continuous medium the optimum location of Wb, for minimum torsional response was specified from Eq. (25.10) in combination with Eq. (25.11). Therefore, taking into account that along the y-direction there are four frames and two walls, the ratio $\overline{\omega}_f^2/\overline{\omega}_w^2$, was found equal to 0.15, 0.27 and 0.38 for Models 1, 2 and 3 respectively. Relatively similar values for the same ratio, were obtained from the numerical analyses (SAP2000), i.e.: 0.17, 0.31 and 0.42. Accordingly, the corresponding optimum locations, x, of Wb are as follows (also shown in a normalized form $\overline{x} = x/r$)

$$\begin{array}{lll} & \text{Continuous medium method} & \text{Stiffness matrix method} \\ \text{Model 1}: & x = 3.725\,\text{m}\ (\overline{x} = 0.44) & x = 3.955\,\text{m}\ (\overline{x} = 0.47) \\ \text{Model 2}: & x = 5.105\,\text{m}\ (\overline{x} = 0.61) & x = 5.565\,\text{m}\ (\overline{x} = 0.66) \\ \text{Model 3}: & x = 6.370\,\text{m}\ (\overline{x} = 0.76) & x = 6.830\,\text{m}\ (\overline{x} = 0.81) \end{array} \tag{25.13}$$

Figure 25.5 shows the complete torsional response of the investigated structural models, for any location of Wb, under the Erzincan excitation. The response is indicated by the peak values of base shears and torques for both the elastic (black lines) and inelastic (colored lines) systems. Normalized peak values of base shears in

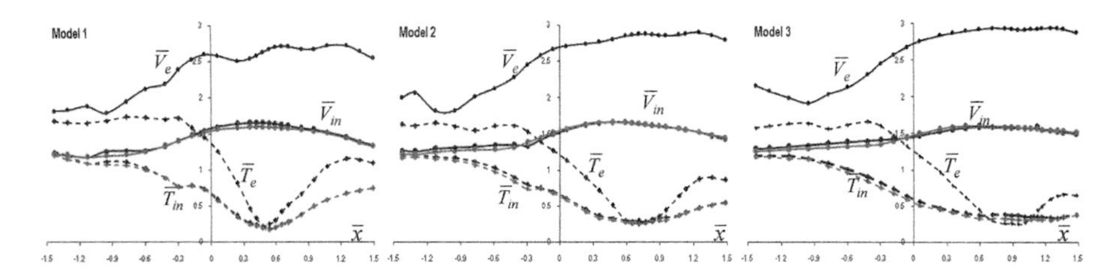

Fig. 25.5 Normalized base shears $(\overline{V}_e, \overline{V}_{in})$ and torques $(\overline{T}_e, \overline{T}_{in})$ of the analyzed models

the y-direction (elastic: $\overline{V}_e = V_e/V_d$ and inelastic: $\overline{V}_{in} = V_{in}/V_d$) are shown by solid lines and normalized peak values of base torques (elastic: $\overline{T}_e = T_e/rV_d$ and inelastic: $\overline{T}_{in} = T_{in}/rV_d$) by dotted lines. The red lines of $\overline{V}_{in}$ and $\overline{T}_{in}$ represent the response of the inelastic system when the gravity loads are neglected and the green ones the response when their effect has been taken into account (prior to the application of the assumed ground motion).

Considering the behavior of the elastic models, it may be seen that the locations of Wb, predicting minimum torsion according to Eq. (25.13), are within the range of coordinates $\overline{x}$, where the curves of $\overline{T}_e$ in Fig. 25.5 obtain minimum values. This is clear for model 1, while for the other models, where a wider range of values of $\overline{x}$ produce minimum torsion, Eq. (25.13) predicts the location closer to CM. It is worth mentioning here that when the elastic torsional response in Fig. 25.5 obtains relatively small values, this is accompanied by increased values of the base shear, $\overline{V}_e$. In general, the response data of the inelastic systems (base shears and torques) are of a reduced magnitude and smoother than those of the elastic ones. With the exception of the range of $\overline{x}$, where the torsional response is minimized and almost equal values of base torques appear in both elastic and inelastic systems, the latter systems sustain reduced shears and torques, which may attain low values up to half of the corresponding elastic ones. In models 1 and 2, minimum inelastic torsion is registered in the range of $\overline{x}$ where the elastic torsion is also minimum, but in a flatter form. In model 3, the variation of the inelastic base shears and torques is relatively smaller than those of the other models and they are more or less constant when the wall Wb 'moves' to the right of CM ($\overline{x} \geq 0$).

25.6 Conclusions

Dual wall-frame concrete buildings are shown to have a rather stable yield displacement for a given wall aspect ratio. The yield displacement depends only on the shear force allocated to the frame sub-system and can be predicted by a simple equation, which incorporates the widely accepted concept that stiffness and strength are interrelated in R.C. elements. The results signify the importance of the yield point spectra (YPS). As the yield displacement is more or less independent of the overall lateral strength of the building, the use of YPS allows the designer to select the strength which satisfies the desired displacement performance. In any case dual building systems may effectively withstand strong ground motions when they respond in a translation mode. A simple procedure is presented for designing building systems to sustain minimum torsional response. It is based on the condition that the modal center of rigidity (m-CR) should be as close as possible to the mass axis of the building. This design oriented procedure is demonstrated in typical ten-story dual systems, whose seismic behavior is investigated under the ground excitation of Erzincan-NS 1992.

Acknowledgement The author wishes to acknowledge the financial support for the dissemination of this work from the Special Account for Research of ASPETE through the funding program "Strengthening research of ASPETE faculty members".

References

Black E, Aschheim M (2000) Seismic design and evaluation of multistory buildings using yield point spectra. Mid-America Earthquake Center, University of Illinois, Urbana

Chopra AK (ed) (2008) Dynamics of structures, 3rd edn. Prentice Hall, Englewood Cliffs

Clough RW, Penzien J (1993) Dynamics of structures, 2nd edn. McGraw-Hill International Eds, New York

EAK (2000) Greek code for earthquake resistant design. Ministry of Environment, City Planning and Public Works. Greece (in Greek)

Fajfar P, Gaspersic P (1996) The N2 method for the seismic damage analysis of RC buildings. Earthq Eng Struct Dyn 25:31–46

Garcia R, Sullivan TJ, Corte GD (2010) Development of a displacement-based design method for steel-RC wall buildings. J Earthq Eng 14(2):252–277

Georgoussis GK (2009) An alternative approach for assessing eccentricities in asymmetric multistory structures, 1: elastic systems. Struct Design Tall Spec Build 18(2):181–202

Georgoussis GK (2010) Modal rigidity center: its use for assessing elastic torsion asymmetric buildings. Earthq Struct 1(2):163–175

Georgoussis GK (2014) Modified seismic analysis of multistory asymmetric elastic buildings and suggestions for minimizing the rotational response. Earthq Struct 7(1):039–052

Georgoussis GK (2015) Minimizing the torsional response of inelastic multistory buildings with simple eccentricity. Can J Civ Eng 42(11):966–969

Georgoussis GK (2016a) An approach for minimum rotational response of medium-rise asymmetric structures under seismic excitations. Adv Struct Eng 19(3):420–436

Georgoussis GK (2016b) Preliminary structural design of wall-frame systems for optimal torsional response. Int J Concr Struct Mater. https://doi.org/10.1007/s40069-016-0183-2

Gupta B, Kunnath SK (2000) Adaptive spectra-based pushover procedure for seismic evaluation of structures. Earthquake Spectra 16(2):367–391

Heidebrecht AC (1975) Dynamic analysis of asymmetric wall- frame buildings. ASCE, National Structural Engineering Convention

Heidebrecht AC, Stafford SB (1973) Approximate analysis of tall wall-frame structures. J Struct Div ASCE 2:169–183

Newmark NM, Rosenblueth E (1971) Fundamentals of earthquake engineering. Prentice-Hall, Englewood Cliffs

Paulay T (2002) An estimation of displacement limits for ductile systems. Earthq Eng Struct Dyn 31:583–599

Paulay T (2003) Seismic displacement capacity of ductile reinforced concrete building systems. Bull NZ Natl Soc Earth Eng 36(1):47–65
Paulay T, Priestley MJN (1992) Seismic design of reinforced and masonry buildings. Wiley Interscience, New York
Priestley MJN (1998) Brief comments on elastic flexibility of RC frames and significance to seismic design. Bull NZ Natl Soc Earth Eng 31(4):246–259
Priestley MJN (2000) Performance based seismic design. 12 WCEE
Priestley MJN, Kowalsky MJ (1998) Aspects of drift and ductility capacity of rectangular walls. Bull NZ Natl Soc Earth Eng 31(2):73–85
Priestley MJN, Calvi GM, Kowalsky MJ (2007) Direct displacement based design of buildings. IUSS Press, Pavia
Tjhin TN, Aschheim MA, Wallace JW (2004) Yield displacement estimates for displacement-based seismic design of ductile reinforced concrete structural wall buildings. 13 WCEE, Vancouver, Canada

Chapter 26
Dynamic Resistance of Residential Masonry Building with Structural Irregularities

F. Pachla and T. Tatara

Abstract Kinematic loads resulting from earthquakes or human induced events can act on engineering structures. Poland is not an active seismic region. In south part of Poland there are located quarries and mines. Surface and underground exploitation of mineral resources results in regional seismic phenomena. Seismic waves originating from these phenomena are travelling towards surface and in form of surface wave can impact the structures. Mining-related surface vibrations show a lot of similarities with earthquakes but also some differences. In previous years the main load used in design procedure of buildings in mining regions in Poland were dead and live load, technology load and gusts of wind. The structures were not designed to dynamic loads resulting from mining-related vibrations of ground. Dynamic loads acting on buildings result in extra inertia forces which load those structures. This specially concerns to residential masonry objects which present group of structures frequently occurring in mining areas. In many cases, these buildings are characterized by irregularities in their construction, and this significantly reduces their dynamic resistance. The paper presents the selected problems of analyzing selected masonry building with irregularities in structure using finite element method (FEM). The analysis relates to dynamic effects caused by surface vibrations. The selected problems with assuming the kinematic excitations and soil-structure interaction are presented as well as some aspects of constitutive modelling of masonry structures. The paper also includes comparison of earthquakes and mining-related tremors. The differences in frequency domain and significant duration of intensive phase are also discussed.

Keywords Mining tremor · Masonry · Irregular structure · Seismic

F. Pachla (✉) · T. Tatara
Institute of Structural Mechanics, Cracow University of Technology, Cracow, Poland
e-mail: fpachla@pk.edu.pl; ttatara@pk.edu.pl

D. Köber et al. (eds.), *Seismic Behaviour and Design of Irregular and Complex Civil Structures III*, Geotechnical, Geological and Earthquake Engineering 48,
https://doi.org/10.1007/978-3-030-33532-8_26

26.1 Introduction

In areas of natural occurrence of earthquakes and mining tremors there are a number of buildings made in traditional technology (Pachla 2011; Tatara 2012). These are primarily multi-family buildings or masonry with ceramic small size elements. Due to architectural solutions, they are often irregularly shaped buildings.

Structural irregularity in buildings is the result of a combination of irregularity in plan and in elevation (De Stefano and Pintucchi 2008). Many uncertain factors are important in seismic research of existing structures. In case of buildings with irregularities crucial role can play an inner variability of the mechanical properties of materials what can also become source of irregularity (De Stefano et al. 2013a, b). Some works refer to an evaluation of seismic behaviour of existing RC buildings with irregularities according to current European Technical Code (La Brusco et al. 2016).

Action of kinematic force at the level of foundation of masonry buildings results in additional horizontal inertia forces (Dulińska 2006; Pachla 2011;Tatara 2012), which in many cases have not been considered in the design process. The masonry bearing walls of structure are primarily designed to carry compressive loads. Tensile or shear strength of masonry wall is much lower than the compressive strength.

In the case of such loads, cracks often occur in bearing elements (Zembaty et al. 2007; Pachla 2011; Tatara 2012). This applies to masonry elements, which are the weakest link in the entire structure. In the case of existing buildings, it is often necessary to determine dynamic resistance. Then, nonlinear analysis is often made in the time domain using the most intense tremors recorded in a mining area (Dulińska 2006; Pachla 2011; Tatara et al. 2017) or using Response Spectrum Analysis (Pachla 2011; Maciag et al. 2016).

The paper deals with the results of such analysis in the case of a multi-family building. In addition, a comparison of mining shocks with a selected earthquake was made. The equations between these phenomena were pointed out. This is particularly important since the seismic engineering experience is used in the analysis of buildings subjected to mining shocks.

26.2 The Dynamic Model of the Building

The building discussed in the paper is made of masonry and is supported by a foundation slab. It's a four story multi-residential building with monolithic reinforced concrete basement. The walls are made of ceramic blocks with thickness of 24 cm. According to the design project, the slabs over each floor are made of reinforced concrete, with thickness of 15 cm. The roof is made of wooden rafters. In plane, the building is like L-shape with length of 30 m and width of 20 m. The height of the building in ridge measured from the foundation level is 16 m. Plan of the building's floor is presented in Fig. 26.1a.

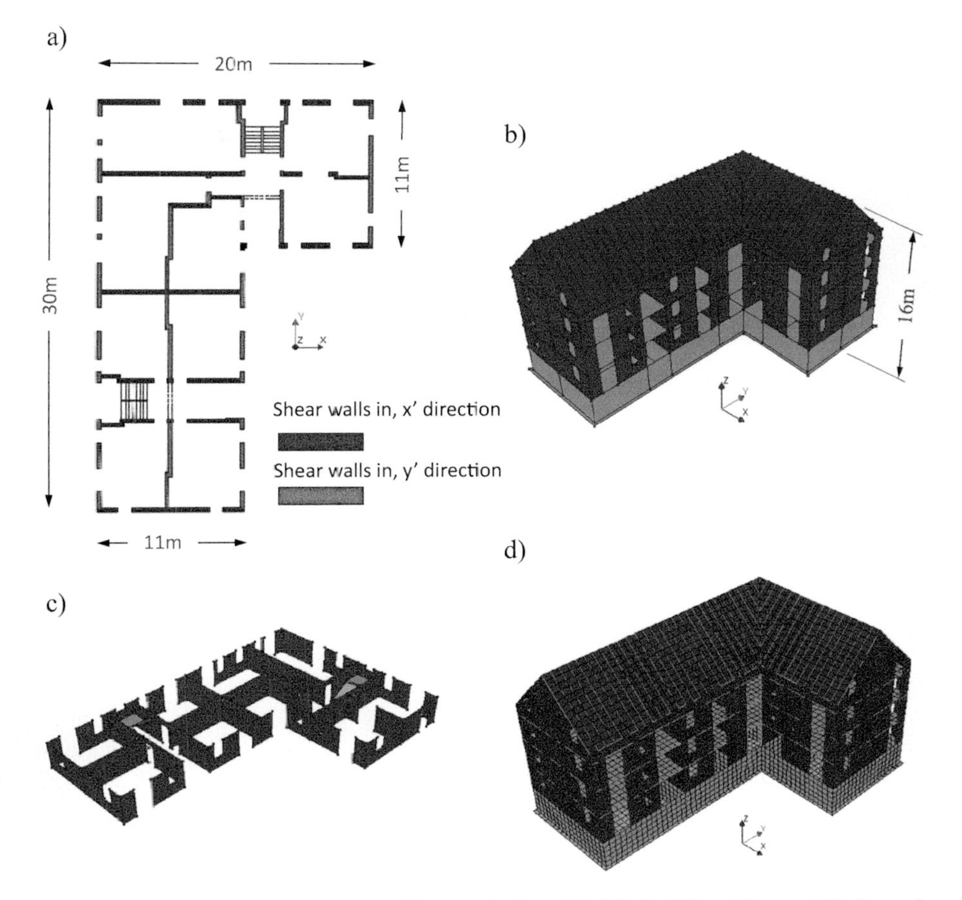

Fig. 26.1 (**a**) plan view of the building, (**b**) numerical model of the building, (**c**) repeatable floor of the building, (**d**) finite element discretization of the model

The general view of numerical model of the analyzed building is presented in Fig. 26.1b. The shear walls in perpendicular direction, which creates the structure of the building, are shown in Fig. 26.1. Each floor of the structure is repeatable and it's presented in Fig. 26.1c. The dynamic studies were focused on adopting a numerical model of the building under interest.

The adopted model was built using the FEM code Diana (TNO Diana User's Manual release 10.2 2017) and documentation of the building. Because of its irregularity the 3D model of the building was analysed (Fig. 26.1d). All structural elements that can influence on the stiffness were included into the model. The stiffness and the mass of the structure were included in the 3D FEM model.

The combination of dead and technological load was assumed according to code (PN-B-02170:2016-12 2016) by formula (26.1):

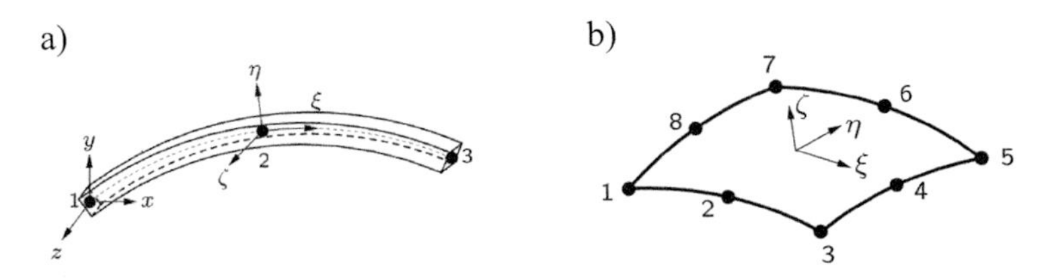

Fig. 26.2 Assumed finite elements: (**a**) CL18B 3 nodes curved beam element, (**b**) CQ40L eight nodes curved layered shell element for walls and slabs

Table 26.1 Material properties for masonry

	Value	Unit
Young modulus (bed joint direction)	3.0	GPa
Young modulus (orthogonal to bed joint direction)	6.0	GPa
Shear modulus	1.875	GPa
Mas density	1800	kg/m^3
Bed-joint tensile strength	0.2	MPa
Compressive strength	6.0	GPa
Fracture energy in compression	5.0	kN/m
Angle between diagonal stepped crack and bed joint	30	deg
Factor to strain at compressive strength	2.3	–
Friction angle	37	deg
Cohesion	0.4	MPa
Fracture energy shear	20	N/m
Unloading factor	0.8	–

$$Q_k = Q_k{}' + 0.6 \cdot Q_k{}'' \tag{26.1}$$

where:

$Q_k{}'$ – dead load
$Q_k{}''$ – technological load

The global coordinate system shown in Fig. 26.1b. It allows to describe geometry. Loads and interpreting the results of the dynamic analysis.

While building the model, curved beam elements for the RC columns and layered shell element for modelling the slabs and walls were implemented (Fig. 26.2).

The model consists of 37,434 finite elements after discretization. Nonlinear Masonry engineering model was assumed for masonry elements. The columns and slabs, as well as the basement construction, all from concrete were modelled as linear elastic. The wooden structure element like rafters and wooden plates on roof were also modelled as linear elastic. The soil influence was modelled using Coulomb friction law. The material properties are presented in Tables 26.1, 26.2, 26.3 and 26.4.

Table 26.2 Material properties for concrete slabs, beams and columns

Linear elastic model	Value	Unit
Young modulus	30	GPa
Poisson ratio	0.2	–
Mas density	2500	kg/m^3

Table 26.3 Material properties for timber roof beams and plates

Linear elastic model	Value	Unit
Young modulus	10	GPa
Poisson ratio	0.3	–
Mas density	600	kg/m^3

Table 26.4 Material properties for soil interface

Coulomb friction law	Value	Unit
Normal stiffness	0.1	GN/m^3
Shear stiffness	0.1	GN/m^3
Cohesion	0.1	MPa
Friction angle	30	deg
Dilatancy angle	5	deg

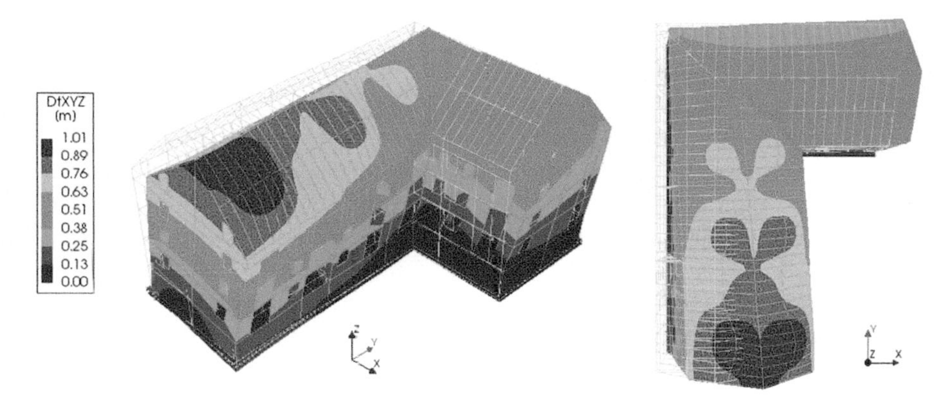

Fig. 26.3 1st mode shape ($f_{1,x}$ = 6 10 Hz), isometric and top view

26.3 Dynamic Analyses

26.3.1 Natural Frequencies

Natural frequencies of the 3D model and the corresponding mode shapes are shown in Figs. 26.3, 26.4 and 26.5. The range of calculated natural frequencies is dense (the first natural horizontal frequencies in 'x', and 'y' direction are equal 6.10, 6.85 Hz respectively and the first torsional frequency equals 8.24 Hz).

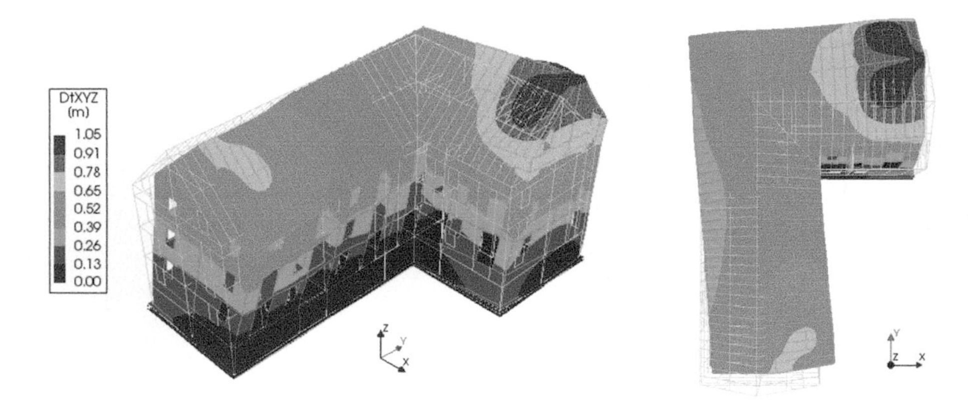

Fig. 26.4 2nd mode shape ($f_{2.y} = 6.85$ Hz), isometric and top view

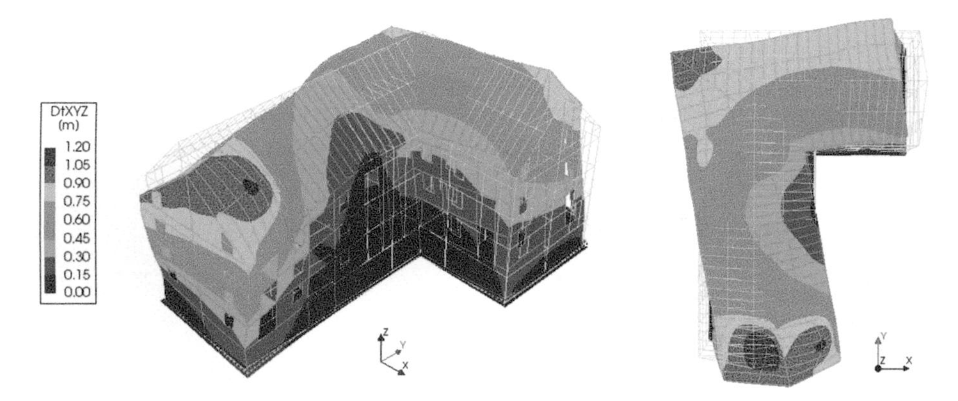

Fig. 26.5 3rd mode shape ($f_{3.tors} = 8.24$ Hz), isometric and top view

26.3.2 Assumed Kinematic Loads and Dynamic Analysis

Horizontal components of free-field acceleration vibrations originating from representative earthquake and rockburst were applied in dynamic analysis as kinematic load. The Sitka earthquake occurred on July 30, 1972 and had magnitude Mw = 7.6. The coordinates of the seismological station are the following 57.0576 N and 135.3273 W. Eearthquakes are the result of tectonic plates pushing against one another. Mining tremors are a regional phenomenon, dependent on human activities. Surface free-field vibration records caused by the most intensive mining shock were used also in dynamic calculations. In Poland, occurrence of rockbursts is related with underground or surface exploitation of mineral resources such as hard and brown coal, lignite and copper ore.

Rockbursts originate from excavation process occurring in a part of the deposit. The zones of increased pressure are formed in the rock mass surrounding the excavation zone. The stress state deviates from the normal one. Elastic energy

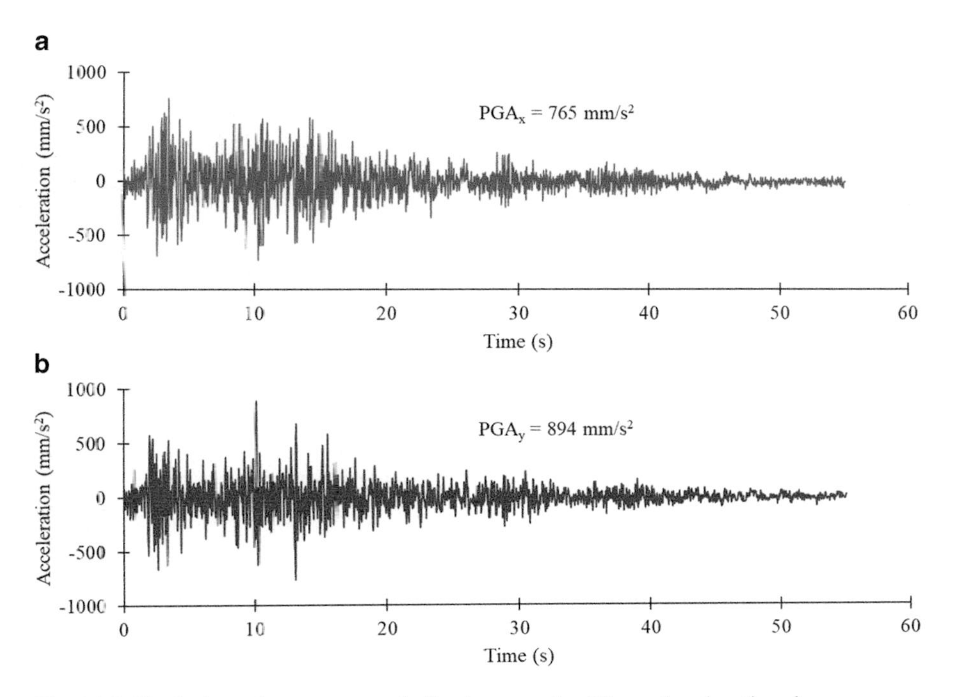

Fig. 26.6 The horizontal components of vibration records of the analysed earthquake

accumulated in rocks, rapidly discharges in a manner similar to an explosion. The sudden relaxation of the rock mass at the hypocentral depths at a focus causes seismic waves that reach the surface due to the elastic energy release. In the surface layer surface waves are formed. They have the greatest impact on buildings, particularly their horizontal components.

One of the most exploited mining areas in Poland is the Legnica-Glogow Copper District (LGOM). In this area rockbursts of energies over 10^7 J occur (Zembaty 2004; Zembaty et al. 2015). The measuring network established in LGOM permit continuous registration surface free - field vibrations. One of the most intensive mining shocks recorded in LGOM took place on 21st May 2006 – comp. Fig. 26.7.

The free-field acceleration vibrations of SITKA earthquake and rockbursts are presented in Figs. 26.6 and 26.7, respectively. Figure 26.8 shows FFT of the horizontal components of mining – related vibrations presented in Fig. 26.7. Comparison of the parameters characterizing records of vibrations applied for dynamic analysis shows some differences between earthquakes and mining tremors and is given in Table 26.5. Arias intensity was used to establish duration of the intensive phase of analysed records.

Generally the most significant differences between earthquakes and mining tremors are the following: (a) duration of the intensive phase of the vibration, (b) frequency characteristics, (c) values of PGA/PGV ratio, (d) PGA, PGV and PGD values. The applied in dynamic analysis mining shock was one of the most

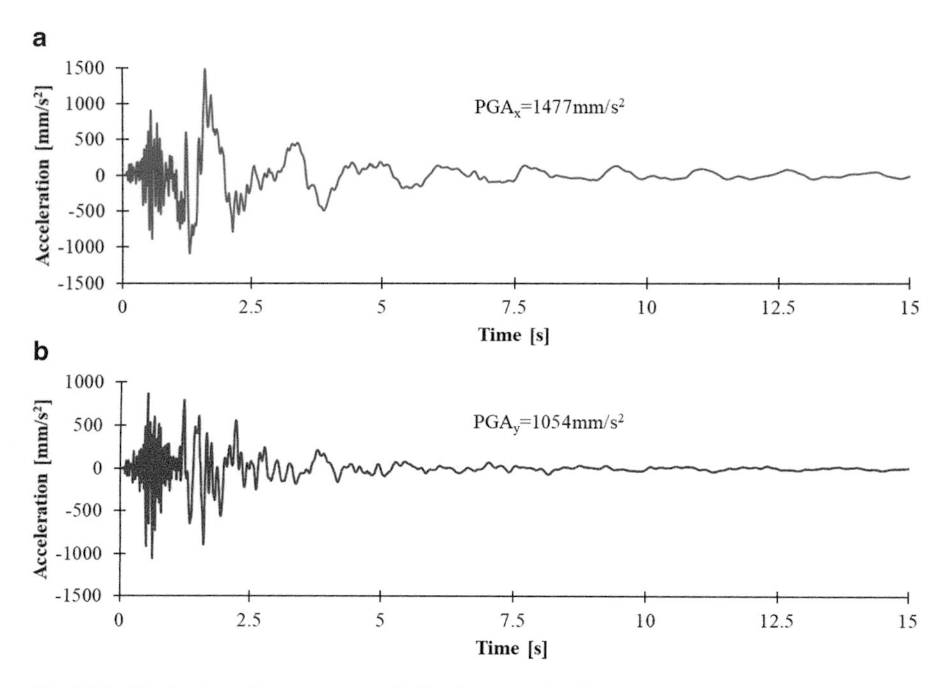

Fig. 26.7 The horizontal components of vibration records of the analysed mining tremor

intensive events in this area. Normally maximal values of horizontal components of vibration acceleration do not exceed 0.2–0.3 g (g – acceleration of gravity) (Maciag et al. 2016).

Two stages of dynamic numerical analysis were conducted. The first stage considered applying incrementally gravity loads as quasi static. The regular incremental iterative Newton method with a constant time step of 0.005 s was used to solve the nonlinear static equilibrium equations in time for gravity loads. After calculation concerning application the gravity load, the nonlinear time history analysis was performed for horizontal earthquake components applied at the base of the structure. The selected value of time step fulfils convergence condition to ensure the accuracy of the results. The quasi-Newton (secant) method was applied as the iterative method to solve the nonlinear dynamic equilibrium equations. This method uses the Broyden – Fletcher – Goldfarb – Shanno (BFGS) stiffness update method. In this method the last obtained secant stiffness matrix is stored and used as the initial stiffness matrix at the first iteration of the next time step. The relative norm of the last displacement increment vector is used as the convergence criterion of the solution.

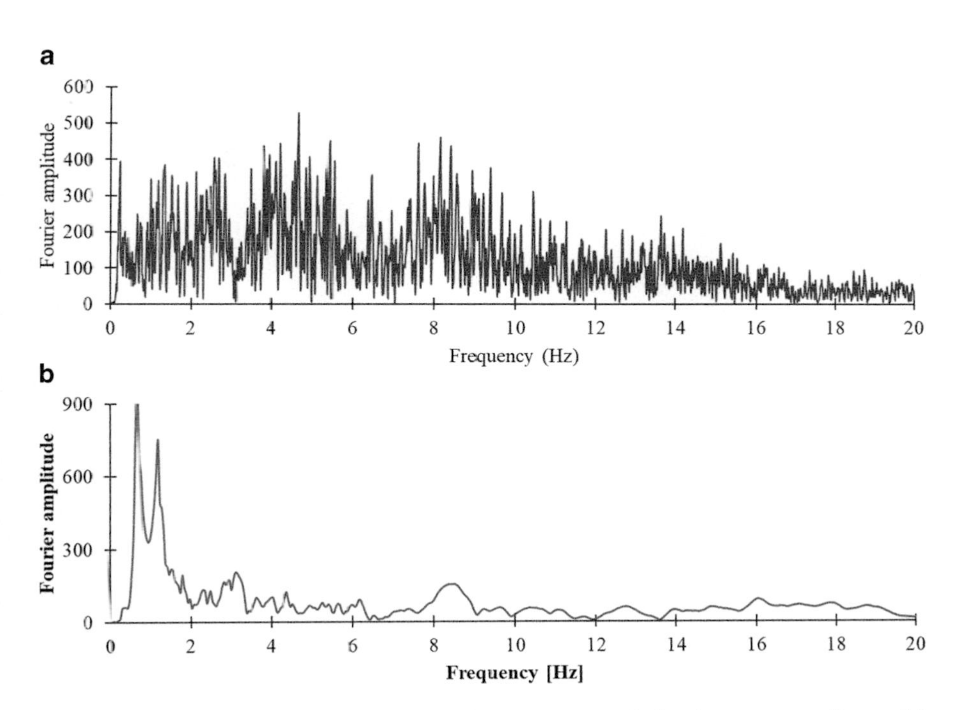

Fig. 26.8 Fourier spectrum for the most intensive components of vibrations corresponding to: (**a**) Fig. 26.6b, (**b**) Fig. 26.7a

Table 26.5 Material properties for soil interface

No.	PGAx (mm/s²)	PGAy (mm/s²)	PGVx (mm/s)	PGVy (mm/s)	PGAx/PGVx (1/s)	PGAy/PGVy (1/s)	tha (s)	thv (s)
Sitka	765	894	74.2	67	10.3	13.3	27.1	28.9
Lgom	1054	1477	48.2	180.2	21.9	8.2	7.7	8.2

26.3.3 Results of the Analysis

The combination of the dead load and the dynamic load (base acceleration) was considered. The global response of the model for the base excitations was analyzed. It can be seen that the model response corresponds to first three mode shapes of the natural frequencies, with a tendency to rotate around. This is observed for all analyzed kinematic excitations.

The failure of the model and crack propagation is presented in Figs. 26.9, 26.10, 26.11 and 26.12. As seen, the cracks (shown using the maximum crack widths) concentrate near the windows openings. Major cracks occur in the shear walls in ‘x’ direction. These walls transfer the seismic horizontal forces. Though the short time of intensive phase of the rockburst from Fig. 26.7, the failure range is significant but

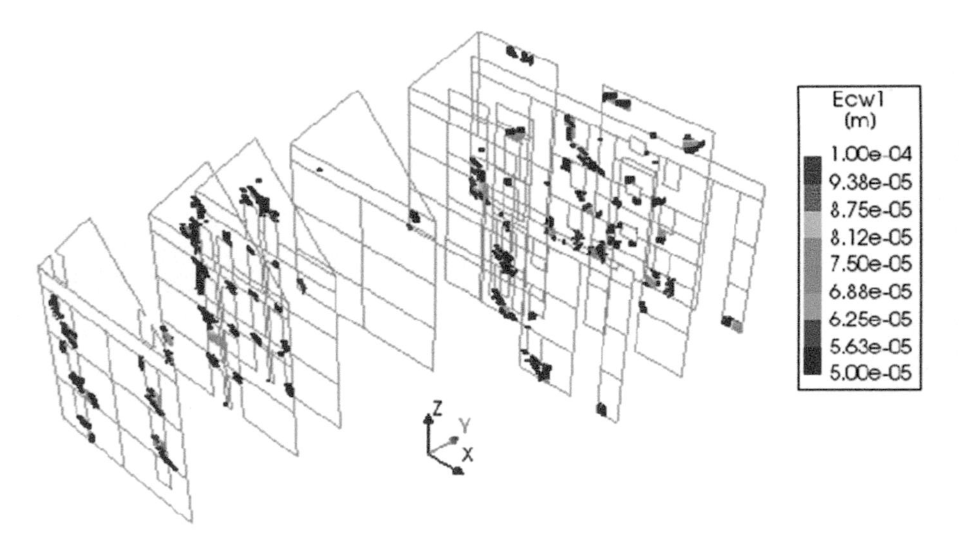

Fig. 26.9 Crack width in the 'x' shear walls due to SITKA earthquake

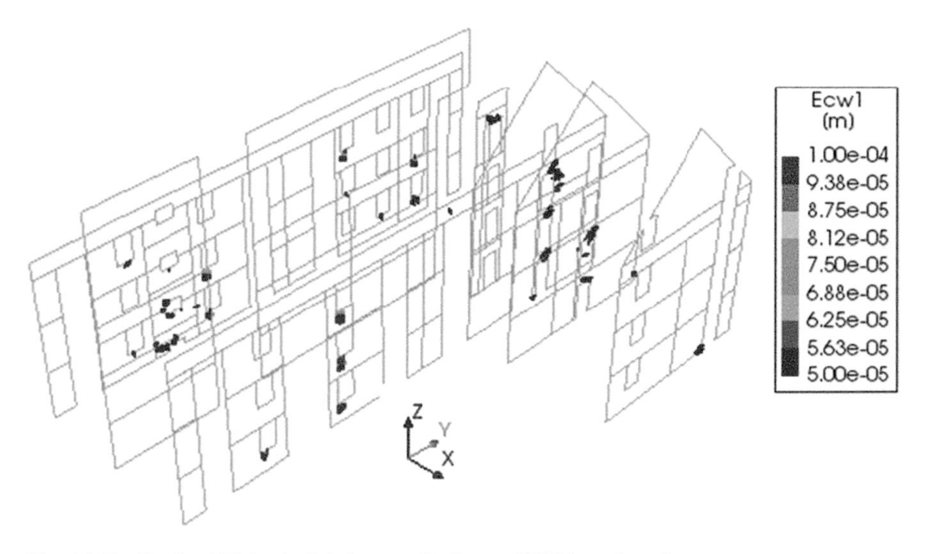

Fig. 26.10 Crack width in the 'y' shear walls due to SITKA earthquake

safe for the structure. An emergency state is not reached. The envelope of the cracks pattern is presented in Fig. 26.7. Despite the damages, the building can be safely used. It is required to carry out renovation work to restore the original condition of the building (Figs. 26.9 and 26.10). In the case of SITKA earthquake from Fig. 26.6 though of much smaller values of PGA (half smaller), but longer time of duration of intensive phase, damage range is very similar (Figs. 26.11 and 26.12). This leads to

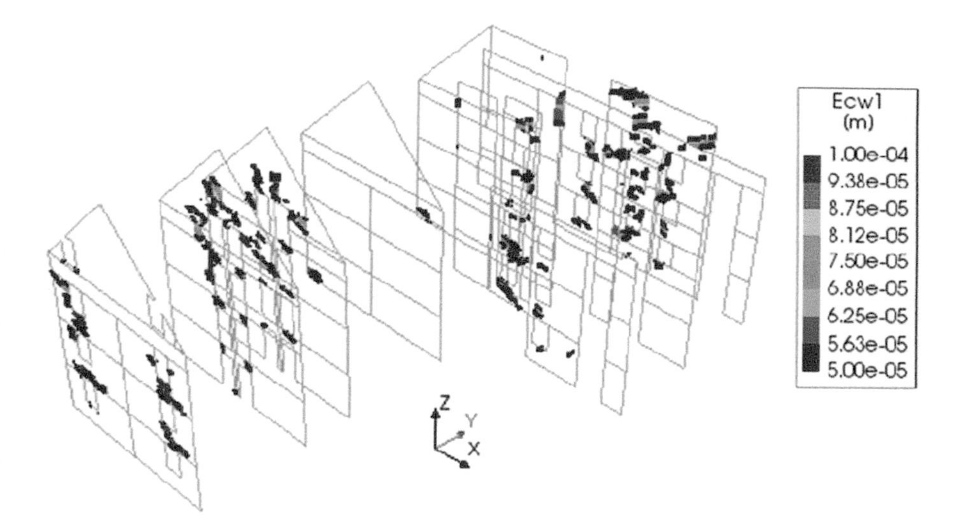

Fig. 26.11 Crack width in the 'x' shear walls due to LGOM mining tremor

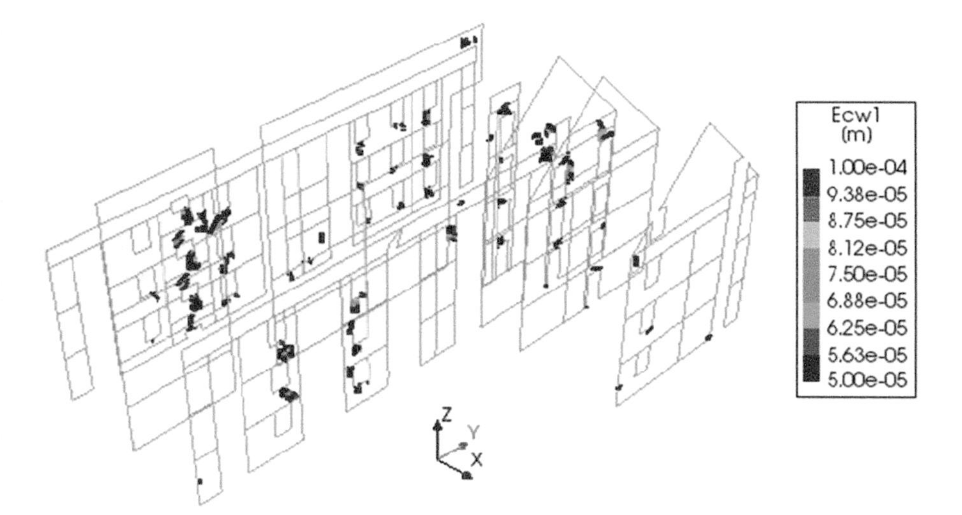

Fig. 26.12 Crack width in the 'y' shear walls due to LGOM mining tremor

the conclusion that the duration of intensive phase of seismic event is an important factor in the response of structures. This also indicates that even relatively weak earthquake because of long time of intensive phase can cause the initiation of the cracks in the structure with irregularities.

26.4 Conclusions

Two types of base accelerations were assumed. Excitations were selected because of their basic characteristics as duration of the intensive phase and PGA values. An irregular masonry building was subjected to all selected kinematic excitations. Nonlinear time history analysis were made. According to the size of the cracks, that appear during the intensive phase of vibrations, the construction stresses were identified. This kind of dependence was fully described in the early nineties (Paulay and Priestley 1992). The calculations results show the significant influence of intensive phase of duration on the response of the building with structural irregularities. Comparing calculated results for the mining shock and the SITKA earthquake it can be observed that the significant cracks appear in the structural elements, but it doesn't lead to emergency state of the structure. Selected excitations differ with the PGA values as well as the duration of intensive phase. It is confirming the thesis of significant influence of intensive phase duration on the response of building with structural irregularities. Earthquake excitation has long-term duration of intensive phase with values of the PGA which are about half lower than for LGOM mining shock. The size of damages in the analysed cases is similar. The literature overview confirms the dependence between long-term vibrations and damages occurring in the residential buildings for the mining shocks (Zembaty et al. 2007; Pachla 2011; Tatara 2012).

References

De Stefano M, Pintucchi B (2008) A review of research on seismic behaviour of irregular building structures since 2002. Bull Earthq Eng 6(2):285–308

De Stefano M, Tanganelli M, Viti S (2013a) On the variability of concrete strength as a source of irregularity in elevation for existing RC building: a case study. Bull Earthq Eng 11 (5):1711–1726

De Stefano M, Tanganelli M, Viti S (2013b) Effect of the variability in plan of concrete mechanical properties on the seismic response of existing RC framed structures. Bull Earthq Eng 11 (4):1049–1060

Dulińska J (2006) Odpowiedź dynamiczna budowli wielopodporowych na nierównomierne wymuszenie parasejsmiczne pochodzenia górniczego (Dynamic response of multi-support structures to unequal excitation of mining tremors). Scientific Publisher of the Cracow University of Technology, Kraków

La Brusco et al (2016) Seismic assessment of an existing irregular RC building according to Eurocode 8 methods. In: Zembaty Z, De Stefano V (eds) Seismic behaviour and design of irregular and complex civil structures II, geotechnical, geological and earthquake engineering 40. Springer, Cham, pp 135–147

Maciag E, Kuzniar K, Tatara T (2016) Response spectra of the ground motion and building foundation vibrations excited by Rockbursts in the LGC region. Earthq Spectra 32 (3):1769–1791

Pachla F (2011) Analiza metod oceny wpływu wstrząsów górniczych na budynki murowe na przykładzie LGOM (Analysis of the methods concerning an estimation of dynamic resistance of masonry buildings due to mining - related rockbursts at LGOM region). Dissertation, Cracow University of Technology

Paulay T, Priestley MJN (1992) Seismic design of reinforced concrete and masonry buildings. Wiley, New York

PN-B-02170:2016-12 (2016) Ocena szkodliwości drgań przekazywanych przez podłoże na budynki (Assessment of harmfulness of vibration transmitted by the ground on buildings), Polish Standardization Committee, Warszawa

Tatara T (2012) Odporność dynamiczna obiektów budowlanych w warunkach wstrząsów górniczych (Dynamic resistance of building structures under mining shock conditions), Scientific Publisher of the Cracow University of Technology, Kraków

Tatara T, Pachla F, Kubon P (2017) Experimental and numerical analysis of an industrial RC tower. Bull Earthq Eng 15(5):2149–2171

TNO Diana User's Manual (2017) Release 10.2

Zembaty Z (2004) Rockburst induced ground motion - a comparative study. Soil Dyn Earthq Eng 24(1):11–23

Zembaty Z, Jankowski R, Cholewicki A et al (2007) Trzęsienia ziemi w Polsce w 2004 roku (Earthquakes in Poland in 2004). Czasopismo Techniczne 2-B:115–126

Zembaty Z, Kokot S, Bozzoni F et al (2015) A system of mitigate deep mine tremors effects in the design of civil infrastructure. Int J Rock Mech Min Sci 74:81–90

Chapter 27
Effect of Mass Irregularity on the Progressive Collapse Potential of Steel Moment Frames

Gholamreza Nouri and Mohammad Reza Yosefzaei

Abstract Progressive collapse is a phenomenon in which a minor damage leads to total failure of the structure or the collapse of large parts of it. In this paper, influence of mass irregularity in height of the structures on the progressive collapse potential was investigated. To this end, four-, eight- and twelve-story steel moment resisting frame structures are designed and progressive collapse potential was evaluated using alternate load path method, recommended by the American General Service Administration 2003. Three dimensional buildings studied in the current paper were modeled using finite element method in ABAQUS software. Results revealed that in the cases with column removal in the first floor, increasing the number of floors, decreases progressive collapse potential. Maximum dynamic displacement under the removed column in regular 4-story building is about 1.41 times larger than that of regular 8-story building and about 2.16 times larger than that of regular 12-story building. Moreover, comparing performance of regular and irregular buildings showed that regardless location of column removal and story number, mass irregularity in height increases progressive collapse potential. Maximum dynamic displacement due to removed column in irregular buildings subjected to column removal in penultimate floor is 19, 20 and 22 percent larger than that of regular one for four-, eight- and twelve-story buildings respectively.

Keywords Progressive collapse · Alternate load path method · Irregular buildings · Steel moment frame · Column removal · Dynamic nonlinear analysis

---

G. Nouri (✉) · M. R. Yosefzaei
Department of Civil Engineering, Faculty of Engineering, Kharazmi University, Tehran, Iran
e-mail: r.nouri@khu.ac.ir

D. Köber et al. (eds.), *Seismic Behaviour and Design of Irregular and Complex Civil Structures III*, Geotechnical, Geological and Earthquake Engineering 48,
https://doi.org/10.1007/978-3-030-33532-8_27

27.1 Introduction

In recent years, the problem of the progressive collapse of the structure has attracted the attention of many researchers. Progressive collapse is defined as an abnormal incident causing minor damage to the building. In buildings incapable of reaching a new statically stable configuration following the damage, the incident could lead to the failure of a major part of the building or its total collapse. In progressive collapse, the structural system primarily transmits the overload induced by the column removal through flexural mechanism of the beams connected to top of the removed column and then undergoes large deformations to form catenary action which mitigates collapse propagation. Disproportionate initial damage of a building as compared to its final destruction condition may be outlined as the main feature of progressive collapse. In 1968, destruction of a large part of a 22-story building in the UK due to the explosion in the eighteenth floor triggered the research into progressive collapse problem. The tragic terrorist attack on World Trade Center towers on September 11, 2001, rendering their total collapse and demolition, was the turning point in the development of studies in the area of progressive collapse that led to the development of codes and recommendations in the design practice for progressive collapse.

Blast load is one of the main causes of triggering progressive collapse in buildings. Very short time period and duration are among the most prominent features of explosive loads. Given their specifications and reviewing the past research, it may be observed that explosive loads, unlike seismic loads, do not stimulate the lateral load bearing system of the structure; but considering limited time period of the loads, they shock the main load bearing elements of the structure, causing initial damages to the building (Lee et al. 2009). Extreme fires could be considered as a triggering event to begin progressive collapse in buildings as well. An analysis on progressive collapse in high-rise concrete frame buildings revealed that flexural failure of the peripheral columns nearly 7 hours after being exposed to fire is the main cause of progressive collapse initiation (Lu et al. 2017). Most of the codes related to the assessment of progressive collapse potential have adopted the Alternate load Path Method (APM) recommended by the Ministry of Defense and the US General Services Administration's instructions (GSA 2003). The APM is an incident-independent technique which means that it does not consider the preliminary cause of failure; rather, it evaluates the structural response after the removal of one of the main load bearing elements. The method is based on the removal of corner or middle columns. In the APM, linear static, nonlinear static and dynamic methods may be applied. Research shows that nonlinear dynamic method yields more accurate response compared to the static one (Pretlove et al. 1991).

Optimal performance of moment frame structures against progressive collapse is due to the extra strength in the moment frame system as designed to withstand lateral loads and high degree of redundancy which provides multiple load transmission paths (Kim et al. 2011). The progressive collapse of two-dimensional eccentrically and concentrically braced steel frames using the APM in a ten-story building were studied and were concluded that although both systems utilize braced frame to resist lateral forces, eccentric frames are less vulnerable to progressive collapse than concentric ones (Khandelwal et al. 2009). Evaluating the progressive collapse potential in tall buildings with steel plate shear walls indicates a greater resistant compared to moment frame and braced frames (Mashhadiali et al. 2016). Conducting progressive collapse analysis for a twenty-story building in various sudden column removal scenarios reveals that the building is more vulnerable when two columns are removed. The reason could be attributed to wider area being affected by removing two columns (Fu 2009). Assessing structural resistance against progressive collapse shows that nonlinear dynamic analysis provides more accurate and generally higher responses, while using linear analysis leads to a more conservative result (Kim and Kim 2009). Study the structural behavior of high-rise reinforced concrete buildings with shear wall lateral resistant system under progressive collapse condition indicates that a weak frame coupled with strong shear walls is not strong enough to withstand removal of a key element (Ren et al. 2014). Evaluation of the progressive collapse potential in earthquake-resistant RC building using GSA linear regression technique, nonlinear static, and nonlinear dynamic analyses reveal that buildings are less likely to undergo progressive collapse when using linear static method proposed by the GSA (Tsai and Lin 2008). Using the APM to analyze three-dimensional moment frames with and without including slab floors in the model reveals the importance of introducing slab in the model in order to assess the progressive collapse potential (Qian et al. 2015). Using end-plate beam-to-column connections, Vierendeel truss in top floors, and fin-plate connection increase joint rotational strength and mitigate progressive collapse potential in composite structures (Jeyarajan et al. 2015).

Steel moment frame is one of the most widely used structural systems in moderate to high seismicity zones. The building system is of interest due to its high ductility and considerable lateral resistance. The building system is also preferred due to the architectural forms it provides. In the this paper, progressive collapse potential in buildings considering both regular and irregular mass distribution in height was evaluated using APM. The buildings are designed according to Iranian Code of Practice for Seismic Resistant Design of Buildings and the evaluation of progressive collapse is conducted using ABAQUS software.

27.2 Modeling and Analysis Method

The models studied in this research include steel moment frames having 4-, 8-, and 12-stories, respectively. Beam element of ABAQUS (Beam – B31) was used in 3-D models of beams and columns. Recent study was showed the accuracy of that element for the modeling of the structures in progressive collapse analysis (Liu 2010). Due to very short column removal time, material hardening is not included and elastoplastic behaviour was assumed for steel. Rayleigh-type damping with the coefficient of 3.5% were assumed.

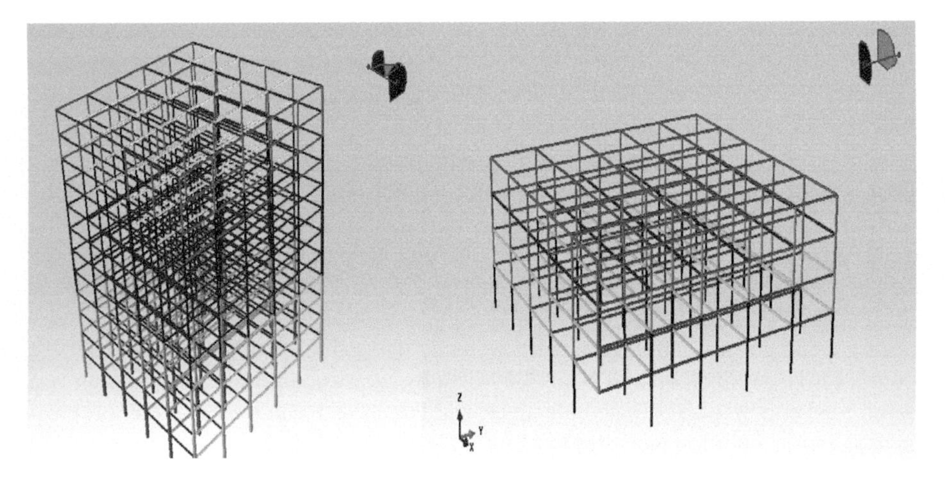

Fig. 27.1 12 and 4-Story model simulated in ABAQUS software

Because of the lack of cyclic behavior in the analysis of the alternative load path method, hysteresis behavior is not modeled. Geometric nonlinearity is included as well as material nonlinearity in the analysis. Moment-resisting frame is a rectilinear assemblage of beams and columns, with the beams rigidly connected to the columns. Failure modes defined in the simulated models include material rupture and element buckling. Mesh size depends on element position; a finer mesh is used around the removed column.

Explicit dynamic analysis method is used to solve the governing differential equations. Its computational efficiency for analyses of extremely discontinuous events with relatively short dynamic response time makes it suitable for the case under investigation. The automatic time increment (based the highest element frequency of the whole model) has been preferred as it accounts for changes in the stability limit with a global time estimator.

Modeling assumptions and analysis method were verified with the experimental test results (Chen et al. 2011). Vertical displacement in location of removal column showed that the modeling results are in good agreement with the experimental ones. The maximum vertical deformation of the model is 4.1 cm and in the experimental sample is about 3.9 cm that the error rate is about 5%. The 12 & 4-story model is depicted in Fig. 27.1.

All three structures have similar plans. Each plan includes five steel moment frames in two orthogonal directions. In spans without column removal, a load combination equal to dead load plus one fourth of the live load is applied, and in spans with column removal, a load combination twice the former is applied. In total, six buildings with and without mass irregularity are designed according to Iranian Code of Practice for Seismic Resistant Design of Buildings (Standard No.2800). To assess the impact of mass irregularity, a live surcharge is applied on the third, seventh and eleventh floors. The following parameters are used in the design of the considered buildings.

Table 27.1 Beam I sections used in the models

Beam section label	B-1	B-2	B-3	B-4	B-5	B-6	B-7
Height (cm)	30	30	30	30	40	40	40
Flange width (cm)	20	20	25	25	25	25	30
Flange thickness (cm)	1	1.5	1.5	2	1.5	2	2
Web thickness (cm)	0.8	1	1	1	1	1	1

Table 27.2 Column Box sections used in the model

Column label	C-1	C-2	C-3	C-4	C-5	C-6	C-7	C-8	C-9	C-10	C-11
Width (cm)	30	30	30	30	40	40	40	50	50	60	60
Thickness	0.8	1	1.5	2	1	1.5	2	1.5	2	2	2.5

- Used steel A36 (according to ASTM, (Es = 206000Mpa, fy 331 MPa)).
- Lateral resistant system: OMRF (ordinary moment resisting frame).
- Square-shaped plan consisting comprising six spans of 5 m length.
- The first story is 3.5 m high and other stories are 3.2 m high.
- To include mass irregularity in height, a live load equal to 750 Kg/m^2 is applied on the third, seventh and eleventh floors.
- Building is located in a region with very high seismicity level and design acceleration is taken as 0.35 g.
- Soil type classification of the site is type C according to IBC 2006.

Section properties used in beams and columns are given in the Tables 27.1 and 27.2.

27.2.1 Loading and Boundary Conditions

Removal of the column is simulated by running 12 s time-history analysis which is done by applying a load in which loads opposite to those of the equivalent column loads are scaled from zero to one. In order to exclude vibrations induced by removing column, loading is increased steadily to reach removal column load in a 5 s (Fig. 27.2). In order to damp out vibrations induced by the first loading stage, loading is continued for a more 2.5 s constantly and then it is dropped to zero in a time period equal to one-tenth of the building fundamental vertical vibration period. In the current research this time period was considered 0.01 s.

27.2.2 Alternate Load Path Analysis Cases

To assess progressive collapse potential of the selected structures, different scenarios of sudden removal of columns compatible with GSA code recommendations were

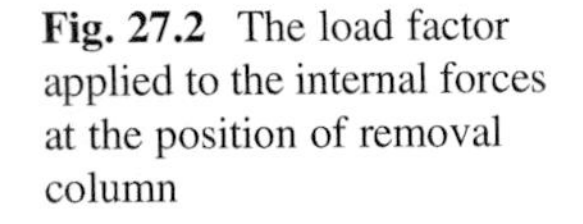

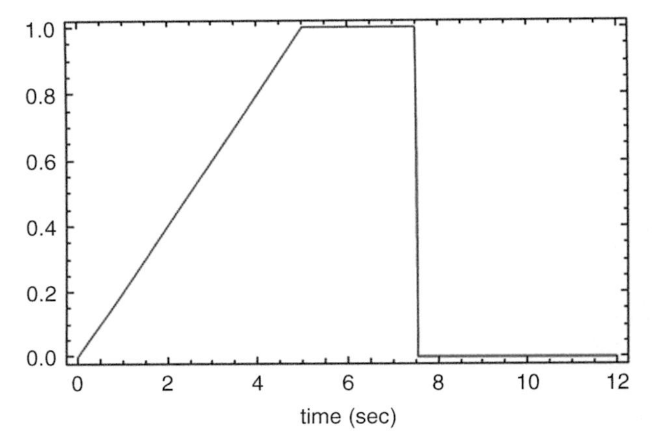

Fig. 27.2 The load factor applied to the internal forces at the position of removal column

considered. Based on the symmetry of the plan and GSA code recommendations, corner column in the first and penultimate floors are chosen to be eliminated. All assumed analyses are listed in Table 27.3. Abbreviations used in the table are as follows:

M- represents type of the structural system (M for moment resisting frame), n- Number of floors, R or I- R represents mass regularity and I shows mass irregularity.

27.3 Numerical Results

In Figs. 27.3 and 27.4 time history of vertical displacement at the joint due to sudden column removal of the first floor are shown. As it is observed, the models that have mass irregularity show larger displacement compared to the regular ones. For example, maximum dynamic deflection at the joint in irregular buildings subjected to sudden column removal in the first floor is about 27, 6 and 15% larger than that of a regular one for four-, eight- and twelve-story buildings respectively. This indicates that mass irregularity in height, makes building much more vulnerable to progressive collapse.

In addition, for column removal in the first floor, results showed that increasing the number of floors decreases vertical displacement of removed column point. For example, maximum dynamic displacement at the point of removed column in 4-story regular building is about 1.41 times larger than that of regular 8-story building and about 2.16 times larger than that of regular 12-story building. The same results were observed for irregular buildings with sudden column removal in the first floor. 4-story irregular building has maximum dynamic displacement under the removed column 1.68 times larger than that of irregular 8-story building and 2.39 times larger than that of 12-story building. In other words, increasing the number of floors decreases progressive collapse potential due to column removal in the first floor. This could be attributed to the fact that increasing beam dimensions as a result

Table 27.3 Different analysis cases

Floor of the removed column	1	1	1	1	1	1	3	7	11	3	7	11
Number of stories and regularity status	M4R	M8R	M12R	M4I	M8I	M12I	M4R	M8R	M12R	M4I	M8I	M12I
Analysis label	APM1	APM2	APM3	APM4	APM5	APM6	APM7	APM8	APM9	APM10	APM11	APM12

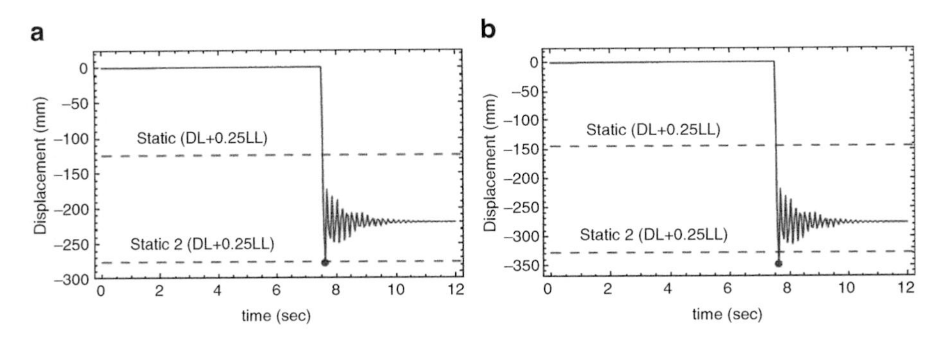

Fig. 27.3 Vertical deflection time history at the joint, subjected to sudden column removal in the first floor 4-story (**a**) regular building -APM1-, (**b**) irregular building- APM4

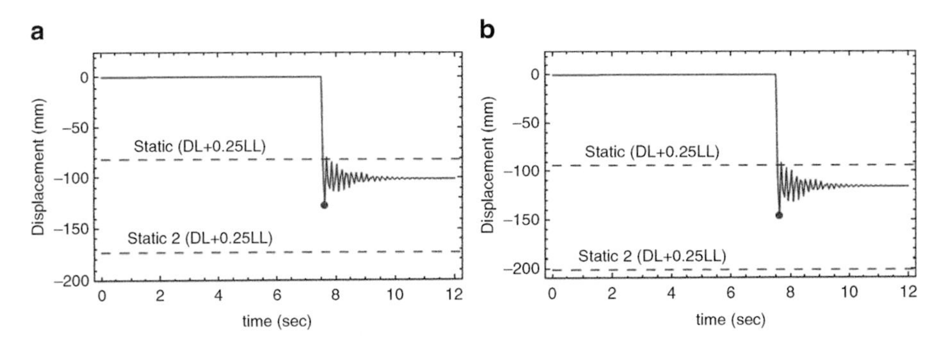

Fig. 27.4 Vertical deflection time history at the joint, subjected to sudden column removal in the first floor 12-story building (**a**) regular building –APM3-, (**b**) irregular building- APM6

of increasing the number of building floors, rises flexural strength of the elements transferring load from removed column to adjacent ones as well as increasing the number of transfer paths. Hence, vertical displacement due to column removal decreases. Resisting mechanism against progressive collapse is flexural strength of the beams connected to the top of the removed column. This conclusion holds for both regular and irregular models.

It is evident from the past events like World Trade Center attack that structural damage in top floors could lead to the stories falling over the other and cause total collapse of the structure. In order to evaluate progressive collapse potential following column removal in top floors, corner column in penultimate floor is eliminated.

Referring to Figs. 27.5 and 27.6, vertical displacement at the removed column node is no longer a function of floor numbers and depends on the flexural strength of the beams connected to the top of the removed column. Since identical beam sections are used in top floors of all modeled buildings, vertical displacement under the removed column in both regular and irregular models nearly are same. In addition, Vertical displacement time history due to column removal in penultimate floor is much larger than those due to column removal in the first floor. This indicates likeliness of propagation of plastic region at beams because of the large

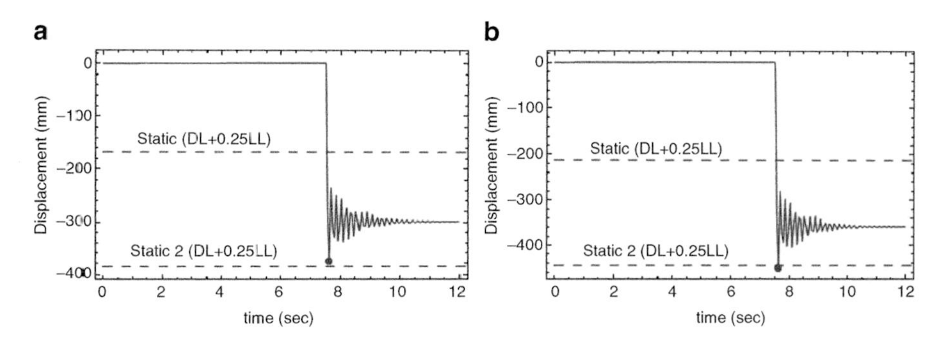

Fig. 27.5 Vertical deflection time history at the joint, subjected to sudden column removal in the penultimate floor 4-story building (**a**) regular building –APM7-, (**b**) irregular building- APM10

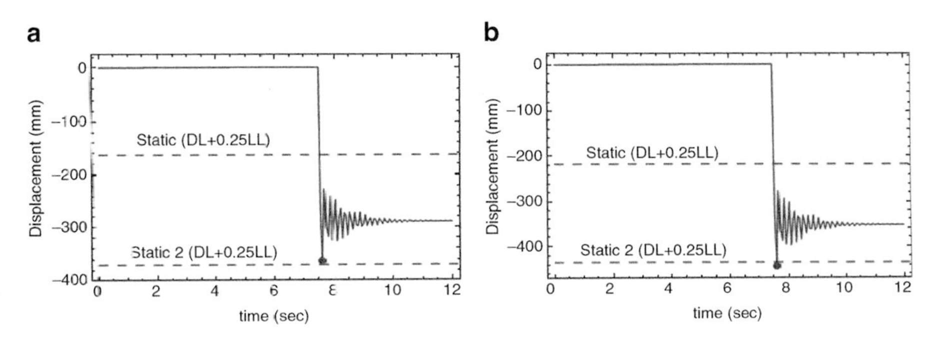

Fig. 27.6 Vertical deflection time history at the joint, subjected to sudden column removal in the penultimate floor 12-story building (**a**) regular building –APM9-, (**b**) irregular building- APM12

deformations. Therefore, in order to calculate vertical displacement it is needed to define nonlinear materials. It is worth mentioning that considering material nonlinearity causes permanent deformations in the structure.

Although in the present research no structural collapse occurs due to column removal in penultimate floor, vertical displacement at the point of the removed column is significantly large. Occurrence of plastic strains in top floor beams necessitates to design them highly ductile and capable of undergoing large deformations. It is also needed to control them withstand axial forces due to catenary action of progressive collapse.

Another result is that mass irregularity in height increases progressive collapse potential in cases with column removal in penultimate floor as well. Results showed that maximum dynamic displacement under removed column in irregular buildings subjected to column removal in penultimate floor is 19, 20 and 22% larger than that of regular one for four-, eight- and twelve-story buildings respectively. In other words, for irregular buildings, vertical displacement is, averagely, 20% larger than that of a regular one. The reason could be attributed to added live surcharge to the floor just above the removed column and the lack of enough load alternate paths to dissipate suddenly exerted forces.

Static and dynamic analyses results showed that it is not generally true to extract a coefficient to convert static analysis results to a dynamic one. Location of the removed column and the amount of nonlinear deformations influence that coefficient greatly. For models with column removal in the first floor, the coefficient is smaller than 2 and in models with column removal in penultimate floor the coefficient is close to 2. Results of the alternate load path method analyses were listed in Table 27.4 for all cases.

27.4 Conclusion

In this paper, mass irregularity effect on the progressive failure potential was investigated. The models studied in this study are 4, 8 and 12 floors building. Dynamic analysis was applied on the 3D finite element models.

Results showed that in models with column removal in the first floor (APM1 to APM6), by increasing the number of floors, vertical displacement at the point of removed column decreases. For example, maximum dynamic displacement in regular 4-story building is about 1.41 times larger than that of regular 8-story building and about 2.16 times larger than that of regular 12-story building. The same results are observed for irregular buildings. 4-story irregular building has maximum dynamic displacement beneath the removed column 1.68 times larger than that of irregular 8-story building and 2.39 times larger than that of 12-story building. This shows that increasing the number of floors, decreases progressive collapse potential due to column elimination in lower floors. The reason is that by increasing the number of floors and consequently beams dimensions to resist lateral forces, flexural strength of elements transferring load from the removed column to adjacent ones increases.

- Results of analyses for the models of APM7 to AMP12 (column elimination in penultimate floor) reveal that vertical displacement under the removed column in these models do not depend on the number of floors and is a function of flexural strength of the beams connected to the top of the removed column.
- Column removal in penultimate floor causes large vertical displacement, therefore beams connected to the removed column need to be highly ductile and capable of undergoing large deformations. It is also needed to control them withstand axial forces due to catenary action of progressive collapse.
- Mass irregularity in height increases progressive collapse potential totally. Maximum dynamic deflection under removed column in irregular buildings subjected to sudden column removal in the first floor is about 27, 6 and 15 percent larger than that of a regular one for four-, eight- and twelve-story buildings respectively.
- Static and dynamic analyses results show that it is not generally true to extract a coefficient to convert static analysis results to a dynamic one. Location of the removed column and the amount of nonlinear deformations influence that coefficient.

Table 27.4 Maximum vertical displacement at the point of the removed column in different analysis cases

Analysis case	APM1	APM2	APM2	APM4	APM5	APM6	APM7	APM8	APM9	APM10	APM11	APM12
Δ_{max} (mm)Static (DL + 0.25LL)	124	96	81	143	108	93	167	178	162	212	224	218
Δ_{max} (mm) Static 2(DL + 0.25LL)	276	220	173	327	235	201	385	404	369	445	465	434
Δ_{max} (mm) Dynamic (DL + 0.25LL)	273	194	126	348	207	145	376	390	363	450	471.25	443

References

Chen J, Huang X, Ma R, He M (2011) Experimental study on the progressive collapse resistance of a two-story steel moment frame. J Perform Constr Facil 26(5):567–575

Fu F (2009) Progressive collapse analysis of high-rise building with 3-D finite element modeling method. J Constr Steel Res 65(6):1269–1278

GSA, U (2003) Progressive collapse analysis and design guidelines for new federal office buildings and major modernization projects. General Services Administration, Washington, DC

Jeyarajan S, Liew JR, Koh C (2015) Progressive collapse mitigation approaches for steel-concrete composite buildings. Int J Steel Struct 15(1):175–191

Khandelwal K, El-Tawil S, Sadek F (2009) Progressive collapse analysis of seismically designed steel braced frames. J Constr Steel Res 65(3):699–708

Kim J, Choi H, Min KW (2011) Use of rotational friction dampers to enhance seismic and progressive collapse resisting capacity of structures. Struct Design Tall Spec Build 20 (4):515–537

Kim J, Kim T (2009) Assessment of progressive collapse-resisting capacity of steel moment frames. J Constr Steel Res 65(1):169–179

Lee K, Kim T, Kim J (2009) Local response of W-shaped steel columns under blast loading. Struct Eng Mech 31(1):25–38

Liu J (2010) Preventing progressive collapse through strengthening beam-to-column connection, part 2: finite element analysis. J Constr Steel Res 66(2):238–247

Lu X, Li Y, Guan H, Ying M (2017) Progressive collapse analysis of a typical super-tall reinforced concrete frame-core tube building exposed to extreme fires. Fire Technol 53(1):107–133

Mashhadiali N, Gholhaki M, Kheyroddin A, Zahiri-Hashemi R (2016) Analytical evaluation of the vulnerability of framed tall buildings with steel plate shear wall to progressive collapse. Int J Civil Eng 14(8):595–608

Pretlove A, Ramsden M, Atkins A (1991) Dynamic effects in progressive failure of structures. Int J Impact Eng 11(4):539–546

Qian K, Li B, Zhang Z (2015) Testing and simulation of 3D effects on progressive collapse resistance of RC buildings. Mag Concr Res 67(4):163–178

Ren P, Li Y, Guan H, Lu X (2014) Progressive collapse resistance of two typical high-rise RC frame shear wall structures. J Perform Constr Facil 29(3):04014087

Tsai M-H, Lin B-H (2008) Investigation of progressive collapse resistance and inelastic response for an earthquake-resistant RC building subjected to column failure. Eng Struct 30 (12):3619–3628

Part III
Seismic Control and Monitoring of Irregular Structures

Chapter 28
On the Response of Asymmetric Structures Equipped with Viscous Dampers Subjected to Simultaneous Translational and Torsional Ground Motion

Jafar Kayvani, Gholamreza Nouri, Shahin Pakzad, and Morteza Tahmasebi Yamchelou

Abstract Past experience of earthquakes has shown the importance of asymmetry effects on the severity of the damage sustained by structures. In order to mitigate torsional response of asymmetric structures under seismic loads, employing passive energy dissipating equipment, such as viscous dampers, has been investigated by many researchers during recent years. One decisive factors contributing to the seismic behaviour of asymmetric structures is the torsional component of strong ground motion. The main purpose of this research is to compute a damping eccentricity which minimizes torsional response of asymmetric structures. To this end, several one-storey, three-dimensional steel moment frames with various stiffness and strength eccentricities are studied. Due to the limitations in recording torsional component of earthquakes, translational components are used to develop torsional counterparts. Models were analyzed using time history analysis method under 7 records of strong ground motions. Results indicate that incorporating viscous dampers could not only mitigate structural response, but it also could suppress torsional deflections due to asymmetry in structures. Results showed by increasing the asymmetry of the structure, optimal damping eccentricity should be shifted towards the flexible edge to reduce the displacement difference considerably.

Keywords Asymmetric structures · Torsional response · Damping eccentricity · Translational and torsional ground motion

J. Kayvani · G. Nouri (✉) · S. Pakzad · M. T. Yamchelou
Department of Civil Engineering, Faculty of Engineering, Kharazmi University, Tehran, Iran
e-mail: r.nouri@khu.ac.ir

D. Köber et al. (eds.), *Seismic Behaviour and Design of Irregular and Complex Civil Structures III*, Geotechnical, Geological and Earthquake Engineering 48,
https://doi.org/10.1007/978-3-030-33532-8_28

28.1 Introduction

Behaviour of buildings during the past earthquakes indicates that asymmetric structures are more vulnerable to damage than symmetric ones and undergo more serious deteriorations. A structure is deemed to be asymmetric when its centre of mass does not coincide with its centre of stiffness. Asymmetric structures exhibit torsional displacement in addition to the lateral displacements, and therefore, experience torsional moments in addition to the lateral forces. The origin of these torsional excitations could be ascribed to various factors, such as torsional nature of ground motions with respect to the vertical axis of the structure, noncoincidence of the centre of mass and the centre of stiffness due to asymmetry in the arrangement of lateral resisting elements, and irregular distribution of mass. Investigating the behaviour of such structures during earthquakes reveals that the inflicted damage is mainly due to the excessive torsional moments and displacements emerging from asymmetry of the diaphragms, which causes damage in structural and nonstructural elements, especially, at exterior edges of the structure. If the distribution of mass, stiffness, and strength is not uniform, earthquake-induced loads and lateral resisting forces do not coincide. Subsequently, torsional moments are developed in case the diaphragm is rigid or semi-rigid. During recent years, controlling the structural response by making use of energy-dissipating systems has become a practical approach for protecting structures against wind and earthquake loads. Making use of viscous damper is an effective approach to controlling torsional response of asymmetric structures, provided that an optimal arrangement of dampers is achieved. One of the advantageous usages of viscous damper is for circumstances in which the architectural limitations increase the stiffness and strength asymmetry significantly. In such cases, a proper distribution of supplementary dampers can mitigate the torsional response in the edges.

Investigation of inelastic behaviour of asymmetric structures was carried out by modelling an asymmetric one-storey building (Kan and Chopra 1976). They concluded that the effect of the torsional moment in the nonlinear region significantly depends on the ratio of uncoupled torsional frequency to the translational frequency (Ω). Study of the seismic behaviour of linear asymmetric one-story buildings with supplemental damping revealed that changing the arrangement and location of supplemental damping devices throughout the plan had a minor effect on modal parameters, such as the period and the modal participation factor; while, dynamic amplification factor and modal damping ratio were highly affected (Goel 2000). Investigating the torsional effects on the displacement of ductile buildings indicated that both strength and stiffness eccentricity affect the torsional behaviour of the structure; however, the effect of strength eccentricity is more considerable during the ductile response of the structure (Crisafulli et al. 2004).

Torsional ground motions occurring during an earthquake could affect the seismic response of structures, particularly, when the center of stiffness and the center of mass do not coincide. Properties of rotational ground motion using translational records in Chiba dense array were estimated and effect of torsional ground motion on the structural response have been studied by several scholars (Ghayamghamian and Nouri 2007; Ghayamghamian et al. 2009; Nouri et al. 2010, 2016). Their findings indicate that for the most structural periods in the engineering practice assuming 5% accidental eccentricity is on the safe side. Although, for structures with periods shorter than 0.3 s, the inclusion of torsional components increases the building displacement up to four times. The concept of torsional equilibrium was used to determine the arrangement of the dampers which optimize the torsional behaviour of asymmetric structures (De la Llera et al. 2005). Their investigation on one- and multi-story building structures revealed that such arrangement of dampers minimizes the diaphragm rotation, as well as the lateral displacement of the structure. The would be an optimal position for the centre of strength to mitigate ductility demand of one-storey models, which have lateral resisting elements in both orthogonal directions (De Stefano et al. 1993).

This research investigates torsional effects in asymmetric structures subjected to earthquake excitation, with the inclusion of rotational components, and seeks to improve the structural performance of such structures through the optimal arrangement of supplementary dampers. The main focus of this research is the effects of rotational component of the earthquake on the response of asymmetric one-story structures. Investigated responses include the displacement difference between the stiff and flexible edges, diaphragm rotation, and base shear. To calculate the response of an asymmetric structure subjected to earthquake excitations, nonlinear time history analysis was performed.

28.2 Theoretical Framework

28.2.1 Determining the Stiffness and the Strength Eccentricity

A structural plan similar to the one depicted in Fig. 28.1 is considered. m lateral resisting elements are considered along X axis, and n elements along Y axis. Stiffness and strength of the i-th element in X and Y directions are denoted by k_{xi}, f_{xi}, k_{yi}, and f_{yi}, respectively. The origin of the coordination system is the center of the mass of the diaphragm. To create stiffness eccentricity, dimension of beams and columns are increased at the stiff (left) edge and decreased at the flexible (right) edge of the structural plan. Total stiffness and strength along X and Y directions are as Eq. 28.1.

$$k_x = \sum_1^m k_{xi}; k_y = \sum_1^n k_{yi}; f_x = \sum_1^m f_{xi}; f_y = \sum_1^n f_{yi} \qquad (28.1)$$

The torsional stiffness of the structure can be stated as Eq. 28.2.

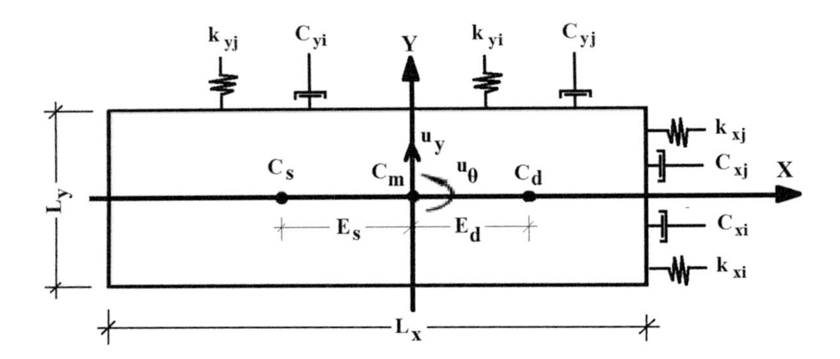

Fig. 28.1 Structural diaphragm representing lateral resisting elements

$$k_\theta = \sum_1^m \left(k_{xi}y_i{}^2\right) + \sum_1^n \left(k_{yi}x_i{}^2\right); f_\theta = \sum_1^m \left(f_{xi}y_i{}^2\right) + \sum_1^n \left(f_{yi}x_i{}^2\right) \tag{28.2}$$

Where x_i and y_i denote the distance of the element i from X and Y axes, respectively. Absolute and normalized eccentricities are defined as:

$$E_{sx} = \frac{1}{k_y}\sum_1^n x_i k_{yi}; e_{sx} = \frac{E_{sx}}{L_x} = \frac{1}{L_x k_y}\sum_1^n x_i k_{yi} \tag{28.3}$$

$$E_{sy} = \frac{1}{k_x}\sum_1^m y_i k_{xi}; \quad e_{sy} = \frac{E_{sy}}{L_y} = \frac{1}{L_y k_x}\sum_1^m y_i k_{xi} \tag{28.4}$$

In the Eqs. 28.3 and 28.4 L_x and L_y denote floor dimensions along X and Y axes, respectively (Sarvghad-Moghadam 1998).

28.2.2 Determining Uncoupled Translational and Torsional Frequencies

One key parameter affecting the behaviour of asymmetric structures is the ratio of uncoupled torsional frequency to translational frequency. Uncoupled torsional frequency can be calculated by determining the torsional stiffness of the structure about the centre of stiffness, which is stated by Eq. 28.5.

$$k_{\theta_{cs}} = k_\theta - E_{sx}^2 k_y - E_{sy}^2 k_x \tag{28.5}$$

So the uncoupled torsional frequency can be expressed as:

$$\omega_\theta = \sqrt{\frac{k_{\theta_{cs}}}{I_{CM}}} I_{CM} = {}^{m}\!/_{12}\left(L_x^{\,2} + L_y^{\,2}\right) = m.\rho_m^2 \tag{28.6}$$

Where ρ_m^2 denotes mass gyration radius and I_{CM} is the mass moment inertia. If the ratio of uncoupled torsional frequency to uncoupled lateral frequency (Ω) is greater than unity, the structure is torsionally stiff, if it is equal to unity the structure is torsionally coupled, otherwise the structure is torsionally flexible.

28.2.3 Determining Damping Eccentricity

If C_{xi}, C_{yi} and $x_i$and y_i are the damping coefficients and distance of the damper from the center of mass in the *x* and *y* directions, total damping along *x* and *y* axes and torsional damping are calculated as Eq. 28.7.

$$c_x = \sum_i c_{xi},\, c_y = \sum_i c_{yi},\, c_\theta = \sum_i c_{xi} y_i^{\,2} + \sum_i c_{yi} x_i^{\,2} \tag{28.7}$$

Similar to the definition used for the stiffness eccentricity, damping eccentricity (E_{dx}, E_{dy}) and normalized damping eccentricity (e_{dx}, e_{dy}) along *x* and *y* axes can be written as Eqs. 28.8 and 28.9.

$$E_{dx} = \frac{1}{c_y} \sum_1^n x_i c_{yi},\, e_{dx} = \frac{E_{dx}}{L_x} = \frac{1}{L_x c_y} \sum_1^n x_i c_{yi} \tag{28.8}$$

$$E_{dy} = \frac{1}{c_{yx}} \sum_1^n y_i c_{xi},\, e_{dy} = \frac{E_{dy}}{L_y} = \frac{1}{L_y c_x} \sum_1^m y_i c_{xi} \tag{28.9}$$

One of the major parameters in assessing the response of structures equipped with supplementary viscous dampers is the torsional gyration radius. To calculate this parameter, it is required to determine the torsional damping about the centre of damping:

$$c_{\theta_{csa}} = c_\theta - E_{dx}^{\,2} c_y - E_{dy}^{\,2} c_x \tag{28.10}$$

The damping gyration radius can be expressed as:

$$\rho_{sd,x} = \sqrt{\frac{c_{\theta_{csd}}}{c_x}}, \rho_{sd,y} = \sqrt{\frac{c_{\theta_{csd}}}{c_y}} \tag{28.11}$$

So, for a one-story building the damping matrix was described as Eq. 28.12:

$$[C_{sd}] = \begin{bmatrix} c_x & 0 & -e_{dy}c_x \\ 0 & c_y & e_{dx}c_y \\ -e_{dy}c_x & e_{dx}c_y & c_\theta \end{bmatrix} \tag{28.12}$$

28.3 The Model Specification

In order to evaluate the effect of dampers on modifying the torsional behaviour of irregular structures, the optimal arrangement of dampers in a simplified structural model is studied. In this research, a one-story building with one-way stiffness, strength, and damping eccentricity is considered. The basic assumptions made in this research are as follows.

The model is a one-story building with an ordinary moment frame as the lateral resisting system. All frames are assumed to be fixed at the base and the soil-structure interaction is ignored. Translational and torsional components of the earthquake are taken into account. Dampers are assumed to be ideally viscous with linear force-velocity behaviour. The uncertainty in the determination of damping ratio of dampers and the variations of damping during the earthquake excitation is not considered.

The story height is 3.2 m and the structure is symmetric along both X and Y directions. The plan of studied structure is illustrated in Fig. 28.2.

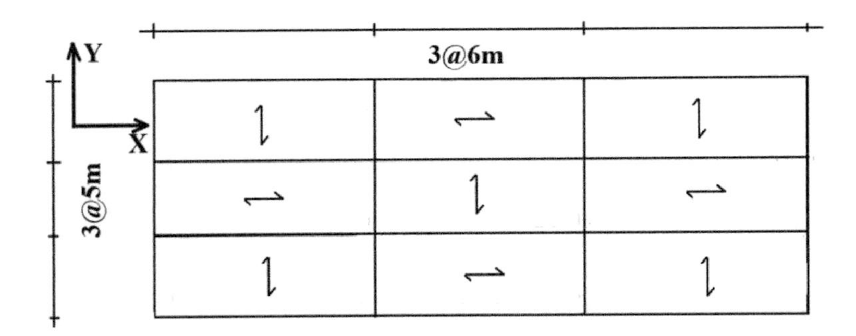

Fig. 28.2 One-storey building configuration in the plan

Dead load and live load are assumed 650 kg/m^2 and 200 kg/m^2, respectively. The building is located in a region with high level of seismicity ($A = 0.35g$) according to Iranian code of practice for seismic resistant design of buildings and the site soil is of type II.

Six asymmetric models are generated by changing the section area of beams and columns along the y axis of the base model. The amount of increase and decrease in cross section of beams and columns are adjusted so the lateral stiffness of all models along x axis are the same. Nonlinear static pushover analysis is performed to determine the strength and stiffness of frames.

In order to create various damping eccentricities, linear viscous dampers are added to the frames along Y-axis as bracing. Therefore, it is assumed that in X direction no damper is installed and eccentricity of the damping is one-way just like the stiffness eccentricity. No strength and stiffness is considered for dampers and it is assumed that they function ideally viscous and do not affect the strength and stiffness of the frame.

As changing the parameters of dampers influences the structural response, the total damping ratio of dampers in Y direction is taken constant to increase the accuracy of the results. The value of lateral damping ratio for a structure equipped with dampers depends on the amount of structural response reduction, such as relative displacement and base shear. Indices for relative story displacement, story acceleration, base shear, and overturning moment in two cases of being equipped with dampers and without dampers could be found in the literature (Singh and Moreschi 2001). They suggested that a 40% decrease in relative story displacement can be a good index for determining the value of overall lateral damping ratio.

The same approach is adopted in this study and total damping ratio along Y axis is adjusted so that the reduction in relative displacement of the center of mass for asymmetric building is 40%. By considering a 5% inherent damping ratio for the structure, the total damping ratio is taken as 20%. Seven damping eccentricities of 0, ± 0.167, ± 0.33, and $\pm$ 0.5 were assumed in this study. For damping eccentricity equal to zero, the maximum damping gyration radius is obtained, and leads to the smallest response in both soft and stiff edges of the structure, which agrees with the results of a previous study (Goel 1998).

28.4 Seismic Assessments

Earthquake records for performing nonlinear time history analysis are listed in Table 28.1 and critical incidence angle to obtain the most critical responses (lateral displacement, diaphragm rotation, and base shear) is determined. Time derivation method was used to generate rotational components of the selected earthquake records (Ghafory-Ashtiany and Singh 1986). In this research, seven earthquakes recorded at sites located on a type C soil, according to NEHRP, was selected (Provisions 1997). The magnitude of selected earthquakes is between 6 and 7.6. All selected earthquake records are far-field type with a minimum distance of 25 km

Table 28.1 Characteristics of selected earthquake records

ID	Earthquake	Year	Magnitude	Duration (sec)	PGA (g)		Site	Distance (km)
					E	N		
E01	Chi-Chi	1999	7.6 M	35	0.301	0.413	TCU047	33
E02	Northridge	1994	6.7 M	20	0.222	0.256	LA-century	25.7
E03	Manjil	1990	7.4 Mw	25	0.131	0.184	Qazvin	49.47
E04	Kern County	1952	7.4 Mw	25	0.156	0.178	Taft	41
E05	Imperial Valley	1979	6.5 M	40	0.157	0.169	Cerro Prieto	26.5
E06	N. Palm Spring	1986	6 M	20	0.239	0.250	San Jacinto	25
E07	Cape Mendocino	1992	7.1 M	40	0.229	0.189	Shelter Cove	33.8

Table 28.2 Critical angle of selected earthquake records

Earthquake ID	E01	E02	E03	E04	E05	E06	E07
Critical incidence angle	0	60	30	60	0	60	0

from the seismic source. For time history analysis, earthquake records were scaled to 0.15 g, 0.35 g, 0.55 g, and 0.75 g. Table 28.1 presents the characteristics of selected earthquake records.

Strong ground motion can be applied to the structure in two principal directions. Although, it has been shown that for a particular incidence angle structural response can be much greater in comparison with those of principal directions. Considering this, firstly, the ground motion was applied to the structure in principal directions and then the incidence angle was increased by 5 degrees in each stage and the corresponding responses were calculated to find the critical incidence angle. For the selected ensemble of earthquakes, critical incidence angles are listed in Table 28.2.

28.5 Earthquake Rotational Components

The ground motion comprises three translational and three rotational components. The rotational component comprises two rocking components (about horizontal axes) and one torsional component (about the vertical axis). To generate rotational components of the earthquake, time derivation method was adopted here. The following expression for generating rotational components of the earthquake using translational ones can be found in the literatue (Ghafory-Ashtiany and Singh 1986):

$$\ddot{\psi}_k(t) = -\frac{1}{2c_j}\frac{d}{dt}\left[\ddot{X}_j(t) - \ddot{X}_i(t)\right] \tag{28.13}$$

Where $\ddot{\psi}_k(t)$ is the rotational component about k axis, $\ddot{X}_j(t)$ and $\ddot{X}_i(t)$ denote translational components along i and j directions, c_j stands for shear wave velocity along x_j direction, and x_i, x_j, and x_k represent principal axes.

28.6 Numerical Results

Seven models with different damping eccentricities of 0, ±0.167, ±0.33, and ± 0.5 were created and nonlinear time history analysis was performed using earthquake records scaled to 4 different PGAs. Three different response parameters were calculated for the aforementioned cases. Furthermore, to study the effect of earthquake rotational components on structural response all models were subjected to Northridge and Kern County earthquake records without rotational components. As the obtained responses followed the same trend, only the results for Chi-Chi earthquake are presented. The first examined response was the displacement difference between flexible and stiff edges. (e_s) and (e_r) are the stiffness and strength eccentricities, respectively.

Figure 28.3 illustrates the displacement difference between the flexible (right) edge and stiff (left) edge versus different damping eccentricities. As it is observed in Fig. 28.3a (The symmetric model) optimal damping eccentricity to achieve the greatest reduction in the displacement difference between flexible and stiff edges shifts towards the stiff edge.

As the strength and stiffness eccentricities move towards the stiff edge of the structure (Fig. 28.3b), displacement of flexible edge increases and that of the stiff edge decreases. Therefore, by increasing the asymmetry of the structure, optimal damping eccentricity shifts towards the flexible edge to cause the most reduction in the displacement difference between flexible and stiff edges. Yet, the maximum

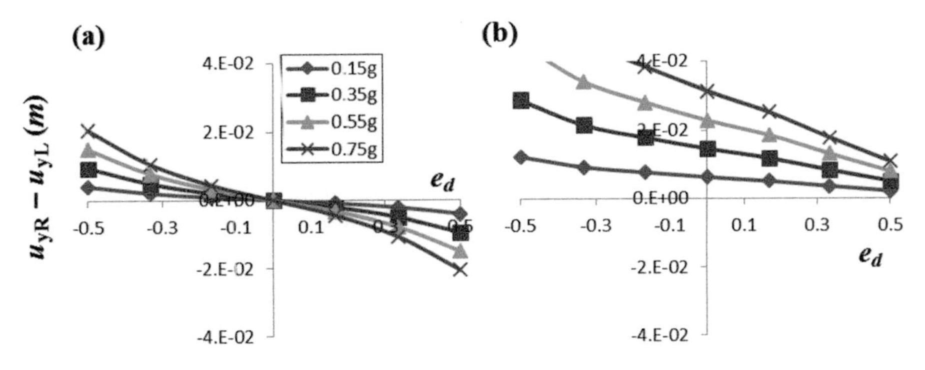

Fig. 28.3 Maximum displacement difference between stiff and flexible edges versus damping eccentricity under Chi-Chi earthquake: (**a**) $e_r = 0$, $e_s = 0$; (**b**) $e_r = -0.24$, $e_s = -0.22$

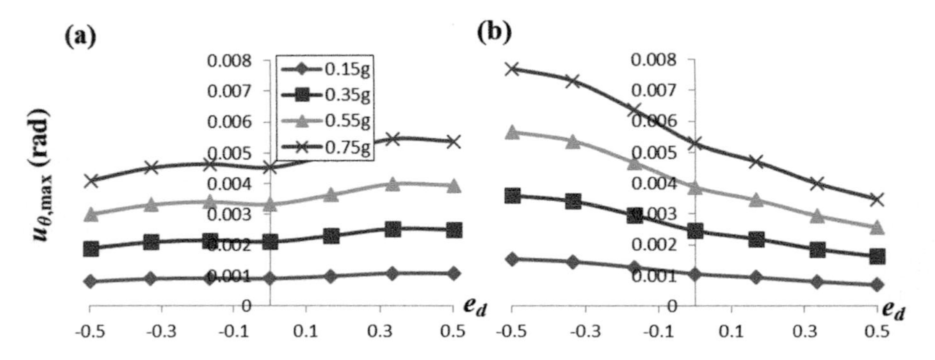

Fig. 28.4 Diaphragm rotation versus damping eccentricity under Chi-Chi earthquake: (**a**) $e_r = 0$, $e_s = 0$; (**b**) $e_r = -0.24$, $e_s = -0.22$

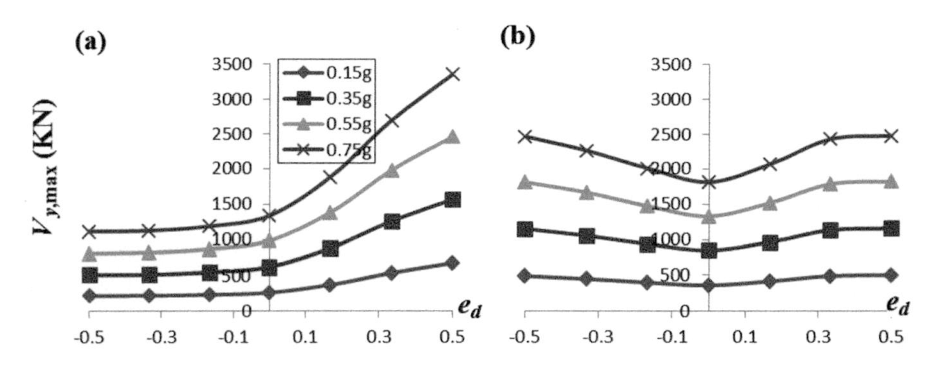

Fig. 28.5 Base shear versus damping eccentricity under Chi-Chi earthquake: (**a**) $e_r = 0$, $e_s = 0$; (**b**) $e_r = -0.24$, $e_s = -0.22$

damping eccentricity cannot balance the lateral displacement, and in order to reduce the displacement difference between flexible and stiff edges, greater damping eccentricity or higher damping capacity is required. Figure 28.4 shows the maximum diaphragm rotation versus different damping eccentricities for various strength and stiffness eccentricities. As it is observed in Fig. 28.4a, in the symmetric model, to obtain the most reduction in diaphragm rotation, the damping eccentricity needs to be shifted towards the stiff edge. But, by moving the strength and stiffness eccentricities towards the stiff edge of the structure (Fig. 28.4b), optimal damping eccentricity increases and shifts towards the flexible edge of the structure. In other words, maximum diaphragm rotation occurs when the damping eccentricity lies at the right (flexible) edge of the structure. Another important point that can be inferred from the graph is that the variations of optimal damping eccentricity to reduce the diaphragm rotation are similar to that of the displacement difference between the stiff and flexible edges.

In Fig. 28.5, base shear versus damping eccentricity for two different strength and stiffness eccentricities are illustrated. As it is seen in the Fig. 28.5, neglecting the effect of damper arrangement, by shifting the strength and stiffness eccentricity

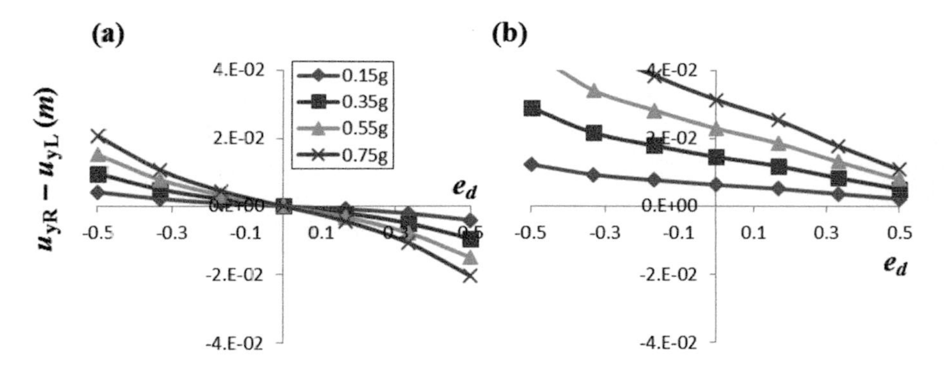

Fig. 28.6 Maximum displacement difference between stiff and flexible edges versus damping eccentricity under Northridge earthquake without including rotational components: (**a**) $e_r = 0$, $e_s = 0$; (**b**) $e_r = -0.24$, $e_s = -0.22$

towards the stiff edge of the structure, base shear decreases. Moreover, the optimal damping eccentricity to achieve the most reduction in base shear occurs when strength and stiffness eccentricities lie on the stiff side of the structure.

28.6.1 *Effects of Rotational Component of the Earthquake*

To examine the effect of including rotational components of the earthquake on the structural performance of the studied structure, response parameters studied in the previous section were calculated.

Figure 28.6 illustrates the displacement difference between stiff and flexible edges under Northridge earthquake without including the rotational component of the earthquake. As it is observed in Fig. 28.6a, in a symmetric structure which is subjected only to translational components of the earthquake, optimal damping eccentricity is equal to zero and by increasing the asymmetry towards the left edge of the structure, it moves towards the flexible edge (Fig. 28.6b). This agrees well with the results of previous studies (Goel 1998; Mansoori and Moghadam 2009). According to the results, ignoring the rotational component can cause an error equal to 97% and 350% in optimal damping eccentricity for $e_r = -0.05$, $e_s = -0.05$ and $e_r = -0.08$, $e_s = -0.05$, respectively.

Figure 28.7 shows diaphragm rotation under Northridge earthquake without including rotational components. In Fig. 28.7a (symmetric structure) optimal damping eccentricity is zero, but by increasing the strength and stiffness eccentricities towards the stiff edge of the structure, the optimum damping eccentricity increases and shifts towards the flexible edge (Fig. 28.7b). Finally, the base shear experienced under Northridge earthquake without including rotational components is shown in Fig. 28.8. Results reveal that the optimal damping eccentricity lies at the stiff edge no matter the rotational component of the earthquake is included or not.

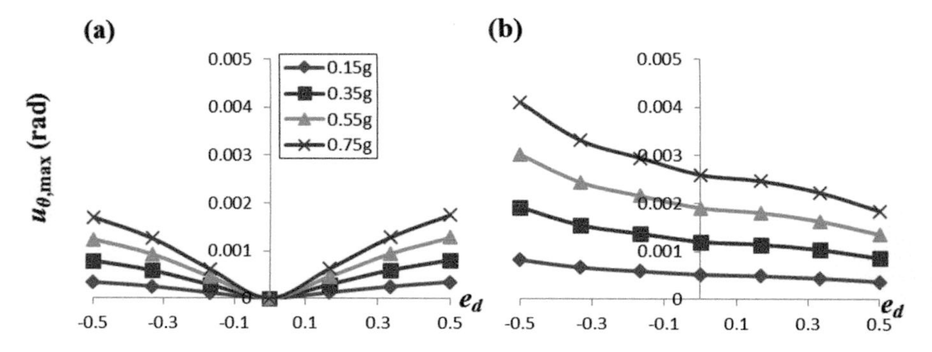

Fig. 28.7 Diaphragm rotation versus damping eccentricity under Northridge earthquake without including rotational components: (**a**) $e_r = 0$, $e_s = 0$; (**b**) $e_r = -0.24$, $e_s = -0.22$

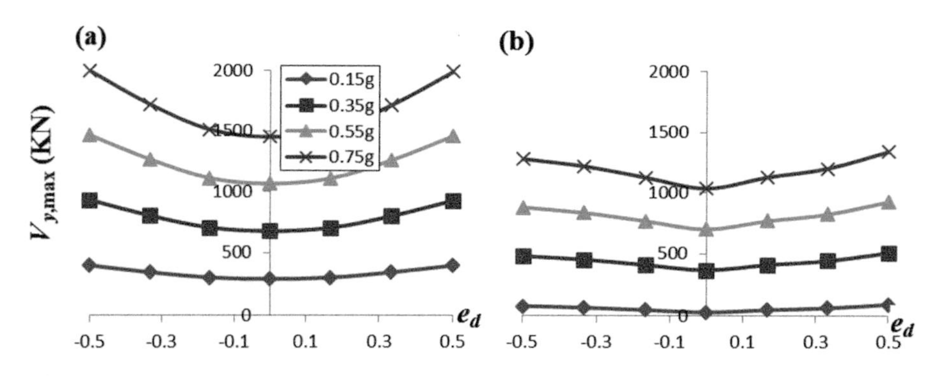

Fig. 28.8 Base shear versus damping eccentricity under Northridge earthquake without including rotational components: (**a**) $e_r = 0$, $e_s = 0$; (**b**) $e_r = -0.24$, $e_s = -0.22$

28.7 Conclusions

In this research, seismic response of asymmetric structure, equipped with supplementary dampers, subjected to earthquake excitation including rotational components is studied.

Investigated response parameters include displacement difference between the stiff and flexible edges, diaphragm rotation, and base shear. Results showed that by increasing the asymmetry of the structure, optimal damping eccentricity should be shifted towards the flexible edge to cause the most reduction in the displacement difference between flexible and stiff edges. The same trend is observed for diaphragm rotation. That is, maximum diaphragm rotation occurs when the damping eccentricity lies at the right (flexible) edge of the structure. But, for the base shear, increasing the strength and stiffness eccentricity towards the stiff edge reduces base shear and damping eccentricity should be increased towards the stiff edge to yield maximum base shear reduction. Evaluating the same response parameters for the structure under strong ground motion without rotational components indicate that an error equal to 350% could be raised in determining the optimal damping eccentricity. Similar to the previous case, wherein rotational components were included, by increasing the strength and damping eccentricity towards the stiff edge of the structure, the damping eccentricity shifted towards the flexible edge of the structure to yield maximum reduction in the displacement difference between stiff and flexible edges and diaphragm rotation.

References

Crisafulli F, Reboredo A, Torrisi G (2004) Consideration of torsional effects in the displacement control of ductile buildings. Paper presented at the 13th world conference on earthquake engineering, Vancouver, BC, Canada, paper

De la Llera JC, Almazán JL, Vial IJ (2005) Torsional balance of plan-asymmetric structures with frictional dampers: analytical results. Earthq Eng Struct Dyn 34(9):1089–1108

De Stefano M, Faella G, Ramasco R (1993) Inelastic response and design criteria of plan-wise asymmetric systems. Earthq Eng Struct Dyn 22(3):245–259

Ghafory-Ashtiany M, Singh MP (1986) Structural response for six correlated earthquake components. Earthq Eng Struct Dyn 14(1):103–119

Ghayamghamian M, Nouri G (2007) On the characteristics of ground motion rotational components using Chiba dense array data. Earthq Eng Struct Dyn 36(10):1407–1429

Ghayamghamian MR, Nouri GR, Igel H, Tobita T (2009) The effect of torsional ground motion on structural response: code recommendation for accidental eccentricity. Bull Seismol Soc Am 99 (2B):1261–1270

Goel RK (1998) Effects of supplemental viscous damping on seismic response of asymmetric-plan systems. Earthq Eng Struct Dyn 27(2):125–141

Goel RK (2000) Passive control of earthquake-induced vibration in asymmetric buildings. Paper presented at the proceeding of 12th world conference on earthquake engineering

Kan CL, Chopra AK (1976) Coupled lateral torsional response of buildings to ground shaking

Mansoori M, Moghadam A (2009) Using viscous damper distribution to reduce multiple seismic responses of asymmetric structures. J Constr Steel Res 65(12):2176–2185

Nouri GR, Ghayamghamian MR, Hashemifard M (2010) A comparison among different methods in the evaluation of torsional ground motion. J Iran Geophys 4:32–44

Nouri G, Ghayamghamian M, Hashemifard M (2016) Evaluation of torsional component of ground motion by different methods using dense array data. In: *Seismic behaviour and design of irregular and complex civil structures II*. Springer, Cham, pp 25–34

Provisions, B. P. O. I. S. S (1997) NEHRP recommended provisions for seismic regulations for new buildings and other structures: provisions (Vol. 302). Fema

Sarvghad-Moghadam A (1998) Seismic torsional response of asymmetrical multi-storey frame buildings

Singh MP, Moreschi LM (2001) Optimal seismic response control with dampers. Earthq Eng Struct Dyn 30(4):553–572

Chapter 29
Base Isolation as an Effective Tool for Plan Irregularity Reduction

R. Volcev, N. Postolov, K. Todorov, and Lj. Lazarov

Abstract Irregularity in plan can significantly increase the vulnerability of structures exposed to strong earthquakes. Application of base isolation systems is one of the possible solutions to avoid the negative effects from irregularity of structures in plan. In order to investigate the advantages of the application of these systems, a detailed evaluation of the behaviour of four five story reinforced concrete frame structures was performed. All structures are rectangular in plan and are analysed as fixed base and base isolated. The fixed base models differ between them according to the degree of irregularity. Because base isolated models have a significantly lower eccentricity between the centre of stiffness and the centre of masses, these models are regular in plan. Total obtained displacements in base isolated models are higher, but main part of them are concentrated at the level of the isolation system which results with lower interstorey drifts and small floor rotations.

Keywords Eurocode 8 · Plan irregularity · Base isolation · Dynamic analysis

29.1 Introduction

Irregular distribution of strength, stiffness and mass in building plan, can lead to coupling of torsional and translational oscillations, which in some cases may be uncontrollable and potentially very dangerous. The application of systems for base isolation is one of the possible solutions that can be used for reduction of these kind of negative effects. Despite the large number of researches carried out in the field of the application of base isolation systems for plan irregular structures in the last few decades, research community have not reached a consensus on many topics in this area. Summary of the most relevant findings on the torsional response of base

R. Volcev (✉) · N. Postolov · K. Todorov · L. Lazarov
Faculty of Civil Engineering, University "Ss. Cyril and Methodius", Skopje, Republic of Macedonia
e-mail: volcev@gf.ukim.edu.mk; postolov@gf.ukim.edu.mk; todorov@gf.ukim.edu.mk; lazarov@gf.ukim.edu.mk

D. Köber et al. (eds.), *Seismic Behaviour and Design of Irregular and Complex Civil Structures III*, Geotechnical, Geological and Earthquake Engineering 48,
https://doi.org/10.1007/978-3-030-33532-8_29

isolated structures can be found in Tena-Colunga and Gómez-Soberón (2002), De Stefano and Pintucchi (2008). The most common analysed parameters which have an influence on the behaviour of the base isolated plan irregular structure are: eccentricity of the mass and the stiffness of the superstructure, eccentricity of base isolation system, type and stiffness of the system for isolation, location of isolation interface, etc. Nagarajaiah et al. in two companion papers investigate a torsional coupling in base-isolated structures with inelastic elastomeric isolation system, (Nagarajaiah et al. 1993) and with sliding isolation system (Nagarajaiah et al. 1993a) due to bidirectional lateral ground motions. They examined the influence that numerous system parameters have on the lateral torsional response of base isolated structures, including the stiffness eccentricity in the superstructure and eccentricity in the isolation system. It is shown that the application of base isolation with elastomeric bearings significantly reduce the total response of the superstructure. Despite that, the torsional amplification, which mainly depend upon the eccentricity in the isolation system and the superstructure and from the torsional and lateral flexibility, can be significant. The nonlinear response of torsionally coupled systems with base isolation exposed to random ground motions have been studied by Jangid and Datta (1994). From the obtained results they concluded that the superstructure eccentricity does not have a significant impact on the displacement of the isolation system. They also concluded that the isolator eccentricity reduces the effectiveness that it has on torsional deformation as well as at the reduction of superstructure displacement. With use of nonlinear dynamic analyses, the torsional response of base isolated structures which have isolation eccentricity have been investigated by Tena-Colunga and Zambrana – Rojas (2004). Among other relevant issues, they concluded that the torsional response has negative implication on the design of the isolation system. In general, with the increase of the eccentricity, the amplification factors of the asymmetric system for the maximum isolator displacement increases with respect to the symmetric system. Kilar and Koren (2009) investigate the influence of distribution of base isolation system with lead rubber bearings under asymmetric four storey RC frame building. From the results obtained by the 3D nonlinear dynamic analyses can be concluded that all considered dispositions of bearings, significantly reduce the adverse torsional effects. Stiffness eccentricity of base isolators as well as mass eccentricity of a superstructure have been investigated by Khoshnoudian and Imani Azad (2011). From the presented results they concluded that the effects of near-fault ground motions which are bidirectional would amplify torsional intensification in comparison with the unidirectional ones in a bilinear base isolation system. The effects on seismic response in relation of mass eccentricity on asymmetrical structures with base isolation system composed by TCFP bearings and exposed to near field ground motions have been investigated by Tajammolian et al. (2016). The obtained results show that the mass eccentricities do not have a significant effect on the displacement of the isolator. Opposite to the displacement, the torsional effects of mass eccentricity increase the base shear and amplify the roof acceleration of the analysed models, compared with the symmetric superstructure. It is also concluded the base shear impact is due to the eccentricity that the structure has in the direction of the earthquake motion, while the roof acceleration and the isolator displacement have mostly influenced by the eccentricity perpendicular to the earthquake path.

29.2 Criteria for Regularity in Plan

Most common reasons for the damage of plan irregular structures are the large torsional effects that occur in the event of an earthquake. Therefore, in modern seismic codes for designing of seismically resistant structures, irregularity has implication in relation to seismic design, i.e. it is reflected on the mathematical model of structure, the method of analysis, the intensity of the loads etc. In that direction, the position and functionality of the building, the dimensions in plan and elevation, the disposition of the bearing elements, the materials that would be used during construction etc. are of great importance.

According to Eurocode 8 (CEN 2004), specific criteria should be met in order to consider structure as regular. Some of these conditions are qualitative, and can be checked in the preliminary design stage, but some of them, which are related to the determination of torsional radius or eccentricity between the centre of stiffness and centre of mass, are quantitative and need to be determined additionally. In order to obtain satisfactory seismic behaviour, the mass and lateral stiffness at conventional designed structures need to have approximately symmetrical distribution in plan with respect to two orthogonal axes.

29.3 Principle of Base Isolation System

Base isolation is a technique for passive structural control that has been used for protection of structures from the damaging effects of earthquake, Naeim and Kelly (1999). Passive systems do not require any additional energy source to operate and they are activated by the earthquake input motion only. The basic principle of base isolation is to isolate the structure from the ground, in order to avoid damages and to enable the structure to withstand severe earthquakes in elastic domain. The link between the base ground and the superstructure is achieved by installing bearings with certain characteristics (stiffness and damping), providing the reduction of the inertial force, whereby it partially or wholly constrain the diffusion of seismic waves through the body of the structure. The installation of isolators in building at base level significantly increases the fundamental period of the structure, modifies the shape of fundamental mode and increases the damping, which significantly reduces the risk of damage to structural elements and leads to better seismic performance of the building.

29.4 Numerical Examples

29.4.1 Description of Analysed Structures

The analysed structures are spatial five storey RC structures with different level of irregularity. All four structures are composed of 3 frames in direction X and 5 frames in direction Y, all at distance of 5 m and storey height 3 m. The columns of the first two stories are 50/50 cm and 45/45 cm on the above 3 stories. All beams have dimensions 40/45 cm. The thickness of the slab is 15 cm. The first structure is regular. In the other three structures, the column, at the middle axis of the first frame in Y direction is replaced with a RC wall with dimensions 120/40_cm, 160/40 cm and 200/40 cm respectively, Fig. 29.1.

In the mathematical models, all beams are modelled taking into account the slab contribution. The dimensions of sections are determined in reference with Eurocode 2. Width of flange of external beams is 110 cm, while for internal beams it is 180 cm. The structures are loaded with uniform distributed loads which are applied on the beam elements. The value of the permanent load at the top of the structures is 14 kN/m on internal beams and 10 kN/m on external beams and at the other stories 26 kN/m and 17 kN/m respectively. All the stories are loaded with live load of 5 kN/m on internal beams and 2.5 kN/m on external beams.

The elastomeric bearings that are used for the base isolated structures are made by the Italian company FIP Industriale. Total horizontal effective stiffness of the isolation system is determined from the requirement that the period of base isolated structure should surpass 3 times the period of the superstructure. The fundamental periods of vibration for analysed structures are in range of 0.64 s–0.67 s. This values are obtained for models of structures in which cracking of elements is included, i.e. determined bending and shear stiffness is 50% from the stiffness of sections without cracks, in accordance with recommendations of Eurocode 8.

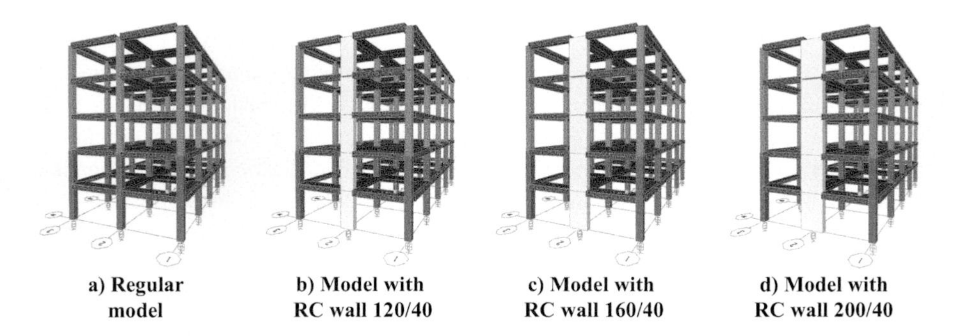

Fig. 29.1 Mathematical models. (**a**) Regular model, (**b**) Model with RC wall 120/40. (**c**) Model with RC wall 160/40. (**d**) Model with RC wall 200/40

$$K_{eff} = \frac{(2\pi)^2 \cdot M}{T_{eff}^2} = \frac{(2\pi)^2 \cdot 1109.6}{(2 \div 2.5)^2} = (7001.7 \div 10940.2) kN/m \qquad (29.1)$$

In case of installation of one isolator below each column, required effective stiffness of one isolator will be 470–730 kN/m. In compliance with the required stiffness, isolator type SI-N 350/125, with effective secant stiffness $K_{eff} = 620$kN/m and maximal horizontal displacement of 250 mm was selected.

$$T_{eff} = 2\pi\sqrt{\frac{M}{K_{eff}}} = 2\pi\sqrt{\frac{1109.6}{15 \cdot 620}} = 2.17s. \qquad (29.2)$$

All analysed structures satisfy the requirement $3T_f < T_{eff} < 3$ s. Selected isolator is with diameter 350 mm, height 213 mm, total thickness of elastomer 125 mm, weight 1.35 kN and vertical stiffness 480,000 kN/m. Maximal shear force that isolator can bear is 155 kN.

29.4.2 *Irregularity in Plan*

As the four analysed structures have ideal symmetry of lateral stiffness and mass distribution with respect to X axis, the structures are analysed for seismic action in Y direction and the provisions for verifying the regularity in plan are considered with respect to Y axis. For the regular structure the centre of mass corresponds with the centre of stiffness. The eccentricity between centre of stiffness and centre of mass, for fixed and base isolated models, at all storey levels are presented in Fig. 29.2.

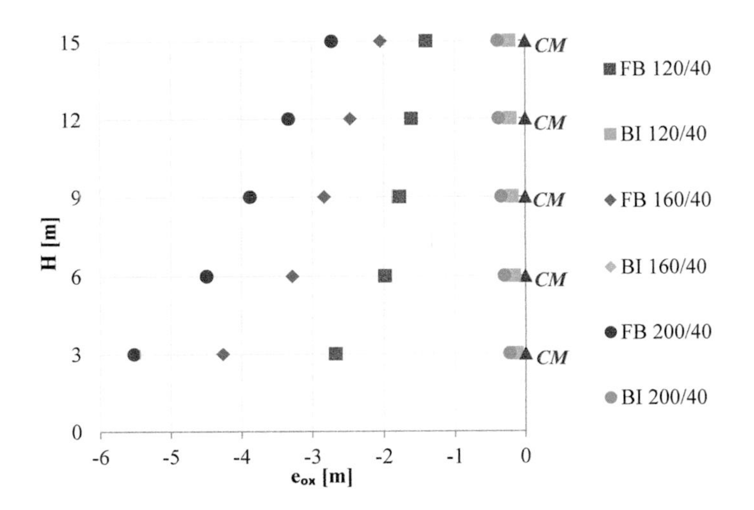

Fig. 29.2 Eccentricity of CS in relation to CM per floors

From the presented results it is noticed that as the length of RC wall increases the eccentricity between the center of stiffness and center of mass is also increasing. For fixed models, the eccentricity is in range of 1.40–2.67 m and from 2.72 m to 5.52 m for the model with length of the RC wall 120 cm and 200.cm, respectively. Whereupon, the eccentricity has highest value at the first storey and it decreases as going in height. At the base isolated models, the eccentricities are lower up to 25 times in relation to the fixed models.

The four analysed structures are rectangular in plan, with dimensions B/L = 10 m/20 m, the storey masses are equally distributed, and the radius of gyration of the storey masses, used also for verifying the regularity in plan, can be calculated as:

$$l_s = \sqrt{\frac{L^2 + B^2}{12}} = \sqrt{\frac{20^2 + 10^2}{12}} = 6.46m \quad (29.3)$$

The condition that controls the slenderness of the building in plan($\lambda = L_{max}/L_{min} \leq 4$) is satisfied, where L_{max} and L_{min} are respectively the larger and the smaller in plan dimension of the building, measured in orthogonal directions.

The necessary results for verification of the remaining two conditions are presented in Table 29.1. From the results it can be noticed that the condition where the torsional radius must be higher than the radius of gyration of the storey masses, is satisfied for all of the analysed models, fixed and base isolated, at every storey level.

The structures with RC wall, for the fixed base models, do not comply the condition which control the ratio between the eccentricity and 30% of the torsional radius. Therefore, these buildings are classified as irregular in plan. This type of irregularity occurs due to unsymmetrical distribution of the lateral stiffness in plan, which depend on the dimensions of the RC wall. For the structure with RC wall 120/40 cm this condition is not satisfied only at the first storey, where the largest eccentricity occurs. For the structure with RC wall 160/40 cm the condition is not satisfied for the first three stories. The structure with RC wall 200/40 cm does not comply this condition at any storey level. The stiffness of the isolation system, for base isolated models, has large impact on the eccentricity in plan. The largest reduction of the eccentricity occurs at the first story, for the structure with highest degree of irregularity in plan and is round 25 times and the smallest reduction is round 6 times, at the highest storey of the structure with lowest degree of irregularity in plan.

From the presented results in Table 29.1 it can be noted that all models with base isolation comply the condition $e_{ox,i} < 0.3 \cdot r_{x,i}$ at all levels of the structures, so they are described as regular in plan. For assumption of linear increasing, the base isolated models will reach up the regularity limit for approximate length of RC wall of 1750 cm, which cannot be fulfilled for this type of structure, because the dimension of the structure in plan in that direction is 1000 cm. This is indicator of the advantages of the base isolated structures in case of regularity in plan.

Table 29.1 Control of criteria for structural regularity in plan

Fixed base models									Base isolated models					
	Floor	Level	$\lvert e_{0x,i} \rvert$	</>	$0,3r_{x,i}$	$r_{x,i}$	</>	l_s	$\lvert e_{0x,i} \rvert$	</>	$0,3r_{x,i}$	$r_{x,i}$	</>	l_s
Structures with RC wall 120/40	1	3	2.67	>	2.48	8.28	>	6.46	0.13	<	2.46	8.18	>	6.46
	2	6	1.98	<	2.50	8.34	>	6.46	0.16	<	2.47	8.22	>	6.46
	3	9	1.78	<	2.48	8.28	>	6.46	0.19	<	2.46	8.20	>	6.46
	4	12	1.61	<	2.48	8.28	>	6.46	0.21	<	2.47	8.22	>	6.46
	5	15	1.40	<	2.51	8.35	>	6.46	0.23	<	2.47	8.23	>	6.46
Structures with RC wall 160/40	1	3	4.26	>	2.38	7.92	>	6.46	0.19	<	2.47	8.22	>	6.46
	2	6	3.28	>	2.46	8.19	>	6.46	0.24	<	2.47	8.23	>	6.46
	3	9	2.83	>	2.52	8.41	>	6.46	0.27	<	2.47	8.22	>	6.46
	4	12	2.46	<	2.53	8.45	>	6.46	0.30	<	2.47	8.25	>	6.46
	5	15	2.04	<	2.50	8.32	>	6.46	0.31	<	2.48	8.26	>	6.46
Structures with RC wall 200/40	1	3	5.51	>	2.22	7.40	>	6.46	0.22	<	2.46	8.20	>	6.46
	2	6	4.49	>	2.36	7.85	>	6.46	0.30	<	2.47	8.23	>	6.46
	3	9	3.87	>	2.42	8.06	>	6.46	0.34	<	2.47	8.23	>	6.46
	4	12	3.33	>	2.49	8.30	>	6.46	0.37	<	2.48	8.26	>	6.46
	5	15	2.72	>	2.48	8.26	>	6.46	0.39	<	2.48	8.27	>	6.46

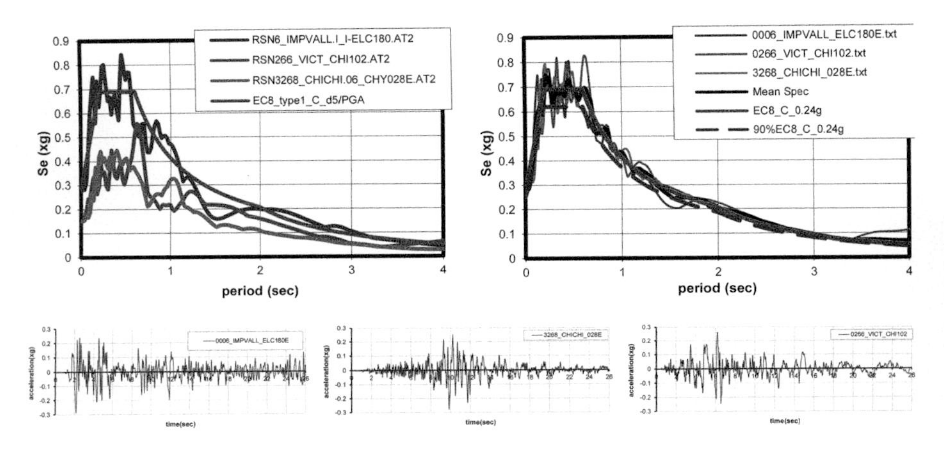

Fig. 29.3 Elastic4 spectrum and ground motion acceleration time history

29.4.3 Seismic Action

Response of structures is determined with application of dynamic time history analysis, whereupon 3 input motions from previous earthquakes (Imperial Valley – El Centro 1940, Chi – Chi Taiwan 06 – CHY028 1999, Victoria Mexico – Chihuahua 1980) are used.

In order to satisfy the Eurocode 8 requirements for ground motion selection, original registrations of selected earthquakes are scaled by amplitudes and frequency. Target spectral ordinates are determined for spectrum Type 1, soil class C, and peak ground accelerations of 0.24 g. Original acceleration spectra and spectra obtained from scaled input motion, as well as the acceleration time history are presented in Fig. 29.3. Base isolated structures was analysed for three levels of seismic hazard, which corresponding to PGA of 0.12 g, 0.24 g and 0.36 g. All analysed structures are exposed to seismic action only in one direction, perpendicular to eccentricity.

29.4.4 Comparison of Results

In this section the most important results obtained from performed analysis are presented. The range of stories displacements for fixed and base isolated models of the structures are given at Fig. 29.4. From the graphs it can be noticed that the maximal displacement at the top of the structures occur for the base isolated models. Displacements of base isolated regular structure are larger for 9.2 cm, i.e. 65%, and in structure with RC wall 200/40 the variation is 3 cm, i.e. 14%. With increasing of degree of irregularity, the maximal displacements at fixed models are also increasing, while for the base isolated models this increase is insignificant.

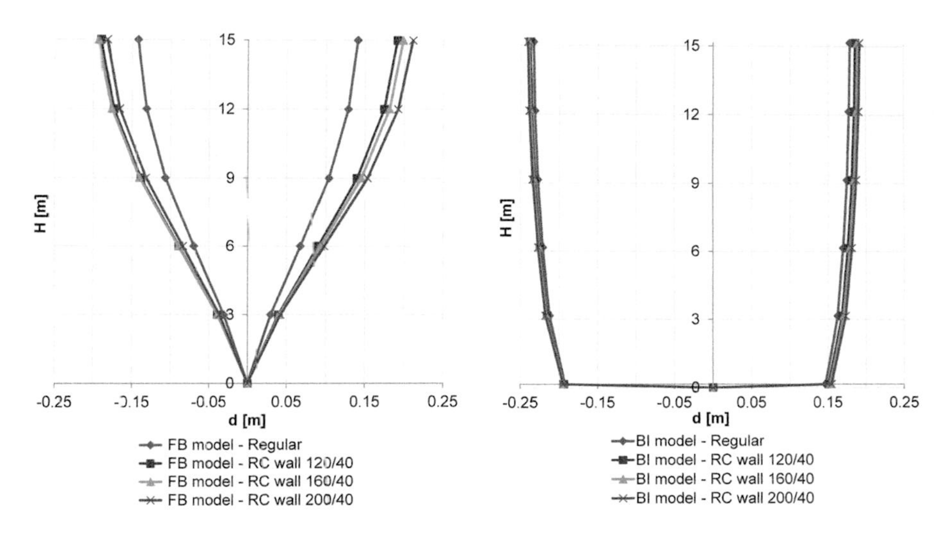

Fig. 29.4 Range of displacements along the height

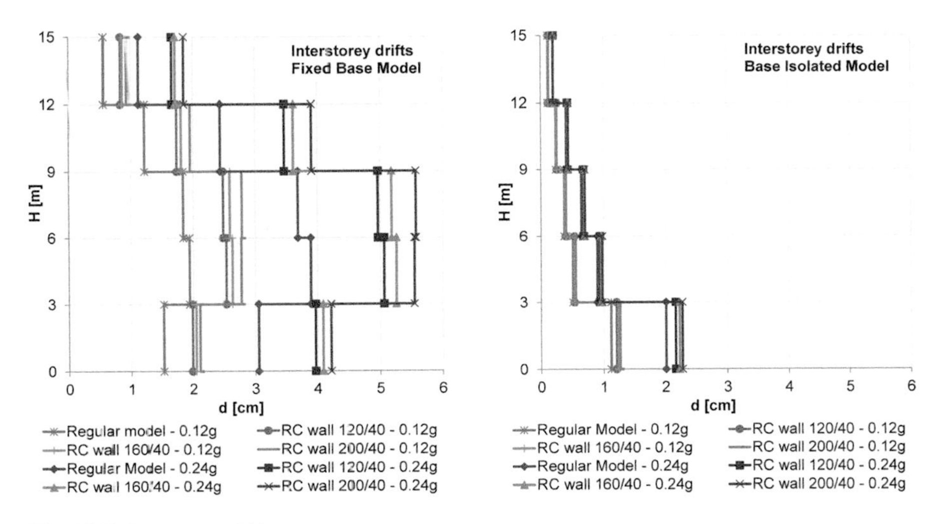

Fig. 29.5 Interstorey drifts

In contrast to total displacements, interstorey drifts for the base isolated structures are significantly less than those obtained for fixed base structures. This leads to significantly lower risk of damages of structural and non-structural elements.

At Fig. 29.5 the distribution of interstorey drifts for fixed base and base isolated models obtained for PGA 0.12 g and 0.24 g are presented. Increasing the level of irregularity results with the increasing of interstorey drifts, which are extreme at the flexible side of the structure. Maximal interstorey drifts obtained for PGA of 0.24 g is equal to 5.57 cm and occurs at the third storey for the fixed base model with RC wall 200/40 cm. This maximal interstorey drift is almost two and a half times greater compared to the maximal interstorey drift of the respective base isolated structure. Maximal interstorey drifts of the base isolated structure occur at the first storey and are equal to 2.27 cm. For all structures the maximal values are obtained from analysis with Victoria Mexico input motion, except at regular fixed base structure where maximal interstorey drifts are obtained from Chi Chi Taiwan input motion.

The figures below present the displacements in plan obtained from the analysis of three input motions for PGA 0.24 g. The presented displacements in plan are at the highest storey of the four analysed structures, for fixed and base isolated models. There are 6 representative cases of response that are observed: minimal and maximal displacement of the highest storey for the left edge of the structure, minimal and maximal displacement of the highest storey for the right edge of the structure and minimal and maximal rotation of the highest storey of the structure in plan (Fig. 29.6).

Because of the symmetry in plan in both direction of the structure and the seismic action in one direction, the regular structure has translational movement only. For the other structures due to the eccentricity between the centre of stiffness and centre of mass, rotation of floor occurs. From the graphs it can be noticed that extreme displacement of the left or right edge of the structure are not always in the phase with the moment when the extreme rotations occurred. Maximal displacements appear from the input ground motion of Victoria Mexico while minimal are obtained from El Centro ground motion. It can be noticed that the displacement on the left edge of the structure, where the RC wall is placed, for fixed based models, are decreasing as the degree of irregularity in plan increases. This leads to increasing of storey rotation, and amplification of torsional effects. From the graphs of the base isolated models it can be noticed that although the absolute displacements of the two edges are increasing, the difference between them is irrelevant, and the rotation of the analysed storey is negligible compare with the same obtained from the fixed model. The minimal and maximal rotation for fixed models almost correspond with the maximal displacement of the right edge of the structure, while at base isolated models maximal rotations are occurring for displacements lower than the maximal.

At Fig. 29.7 the time histories of rotations at first and fifth floor obtained for the structure with RC wall 200/40 cm are presented. For both models, fixed and base isolated, the larger rotation occurs at the top storey. At fixed base structure the maximal rotations at the top storey are 4–5 times larger compared to the base isolated. At the level of first storey, for the structure with RC wall 200/40 cm, the maximal rotations of both models are almost equal.

Hysteresis loops (lateral force – horizontal displacement) for isolators at the left and right edge of the structure obtained for the structure with RC wall 200/40 from performed non-linear dynamic analysis of Victoria Mexico input motion are presented in Fig. 29.8. The maximal initiated force in isolators for PGA 0.36 g is in range of 155–180 kN, whereof the maximal horizontal displacements are in range of 25.37–29.67 cm. Independent of the level of irregularity of the analysed structures and the level of seismic hazard, the isolators from the left and from the right edge of the structure behave pretty uniformly for all conducted analyses.

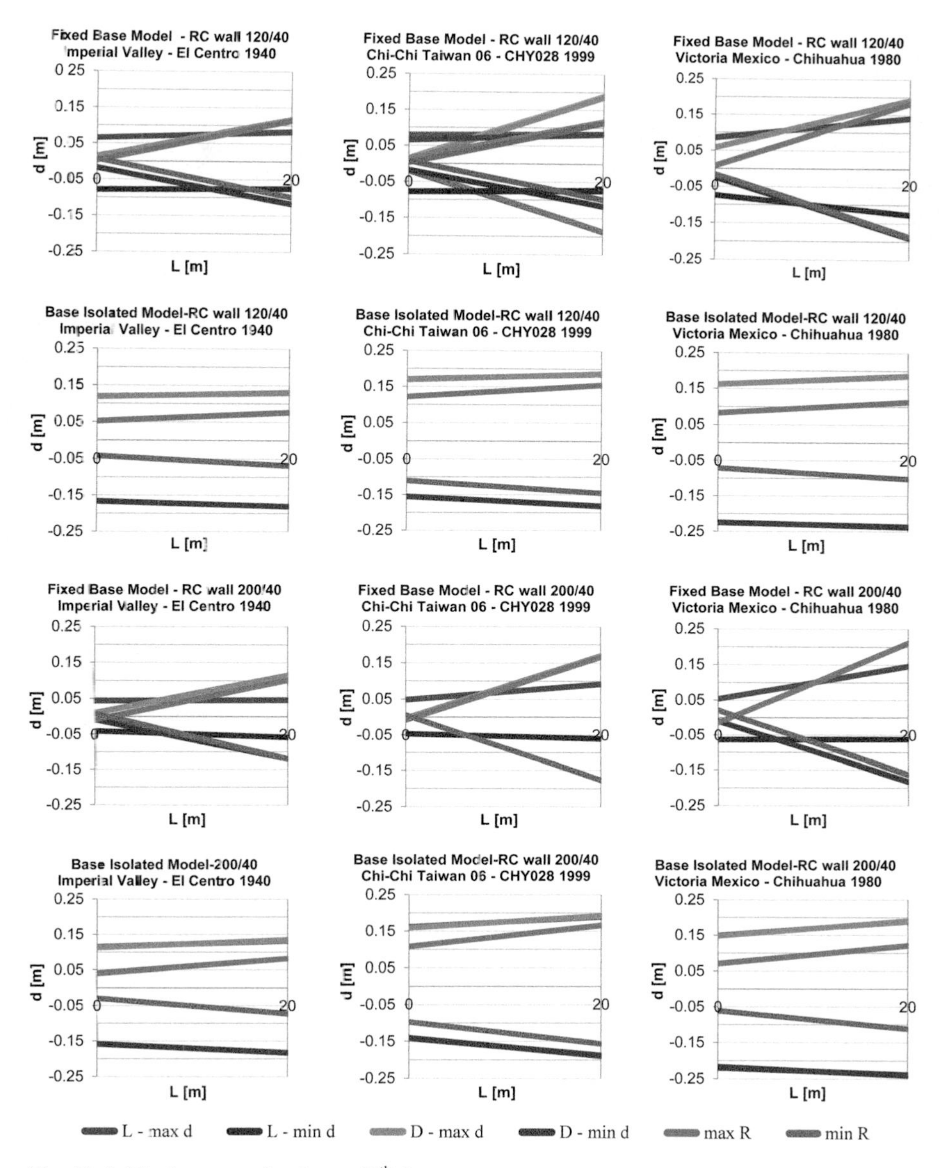

Fig. 29.6 Displacements in plan on Vth floor

29.5 Conclusions

By increasing the length of the RC wall, the eccentricity between the centre of stiffness and the centre of the mass increases. The eccentricity of base isolated structures is up to 25 times smaller compared to the corresponding fixed base ones, which is due to the position and influence of centre of stiffness of the isolation

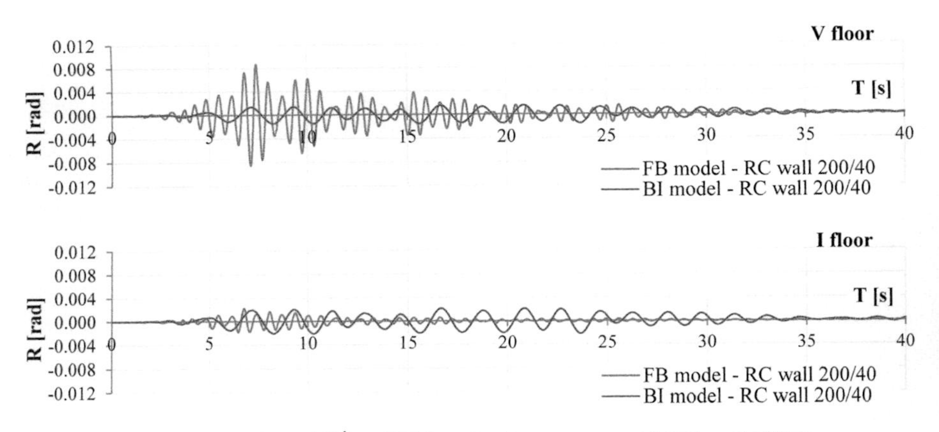

Fig. 29.7 History of rotation of Vth and Ist floor for the structure with RC wall 200/40

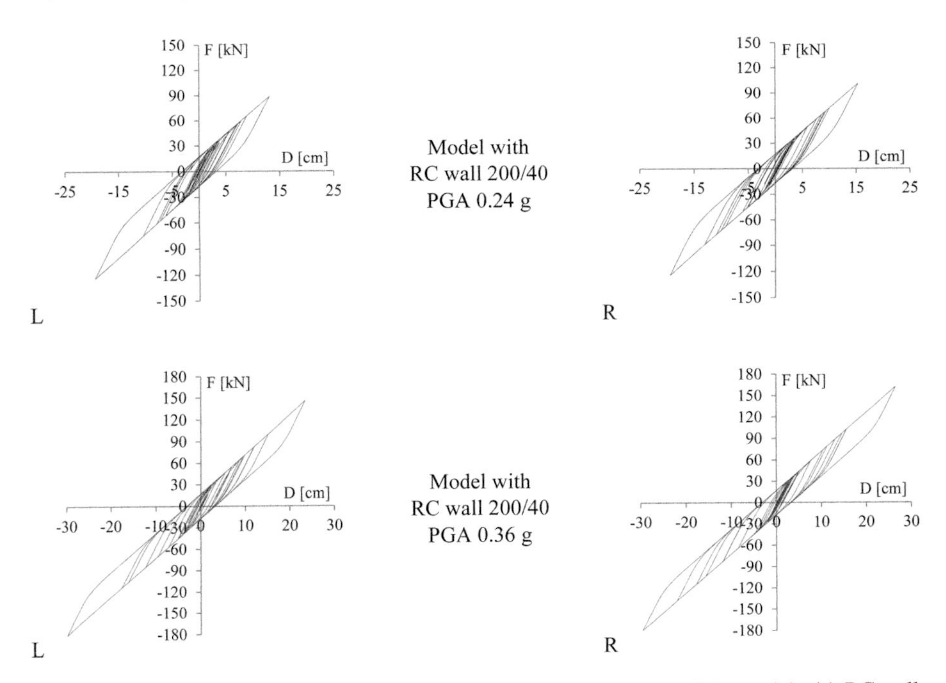

Fig. 29.8 Hysteresis diagrams for isolators from the left and right edge of the model with RC wall 200/40

system. Although the total displacement of base isolated models is higher, interstorey drifts for the base isolated structures are significantly smaller compared to those obtained for fixed base structures. With increasing the degree of eccentricity the displacements at fixed base models significantly increase, while at the base isolated structures they are negligible. Base isolated structures have lower floor rotations comparing with the corresponding fixed base ones. The largest differences are noticed at the upper floors. Low rotations are mainly due to uniformly work of base isolation system. These advantages are indicators for the effectiveness of the base isolation system in relation to regularity, as well as to significantly lower risk of damages of structural and non-structural elements.

References

CEN – European Committee for Standardization (2004) Eurocode 8: design of structures for earthquake resistance – part 1. European standard EN 1998 – 1

De Stefano M, Pintucchi B (2008) A review of research on seismic behavior of irregular building structures since 2002. Bull Earthq Eng 6(2):282–308

Jangid RS, Datta TK (1994) Nonlinear response of torsionally coupled base isolated structure. ASCE J Struc Eng 120(1):1–22

Khoshncudian F, Imani Azad A (2011) Effect of two horizontal components of earthquake on nonlinear response of torsionally coupled base isolated structures. Struct Design Tall Spec Build 2011(20):986–1018

Kilar V, Koren D (2009) Seismic behaviour of asymmetric base-isolated structures with various distributions of isolators. Eng Struct 31(4):910–921

Naeim F, Kelly JM (1999) Design of seismic isolated structures: from theory to practice. Wiley, New York

Nagarajaiah S, Reinhorn AM, Constantinou MC (1993) Torsion in base isolated structures with elastomeric isolation systems. ASCE J Struct Eng 119(10):2932–2951

Nagarajaiah S, Reinhorn AM, Constantinou MC (1993a) Torsional coupling in sliding base isolated structures. ASCE J Struct Eng 119(1):130–149

Tajammolian H, Khoshnoudian F, Mehr NP (2016) Seismic responses of isolated structures with mass asymmetry mounted on TCFP subjected to near-fault ground motions. Int J Civil Eng 14 (8):573–584

Tena-Colunga A, Gómez-Soberón LA (2002) Torsional response of base-isolated structures due to asymmetries in the superstructure. Eng Struct 24(12):1587–1599

Tena-Colunga A, Zambrana-Rojasb C (2004) Torsional response of base-isolated structures due to stiffness assimetries of the isolation system. 13th world conference on earthquake engineering, Vancouver, BC, Canada, August 1–6, 2004, paper no. 2022

Chapter 30
Study on Polymer Elements for Mitigation of Earthquake-Induced Pounding Between Buildings in Complex Arrangements

Barbara Sołtysik, Tomasz Falborski, and Robert Jankowski

Abstract Pounding between neighboring structures during seismic events has been revealed as one of the most commonly observed reasons for severe damage or even total collapse of the adjacent buildings. Therefore, pounding effects have recently become an issue of great interest of many numerical and experimental investigations in many earthquake-prone regions of the world. It has also been observed that the differences in dynamic characteristics is the key reason leading to interaction between colliding, insufficiently separated structures. The problem is much more complicated for complex arrangements of structures, for example, in the case of collisions between few structures in a row. A lot of different approaches have been considered to mitigate earthquake-induced structural pounding. One method is based on placing between the structures some viscoelastic elements acting as bumpers. Another one is stiff linking the structures. It allows the forces to be transmitted between buildings and thus eliminate undesired interactions. The aim of this paper is to present the results of experimental research focused on mitigation of pounding between buildings in complex arrangements by using polymer elements installed between structures. In the present study, three steel models characterized by various dynamic properties and different in-between distances were investigated. Additional masses were mounted at the top of each model in order to obtain different dynamic characteristics. The unidirectional shaking table, available at the Gdansk University of Technology (Poland), was employed to conduct this study. Experimental models were mounted to shaking table platform. The results of the study explicitly show that the approach of using polymer elements can be an effective pounding mitigation technique in the case of complex arrangement of buildings. It may partially or fully prevent from damaging collisions between adjacent buildings during seismic events. It also enhances the dynamic response leading to the reduction in lateral vibrations under different strong ground excitations.

B. Sołtysik (✉) · T. Falborski · R. Jankowski
Faculty of Civil and Environmental Engineering, Gdansk University of Technology, Gdansk, Poland
e-mail: barwiech@pg.gda.pl; tomfalbo@pg.gda.pl; jankowr@pg.gda.pl

D. Köber et al. (eds.), *Seismic Behaviour and Design of Irregular and Complex Civil Structures III*, Geotechnical, Geological and Earthquake Engineering 48,
https://doi.org/10.1007/978-3-030-33532-8_30

Keywords Earthquake-induced pounding · Shaking table investigation · Seismic response · Irregular structures

30.1 Introduction

Due to lack of space and high cost of land in a number of urban areas, designers are forced to take into account a need to build closely-separated structures. That situation may lead to the problem of pounding between neighboring structures as an effect of earthquake excitation. The safety and reliability of steel structures under seismic load is one of the aims during the design process. Meanwhile, earthquake-induced structural pounding may cause some local damage at the points of interactions, it may also lead to substantial destruction, permanent deformations or total collapse of colliding structures (see Kasai and Maison 1997). For example, the 1971 San Fernando earthquake resulted in pounding between the main building of the Olive View Hospital and one of its independent stairway towers. It resulted in its permanent tilting (see Bertero and Collins 1973). The observations and findings made after the 1985 Mexico earthquake reveal that one of the main reasons of damage was related to the pounding effects between the buildings (see Rosenblueth and Meli 1986). A massive pounding damage of school buildings was also noticed after the 1999 Athens earthquake (see Vasiliadis and Elenas 2002).

A major reason leading to pounding between adjacent buildings results from the differences in their dynamic characteristics (see Maison and Kasai 1992; Naderpour et al. 2016; Elwardany et al. 2017). The difference in the natural vibration periods of the structures leads to their out-of-phase vibrations (see Karayannis and Favvata 2005; Jankowski and Mahmoud 2015). Also, the propagation of seismic wave may result in various seismic inputs leading to interactions between decks of bridge structures (see Jankowski 2015).

The effects of structural pounding during seismic excitations have been studied for more than two decades now (see, for example, Anagnostopoulos 1988; Maison and Kasai 1990; Anagnostopoulos and Spiliopoulos 1992; Jankowski 2005; Favvata et al. 2009; Polycarpou et al. 2014; Jankowski and Mahmoud 2016). However, most of the studies have concerned masonry as well as reinforced concrete structures and investigations on steel structures are very limited (see Sołtysik and Jankowski 2013, 2016a, b). Meanwhile, increased displacements observed in steel structures during ground motions due to their flexibility and low damping properties make them more vulnerable to collisions. The aim of the present paper is to show the results of experimental investigation concerning earthquake-induced pounding between models of steel structures in a row. The analysis was focused on mitigation of pounding between structures in such a complex arrangement by using polymer bumpers.

30.2 Experimental Investigation

30.2.1 Experimental Setup

In order to conduct the experimental study, three models of steel structures (each 1000 mm high) with different dynamic parameters were prepared. They were constructed out of four steel columns made from rectangular box section 15 × 15 × 1.5 mm with spacing of 480 mm in the longitudinal direction (load direction) and 571 mm in the transverse one. Additional steel skew bracings of the same cross section were used to prevent transverse and torsional vibrations. To obtain different dynamic characteristics of the models, additional weight (concrete plates with mass of 42.2 kg) was added at the top of each tower. In the experiment, complex configuration of the towers, in which two concrete plates were mounted on the external structures and only one plate at the top of the middle tower, was investigated (Fig. 30.1a). The study was conducted using the middle-sized one-directional shaking table located at the Faculty of Civil and Environmental Engineering, Gdansk University of Technology, Poland (Fig. 30.1b, c). The device is equipped with a platform (dimensions 2000 × 2000 mm) which allows us to test the structural models with a maximum weight of 1000 kg. The platform is connected to the linear actuator which may induce arbitrary movement with a maximum acceleration of 10 m/s^2 and a maximum strength of 44.5 kN. The following equipment for the measurements was used:

- four accelerometers with a mechanical restriction frequency of 4 kHz (three of the sensors were located at the top of each tower and one was placed on the shaking table platform),
- eight-channel amplifier with low pass filter of 100 Hz,
- analogue-digital card to record the measurements.

30.2.2 Free Vibration Test

In the first stage of the study, the free vibration tests were conducted so as to determine the basic dynamic properties of the models of steel structures. Based on the results of free vibration tests, the values of the natural frequencies for all steel towers were identified (see Table 30.1). It can be seen from the table that the obtained frequency values are within the range of frequencies typical for small steel buildings, what somehow validates the experimental models and, based on the experimental results, allows us to draw more general conclusions concerning real steel structures.

Fig. 30.1 Experimental setup – different views: (a) front view, (b) side view, (c) oblique view

Table 30.1 Natural frequencies for free vibration modes (Hz)

	Natural frequency (Hz)
Tower no 1	1.825
Tower no 2	3.257
Tower no 3	1.792

30.2.3 Seismic Test

After conducting the free vibration tests, the steel towers in a row with different in-between gap sizes were tested under the following earthquakes:

- Northridge (17.01.1994, 75% of the nominal amplitude of the EW component, PGA = 6.50 m/s^2),
- San Fernando (09.02.1971, 25% of the nominal amplitude of the N74°E component, PGA = 2.85 m/s^2).

It should be underlined that the earthquake records were scaled down so as to prevent damage to analyzed models of steel structures.

Firstly, the structural response of each tower under earthquake excitation was measured for the in-between gap size equal to 20 mm. After this test, the influence of additional polymer elements with thickness $t_1 = 20$ mm mounted between the towers with distance 40 mm, was studied (see Fig. 30.2). Polymer mass with high damping properties (especially designed flexible elastoplastic two-component grout based on polyurethane resin – see Falborski and Jankowski 2017, 2018) was used for such bumper elements. The examples of the results of the experimental study, for the Northridge and the San Fernando earthquakes, are presented in this paper. The peak values of the structural response acceleration are summarised in Table 30.2. Exemplary acceleration time histories for the Northridge earthquake are also shown in Figs. 30.3, 30.4 and 30.5.

The results of the experimental study clearly confirm the influence of pounding on the response of models of steel structures exposed to ground motions. Moreover it can be seen from Table 30.2, that using polymer elements between structures, which play a role of a bumper, leads to a significant reduction in the dynamic response. For example, in the case of steel tower no 1 under the Northridge earthquake, the use of 20 mm thick polymer elements results in the decrease in the peak value of acceleration from 155.810 m/s^2 to 5.809 m/s^2 (decrease by 96.27%). Similar results are also visible for tower no 2 and tower no 3. For both analysed ground motions, the reduction in peak values of acceleration is between 93.3% and 96.3%.

a)

b)

c)

Fig. 30.2 Experimental model with polymer elements with thickness of 20 mm: (**a**) front view, (**b**) side view, (**c**) oblique view

Table 30.2 Peak values of acceleration for three towers without polymer elements and with polymer elements with thickness $t_1 = 20$ mm (gap size 20 mm)

Tower no	Peak values of acceleration (m/s^2)		
	Without polymer elements	With polymer elements 20 mm	Difference (%)
Northridge earthquake – 17.01.1994 (PGA = 6.50 m/s^2)			
Tower no 1	155.810	5.809	Reduction by 96.27%
Tower no 2	355.001	13.866	Reduction by 96.09%
Tower no 3	105.957	6.179	Reduction by 94.17%
San Fernando earthquake – 09.02.1971 (PGA = 2.85 m/s^2)			
Tower no 1	72.350	4.631	Reduction by 93.60%
Tower no 2	150.551	10.100	Reduction by 93.29%
Tower no 3	111.751	5.705	Reduction by 94.89%

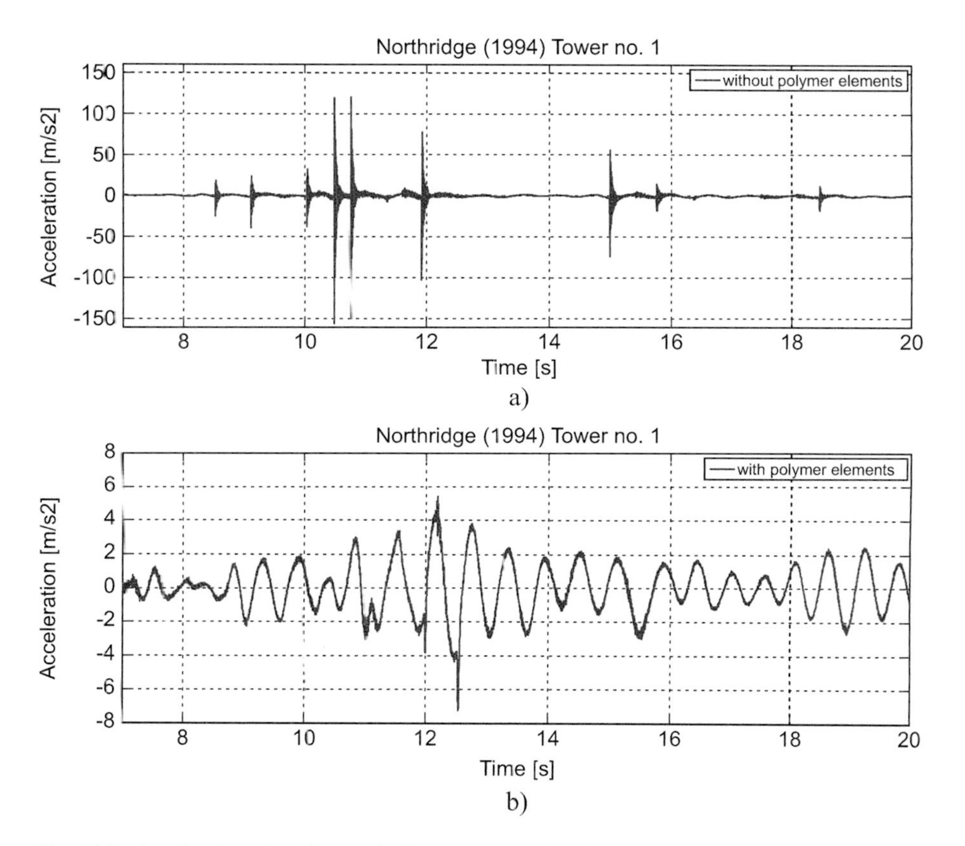

Fig. 30.3 Acceleration time history for Tower no 1 under the Northridge earthquake: (**a**) without polymer elements; (**b**) with polymer elements

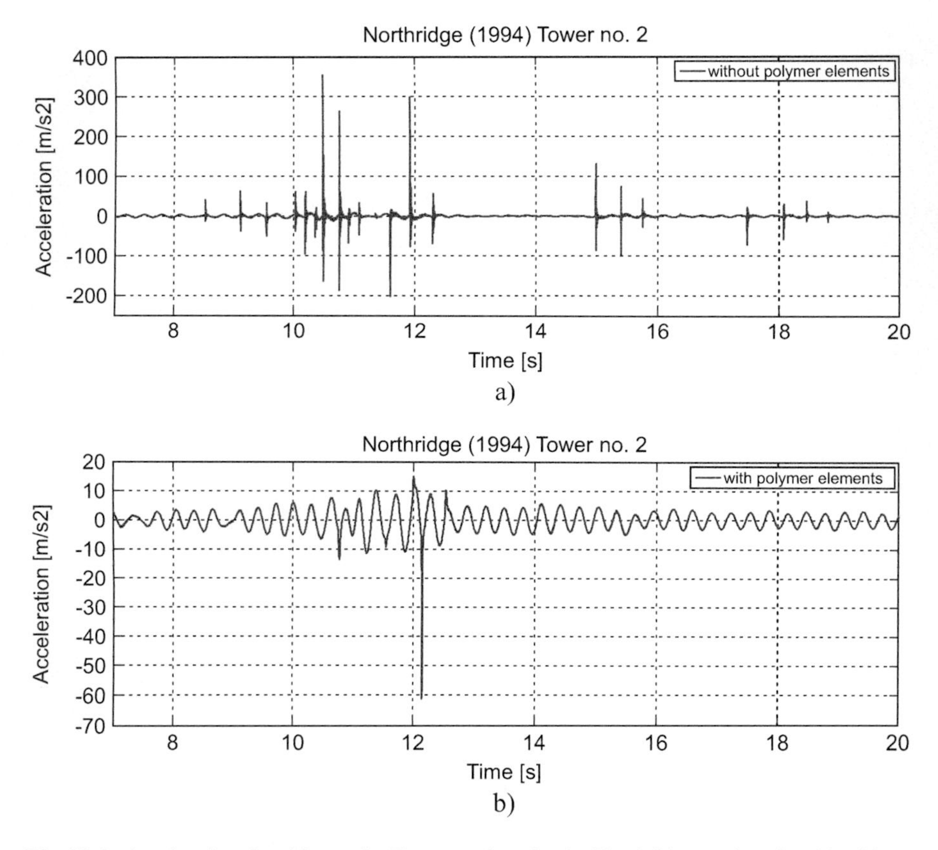

Fig. 30.4 Acceleration time history for Tower no 2 under the Northridge earthquake: (**a**) without polymer elements; (**b**) with polymer elements

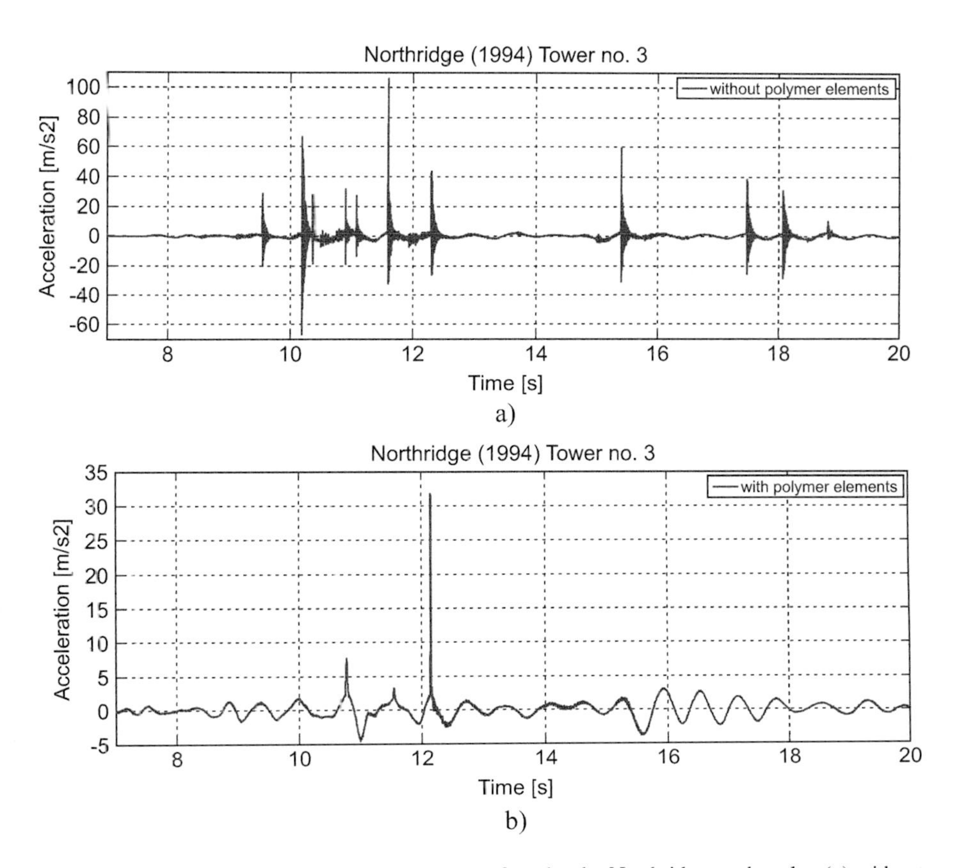

Fig. 30.5 Acceleration time history for Tower no 3 under the Northridge earthquake: (**a**) without polymer elements; (**b**) with polymer elements

30.3 Summary

The results of the shaking table experimental study focused on earthquake-induced pounding between models of three steel structures with different dynamic parameters have been presented in this paper.

The results of the study clearly indicate that pounding may have a significant influence of the response of colliding structures under ground motions. The results also show that the method of using polymer elements can be an effective pounding mitigation technique in the case of complex arrangement of buildings. It allows us to prevent damaging collisions between adjacent structures during earthquakes. It also improves the structural behaviour leading to the reduction in vibrations under different seismic excitations.

References

Anagnostopoulos SA (1988) Pounding of buildings in series during earthquakes. Earthq Eng Struct Dyn 16:443–456

Anagnostopoulos SA, Spiliopoulos KV (1992) An investigation of earthquake induced pounding between adjacent buildings. Earthq Eng Struct Dyn 21:289–302

Bertero VV, Collins RG (1973) Investigation of the failures of the Olive view stairtowers during the San Fernando earthquake and their implications on seismic design. EERC report no 73-26, University of California, Berkeley, USA

Elwardany H, Seleemah A, Jankowski R (2017) Seismic pounding behavior of multi-story buildings in series considering the effect of infill panels. Eng Struct 144:139–150

Falborski T, Jankowski R (2017) Experimental study on effectiveness of a prototype seismic isolation system made of polymeric bearings. Appl Sci 7(8):808. https://doi.org/10.3390/app8030400

Falborski T, Jankowski R (2018) Advanced hysteretic model of a prototype seismic isolation system made of polymeric bearings. Appl Sci 8(3):400. https://doi.org/10.3390/app8030400

Favvata MJ, Karayannis CG, Liolios AA (2009) Influence of exterior joint effect on the inter-story pounding interaction of structures. J Struct Eng Mech 33:113–136

Jankowski R (2005) Impact force spectrum for damage assessment of earthquake-induced structural pounding. Key Eng Mater 293–294:711–718

Jankowski R (2015) Pounding between superstructure segments in multi-supported elevated bridge with three-span continuous deck under 3D non-uniform earthquake excitation. J Earthq Tsunami 9(4), paper no 1550012

Jankowski R, Mahmoud S (2015) Earthquake-induced structural pounding. Springer, Basel

Jankowski R, Mahmoud S (2016) Linking of adjacent three-storey buildings for mitigation of structural pounding during earthquakes. Bull Earthq Eng 14:3075–3097

Karayannis CG, Favvata MJ (2005) Earthquake-induced interaction between adjacent reinforced concrete structures with non-equal heights. Earthq Eng Struct Dyn 34:1–20

Kasai K, Maison B (1997) Building pounding damage during the 1989 Loma Prieta earthquake. Eng Struct 19:195–207

Maison BF, Kasai K (1990) Analysis for type of structural pounding. J Struct Eng 116:957–977

Maison B, Kasai K (1992) Dynamics of pounding when two buildings collide. Earthq Eng Struct Dyn 21:771–786

Naderpour H, Barros RC, Khatami SM, Jankowski R (2016) Numerical study on pounding between two adjacent buildings under earthquake excitation. Shock and vibration 2016, article ID 1504783

Polycarpou PC, Papaloizou L, Komodromos P (2014) An efficient methodology for simulating earthquake-induced 3D pounding of buildings. Earthq Eng Struct Dyn 43:985–1003

Rosenblueth E, Meli R (1986) The 1985 earthquake: causes and effects in Mexico City. Concr Int 8:23–34

Sołtysik B, Jankowski R (2013) Non-linear strain rate analysis of earthquake-induced pounding between steel buildings. Int J Earth Sci Eng 6:429–433

Sołtysik B, Jankowski R (2016a) Earthquake-induced pounding between asymmetric steel buildings. In: Zembaty Z, De Stefano M (eds) Geotechnical, geological and earthquake engineering 40: seismic behaviour and design of irregular and complex civil structures II. Springer, Cham, pp 255–262

Sołtysik B, Jankowski R (2016b) Influence of separation gap on the response of colliding models of steel structures under seismic and paraseismic excitations. In: Kleiber M, Burczyński T, Wilde K, Górski J, Winkelmann K, Smakosz Ł (eds) Advances in mechanics: theoretical, computational and interdisciplinary. CRC Press/Balkema, London, pp 533–536
Vasiliadis L, Elenas A (2002) Performance of school buildings during the Athens earthquake of September 7 1999. 12th European conference on earthquake engineering, paper no 264

Chapter 31
Procedure of Non-linear Static Analysis for Retrofitted Buildings Structures Through Seismic Isolation

Gabriel Dănilă

Abstract Seismic response predictability of irregular buildings structures is a challenging task which engaged many studies. To improve the seismic performance of irregular buildings structures, the base isolation method can be used. By positioning the isolation system components, equilibrated, aside and the other of the mass centre and due to the large flexibility of the isolation system in the horizontal direction, the effect of irregularities is drastically diminished. The paper presents a nonlinear static analysis procedure, applicable to existing buildings structures, retrofitted through seismic isolation. There are performed nonlinear static analyses on a fixed base plan structure, with elevation irregularity and an isolation system is proposed. Based on the capacity curves (F-D curves), of the fixed base structure and of the proposed isolation system, there is computed the F-D curve of the isolated building structure. On the basis of the proposed non-linear static analysis procedure, there is determined the performance point of the seismically isolated building structure.

Keywords Non-linear static analysis · Performance point · Elevation irregular structure · Seismic isolation

31.1 Introduction

Among the non-linear analyses, the most popular one is the non-linear static analysis. Based on this analysis there were developed procedures for seismic assessment of the buildings structures (e.g. the N2 Method – provided in Eurocode 8 (CEN, Comité Européen de Normalisation 2004), the Capacity Spectrum Method (CSM) – presented in ATC40 (ATC, Applied Technology Council 1996) and the Coefficient Method (CM) – given in FEMA356 (FEMA, Federal Emergency Management Agency 2000)). The major advantage of the non-linear static procedures (NSPs) is their simplicity. In the case of irregular structures, this simplicity is

G. Dănilă (✉)
"Ion Mincu" University of Architecture and Urbanism, Bucharest, Romania

D. Köber et al. (eds.), *Seismic Behaviour and Design of Irregular and Complex Civil Structures III*, Geotechnical, Geological and Earthquake Engineering 48,
https://doi.org/10.1007/978-3-030-33532-8_31

becoming "inaccuracy". When using the NSPs for seismically isolated buildings, this "inaccuracy" is drastically diminished due to the isolation system. In the paper it is presented a nonlinear static analysis procedure, applicable to existing buildings structures, retrofitted through seismic isolation. There are performed nonlinear static analyses on a fixed base plan structure, with elevation irregularity. On the basis of the proposed non-linear static analysis procedure, there is determined the performance point of the seismically isolated building structure.

31.2 Description of the Proposed NSP

The NSP, involves prior determination of the capacity curve (force-displacement curve) on the fixed base building structure. The following sequence of steps is considered:

- It is determined the capacity curve ($F_b - D_{s\_fb}^{top}$) on the fixed base building structure;
- An isolated system is proposed and the force-displacement curve of the isolation system is established;
- Knowing the F-D curve of the isolation system and the base shear force, it can be computed the displacement of the isolation system D_{is}, corresponding to each point in the capacity curve;
- Compute the top displacements $D_{s\_is}^{top}$, of the seismically isolated building, using the relation:

$$D_{sis}^{top} = D_{sfb}^{top} + D_{is} \tag{31.1}$$

  Equation (31.1) considers that the total displacement of the isolated building structure is composed of the displacement of the fixed base building structure and the displacement of the isolation system.
- Draw the capacity curve ($F_b - D_{s\_is}^{top}$) of the seismically isolated building structure;
- Convert the capacity curve ($F_b - D_{s\_is}^{top}$) of the seismically isolated building structure into a spectral capacity curve ($S_a - S_d$). The seismically isolated building is considered a single degree of freedom system, without additional convertions.
- Draw the elastic response spectrum, corresponding to the damping with which is creditated the isolated system;
- Overlap the spectral capacity curve to the elastic response spectrum;
- The intersection of the spectral capacity curve with the elastic response spectrum, represents the initial performance point of the isolated structure;
- Assess the effective damping, introduced by the isolation system and by the building structure, corresponding to the initial performance point using the following formula:

$$\xi_{sis} = \frac{1}{4\pi} \cdot \frac{E_D^{is}}{E_S^s + E_S^{is}} \tag{31.2}$$

where: E_D^{is} *is the energy dissipated by the isolation system;* E_S^s *is the elastic strain energy of the building structure and* E_S^{is} *is the elastic strain energy of the isolation system.*

Because the dissipation of the seismic energy is made mainly through the isolation system the hysteretic loop is not affected by any coefficient that takes into account the structural behavior.

- Change the elastic response spectrum according to the effective damping, computed in the previous step;
- The intersection of the spectral capacity curve with the modified elastic response spectrum, represents the performance point of the isolated building structure.

  In computing the performance point no further iterations are needed if the damping ratio falls within 5% of the initial (assumed) damping ratio.
- Draw a radial line from the origin to the performance point. The radial line is the vibration period of the isolated building structure;

In order to apply the procedure, the first vibration period of the seismically isolated building structure must be greater than three times the first vibration period of the fixed base building structure ($T_{is} \geq 3T_{fb}$). This assumption is made to obtain a "rigid solid" behavior of the isolated building during the seismic action.

31.3 Case Study

31.3.1 Description of the Analysed Building Structure

The building structure is a plan frame with three spans of 5 m and nine stories. The ground floor has 4 m height and the stories 1...8 have 3.2 m height. At the 4th storey is presented a drawback, resulting an elevation irregularity. The beams are made of T section, with the web thickness of 30 cm and the cross-sectional height of 60 cm. The columns are made of square cross-section with the edge of 70 cm. Under each column is placed a high damping rubber bearing (Fig. 31.1).

The reinforcement for the structural elements was computed using STAS 10107–0/90 design code which conducted to the following reinforcement ratios of the longitudinal reinforcement:

- Beams:
  - isolation plane, ground floor, story 1 and story 2: 1.37% for the top reinforcement and 0.73% for the bottom reinforcement.
  - story 3, story 4 and story 5: 1.14% for the top reinforcement and 0.73% for the bottom reinforcement.

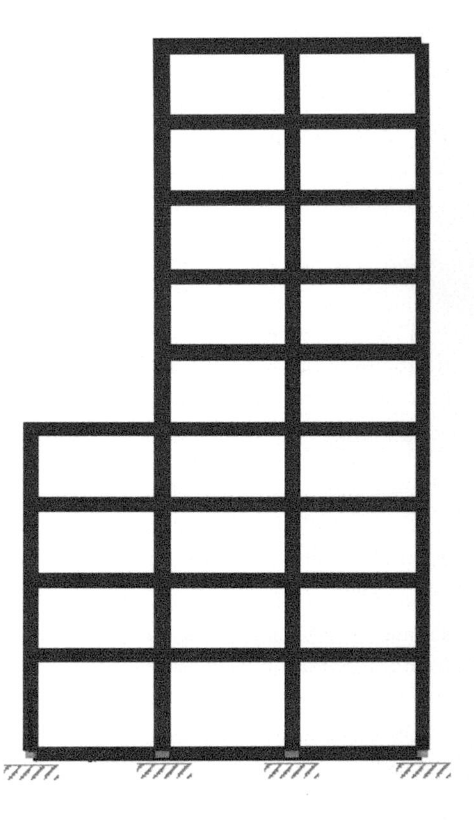

Fig. 31.1 The analysed building structure

– story 6, story 7 and story 8: 0.70% for the top reinforcement and 0.47% for the bottom reinforcement.

- Columns: 0.53% for all the columns

31.3.2 Bearing's Preliminary Design

The preliminary design of the isolation system was carried out, considering the structure a single degree of freedom system. The building structure was isolated to the vibration period of $T_{is} = 3.4$ s, using high damping rubber bearings (HDRB). The damping provided by the isolation system was considered $\xi_{ef} = 20\%$.

The effective stiffness of one bearing k_{ef}, is determined with the Eq. (31.3).

$$k_{ef} = \left(\frac{2\pi}{T_{is}}\right)^2 \cdot \frac{G_{SC}}{n_b \cdot g} = \left(\frac{2\pi}{3.4}\right)^2 \cdot \frac{10802.5}{4 \cdot 9.81} = 940.5\frac{kN}{m} \quad (31.3)$$

Where: G_{SC} is the total weight of the building in the special combination of loads; n_b is the number of the bearings and g is the ground acceleration.

The displacement demand of the isolation systems, d_{dc}, to the design earthquake was determined using Eq. (31.4) (ATC, Applied Technology Council 1996; Universitatea Tehnică de Construcții București 2006).

$$D_D = \left(\frac{T_{is}}{2\pi}\right)^2 \cdot a_g^d \cdot \beta(T_{is}) \cdot \eta = \left(\frac{3.4}{2\pi}\right)^2 \cdot 0.24 \cdot 9.81 \cdot 0.761 \cdot 0.632 = 0.332m \quad (31.4)$$

Where: $a_g{}^d$ is the ground acceleration corresponding to the design earthquake; $\beta(T_{is})$ is the normalized spectral ordinate, corresponding to the vibration period, T_{is} and η is the damping correction factor.

31.3.3 The Seismic Action

The seismic action is described by one recorded accelerogram, corresponding to the N-S component of the March 4, 1977 earthquake and six artificial accelerograms, compatible with the design spectrum for Bucharest. In Fig. 31.2 there are shown the elastic response spectra in $S_a - S_d$ format, corresponding to the seven accelerograms. The response spectra were plotted taking into account the damping $\xi_{ef} = 20\%$.

31.3.4 Performance Point Determination

The non-linear static analyses were performed using the SeismoStruct v6.0 (SeismoStruct [computer software] 2012) computer program, using the mean values of the materials strengths. Due to the elevation irregularity, there ware made non-linear static analysis on both negative and positive directions of the fixed base building structure.

The nonlinear behavior of the structural elements was taken into account using a distributed plasticity model (fiber model). The yielding of the longitudinal reinforcement was considered for specific deformations $\varepsilon \geq 1.93‰$, the concrete cover spalling was considered at specific deformations $\varepsilon b \geq 3.5‰$, crushing of concreted concrete was considered for specific deformations $\varepsilon b \geq 8‰$, and the failure of the longitudinal reinforcements was considered to specific deformations $\varepsilon a \geq 10\%$.

The force – displacement curve of HDRBs was established taking into account a three-linear behavior model, with the increase of the post-elastic stiffness at shear deformations greater than 170%. This increase in stiffness of the HDRBs occurs at shear deformations in between 150% and 200%, due to the rubber crystallization process, which is accompanied by an increase in the dissipated energy (Fig. 31.3).

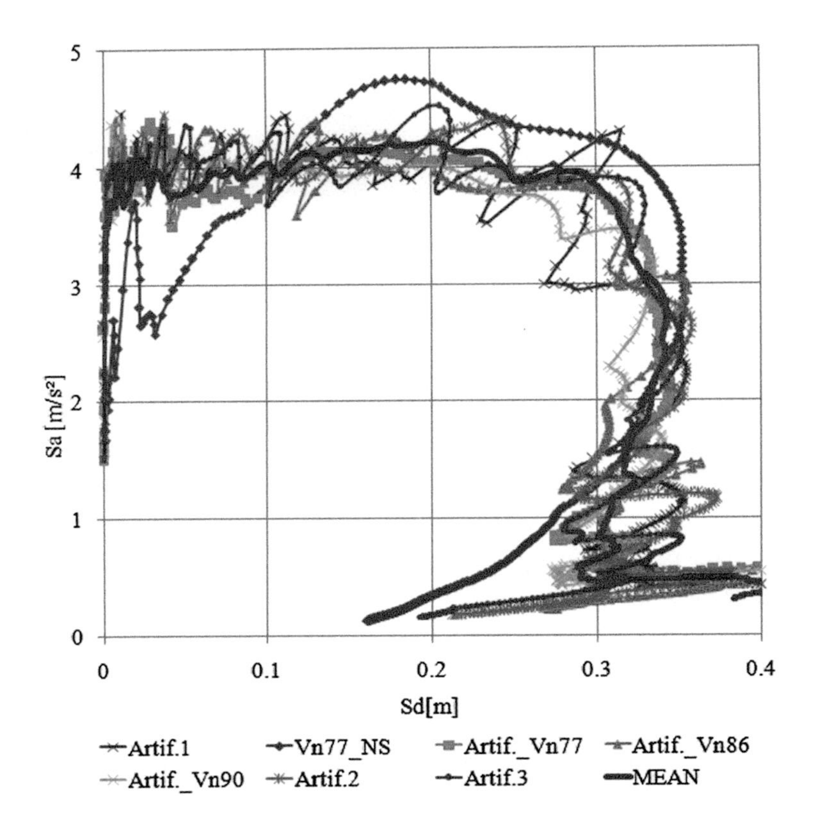

Fig. 31.2 The elastic response spectra

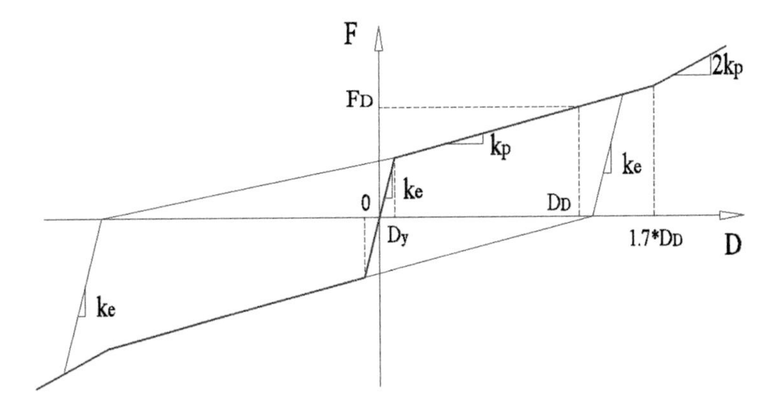

Fig. 31.3 The force-displacement curve of the isolation system

In Table 31.1 there are given the parameters for defining the nonlinear model of the HDRBs used for seismic isolation of the analyzed building structure.

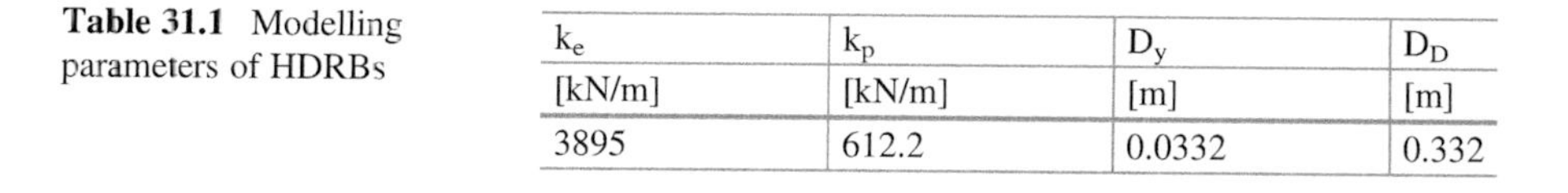

Table 31.1 Modelling parameters of HDRBs

k_e	k_p	D_y	D_D
[kN/m]	[kN/m]	[m]	[m]
3895	612.2	0.0332	0.332

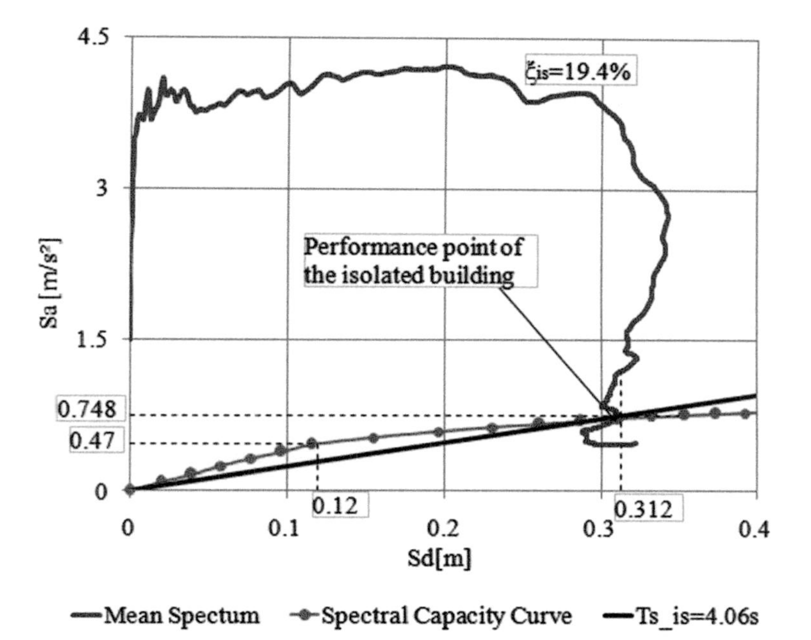

Fig. 31.4 The performance point for the lateral load acting in the positive x-direction

After computing the top displacements $D_{s\_is}{}^{top}$, for the seismically isolated building, there was plotted the capacity curve (F_b–$D_{s\_is}{}^{top}$) of the seismically isolated building structure and converted into a spectral capacity curve (S_a–S_d).

The spectral capacity curve of the seismically isolated building structure was overlap to the mean elastic response spectrum. The performance point was determined by an iterative process, so that the decrease of the mean response spectra reflects the damping introduced by the seismically isolated building structure.

In Fig. 31.4 it is shown the procedure of computing the performance point of the seismically isolated building structure, for the lateral load acting in the positive x-direction. The mean response spectrum was reduced with the following damping ratio (Naiem and Kelly 1997):

$$\xi_{sis}^{xp} = \frac{1}{4\pi} \cdot \frac{E_D^{is}}{E_S^s + E_S^{is}} = 19.4\% \tag{31.5}$$

Where: $E_D{}^{is}$ is the energy dissipated by the isolation system; $E_S{}^s$ is the elastic strain energy of the building structure and $E_S{}^{is}$ is the elastic strain energy of the isolation system.

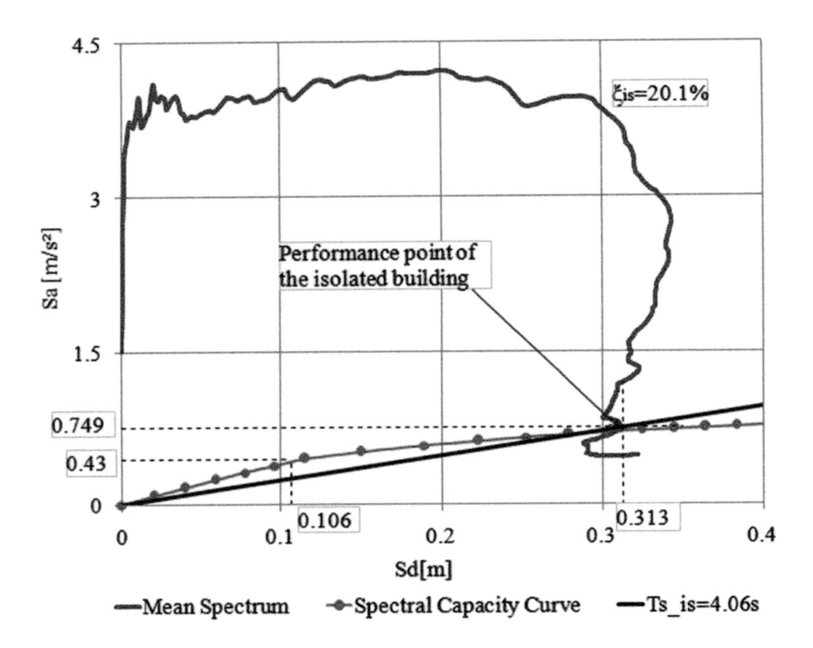

Fig. 31.5 The performance point for the lateral load acting in the negative x-direction

The performance point shows a quasi-elastic behavior of the analysed building structure. Because of the quasi-elastic behavior of the building structure, it can be considered that the energy is dissipaded mainly through the isolation system.

In Fig. 31.5 it is shown the procedure of computing the performance point of the seismically isolated building structure, for the lateral load acting in the negative x-direction.

Also, for negative x-direction, the performance point shows a quasi-elastic behavior of the analysed building structure. Because of the quasi-elastic behavior of the building structure, it can be considered that the energy is dissipaded mainly through the isolation system. The mean response spectrum was reduced with the following damping ratio (Naiem and Kelly 1997):

$$\xi_{sis}^{xn} = \frac{1}{4\pi} \cdot \frac{E_D^{is}}{E_S^s + E_S^{is}} = 20.1\% \tag{31.6}$$

31.4 Conclusions

In this paper, a non-linear static procedure was proposed. The procedure is applicable to existing buildings structures, retrofitted through seismic isolation. Based on this NSP it was analysed an irregular building structure. The performance point was computed for both x-positive and x-negative directions, to account the effect of elevation irregularity. Due to the large flexibility of the isolation system in the horizontal direction, the effect of irregularity is negligible. The base isolation method can be used to improve the seismic performance of irregular buildings structures.

Acknowledgments The author would like to thank to SeismoSoft Company for providing free educational licences of SeismoStruct and SeismoArtif computer programs without which this study could not have been achieved.

References

ATC, Applied Technology Council (1996) Seismic evaluation and retrofit of concrete buildings, vol. 1 and 2, report no. ATC-40. Redwood City, CA

CEN, Comité Européen de Normalisation (2004) Eurocode 8: design of structures for earthquake resistance. Part 1: general rules, seismic actions and rules for buildings. EN 1998–1:2004. Brussels, Belgium

FEMA, Federal Emergency Management Agency (2000) Prestandard and commentary for the seismic rehabilitation of buildings, FEMA356. Washington, DC

Naiem F, Kelly JM (1997) Design of seismic isolated structures. From theory to practice. Wiley, New York/Chichester/Weinheim/Brisbane/Singapore/Toronto

SeismoStruct [computer software] (2012) Pavia: SeismoSoft srl. Available: http://www.seismosoft.com

Universitatea Tehnică de Construcții București (2006) Cod de proiectare seismică P100: Partea I Prevederi de proiectare pentru clădiri. P100-1/2006. București

Chapter 32
Observations of Damage to Uto City Hall Suffered in the 2016 Kumamoto Earthquake

K. Fujii, T. Yoshida, T. Nishimura, and T. Furuta

Abstract The main building of Uto City Hall was constructed in 1965 and severely damaged in the 2016 Kumamoto Earthquake. In the present article, the damage observed is described and discussed. First, observations of damage to the outside of the building are described. Based on the first damage observation, a damage surveillance mission using a mobile rescue robot was planned and carried out. From these observations, it can be concluded that most of the structural damage in the main building was concentrated on the fourth and fifth stories. Some of the columns on the fourth floor had lost the axial strength capacity to sustain vertical loads.

Keywords 2016 Kumamoto Earthquake · Damage observation · Mobile rescue robot

32.1 Introduction

On 14 April 2016, a large earthquake hit Kumamoto Prefecture, Japan. Two days later, another large earthquake occurred, mainly affecting Kumamoto and Oita prefectures. During the second earthquake, Uto City Hall was severely damaged. The main building of Uto City Hall is a five-story reinforced concrete building constructed in 1965. Although external observations were possible, it was too dangerous to enter the main building because of the possibility of total collapse due to aftershocks. To understand the behavior of this building during sequential seismic events, a damage observation of the whole building including interior damage is essential.

K. Fujii (✉)
Department of Architecture, Chiba Institute of Technology, Narashino-shi, Chiba, Japan
e-mail: kenji.fujii@it-chiba.ac.jp

T. Yoshida · T. Nishimura · T. Furuta
Future Robotics Technology Center, Chiba Institute of Technology, Narashino-shi, Chiba, Japan
e-mail: yoshida@furo.org; nishimura@furo.org; furuta@furo.org

D. Köber et al. (eds.), *Seismic Behaviour and Design of Irregular and Complex Civil Structures III*, Geotechnical, Geological and Earthquake Engineering 48,
https://doi.org/10.1007/978-3-030-33532-8_32

In this paper, the damage to the main building of Uto City Hall is described and discussed. Observations of the damage caused by the earthquake to the outside of the building are described. Then, the results of a damage surveillance mission using a mobile rescue robot are described.

32.2 Basic Information

32.2.1 Epicenter of the 2016 Kumamoto Earthquake

Figure 32.1 shows the epicenters of earthquakes that occurred on 14 and 16 April 2016, with magnitudes of M6.5 and M7.3, respectively, according to the Japan Meteorological Agency (JMA) (Japan Meteorological Agency 2016). Table 32.1 shows the basic information of the epicenters for both earthquakes.

The maximum seismic intensity (according to the JMA) of both earthquakes was 7, as recorded in Mashiki, Kumamoto Prefecture. The seismic intensity recorded at Uto, Kumamoto Prefecture was 5.5 on 14 April and 6.2 on 16 April.

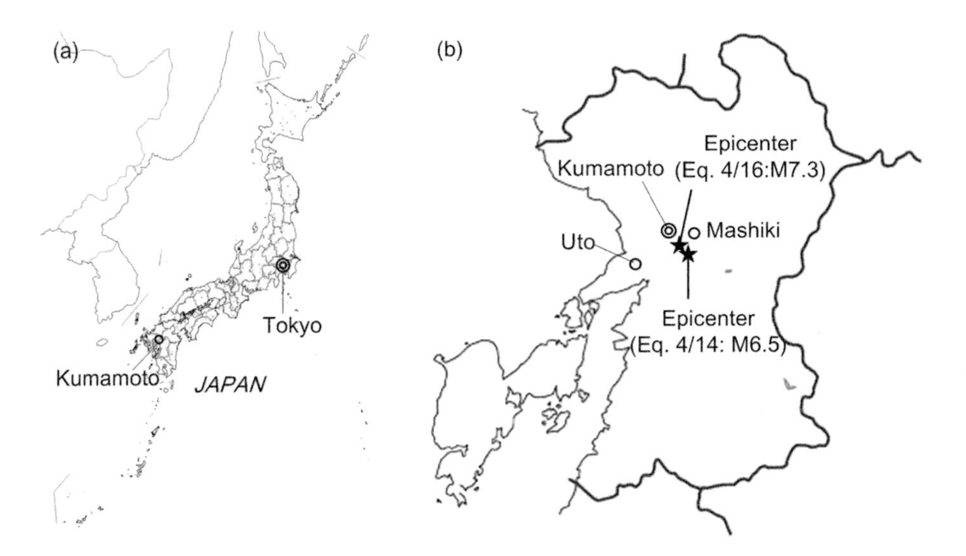

Fig. 32.1 Epicenters of earthquakes on 14 and 16 April 2016

Table 32.1 Basic information on the epicenters (Japan Meteorological Agency)

Event	Date	Latitude	Longitude	Depth	Magnitude
4/14 Earthquake	2016/04/14 21:26:34	32°44.5′N	130°48.5′E	11 km	M6.5
4/16 Earthquake	2016/04/16 01:25:05	32°45.2′N	130°45.7′E	12 km	M7.3

32.2.2 *Uto City Hall*

Figure 32.2 shows an aerial photograph of Uto City Hall. The main building, which suffered the most severe damage during these earthquakes, is a five-story reinforced concrete building. There are several other municipal office buildings on this site. There is also a strong-motion seismograph network (K-NET) station, which is monitored by the National Research Institute for Earth Science and Disaster Resilience.

The construction of the main building was completed in May 1965. The seismic capacity of this building was evaluated in 2003 according to the Guideline for Seismic Capacity Evaluation of Existing Reinforced Buildings (The Japan Building Disaster Prevention Association 2001a). The evaluation results showed that the seismic capacity of this building was insufficient, so it was recommended that the building be rebuilt. When the building was surveyed on the morning of 29 April, access to the area inside the red dotted line shown in Fig. 32.2 was restricted.

Table 32.2 shows the peak acceleration of three components of both earthquakes, recorded at K-NET Uto station (Strong-motion Seismograph Network 2016).

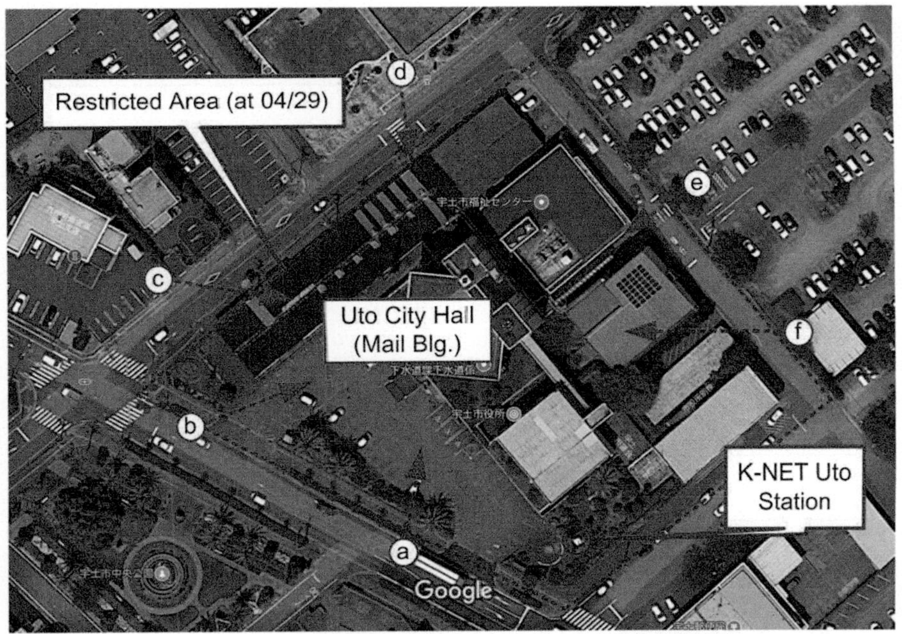

Fig. 32.2 Aerial photograph at Uto City Hall. (From https://www.google.co.jp/maps/@32.6873102,130.658548,154m/data=!3m1!1e3) [Accessed at July 1, 2017]

Table 32.2 Peak accelerations of both earthquakes at K-NET Uto Station

		Peak acceleration (m/s^2)		
	Epicentral distance	NS Comp.	EW Comp.	UD Comp.
4/14 Earthquake	15 km	2.635	3.042	1.987
4/16 Earthquake	12 km	6.515	7.711	4.217

32.3 Building Damage

In this section, observations of the damage to the outside of the building caused by the earthquake are described.

32.3.1 Overview

Figure 32.3 shows photographs of the main building and the locations from which each photograph was taken. As shown in Fig. 32.3a, b and c, the main building is severely damaged on the south and west sides, especially on the third through fifth stories; the fourth story is partially collapsed. Most vertical concrete panels on the south and west side balconies collapsed. The north and east side could not be clearly observed from outside the restricted area owing to other buildings obstructing the view (Fig. 32.3d, e, and f). No severe damage to the other buildings on this site was observed.

32.3.2 Damage to the Main Building

In this section, the damage to the south, west and north elevations are described in detail. Figures 32.4 through 32.5 and 32.6 show the simplified elevations and photographs of each elevation. The observed damage is described below.

South Elevation (Fig. 32.4): At the top of a fifth-story column, large flexural cracks are observed, and the aluminum window frames were deformed vertically (Fig. 32.4a). In the "+"-shaped beam–column joints at the top of the fourth story (Fig. 32.4b) and the third story (Fig. 32.4c), the fallings of the cover concrete were observed.

West Elevation (Fig. 32.5): At the top of the west-side corner column on the fifth story (Fig. 32.5a), some larger flexural cracks at corner were observed. In addition, there were large vertical openings and cracks observed near the "T"-shaped beam–column joint. The fourth story had partially collapsed (Fig. 32.5b), and beams on the fifth floor were inclined (Fig. 32.5c). The middle column on the fourth story was

Fig. 32.3 Views of the main building of Uto City Hall from outside the restricted area. Photographs were taken from (**a**) the south, (**b**) the southwest, (**c**) the west, (**d**) the north, (**e**) the northeast, and (**f**) the east. The locations from which the photographs were taken are shown in Fig. 32.2

severely damaged and the bottom part of column vertically displaced (Fig. 32.5d). Conversely, at the west-side corner column on the third story (Fig. 32.5e), only small flexural cracks at the top were observed.

North Elevation (Fig. 32.6): On the fifth story, some large flexural cracks were observed at the top of the corner column, and large diagonal cracks were observed at the ends of roof-floor beams (Fig. 32.6a). In addition, there were large diagonal cracks at the end of a fifth-floor beam (Fig. 32.6b), at the bottom of a fourth-story column and a fourth-floor beam (Fig. 32.6c).

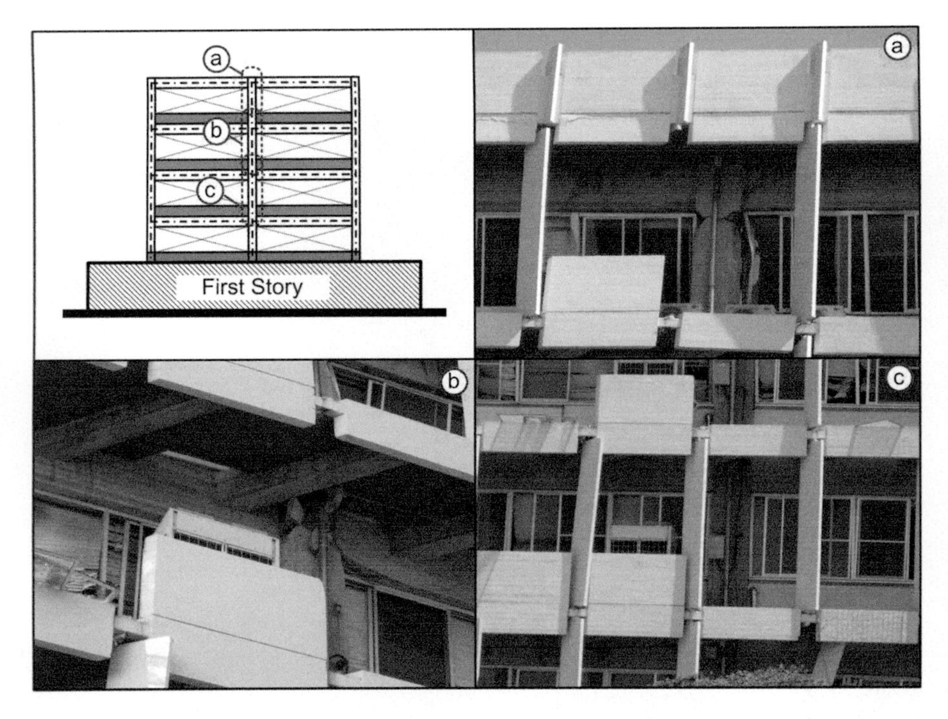

Fig. 32.4 Simplified elevation and photographs of the south elevation. Photographs show damage to (**a**) a fifth-story column, (**b**) a beam–column joint at the top of the fourth story, and (**c**) a third story column

From external observations, it may be concluded that the main damage to this building was limited to the third, fourth and fifth stories on the south and west elevations. The middle column on the fourth story on the west elevation was most severely damaged and almost lost the capacity to bear the vertical load.

32.4 Surveillance of the Main Building by Mobile Rescue Robot

In this section, the results of a damage surveillance mission with a mobile rescue robot developed for the present study are described.

The first point of focus in the surveillance mission was the interior of the fourth floor. As described in the previous section, the fourth story had partially collapsed (from the observation of the west elevation). Therefore, the interior columns on the fourth story were suspected to be severely damaged.

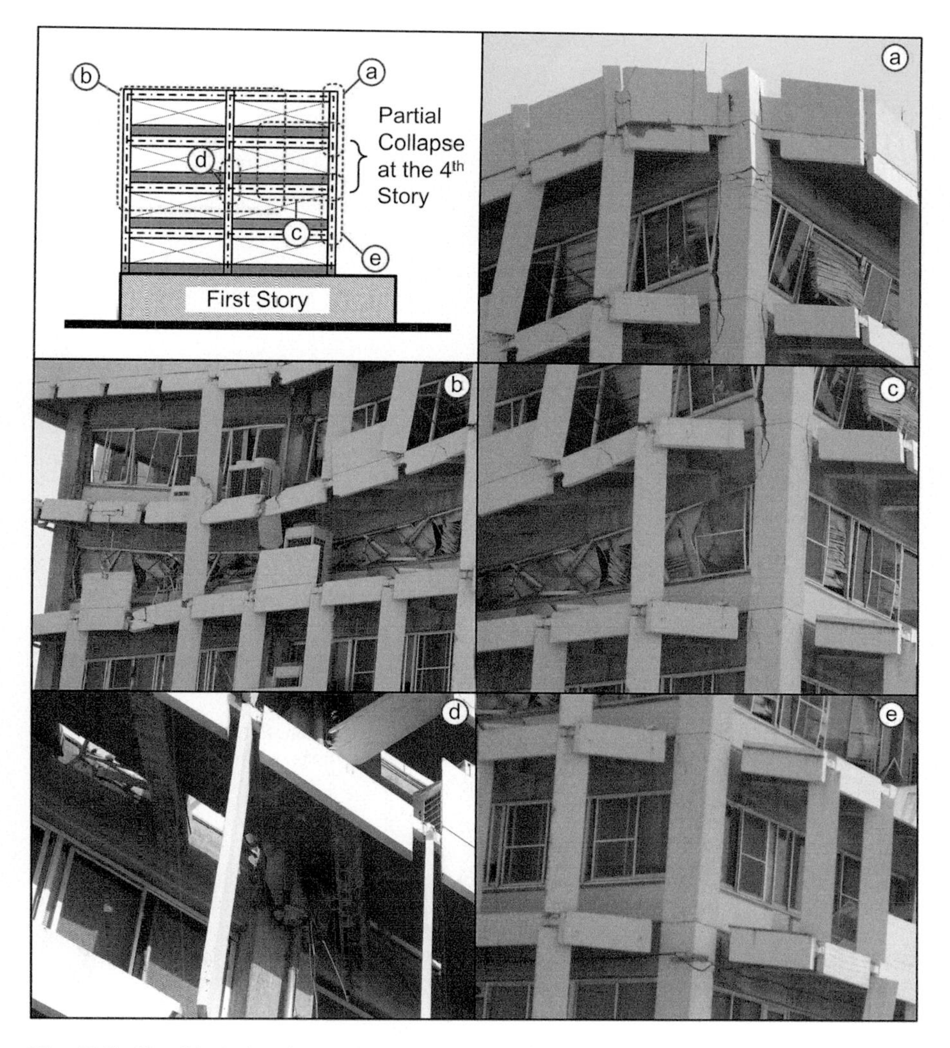

Fig. 32.5 Simplified elevation and photographs of the west elevation. Photographs show (**a**) damage to the corner column on the fifth story, (**b**) partial collapse of the fourth story, (**c**) damage to the corner column on the fourth story, (**d**) damage to the bottom of a fourth story, and (**e**) damage to the corner column on the third story

The second point was the damage of floor slabs. According to the original drawing of the main building, the structure of this building can be divided in two parts (the stair block and the office block), which are connected by floor slabs. If this building had behaved as a unit building structure, the floor slabs were to transfer seismic loads between the two blocks.

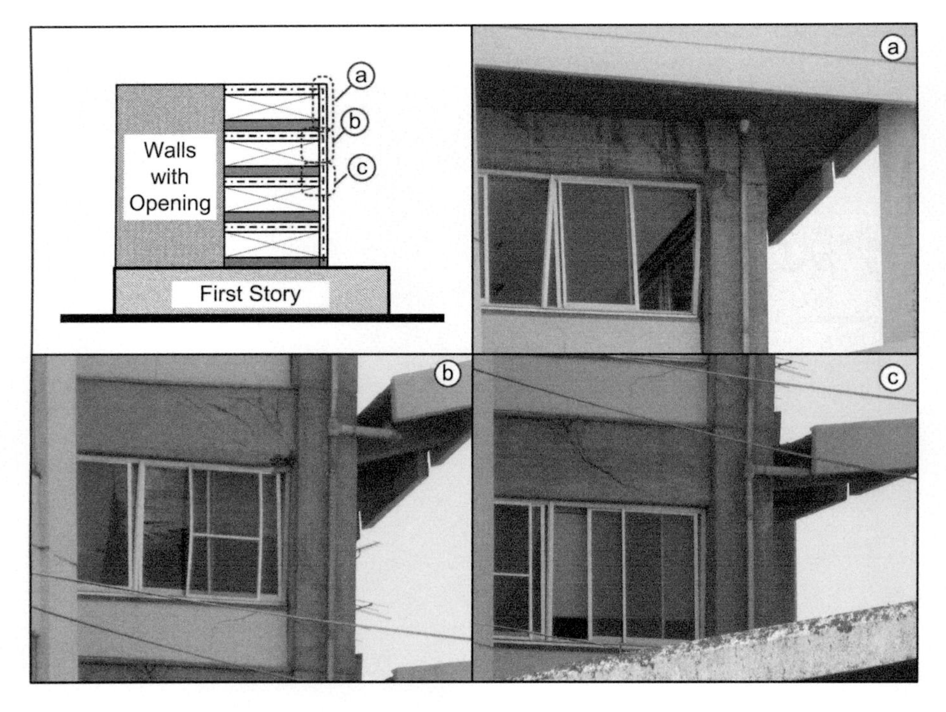

Fig. 32.6 Simplified elevation and photographs of the north elevation. Photographs show damage to (**a**) the top of corner columns and beams on the roof floor, (**b**) a beam on the fifth floor, and (**c**) the bottom of a fourth-story column and a fourth-floor beam

The surveillance mission was planned considering the points above and capacity of batteries used for a mobile rescue robot: the mission time had to be limited to within 2 h.

32.4.1 Mobile Rescue Robot "Sakura-Ichi-Go"

Figure 32.7 shows the mobile rescue robot "Sakura-ichi-go", used in the damage surveillance mission. This robot is an upgraded and compacted version of the rescue robot "Quince" (Nagatani et al. 2013). Quince surveyed inside of the Fukushima Daiichi nuclear power plant, which was damaged by the Great East Japan Earthquake and resulting tsunami in 2011. The robot used in the present study is 420 mm wide, 530–1070 mm long and 870 mm tall, and weighs 46.5 kg. For this surveillance mission, the robot was equipped with two omnidirectional cameras, two wide-angle cameras, and a three-dimensional laser scanner.

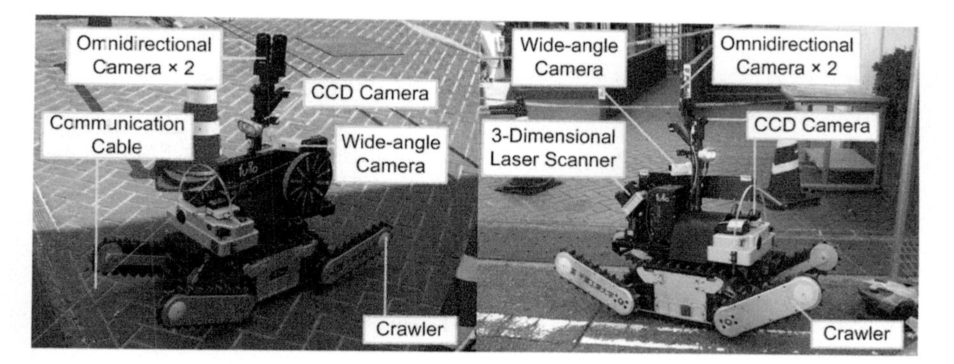

Fig. 32.7 Mobile rescue robot "Sakura-ichi-go"

32.4.2 Route of the Surveillance Mission

Figure 32.8 shows the floor plan of the main building. The surveillance time by the robot was approximately 100 min, with the surveillance robot travelling approximately 250 m.

32.4.3 Surveillance Mission Results

Figure 32.9 shows the interior damage photographed by the robot. Note that characters in each photograph correspond to the place and direction shown in Fig. 32.8. The damage observed by the robot is described below:

First, Second and Third Floor In the office on the first floor, no damage is noticeable. On column A_2-B_2, shown in Fig. 32.9a, no cracks are visible (damage class: less than II). In the stairwell from the first floor to the second floor, spalling of the concrete covering column X_{3B}-Y_6 and diagonal cracks in the shear wall can be seen in Fig. 32.9b (damage class: III). On the second story, diagonal cracks and falling concrete cover from the shear wall in frame Y_5 can be seen in Fig. 32.9c (damage class: III). No damage is visible on other vertical members. On the third floor, there are some flexural cracks at the bottom of column A_2-B_2 (damage class: III), and the finishing materials of the ceiling have fallen to the floor, as shown in Fig. 32.9d.

Fourth Floor At the border between the stair block and the office block (between frames Y_4 and Y_5), the floor slab is severely damaged, and broken pieces of the ceiling and concrete blocks litter the floor, as shown in Fig. 32.9e. Column A_2-B_2 is axially collapsed, as shown in Fig. 32.9f (damage class: V). At the top of this column, the buckling of longitudinal bars and large vertical deformation is visible. The measured height of the ceiling around column A_2-B_2 is about 1500 mm, whereas

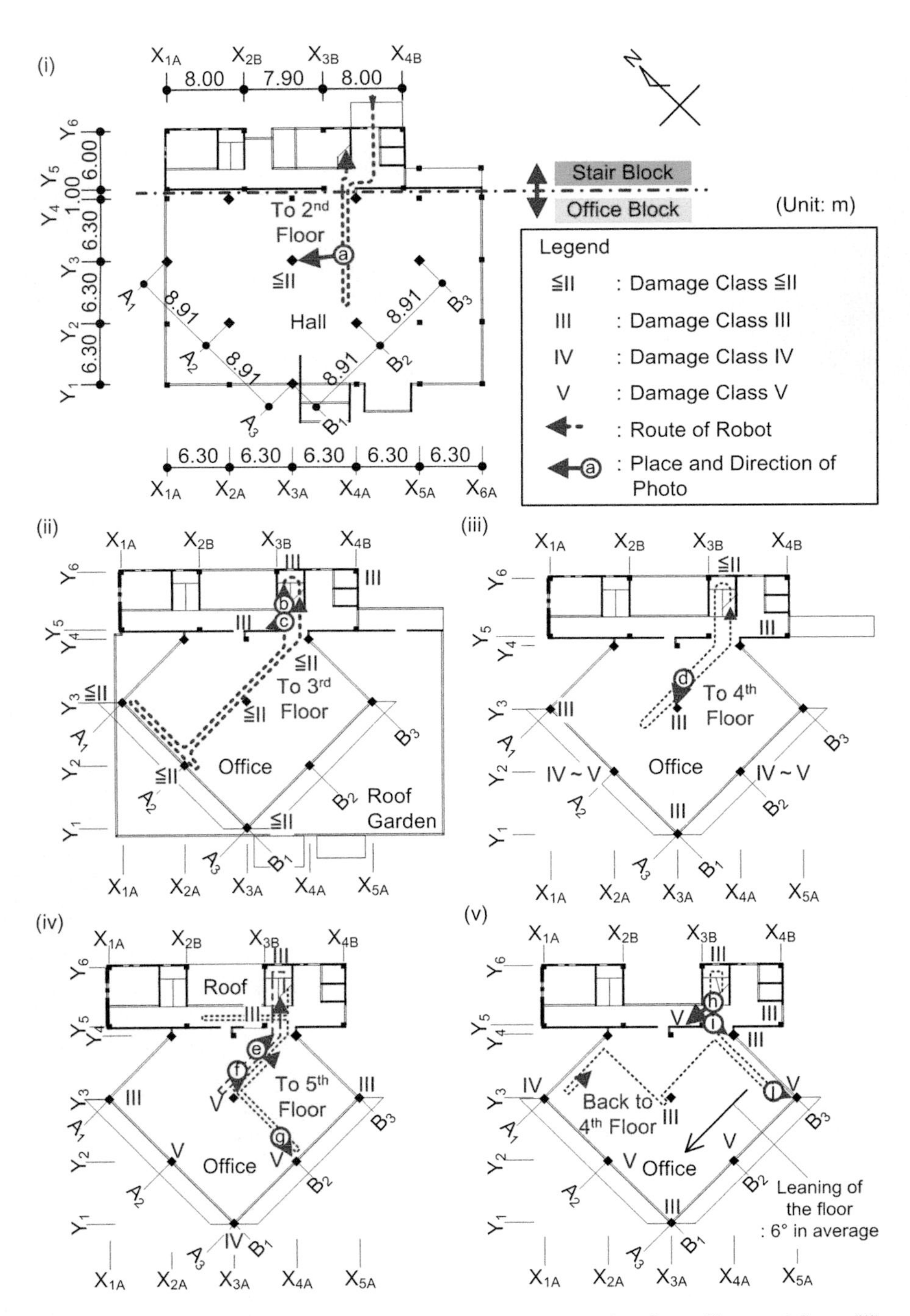

Fig. 32.8 Floor plans of the main building of Uto City Hall. (i) first floor, (ii) second floor, (iii) third floor, (iv) fourth floor, and (v) fifth floor. The red dotted line on each floor plan marks the robot's route. The roman numerals on the floor plans represent the damage class of vertical members, classified according to the Guideline for Post-Earthquake Damage Evaluation and Rehabilitation (The Japan Building Disaster Prevention Association 2001b; Nakano et al. 2004)

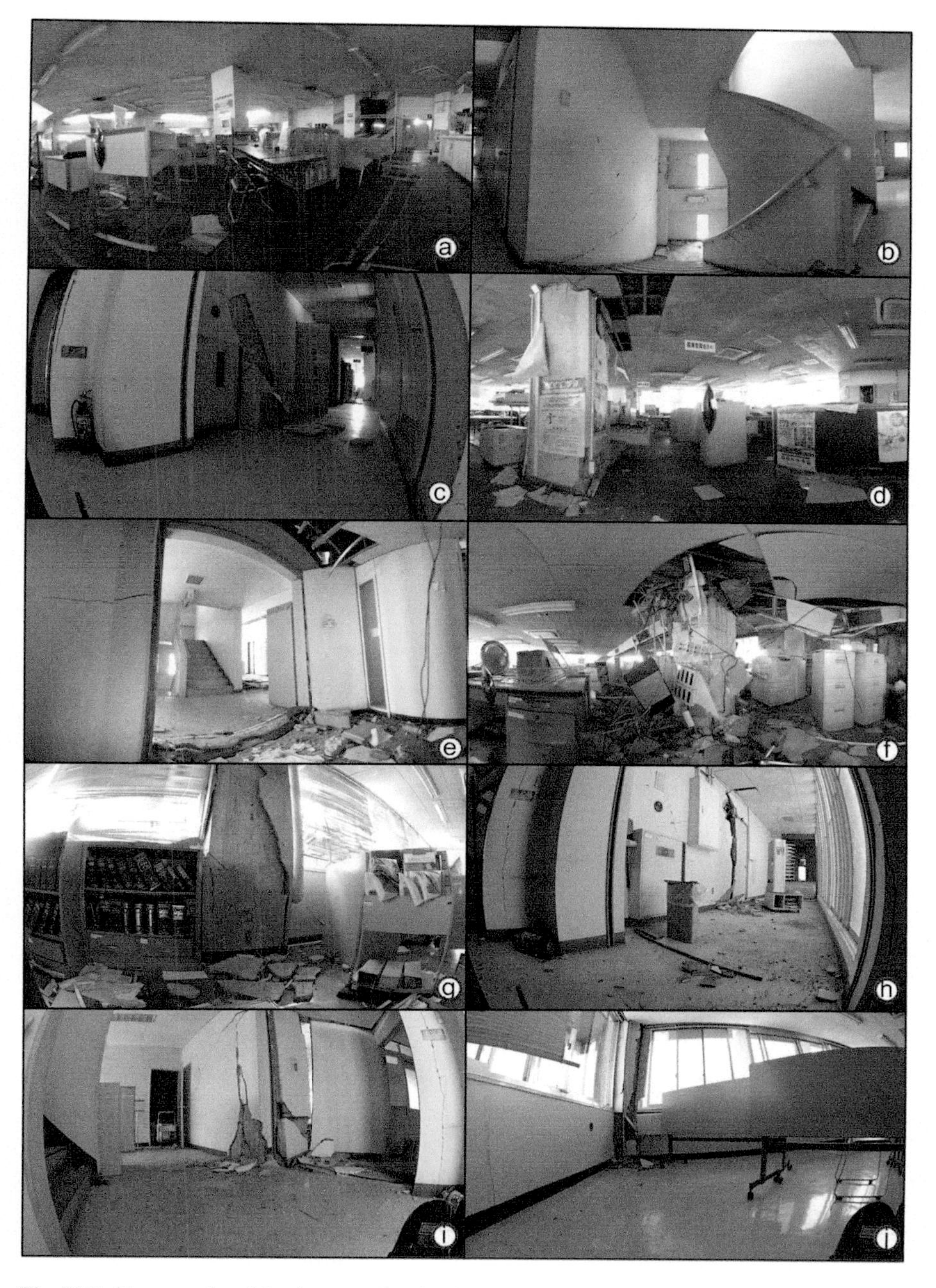

Fig. 32.9 Photographs of the damage taken by the robot inside the main building of Uto City Hall. The locations from which the photographs were taken are shown in Fig. 32.8

the ceiling height of the fourth floor according to the original structural designs is 2600 mm. Therefore, the axial deformation of column A_2-B_2 is approximately 1100 mm. Broken pieces of cover concrete and ceiling materials litter the floor around this column. At column A_3-B_2, shown in Fig. 32.9g, fallen mortar and diagonal cracks are visible (damage class: V. Note that the damage of this column is classified based on not only Fig. 32.9g but also the observation from outside (e.g. Fig. 32.4b).

Fifth Floor At the shear wall in frame Y_5, large diagonal cracks and exposed reinforcing bars are visible around the door in Fig. 32.9h (damage class: V). As on the fourth floor, the floor slab connecting the stair block and the office block is severely damaged, as shown in Fig. 32.9i.

In the office block, the floor slopes down to the west. The slope of the floor was measured by the robot to be 6° on average, which is consistent with the axial deformation of column A_2-B_2 on the third floor. The tangent of inclination angle of the floor slab can be estimated as 1100 mm (column deformation) / 8900 mm (span of beam A_2B_2-A_2B_3) = 0.12. Therefore, the estimated inclination angle is 7.0°, which is close to the angle measured by the robot. At column A_3-B_3, most of the mortar has fallen off and the reinforcing bars near the bottom are exposed and buckled, as shown in Fig. 32.9j (damage class: V).

32.5 Discussion and Conclusions

In this paper, the damage to the main building of Uto City Hall is described and discussed. From external observations, it has been found that the third, fourth and fifth stories of the south and west elevations are severely damaged, whereas in the northeast stair block there is only limited structural damage.

The main findings of the robot surveillance mission are as follows:

In the office block, column A_2-B_2 of the fourth story collapsed axially and the ceiling dropped by about 1100 mm. On the fourth and fifth floors, the concrete floor slabs connecting the stair and office blocks were severely damaged. In the stair block, several of the shear walls on the fifth story were severely damaged. In the lower floors, visibility of damage to structural members is limited.

Based on the results, it may be concluded that most of the structural damage to the main building of Uto City Hall was concentrated in the upper stories. Some columns on the fourth story lost the axial strength necessary to sustain vertical loads.

Acknowledgments First, the authors wish to express respectful regret for those who suffered losses in Kumamoto and Oita prefectures because of these earthquakes, and hope that they can overcome this disaster as soon as possible. The authors wish to thank the staff of "Mr. Sunday," a news talk program broadcast by Fuji Television Network, Japan. This paper is based on an investigation done through cooperation with the staff of "Mr. Sunday." During the investigation, Uto City Hall officials provided original structural designs and other material related to the main buildings. For the preparation of this paper, the photographs published by Google Earth were used.

References

Japan Meteorological Agency (2016) Earthquake information. http://www.jma.go.jp/en/quake/. Accessed 4 May 2016

Nagatani K, Kiribayashi S, Okada Y et al (2013) Emergency response to the nuclear accident at the Fukushima Daiichi nuclear power plants using mobile rescue robots. J Field Robot 30(1):44–63

Nakano Y, Maeda M, Kuramoto H et al (2004) Guideline for post-earthquake damage evaluation and rehabilitation of RC buildings in Japan. In: 13th World Conference on Earthquake Engineering, Vancouver, BC., Canada, 1–6 August 2004

Strong-motion Seismograph Network (K-NET, KIK-NET) Home Page. http://www.kyoshin.bosai.go.jp/kyoshin/. Accessed 4 May 2016

The Japan Building Disaster Prevention Association (2001a) Standard for seismic evaluation of existing reinforced concrete buildings (in Japanese)

The Japan Building Disaster Prevention Association (2001b) Guideline for post-earthquake damage evaluation and rehabilitation (in Japanese)

Chapter 33
Preliminary Evaluation of Seismic Capacity and Torsional Irregularity of Uto City Hall Damaged in the 2016 Kumamoto Earthquake

K. Fujii

Abstract In this paper, the seismic capacity and torsional irregularity of the main building of Uto City Hall are evaluated by using a simple method based on the building's structural drawing. The simplified evaluation method of seismic capacity, which was proposed by Shiga in the 1970s, is based on the wall-area index and the average shear stress in walls and columns. For evaluation of its seismic capacity, the following two cases are considered: the building is assumed to behave as a unit building, and each of the structural blocks responding independently. The evaluation of the torsional parameters, stiffness eccentricity and radius of torsional stiffness with respect to the center of stiffness are based on the sectional area of the columns and walls, which is presented in the Japanese Standard for the seismic evaluation of existing reinforced concrete (RC) buildings. The main findings of this paper are as follows. (a) The seismic capacity of the main building of Uto City Hall is insufficient to survive severe earthquakes. However, the evaluated results of both cases cannot explain the damage observed in upper stories. (b) The ratio of the stiffness eccentricity to radius of torsional stiffness evaluated in each story exceeds 0.15, while the radius ratio of the torsional stiffness with respect to center of stiffness to the gyration of the whole mass above the considered story is smaller than 1. Therefore, the main building of Uto City Hall is sensitive to torsional response: it may be classified as a "torsionally flexible building".

Keywords Seismic capacity evaluation · 2016 Kumamoto Earthquake · Damage · Torsional irregularity · Wall-area index · Column-area index

---

K. Fujii (✉)
Department of Architecture, Chiba Institute of Technology, Narashino-shi, Chiba, Japan
e-mail: kenji.fujii@it-chiba.ac.jp

D. Köber et al. (eds.), *Seismic Behaviour and Design of Irregular and Complex Civil Structures III*, Geotechnical, Geological and Earthquake Engineering 48,
https://doi.org/10.1007/978-3-030-33532-8_33

33.1 Introduction

The main building of Uto City Hall, a five-story reinforced concrete building constructed in 1965, was severely damaged in the 2016 Kumamoto Earthquake. As reported in reference (Fujii et al. 2017), most of the structural damage to this building was concentrated in the upper stories. External observation of the main damage to this building was limited to the third to fifth stories on the south and west elevations. From this point, the following questions arise. (i) Had this building structure had enough seismic capacity to withstand severe earthquake? (ii) Why was most of the structural damage concentrated in the upper stories? (iii) From the damage observation, the torsional response might be significant. Had this building been sensitive to torsional response?

In this paper, the seismic capacity and torsional irregularity of the main building of Uto City Hall are evaluated by using a simple method based on its structural drawing. The simplified evaluation scheme of seismic capacity, which was proposed by Shiga in the 1970s, is based on the wall-area index and the average shear stress in walls and columns (Shiga 1977). Evaluating the torsional parameters, stiffness eccentricity and radius of torsional stiffness with respect to the center of stiffness, is based on the sectional area of the columns and walls, which is presented in the Japanese Standard for the seismic evaluation of existing RC buildings (The Japan Building Disaster Prevention Association 2001).

33.2 The Main Building of Uto City Hall

Figure 33.1 shows the structural plan of the main building of Uto City Hall. The structure of this building can be divided into two structural blocks (office block and stair block). The two blocks are connected only by a concrete slab (thickness: 110 mm).

As shown in this figure, all the structural walls are concentrated in the stair block, whereas in the office block concrete columns are the only vertical members to resist lateral loads. Is should be also noted that in the office block not all frames are oriented in X- or Y-directions: frames $A_1 - A_3$ lie on the axis rotated 45° counter-clockwise from the X-axis, and frames $B_1 - B_3$ are orthogonal to frames A_1.

In the fourth and fifth story, the floor slab between frames Y_4 and Y_5, the border of two blocks, was severely damaged because of this earthquake, as described in reference (Fujii et al. 2017).

Figure 33.2 shows the simplified structural elevation of frame Y_4 and A_2. The height of the first story is different in zones (I) and (II) in the office block: in zone (I), where the number of stories is 5, the story height is 4.4 m, while it is 3.1 m in zone (II), where the number of stories is 1. Note that the story height in the stair block is the same as frame A_2.

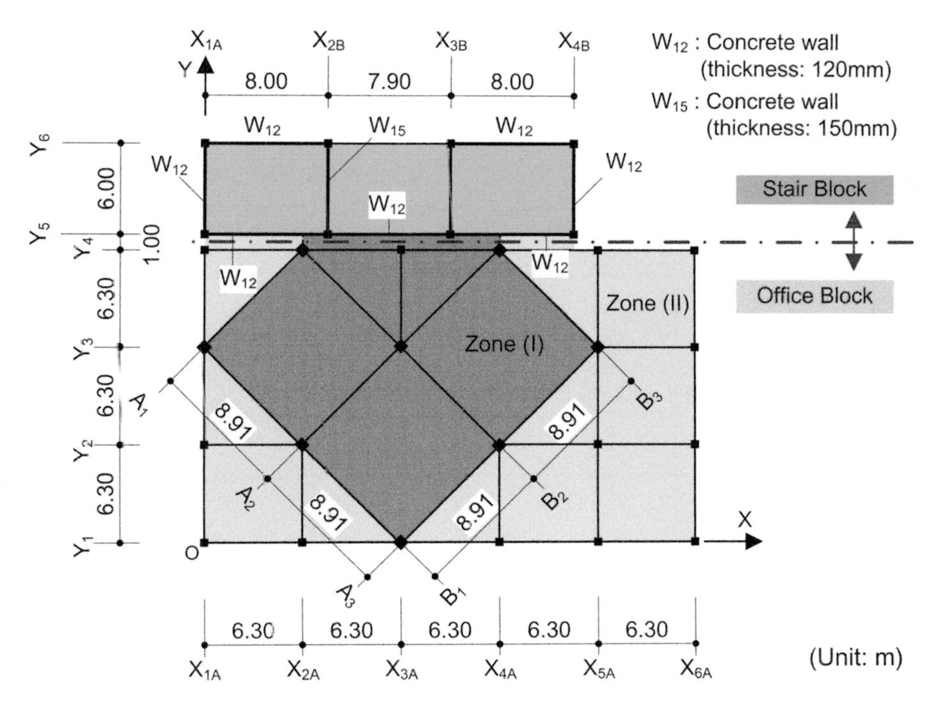

Fig. 33.1 Structural plan of the main building of Uto City Hall

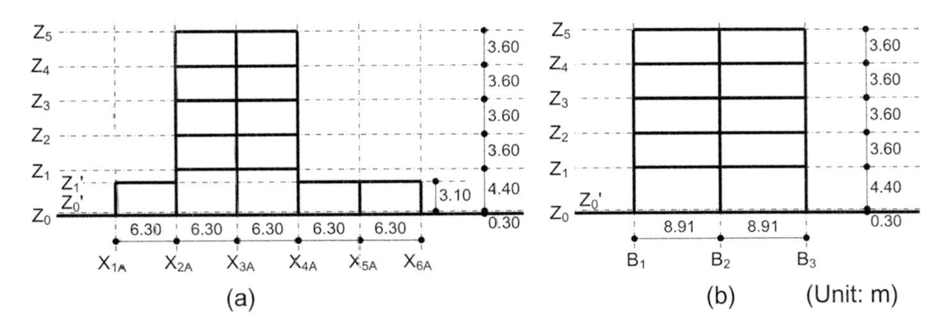

Fig. 33.2 Simplified structural elevation of the main building of Uto City Hall. (**a**) frame Y_4, (**b**) frame A_2

Figure 33.3 shows the sections of column A_1B_1, A_2B_1 and A_2B_2. As shown in this figure, the sectional area of column is drastically reduced from the lower stories to upper stories. This is very common in reinforced concrete buildings constructed before 1981, because at that time the design seismic force in upper stories is smaller than that in the current seismic code of Japan (BCJ 2016).

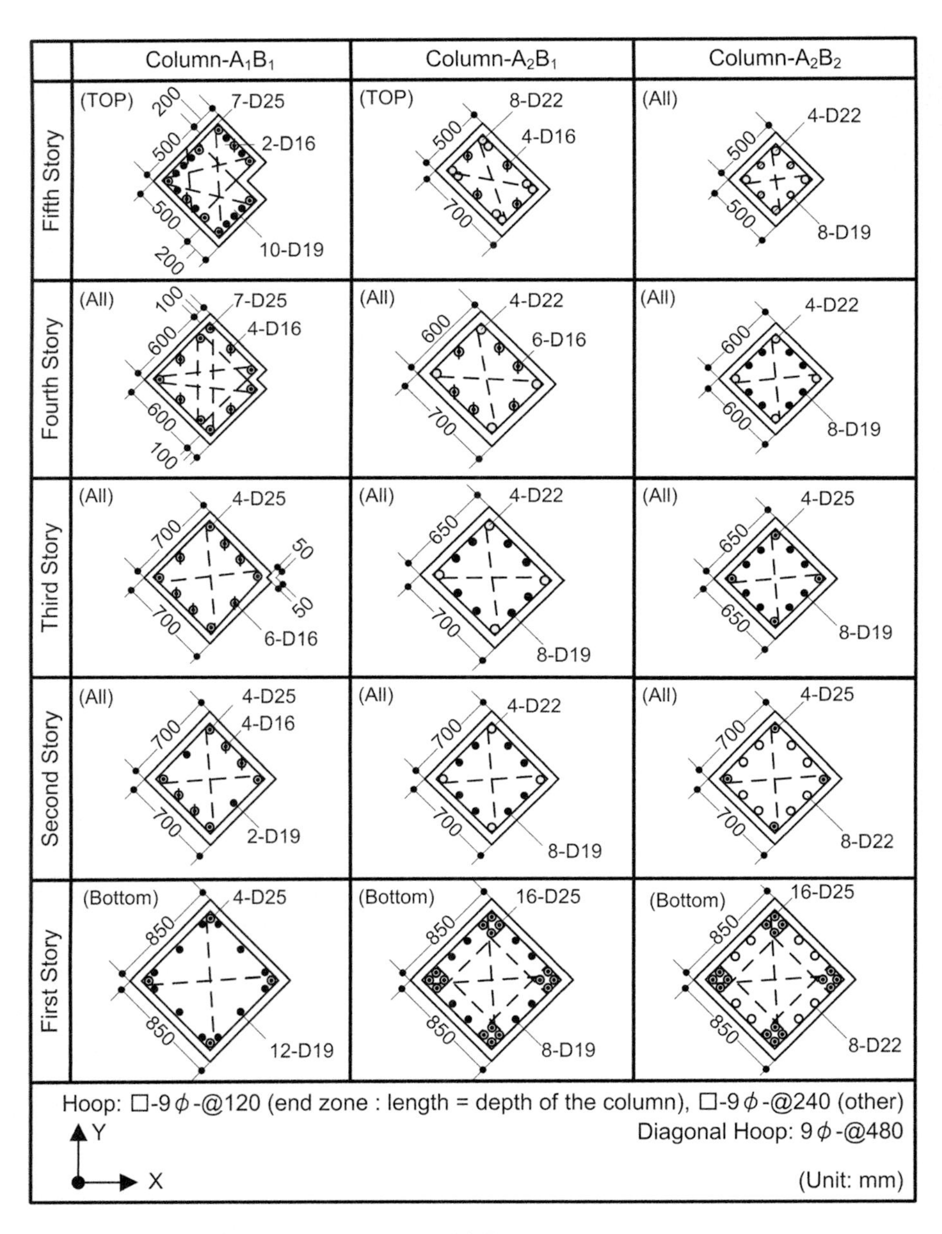

Fig. 33.3 Sections of column A_1B_1, A_2B_1 and A_2B_2

33.3 Seismic Capacity Evaluations

33.3.1 *Description of the Simplified Evaluation Method*

Shiga investigated low-rise reinforced concrete buildings damaged in the 1968 Tokachi-oki Earthquake (Shiga 1977). He explored the relation of earthquake damage and the following parameters, which can be easily obtained from drawings: wall-area index, column-area index, and average shear stress in walls and columns. He had concluded that damaged and undamaged buildings could be significantly distinguished between two parameters, wall-area index and average shear stress in walls and columns. In this study, the seismic capacity evaluation of the main building of Uto City Hall is carried out according to Shiga's method, with some modifications.

In the present study, wall-area index and column-area index of the i-th story, α_{Wi} and α_{Ci}, respectively, are defined by Eq. (33.1).

$$\alpha_{Wi} = \frac{A_{Wi}}{A_i \sum_{j=i}^{N} A_{fj}} \left(\text{unit}: \text{mm}^2/\text{m}^2\right), \alpha_{Ci} = \frac{A_{Ci}}{A_i \sum_{j=i}^{N} A_{fj}} \left(\text{unit}: \text{mm}^2/\text{m}^2\right). \tag{33.1}$$

In Eq. (33.1), A_{Wi} (unit: mm^2) and A_{Ci} (unit: mm^2) are the sum of the sectional area of the walls and columns in the i-th story, respectively, and A_{fj} (unit: m^2) is the area of the j-th floor. Th coefficient A_i is calculated from Eq. (33.2), which is used in the current seismic design code in Japan (BCJ 2016).

$$A_i = 1 + \left(\frac{1}{\sqrt{\alpha_i}} - \alpha_i\right) \cdot \frac{2T}{1+3T}, \alpha_i = \sum_{j=i}^{N} w_j / \sum_{j=1}^{N} w_j. \tag{33.2}$$

In Eq. (33.2), w_j (unit: kN) is the weight of the j-th floor, and T is the natural period of the building that is calculated as a function of the building height H (unit: m).

$$T = 0.02H. \tag{33.3}$$

The average shear stress in walls and columns in the i-th story, τ_{avei}, is calculated from Eq. (33.4), assuming that weight per unit floor area of the building is 10 kN/m^2 and base shear coefficient is 1.0.

$$\tau_{avei} = \frac{A_i \cdot \sum_{j=i}^{N} w_j}{A_{Wi} + A_{Ci}} = 10^4 \times \frac{A_i \cdot \sum_{j=i}^{N} A_{fj}}{A_{Wi} + A_{Ci}} \left(\text{unit}: \text{N/mm}^2\right). \tag{33.4}$$

In the present study, two modifications are made to Shiga's original method. One is that the wall-area index and column-area index are extended for the upper stories in a multi-story building: both indices are divided by A_i coefficient to consider the vertical distribution of lateral seismic forces. The other is that the weight per unit floor area of the building is changed from 1000 kgf/m^2 to 10 kN/m^2, to adjust the SI unit.

In Shiga's investigation, he had concluded that buildings that satisfy either of two conditions, that the wall-area index α_{Wi} is larger than 30×10^2 mm^2/m^2 or the average shear stress in walls and columns τ_{avei} is less than 1.2 N/mm^2, correspond to those that were undamaged or very slightly damaged in the 1968 Tokachi-oki Earthquake. He had also concluded that the buildings within zone A, which is defined by the condition shown as Eq. (33.5), correspond to those whose walls were heavily cracked columns were heavily damaged in shear in case columns were short and shear failure preceded bending failure in 1968 Tokachi-oki Earthquakes (Shiga 1977).

$$1.2A_{Ci} + 3.3A_{Wi} \leq 10^4 \times A_i \cdot \sum_{j=i}^{N} A_{fj}. \tag{33.5}$$

Note that in Eq. (33.5), the average ultimate shear stress of the column is assumed to be 1.2 N/mm^2, while that of wall is assumed to be 3.3 N/mm^2.

33.3.2 Evaluation Cases

In this study, the following two cases are considered for the seismic capacity evaluation of the main building of Uto City Hall. In Case 1, the building is assumed to behave as a unit building, and the evaluation is carried out as if for a single building. In contrast, in Case 2, the stair and office blocks are assumed to behave independently, and the evaluation is carried out as if for two independent buildings. In each case, the X- and Y-directions shown in Fig. 33.1 are evaluated.

33.3.3 Evaluation Results

Figures 33.3 and 33.4 show the evaluation results in each case. In these figures, the zone A is the area corresponding to the most of buildings were heavily damaged while zone C is the area corresponding to the most of buildings were not damaged or only slightly damaged in the 1968 Tokachi-oki Earthquake.

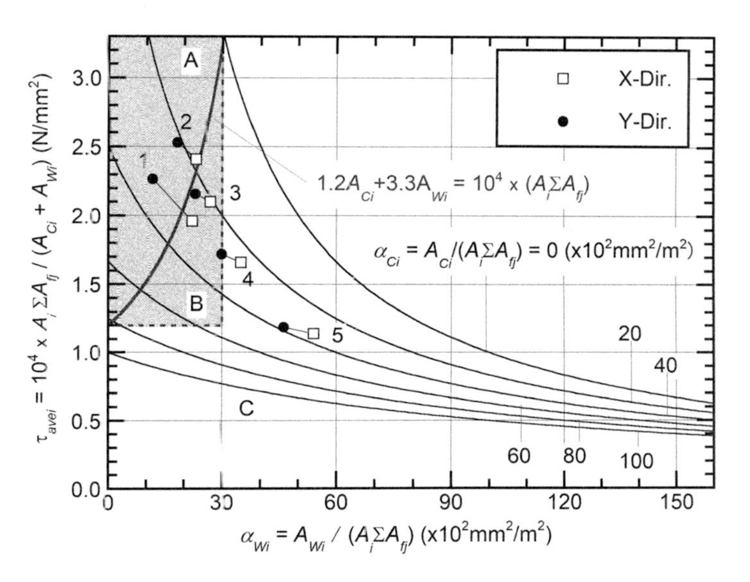

Fig. 33.4 Evaluation results in case 1

33.3.4 Discussions

In Case 1 (Fig. 33.3), the plots of Y-direction in the first and second stories are within zone A, while the plots of the upper stories are in zone B or C. Therefore, it may be concluded that if this building behaved as a united single building, the seismic capacity of this building is insufficient to survive strong earthquakes. However, this result cannot explain the fact that most damage in this building is in the upper stories.

In Case 2 (Fig. 33.4) the plots of office blocks in all stories are within zone A, whereas the plots of stair block in all stories are within zone C. Therefore, it may be concluded that the seismic capacity of the office block is insufficient while that of the stair block is sufficient, under the condition that the two blocks of this building behaved as two independent buildings. However, in the damage observation of this building (Fujii et al. 2017), it was found that the walls in the fifth story of frame Y_5 (stair block) were severely damaged. The results shown in Fig. 33.5 cannot explain this damage. Therefore, the assumption that the stair and office blocks behave independently appears invalid, even though the floor slab at the border of two blocks in the fourth and fifth floors were severely damaged (Fujii et al. 2017).

In conclusion, the seismic capacity of the main building of Uto City Hall is insufficient to survive strong earthquakes. However, neither results can explain the damage of this building observed. The reasons why this simplified evaluation method fails to explain the observed damage are (i) the lateral force distribution coefficient, A_i, is smaller in upper stories because the A_i coefficient cannot reflect the drastic reduction of the sectional area in upper stories, and (ii) the effect of torsion is not considered in this simplified method.

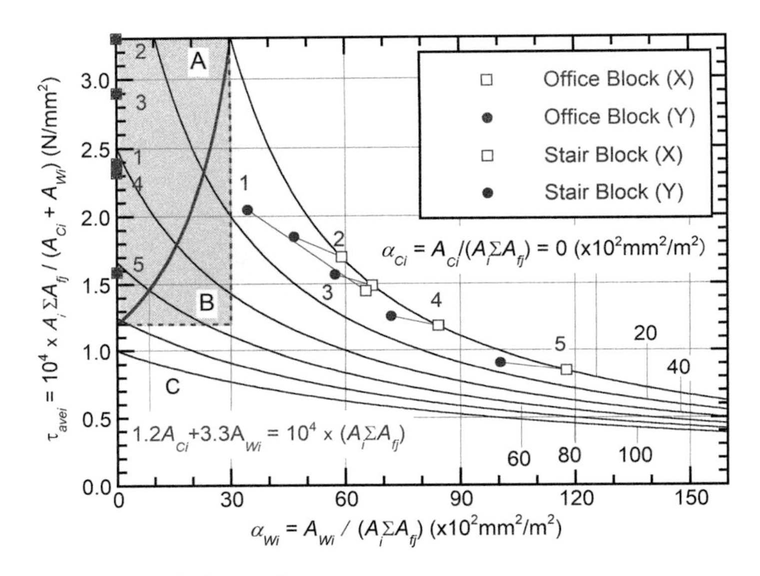

Fig. 33.5 Evaluation results in case 2

33.4 Evaluation of Torsional Irregularity

33.4.1 Description of Calculation Method

In this study, the parameters of torsional irregularity, eccentricity ratio, ratio of gyration of story torsional stiffness with respect to the center of mass, and eccentricity index are calculated according to the standard for seismic evaluation of existing reinforced concrete buildings (The Japan Building Disaster Prevention Association 2001). In this study, those parameters are calculated based on the sectional area of columns and walls as below.

The stiffness index of the j-th frame, S_j, is calculated by Eq. (33.6).

$$S_j = \sum_k a_{Cjk} + \sum_k \alpha_{jk}(1 - \eta_{jk})a_{Wjk}. \tag{33.6}$$

In Eq. (33.6), a_{Cjk} is the sectional area of the k-th column in the j-th frame, and a_{Wjk}, α_{jk}, η_{jk} are the sectional area, stiffness modification factor considering the proportion of wall, and opening ratio, respectively, of the k-th wall in the j-th frame. Figure 33.5 shows the definition of α_{jk}, η_{jk} (Fig. 33.6).

The location of the center of stiffness of each story (x_{Si}, y_{Si}), and the radius of gyration of story torsional stiffness with respect to the center of stiffness, j_{Xi}' and j_{Yi}', respectively, are calculated by using stiffness index S_j.

Let m_i and I_i be the mass and mass moment of inertia of the i-th floor, respectively, and the location of the center of mass of i-th floor is expressed as (x_{Gfi}, y_{Gfi}).

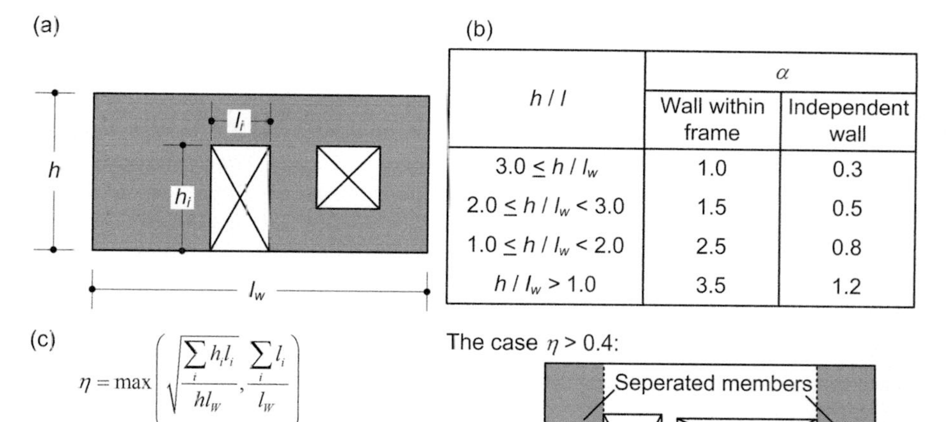

h/l	α	
	Wall within frame	Independent wall
$3.0 \leq h/l_w$	1.0	0.3
$2.0 \leq h/l_w < 3.0$	1.5	0.5
$1.0 \leq h/l_w < 2.0$	2.5	0.8
$h/l_w > 1.0$	3.5	1.2

(c)

$$\eta = \max\left(\sqrt{\frac{\sum_i h_i l_i}{h l_W}}, \frac{\sum_i l_i}{l_W}\right)$$

Note:
If η is larger than 0.4, wall with openings should be modeled as separated members.

The case $\eta > 0.4$:

Seperated members

Fig. 33.6 Definition of α_{jk}, η_{jk} for wall. (**a**) Elevation of wall considered, (**b**) definition of α_{jk}, (**c**) definition of η_{jk}

The location of the center of total mass above the i-th story $(\widehat{x}_{Gi}, \widehat{y}_{Gi})$is calculated from Eq. (33.7).

$$\widehat{x}_{Gi} = \sum_{j=i}^{N} m_j x_{Gfj} / \sum_{j=i}^{N} m_j, \widehat{y}_{Gi} = \sum_{j=i}^{N} m_j y_{Gfj} / \sum_{j=i}^{N} m_j. \tag{33.7}$$

The radius of gyration of mass above the i-th story, r_i, is calculated from Eq. (33.8).

$$r_i = \sqrt{\sum_{j=i}^{N} \left[I_j + m_j\left\{\left(x_{Gfj} - \widehat{x}_{Gj}\right)^2 + \left(y_{Gfj} - \widehat{y}_{Gj}\right)^2\right\}\right] / \sum_{j=i}^{N} m_j}. \tag{33.8}$$

The stiffness eccentricity of the i-th story, e_{Xi} and e_{Yi}, are calculated by Eq. (33.9).

$$e_{Xi} = x_{Si} - \widehat{x}_{Gi}, e_{Yi} = y_{Si} - \widehat{y}_{Gi}. \tag{33.9}$$

The eccentricity indices of the i-th story defined in the current seismic design code of Japan, R_{eXi} and R_{eYi}, respectively, are calculated by Eq. (33.10).

$$R_{eXi} = \left|e_{Yi}/j_{Xi}'\right|, R_{eYi} = \left|e_{Xi}/j_{Yi}'\right|. \tag{33.10}$$

According to Hejal and Chopra (Hejal and Chopra 1987), the classification of systems as either torsionally stiff (TS) or torsionally flexible (TF) systems is based on the ratio of the uncoupled torsional mode to the lateral frequencies $\Omega_{\theta X}$, $\Omega_{\theta Y}$ of the corresponding torsionally balanced system, defined by Eq. (33.11).

$$\Omega_{\theta X} = \omega_{0\theta}'/\omega_{0X}, \Omega_{\theta Y} = \omega_{0\theta}'/\omega_{0Y}. \tag{33.11}$$

In Eq. (33.11), $\omega_{0\theta}'$ is the uncoupled natural circular frequency of rotational oscillation with respect to the center of stiffness. The system $\Omega_{\theta X}$, $\Omega_{\theta Y} > 1$ is classified as a TS system in both the X- and Y-directions (Hejal and Chopra 1987).

For the single-story asymmetric building system (mass: m, mass moment of inertia I, lateral stiffness of system in X- and Y-direction, K_X and K_Y, torsional stiffness with respect to the center of stiffness K_θ'), $\Omega_{\theta X}$, $\Omega_{\theta Y}$ are equal to the radius ratios of gyration of the story torsional stiffness with respect to center of stiffness, j_X'/r and j_Y'/r, as shown in Eq. (33.12).

$$\Omega_{\theta X} = \frac{\omega_{0\theta}'}{\omega_{0X}} = \frac{\sqrt{K_\theta'/I}}{\sqrt{K_X/m}} = \frac{j_X'}{r}, \Omega_{\theta Y} = \frac{\omega_{0\theta}'}{\omega_{0Y}} = \frac{\sqrt{K_\theta'/I}}{\sqrt{K_Y/m}} = \frac{j_Y'}{r}. \tag{33.12}$$

In this study, the system classification as either TS or TF is made based on j_X'/r and j_Y'/r for each direction in each story: the i-th story is classified as TS in X-direction if the ratio j_{Xi}' / r_i is larger than 1, whereas it is classified as TF if j_{Xi}' /r_i is smaller than 1. The i-th story in Y-direction is also classified in the same manner.

33.4.2 Calculated Results

Figure 33.7 shows the location of the center of total mass above the considered story and the center of stiffness of the same, and Fig. 33.8 shows the vertical distribution of three parameters of torsional irregularity.

33.4.3 Discussions

From Fig. 33.7, the center of total mass above the story, G, and the center of stiffness of each story, S, almost lie on the axis of frame X_{3A}; however, the location of S is closer to the stair block than G. This is because all of the walls are in the stair block. Therefore, the eccentricity ratio in X-direction $|e_X/ r|$ is small (0.009–0.030), whereas the eccentricity ratio in Y-direction $|e_Y/r|$ is relatively large (0.219–0.306), as shown in Fig. 33.8a.

The radius ratios of gyration of the story torsional stiffness with respect to S, j_X'/r and j_Y'/r, are smaller than 1, except j_Y'/r in the first story (Fig. 33.8b). In addition, the

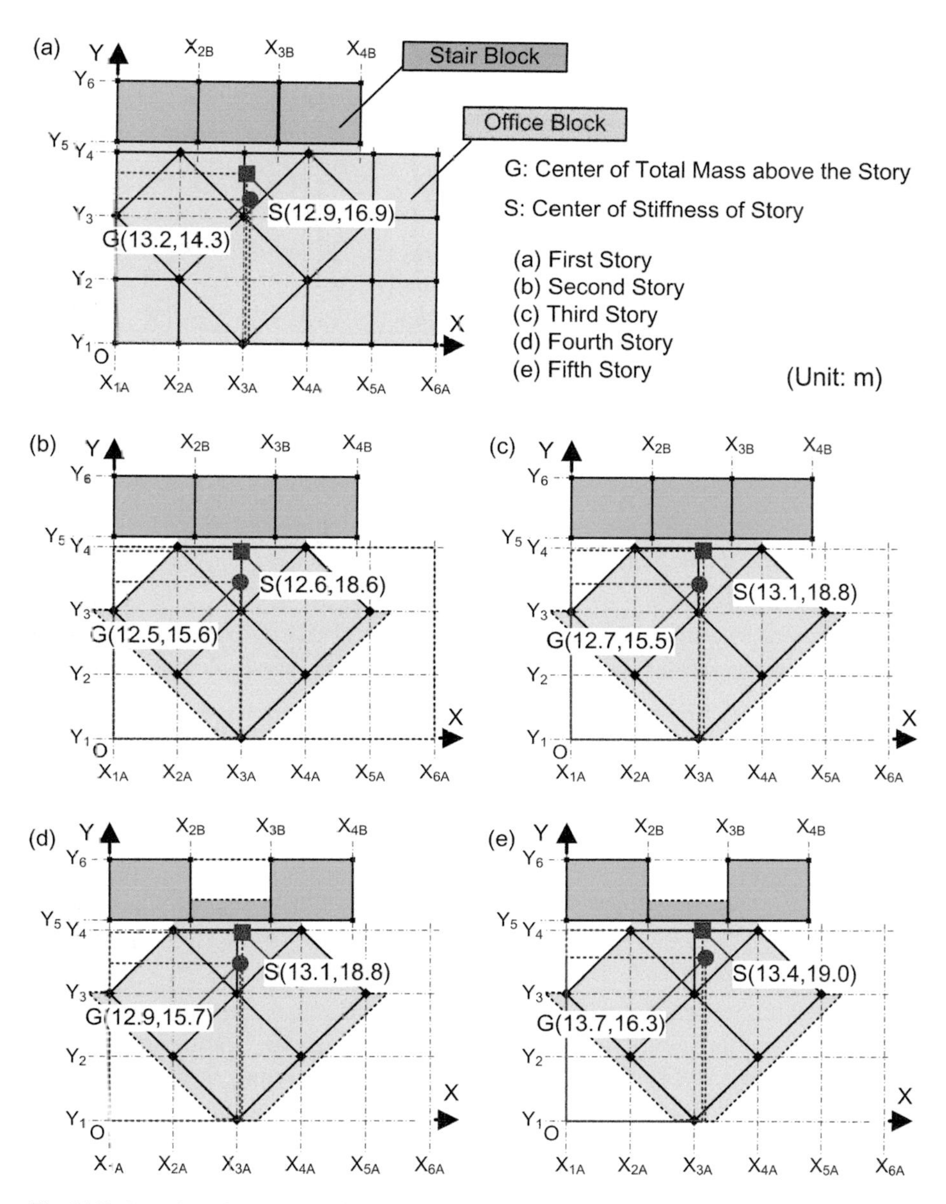

Fig. 33.7 Location of the center of total mass above the considering story and the center of stiffness of the considering story

eccentricity index in X-direction, R_{eX}, is larger than 0.15 in all stories: in particular, R_{eX} is larger than 0.3 in the second to fourth stories (Fig. 33.8c).

Therefore, this building is sensitive to torsional response and is classified as a TF system in all stories in X-direction and the second to fifth stories in Y-direction.

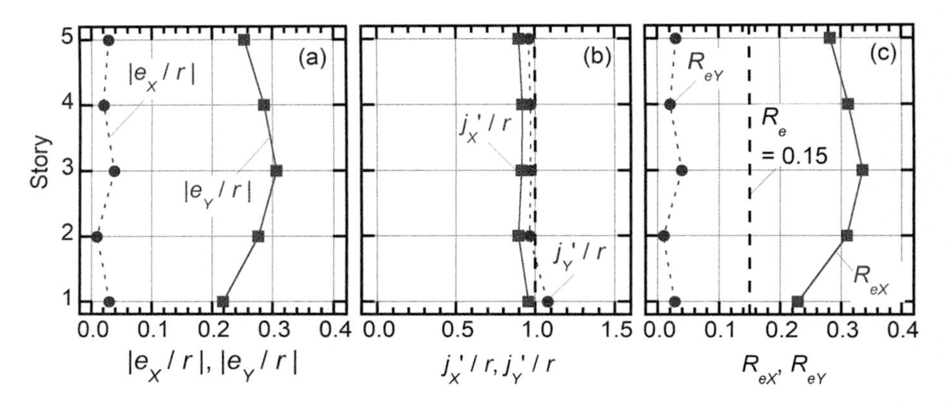

Fig. 33.8 Distribution of parameters of torsional irregularity. (**a**) Eccentricity ratio, (**b**) radius ratio of gyration of story torsional stiffness with respect to the center of stiffness, (**c**) eccentricity index

33.5 Discussions and Conclusions

In this paper, the seismic capacity and torsional irregularity of the main building of Uto City Hall were evaluated by using a simple method based on structural drawings. The main findings of this paper are as follows.

(a) The seismic capacity of the main building of Uto City Hall is insufficient to survive severe earthquakes. However, the evaluated results of both cases cannot explain the damage observed in upper stories.

(b) The ratio of the stiffness eccentricity to radius of torsional stiffness evaluated in each story exceeds 0.15, whereas the radius ratio of the torsional stiffness with respect to center of stiffness to the gyration of whole mass above the considered story is smaller than 1. Therefore, the main building of Uto City Hall is sensitive to torsional response: it may be classified as a "torsionally flexible building."

Note that further detailed investigations, such as a nonlinear time-history analysis of the frame building model, are needed to explain the seismic behavior of the main building of Uto City Hall during sequential seismic events. Seismic response evaluation of this building by using several nonlinear static procedures is also attractive for the validation of these procedures.

Acknowledgments The author wishes to thank the Uto City Hall officials who provided the original structural designs and other material related to the main buildings.

References

BCJ (2016) The building standard law of Japan on CD-ROM, The Building Center of Japan, Tokyo

Fujii K, Yoshida T, Nishimura T et al (2017) Observation of damage to Uto City Hall in the 2016 Kumamoto Earthquake. In: 8th European workshop on the seismic behaviour of irregular and complex structures, Bucharest, Romania, 19–20 October 2017 (submitted)

Hejal R, Chopra AK (1987) Earthquake response of torsionally-coupled buildings. Earthquake Engineering Research Center, report no. UCB/EERC-87/20, College of Engineering, University of California at Berkeley

Shiga T (1977) Earthquake damage and the amount of walls in reinforced concrete buildings. In: 6th world conference on earthquake engineering, New Delhi, India, 9–14 January 1977

The Japan Building Disaster Prevention Association (2001) Standard for seismic evaluation of existing reinforced concrete buildings (in Japanese)

Index

D. Köber et al. (eds.), *Seismic Behaviour and Design of Irregular and Complex Civil Structures III*, Geotechnical, Geological and Earthquake Engineering 48,
https://doi.org/10.1007/978-3-030-33532-8